ISSUES FOR DEBATE
IN SOCIOLOGY

SELECTIONS FROM CQ RESEARCHER

D1511850

$SAGE | PINE FORGE

Los Angeles | London | New Delhi
Singapore | Washington DC

For information:

Pine Forge Press
An Imprint of SAGE Publications, Inc.
2455 Teller Road
Thousand Oaks, California 91320
E-mail: order@sagepub.com

SAGE Publications Ltd.
1 Oliver's Yard
55 City Road
London EC1Y 1SP
United Kingdom

SAGE Publications India Pvt. Ltd.
B 1/I 1 Mohan Cooperative Industrial Area
Mathura Road, New Delhi 110 044
India

SAGE Publications Asia-Pacific Pte. Ltd.
33 Pekin Street #02-01
Far East Square
Singapore 048763

Printed in the United States of America

Library of Congress Cataloging-in-Publication Data

Issues for debate in sociology: selections from CQ researcher.
    p. cm.
Includes bibliographical references.
ISBN 978-1-4129-7860-6 (pbk.)
    1. Sociology. 2. Social problems. I. CQ researcher.

HM585.I87 2010
301—dc22   2009026672

This book is printed on acid-free paper.

10   11   12   13   10 9 8 7 6 5 4 3 2

Acquisitions Editor:	David Repetto
Editorial Assistant:	Nancy Scrofano
Production Editor:	Laureen Gleason
Typesetter:	C&M Digitals (P) Ltd.
Cover Designer:	Candice Harman
Marketing Manager:	Jennifer Reed Banando

Contents

INSTITUTIONS

Annotated Contents

SOCIAL STRUCTURE, PROCESSES, AND CONTROL

Celebrity Culture: Are Americans Too Focused on Celebrities?

In early February, North Korea's leader bragged about his nuclear arsenal, the lagging U.S. dollar started climbing and the Prince of Wales announced his engagement. But the serious-minded readers of Bloomberg News were most interested in Charles and Camilla. Americans have an insatiable appetite for celebrity news, and the juicier the better — from Brad and Jennifer's breakup to Michael Jackson's trial to Martha Stewart's jail term. Some observers say it's harmless to follow the lives of celebrities. Indeed, they even say we are genetically programmed to care, and that the heavy focus on celebrities simply reflects that interest. But media critics say celebrity coverage is squeezing out legitimate news and that, as a result, the United States is becoming a nation that knows more about the "Battle of the Network Stars" than the battle for Baghdad. With less attention being paid to informing citizens about government and the world around them, the critics warn, a cornerstone of a democratic society — an informed populace — is being put in jeopardy.

Teaching Values: Do School-Based Programs Violate Parents' Beliefs?

Thousands of schools in more than a dozen states are participating in a new movement to teach values in schools. Leaders of the "character education" movement point to moral decline among America's youth — evidenced by rising rates of teen pregnancy and youth crime — as the main reason schools should teach values. The programs, which vary greatly depending on the school, have provoked relatively little controversy locally. However, both conservative Christians and civil libertarians see the potential for schools to impose ideologies contrary to parents' values. Leading character-education advocates contend that schools can teach such basic values as respect and responsibility without wading into controversial areas like abortion, sex education and homosexuality.

Cyber Socializing: Are Internet Sites Like MySpace Potentially Dangerous?

Internet socializing has become hugely popular, and Web sites that help people meet potential dates, find new friends and keep track of old ones are big business. Hundreds of sites attract tens of millions of users, and more sites come online daily. Born along with the Internet in the early 1970s, online socializing has helped people worldwide link to others with common interests for conversation and support. Nevertheless, new social-networking sites like Facebook and MySpace raise more troubling privacy issues than traditional Internet chat rooms. Visitors to such sites can access not only individuals' posted profiles but also profiles of their friends. Parents and law enforcement agencies worry that predators can use the information to contact vulnerable teens. Some states are considering requiring tighter security and confidentiality, and a bill introduced in the House of Representatives would require schools and libraries to block teenagers from the sites.

Closing Guantánamo: Can Obama Close the Detention Camp Within One Year?

President Obama on his second full day in office ordered the closing of the Guantánamo detention camp within a year. The facility at the U.S. Naval Station in Cuba has been controversial ever since President George W. Bush decided in late 2001 to use it to hold suspected enemy combatants captured in Afghanistan and elsewhere. Both Obama and Republican candidate John McCain promised during the presidential campaign to close the facility if elected. But that poses many difficult issues about the camp's remaining 241 prisoners. The government wants to send many to other countries — with few takers so far — but worries that some may resume hostile activities

against the United States. Some may be brought to the U.S. for trial, but those prosecutions would raise a host of uncharted legal issues. Meanwhile, opposition already has surfaced to any plans for housing detainees in the United States. And human-rights advocates worry the Obama administration may continue to back some form of preventive detention for suspected terrorists.

INEQUALITY

Middle-Class Squeeze: Is More Government Aid Needed?

Millions of families who once enjoyed the American dream of home ownership and upward financial mobility are sliding down the economic ladder — some into poverty. Many have been forced to seek government help for the first time. The plunging fortunes of working families are pushing the U.S. economy deeper into recession as plummeting demand for goods and services creates a downward economic spiral. A consumption binge and growing consumer debt beginning in the 1990s contributed to the middle-class squeeze, but the bigger culprits were exploding prices for necessities such as housing, medical care and college tuition, cuts in employer-funded benefits and, some say, government policies that favored the wealthy. President Barack Obama has promised major aid for the middle class, and some economists are calling for new programs — most notably national health coverage — to assist working Americans.

Debating Hip-Hop: Does Gangsta Rap Harm Black Americans?

Since exploding from the streets of New York in the 1970s, the cultural phenomenon known as hip-hop has morphed from hard-driving dance numbers into sex- and violence-filled "gangsta rap" — and a record label goldmine. Gangsta lyrics have sparked periodic outbreaks of indignation, but the outrage intensified after white shock jock Don Imus was fired in April for describing black female athletes in the degrading terms used commonly by hip-hop performers. African-American leaders, including Bill Cosby, Oprah Winfrey and the Rev. Al Sharpton, claim the genre's glorification of thug culture — often for the entertainment of white youths — drags down the black community. In response, a few top hip-hop figures have called for cleaning up gangsta content. Meanwhile, a school of socially conscious hip-hop remains vibrant, embraced by political activists, school reformers and artistic innovators who call it an inspiration no matter what happens to the gangsta style.

Gender Pay Gap: Are Women Paid Fairly in the Workplace?

More than four decades after Congress passed landmark anti-discrimination legislation —including the Equal Pay Act of 1963 — a debate continues to rage over whether women are paid fairly in the workplace. Contending that gender bias contributes to a significant "pay gap," reformists support proposed federal legislation aimed at bringing women's wages more closely in line with those of men. Others say new laws are not needed because the wage gap largely can be explained by such factors as women's choices of occupation and the amount of time they spend in the labor force. Meanwhile, a class-action suit charging Wal-Mart Stores with gender bias in pay and promotions — the biggest sex-discrimination lawsuit in U.S. history — may be heading for the Supreme Court. Some women's advocates argue that a controversial high-court ruling last year makes it more difficult to sue over wage discrimination.

Women's Rights: Are Violence and Discrimination Against Women Declining?

Women around the world have made significant gains in the past decade, but tens of millions still face significant and often appalling hardship. Most governments now have gender-equality commissions, electoral gender quotas and laws to protect women against violence. But progress has been mixed. A record number of women now serve in parliaments, but only 14 of the world's 193 countries currently have elected female leaders. Globalization has produced more jobs for women, but they still constitute 70 percent of the world's poorest inhabitants and 64 percent of the illiterate. Spousal abuse, female infanticide, genital mutilation, forced abortions, bride-burnings, acid attacks and sexual slavery remain pervasive in some countries, and rape and sexual mutilation have reached epic proportions in the war-torn Democratic Republic of the Congo. Experts say without greater economic, political and educational equality, the plight of women will not improve, and society will continue suffering the consequences.

INSTITUTIONS

Future of Marriage: Is Traditional Matrimony Going Out of Style?

In the past 40 years, the nation's marriage rate has dropped from three-quarters of American households to slightly over half. Moreover, nearly 50 percent of all U.S. marriages now end in divorce, and the number of households with unmarried couples has risen dramatically. Some scholars say that although traditional marriage will not disappear entirely, it will never again be the nation's pre-eminent social arrangement. In the future, they say, the United States will look more like Europe, where couples increasingly are opting to cohabit rather than marry. But other experts argue that the recent decrease in the divorce rate and other positive trends point to a brighter future for marriage. Meanwhile, actions by a number of state courts and local officials in favor of same-sex unions have helped ignite a debate over the issue and prompted conservatives to push for a constitutional amendment banning gay marriage.

Student Aid: Will Many Low-Income Students Be Left Out?

With a record number of students hoping to attend college next year — and fees higher than ever — finding a way to pay the bills will be tough for many. Congress and the Bush administration made common cause in 2007 to increase federal Pell Grants for students and reduce some student-loan interest rates. Nevertheless, critics say the increases won't go far enough. To help middle-class families, states increasingly offer merit-based grants for college aid. But with merit scholarships replacing need-based aid, low-income and minority students — who often don't have the grades for scholarships — are finding their college dreams harder to realize. Meanwhile, longtime concern that private lenders rake in excess profits from their high-interest student loans has reached new heights. Investigations of student lending are being conducted in several states, even as universities and lenders settle allegations of loan fraud with New York's attorney general.

Religious Fundamentalism: Does It Lead to Intolerance and Violence?

People around the world are embracing fundamentalism, a belief in the literal interpretation of holy texts and, among the more hard-line groups, the desire to replace secular law with religious law. At the same time, deadly attacks by religious extremists in India, Uganda, Somalia and Nigeria are on the rise — and not just among Muslims. Meanwhile, political Islamism — which seeks to install Islamic law via the ballot box — is increasing in places like Morocco and in Muslim communities in Europe. Christian evangelicalism and Pentacostalism — the denominations from which fundamentalism derives — also are flourishing in Latin America, Africa, Central Asia and the United States. Ultra-Orthodox Jewish fundamentalists are blamed for exacerbating instability in the Middle East and beyond by establishing and expanding settlements on Palestinian lands. And intolerance is growing among Hindus in India, leading to deadly attacks against Christians and others. As experts debate what is causing the spread of fundamentalism, others question whether fundamentalists should have a greater voice in government.

The Obama Presidency: Can Barack Obama Deliver the Change He Promises?

As the 44th president of the United States, Barack Hussein Obama confronts a set of challenges more daunting perhaps than any chief executive has faced since the Great Depression and World War II. At home, the nation is in the second year of a recession that Obama warns may get worse before the economy starts to improve. Abroad, he faces the task of withdrawing U.S. forces from Iraq, reversing the deteriorating conditions in Afghanistan and trying to ease the Israeli-Palestinian conflict. Still, Obama begins his four years in office with the biggest winning percentage of any president in 20 years and a strong Democratic majority in both houses of Congress. In addition, as the first African-American president, Obama starts with a reservoir of goodwill from Americans and people and governments around the world. But he began encountering criticism and opposition from Republicans in his first days in office as he filled in the details of his campaign theme: "Change We Can Believe In."

HPV Vaccine: Should It Be Mandatory for School Girls?

A new vaccine that prevents infections from a sexually transmitted disease (STD) that causes cervical cancer is being hailed as a major achievement in women's health.

The human papillomavirus (HPV) vaccine, Gardasil, is recommended by the Centers for Disease Control and Prevention for girls ages 11-12, and could be used by females ages 9-26. Some state lawmakers moved quickly to make inoculations mandatory for school attendance to ensure vaccine access regardless of socioeconomic status. The requirement was approved in the District of Columbia and Virginia. But reactions to an aggressive lobbying campaign by vaccine manufacturer Merck coupled with general concerns about immunization safety stalled efforts to mandate the shots in many states. Conservative groups joined the opposition, saying the vaccine would encourage inappropriate sexual activity and override parental autonomy.

SOCIAL DYNAMICS

Declining Birthrates: Will the Trend Worsen Global Economic Woes?

Nations around the globe worry that low or falling birthrates will cause severe economic problems, including shortages of workers to pay into social security systems to support growing numbers of retirees. While the coming retirement of American baby boomers engenders concern, the United States is exceptional among major industrialized Western nations because its birthrate produces enough children to maintain the population as elderly people die. Most of Europe as well as Japan and China are well below population replacement levels. The current global economic downturn could worsen the situation by forcing young couples to postpone having children until the economy improves. Meanwhile, governments are casting about for solutions, such as cutting spending on the elderly, requiring workers to stay on the job longer before drawing benefits and offering cash bonuses to families to encourage them to have more children.

Rapid Urbanization: Can Cities Cope With Rampant Growth?

About 3.3 billion people — half of Earth's inhabitants — live in cities, and the number is expected to hit 5 billion within 20 years. Most urban growth today is occurring in developing countries, where about a billion people live in city slums. Delivering services to crowded cities has become increasingly difficult, especially in the world's 19 "megacities" — those with more than 10 million residents. Moreover, most of the largest cities are in coastal areas, where they are vulnerable to flooding caused by climate change. Many governments are striving to improve city life by expanding services, reducing environmental damage and providing more jobs for the poor, but some still use heavy-handed clean-up policies like slum clearance. Researchers say urbanization helps reduce global poverty because new urbanites earn more than they could in their villages. The global recession could reverse that trend, however, as many unemployed city dwellers return to rural areas. But most experts expect rapid urbanization to resume once the economic storm has [**Dave/Nancy: This paragraph was incomplete in the original version of the article. Would you be able to complete this sentence, or should we delete it altogether?**]

Reducing Your Carbon Footprint: Can Individual Actions Reduce Global Warming?

As climate change rises closer to the top of the government's policy agenda — and an economic crisis intensifies — more and more consumers are trying to change their behavior so they pollute and consume less. To reduce their individual "carbon footprints," many are cutting gasoline and home-heating consumption, choosing locally grown food and recycling. While such actions are important in curbing global warming, the extent to which consumers can reduce or reverse broad-scale environmental damage is open to debate. Moreover, well-intentioned personal actions can have unintended consequences that cancel out positive effects. To have the greatest impact, corporate and government policy must lead the way, many environmental advocates say.

Socially Responsible Investing: Can Investors Do Well by Doing Good?

Socially responsible investing, which combines financial goals with the aim of improving society through stock screening, shareholder activism and other methods, has grown into a multi-trillion-dollar industry. Concerns about climate change, worker rights and other issues are prompting big institutional accounts as well as small investors to put more and more emphasis on social, environmental and corporate governance factors in weighing investment decisions. But critics say stock-screening methods used by mutual funds are subjective and that socially responsible investments tend not to perform as well as conventional ones. Some of the harshest criticism has been directed at public pension funds using social-investing approaches, such as the California State Teachers' Retirement System, which uses a "double bottom line" approach to investing.

Preface

A re Internet sites like MySpace potentially dangerous? Does gangsta rap harm black Americans? Is traditional matrimony going out of style? Can Barack Obama deliver the change he promises? Can individual actions reduce global warming? These questions and many more are addressed in a unique selection of articles for debate offered exclusively through *CQ Researcher,* CQ Press and SAGE. This collection intended for introductory sociology courses aims to promote in-depth discussion, facilitate further research and help students formulate their own positions on crucial issues.

This first edition includes seventeen up-to-date reports by *CQ Researcher,* an award-winning weekly policy brief that brings complicated issues down to earth. Each report chronicles and analyzes current issues in our society. This collection was carefully crafted to cover a range of issues including celebrity culture, cyber socializing, women's rights, student aid, the Obama Presidency and much more. All in all, this reader will help your students gain a deeper, more critical perspective of timely and important issues.

CQ RESEARCHER

CQ Researcher was founded in 1923 as *Editorial Research Reports* and was sold primarily to newspapers as a research tool. The magazine was renamed and redesigned in 1991 as *CQ Researcher.* Today, students are its primary audience. While still used by hundreds of journalists and newspapers, many of which reprint portions of the reports, the *Researcher's* main subscribers are now high school, college and public libraries. In 2002, *Researcher* won the American Bar Association's coveted Silver Gavel award for magazine excellence for a series of nine reports on civil liberties and other legal issues.

Researcher staff writers — all highly experienced journalists — sometimes compare the experience of writing a *Researcher* report to drafting a college term paper. Indeed, there are many similarities. Each report is as long as many term papers — about 11,000 words — and is written by one person without any significant outside help. One of the key differences is that writers interview leading experts, scholars and government officials for each issue.

Like students, staff writers begin the creative process by choosing a topic. Working with the *Researcher's* editors, the writer identifies a controversial subject that has important public policy implications. After a topic is selected, the writer embarks on one to two weeks of intense research. Newspaper and magazine articles are clipped or downloaded, books are ordered and information is gathered from a wide variety of sources, including interest groups, universities and the government. Once the writers are well informed, they develop a detailed outline, and begin the interview process. Each report requires a minimum of ten to fifteen interviews with academics, officials, lobbyists and people working in the field. Only after all interviews are completed does the writing begin.

CHAPTER FORMAT

Each issue of *CQ Researcher,* and therefore each selection in this book, is structured in the same way. Each begins with an overview, which briefly summarizes the areas that will be explored in greater detail in the rest of the chapter. The next section chronicles important and current debates on the topic under discussion and is structured around a number of key questions. These questions are usually the subject of much debate among practitioners and scholars in the field. Hence, the answers presented are never conclusive but detail the range of opinion on the topic.

Next, the "Background" section provides a history of the issue being examined. This retrospective covers important legislative measures, executive actions and court decisions that illustrate how current policy has

evolved. Then the "Current Situation" section examines contemporary policy issues, legislation under consideration and legal action being taken. Each selection concludes with an "Outlook" section, which addresses possible regulation, court rulings and initiatives from Capitol Hill and the White House over the next five to ten years.

Each report contains features that augment the main text: two to three sidebars that examine issues related to the topic at hand, a pro versus con debate between two experts, a chronology of key dates and events and an annotated bibliography detailing major sources used by the writer.

ACKNOWLEDGMENTS

We wish to thank many people for helping to make this collection a reality. Tom Colin, managing editor of *CQ Researcher,* gave us his enthusiastic support and cooperation as we developed this edition. He and his talented staff of editors and writers have amassed a first-class library of *Researcher* reports, and we are fortunate to have access to that rich cache. We also wish to thank our colleagues at CQ Press, a division of SAGE and a leading publisher of books, directories, research publications and Web products on U.S. government, world affairs and communications. They have forged the way in making these readers a useful resource for instruction across a range of undergraduate and graduate courses.

Some readers may be learning about *CQ Researcher* for the first time. We expect that many readers will want regular access to this excellent weekly research tool. For subscription information or a no-obligation free trial of *CQ Researcher,* please contact CQ Press at www.cqpress .com or toll-free at 1-866-4CQ-PRESS (1-866-427-7737).

We hope that you will be pleased by this edition of *Issues for Debate in Sociology: Selections From CQ Researcher.* We welcome your feedback and suggestions for future editions. Please direct comments to David Repetto, Sr. Acquisitions Editor, Pine Forge Press, an Imprint of SAGE Publications, 2455 Teller Road, Thousand Oaks, CA 91320, or david.repetto@sagepub .com.

—The Editors of SAGE

Contributors

Howard Altman is the courts and cops team leader at the *Tampa Tribune.* He was formerly Mid-Hudson Regional Editor of the *Times Herald-Record,* in Middletown, New York, and editor-in-chief of the *Philadelphia City Paper.* His work has appeared in *The New York Times, Newsday, American Journalism Review,* wired .com. and salon.com, and he is the recipient of more than 50 journalism awards. He graduated from Ithaca College with a BS in communications.

Brian Beary—a freelance journalist based in Washington, D.C.—specializes in European Union (EU) affairs and is the U.S. correspondent for *Europolitics,* the EU-affairs daily newspaper. Originally from Dublin, Ireland, he worked in the European Parliament for Irish MEP Pat "The Cope" Gallagher in 2000 and at the EU Commission's Eurobarometer unit on public opinion analysis. A fluent French speaker, he appears regularly as a guest international-relations expert on television and radio programs. Beary also writes for the *European Parliament Magazine* and the *Irish Examiner* daily newspaper. His last report for *CQ Global Researcher* was "Race for the Arctic."

Thomas J. Billitteri is a *CQ Researcher* staff writer based in Fairfield, Pennsylvania, who has more than 30 years' experience covering business, nonprofit institutions and public policy for newspapers and other publications. His recent *CQ Researcher* reports include "Campaign Finance," "Human Rights in China" and "Financial Bailout." He holds a BA in English and an MA in journalism from Indiana University.

Nellie Bristol is a veteran Capitol Hill reporter who has covered health policy in Washington for more than 20 years. She now writes for *The Lancet, The British Medical Journal* and the *Journal of Disaster Medicine and Public Health Preparedness.* She graduated in American studies from The George Washington University, where she is now working toward a master's degree in public health.

Marcia Clemmitt is a veteran social-policy reporter who previously served as editor in chief of *Medicine and Health,* a Washington industry newsletter, and staff writer for *The Scientist.* She has also been a high school math and physics teacher. She holds a liberal arts and sciences degree from St. John's College, Annapolis, and a master's degree in English from Georgetown University. Her recent reports include "Climate Change," "Controlling the Internet" and "Pork Barrel Politics."

Karen Foerstel is a freelance writer who has worked for the Congressional Quarterly *Weekly Report* and *Daily Monitor, The New York Post* and *Roll Call,* a Capitol Hill newspaper. She has published two books on women in Congress, *Climbing the Hill: Gender Conflict in Congress* and *The Biographical Dictionary of Women in Congress.* Her most recent *CQ Global Researcher* was "China in Africa." She has worked in Africa with ChildsLife International, a nonprofit that helps needy children around the world, and with Blue Ventures, a marine conservation organization that protects coral reefs in Madagascar.

Sarah Glazer, a London-based freelancer, is a regular contributor to the *CQ Researcher.* Her articles on social policy issues have appeared in *The New York Times, The Washington Post, The Public Interest* and *Gender and Work,* a book of essays. Her recent *CQ Researcher* reports include "Fair Trade Labeling" and "Future of Feminism" and "Antisemitism in Europe" in *CQ Global Researcher.* She

graduated from the University of Chicago with a BA in American history.

Kenneth Jost graduated from Harvard College and Georgetown University Law Center. He is the author of the *Supreme Court Yearbook* and editor of *The Supreme Court from A to Z* (both CQ Press). He was a member of the *CQ Researcher* team that won the American Bar Association's 2002 Silver Gavel Award. His previous reports include "Treatment of Detainees" and "War on Terrorism."

Peter Katel is a *CQ Researcher* staff writer who previously reported on Haiti and Latin America for *Time* and *Newsweek* and covered the Southwest for newspapers in New Mexico. He has received several journalism awards, including the Bartolomé Mitre Award for coverage of drug trafficking from the Inter-American Press Association. He holds an AB in university studies from the University of New Mexico. His recent reports include "New Strategy in Iraq," "Prison Reform" and "Real ID."

David Masci specializes in science, religion and foreign policy issues. Before joining The *CQ Researcher* in 1996, he was a reporter at Congressional Quarterly's *Daily Monitor* and *CQ Weekly.* He holds a law degree from The George Washington University and a BA in medieval history from Syracuse University. His recent reports include "Rebuilding Iraq" and "Human Trafficking and Slavery."

Jennifer Weeks is a *CQ Researcher* contributing writer in Watertown, Massachusetts, who specializes in energy and environmental issues. She has written for *The Washington Post, The Boston Globe Magazine* and other publications, and has 15 years' experience as a public-policy analyst, lobbyist and congressional staffer. She has an AB degree from Williams College and master's degrees from the University of North Carolina and Harvard. Her previous *CQ Global Researcher* examined "Carbon Trading."

1

Celebrity Culture

Are Americans Too Focused on Celebrities?

Howard Altman

Billionaire lifestyle entrepreneur Martha Stewart received heavy media coverage after her release from a West Virginia prison on March 4, 2005. The media say they cover celebrities heavily because of strong reader and viewer interest, but critics say excessive coverage of celebrities diverts attention from more serious journalistic pursuits and gives younger readers a distorted view of reality.

From *CQ Researcher*,
March 18, 2005.

Martha is everywhere. For days before and after her release from prison, she is the blazing star around which television, the Internet, newspapers and magazines revolve. There she is, newly svelte and smiling sweetly, leaving prison. Waving girlishly and bussing the pilot on the cheek as she boards a private jet to return to her upstate New York estate. Joking with reporters about not getting cappuccino in prison and missing fresh lemons. Lovingly stroking her handsome horses over the pasture fence. Addressing adoring employees at Martha Stewart OmniMedia and showing off the shawl crocheted for her by a fellow inmate.

Domestic diva, media magnate, hero, outcast, convict, comeback kid and soon-to-be-star of her own reality show — Martha Stewart is among the few people on Earth (along with Jennifer Aniston and Brad Pitt) capable of diverting the media from the all-consuming feeding frenzy of the Michael Jackson child-molestation trial.

In short, Martha is the essence of celebrity — and we can't take our eyes off her.

On a very basic, biological basis, scientists say we humans are hardwired to be fascinated with celebrity, and that our brains receive pleasurable chemical stimuli when we see familiar faces.

"Celebrity journalism has never been hotter," says *Washington Post* media critic Howard Kurtz. "What used to be the realm of *People* magazine and "Entertainment Tonight" now has a foothold in every part of the media business. That's why there are 1,000 journalists camped out in California for the Michael Jackson trial. That's why magazines and newspaper gossip columns breathlessly

Celebrity Coverage Doubled in News Magazines

The percentage of pages in news magazines dedicated to celebrities and entertainment doubled from 1980 to 2003, while coverage of national affairs dropped from 35 percent of all pages to 25 percent.

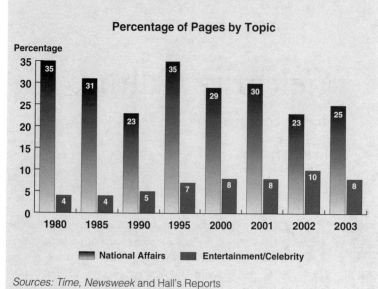

Percentage of Pages by Topic

National Affairs / Entertainment/Celebrity

	1980	1985	1990	1995	2000	2001	2002	2003
National Affairs	35	31	23	35	29	30	23	25
Entertainment/Celebrity	4	4	5	7	8	8	10	8

Sources: Time, Newsweek and Hall's Reports

The constant barrage of celebrity has led more and more people to risk their dignity, and even their lives in some cases, for the crack-like high of their "15 minutes of fame," as artist Andy Warhol famously put it.

Moreover, some researchers argue that as the media dishes out an increasingly rich diet of celebrity hype, less and less attention is paid to informing citizens about government and the world around them — undercutting a cornerstone of a democratic society. Many trace the new emphasis on celebrities to the massive consolidation of the mass media industry, which began in the 1990s when newspapers faced layoffs and drops in circulation and profits. Media companies were gobbled up by mega corporations with a greater commitment to stockholder proftis than to maintaining large, traditionally money-losing news departments.

In many cases, newspapers and broadcast stations owned by family dynasties — with traditionally strong commitments to the local community and relatively low profits — were replaced by huge corporations demanding that news departments produce double-digit profits. As a result, government and foreign news coverage was slashed and often replaced by cheaper-to-produce celebrity gossip, media critics say.

The squeeze on news departments became even more intense when online news outlets began to produce even more competition for viewers' attention.[1]

Yet, as media organizations scale back coverage of government and world events — even the wars in Iraq and Afghanistan — there seems no shortage of resources available for celebrity doings. Celebrity "news" magazine shows have sprouted like mushrooms after a rainstorm. One even devotes a half-hour each day to celebrities' legal problems. Indeed, even as the small army of journalists camps outside the courthouse in California where Michael Jackson is being tried, ABC is debating replacing Ted Koppel's celebrated news show, "Nightline," with more celebrity fluff.

chronicle every breakup by Ben [Affleck] and Jen [Garner], every Britney marriage, every birth to a remotely famous B actress."

Fascination with celebrity has been fueled by an explosion in the number of Internet sites and cable television channels, including 24-hour news shows. As the number of shows and Web sites increased, so did competition for audiences and ad dollars. In turn, that raised the demand for more cheap content, such as the latest celebrity gossip, to fill the burgeoning amounts of broadcast airtime.

"Television, more than any other cultural development, has radically changed our experience of celebrity," says David Blake, a professor of English at the College of New Jersey, in Ewing. "Television has made celebrities both prevalent and ubiquitous, and with the rise of television came a whole new branch of the public relations industry. Public relations once focused on preparing accomplished individuals for the interest and scrutiny that had come to them. Now it involves manufacturing celebrities to meet the culture's seemingly insatiable desire for them."

Part of modern celebrity is the money showered upon true stars. In the eyes of many, Alex Rodriguez, the New York Yankees' third baseman, took on the aura of a Donald Trump when he signed a 10-year, $252 million contract in 2001. Some movie stars make that by working in a few films.

But the fascination with celebrities and their stratospheric earnings has taken its toll. More American teenagers can name the Three Stooges than the three branches of government; more kids know who won the "Battle of the Network Stars" than the Civil War, says comedian and pop-culture commentator Mo Rocca.

Celebrity culture is having other negative impacts on society. According to British researcher Satoshi Kanazawa, of The London School of Economics and Science, children's mental health suffers the more they believe that happiness comes from money, fame and beauty. He found that the human brain was not designed to handle the constant bombardment of celebrity-based stimuli and that we are losing touch with our friends and family as a result. Meanwhile, a study conducted in the United States shows that we are all just a few stressors short of becoming celebrity stalkers.[2] And more and more Americans are seeking plastic surgery, the direct result of people either wanting to look like celebrities or feeling pressured to look younger and better because of the very high beauty bar set by celebrities, says New York plastic surgeon Z. Paul Lorenc.

The outlook for our celebrity-saturated culture, say many media watchers and social scientists, is bleak. "It's already all-Paris-Hilton-all-the-time, or nearly so," says Marty Kaplan, dean of the Annenberg School for Communications at the University of Southern California, "so you don't have to extrapolate that pathology very much to see the future.

"News coverage will continue to shrink; traditional hard news (like politics) will package and present itself even more aggressively as entertainment in order to get attention," Kaplan continues. " 'Journalism' will become an even more important profit center for entertainment conglomerates."

As the amount of news decreases, citizens' ability to stay informed — and thus participate responsibly in democracy — also will diminish, says David T.Z. Mindich, an associate professor of journalism and mass communication at Saint Michael's College, in Colchester, Vt.

As pundits, social scientists and media watchdog groups examine the celebrity culture phenomenon, here are some of the questions they are debating:

Is America's fascination with celebrity bad for society?

Every day, from living-room TVs to supermarket checkout counters, the mass media bombard Americans with

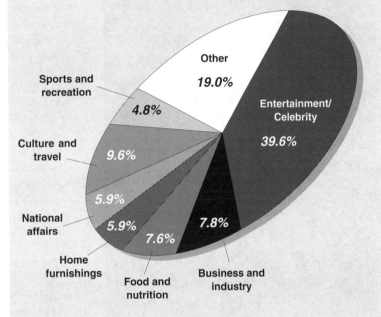

Celebrities Dominate Magazine Covers

Entertainers and other celebrities appeared on the covers of nearly 40 percent of all American magazines in 2004. The next largest category, culture and travel, came in at almost 10 percent, while only 6 percent of covers were related to national affairs.

Other **19.0%**

Entertainment/ Celebrity **39.6%**

Sports and recreation **4.8%**

Culture and travel **9.6%**

National affairs **5.9%**

Home furnishings **5.9%**

Food and nutrition **7.6%**

Business and industry **7.8%**

Note: Percentages do not add to 100 due to rounding.
Source: Hall's Reports

AFP Photo/Robyn Beck

Pop star Michael Jackson arrives at his child-molestation trial in Santa Maria, Calif., being covered by hundreds of media representatives. An explosion in cable television outlets competing for audiences and ad dollars has helped fuel the demand for celebrity news, which is a relatively inexpensive way to fill airtime.

images of celebrities and their rarified lives. But experts have differing opinions on whether it is a good or a bad thing for Americans to be inundated with news about the rich and famous — not only accounts of their privileged lives but also their battles with weight loss, criminal charges, sexual dalliances, drug abuse, broken marriages and problem children.

Perhaps the most obvious downside of celebrity culture is how it has changed whom Americans idolize, says Al Tompkins, group leader for broadcast and online journalism at the Poynter Institute, in St. Petersburg,

Fla.* "Celebrity has taken the place of heroes," he says. "When I ask college and high school students who their heroes are, they usually name celebrities, such as athletes or movie stars, not names that did something heroic or noteworthy."

But Lorenc worries about the danger posed by the impact on people's self-image. "There is tremendous danger" in unchecked celebrity worship, Lorenc says. "A perfect example, is 'I Want A Famous Face' — the MTV television show in which patients come into a doctor's office and say, 'I want to look like Britney Spears,' or 'I want to look like so and so.'

"That shouldn't happen," insists Lorenc, author of *A Little Work: Behind the Doors of a Park Avenue Plastic Surgeon.* "No one should aspire to look like someone else. If I have a patient with a photograph who says, 'I want to look like that,' they don't need me, they need a therapy session. It's very unhealthy to perpetuate that. I won't operate on them."

The danger, he says, is not just that people want to look like specific celebrities but that it perpetuates a worship of youthfulness, and increasingly, Americans are turning to plastic surgery to capture the youth and glamour associated with celebrities. According to the American Society for Aesthetic Plastic Surgery, the number of plastic surgery procedures performed in America increased fourfold from 1997 to 2003 — from slightly more than 2 million to more than 8 million.[3]

"Even celebrities are in a bind," Lorenc says. "They have an image they have to upkeep and are forced to do that with Botox [a botulism neurotoxin injected to eliminate wrinkles]. They have to maintain an image and a lifestyle and an income. Do they influence people? Of course. Patients want to look younger, feel better about themselves."

The youth culture even influences the power elite, he says. "A lot of men from Wall Street say, 'I am competing against men half my age, who are working for a quarter of my price.' We are a youth-oriented culture."

Psychologist James Houran, of Irving, Texas, says celebrity worship is more than skin deep. It is a "gateway drug toward stalking," he cautions.

Houran is the co-creator of the Celebrity Worship Scale, which measures an individual's level of interest in

---

* The nonprofit Poynter Institute owns Congressional Quarterly Inc., the parent company of CQ Press, publisher of the *CQ Researcher*.

celebrities. "Celebrity worship starts off with normal, healthy behavior," he says. "But it can be transformed into more dysfunctional expressions," where people feel a connection to a celebrity that does not exist.

Houran, along with other British and U.S. researchers, found that one-third of Americans suffer from some form of "celebrity worship syndrome" in a study published in February 2002. In its most innocuous form, the condition manifests itself as a sense of emptiness, but the study found it can progress to obsessive thinking and, in rare cases, worsen into behavior — like stalking — that is driven by delusions.[4]

Houran recalls a teenage girl who began injuring herself after learning that punk singer Marilyn Manson, her favorite celebrity, was getting married. "She cut her arms, neck and legs. She was rushed to the hospital. She wanted to be the one to change him. When she was discharged, she realized what she did was extreme. But she still rationalized her obsession, saying, 'I just want him to be happy. If he is happy, I am happy. He is the only person I connect with.' "

Everyone, says Houran, is susceptible. "You don't have to be a stalker to have this [affect] your life, negatively and intensely. Those extreme celebrity worshippers don't start off that way, but the bad news is that it implies there is a stalker in all of us, given the right set of variables."

But not all studies have shown that celebrity worship has a decidedly negative impact. In a study published in March 2004, a group of British researchers found that gossiping about celebrities took up most of the social time of nearly one-third of a sample of 191 English youngsters ages 11 to 16. But these young people were far from being isolated; in fact, researchers found the gossiping children had a stronger network of close friends than their peers who were less interested in celebrities.[5]

The Harvard-educated Rocca, who appears frequently on CNN's "American Morning," believes saturation celebrity coverage has had an inoculating effect on society, particularly young people, and has made college students, in particular, extremely media savvy.

"There is an overwhelming appetite for celebrity and pop culture news across the board in America right now and on campus in particular," Rocca says. "But I have a strange faith in college students. They are both more optimistic and skeptical than everyone else.

"It sounds like a strange contradiction, but they consume all this celebrity news with tongue planted firmly in cheek, I think," Rocca continues. "Nobody is wide-eyed any longer when it comes to celebrity news. When I see college students devouring *Us Weekly*, they know it is all a joke. There is a hunger for something else. When I go to campuses and talk about my interest in presidential history, while a lot of students may not know much, they are hungry for something more substantive than the latest news on the Olsen twins."

Growing up in a celebrity-saturated culture helps turn college students today into experts on how the media work, Rocca says. "I am constantly amazed at how much the average student knows about what goes into making a TV show. Everyone has deconstructed the media, understands the ingredients and understands how the artifice is created. Essentially, students know it is all BS — the work of celebrity publicists and stories they are fed. The students revel in the cheesiness of it."

Conversely, Rocca believes that people who did not grow up with constant celebrity news are more apt to take celebrity news at face value. "I am betting older people were more engrossed by the Laci Peterson [murder] story," he says. "That was essentially tabloid trash. It had no relevance to people's lives. College kids . . . can draw a distinction between legitimate news, say the tsunami or Iraq, and soap operas that masquerade as news, like the Laci Peterson story."

Moreover, says Dan Kennedy, media critic at the *Boston Phoenix*, some heavily played celebrity stories can help make this a better country. "The coverage of the O. J. Simpson murder trial actually helped foster a national conversation about race and celebrity that otherwise would not have taken place, totally apart from the fact that he got away with murder," Kennedy says.

In fact, Kennedy thinks that today's media consumers are more sophisticated than in the past, and thus less obsessed with celebrity. "Large segments of society have always lived vicariously through celebrities," Kennedy observes. "It's not healthy, but it's ever been thus. In the 1860s, the wedding of Charles Stratton and Lavinia Warren — better known as General and Mrs. Tom Thumb — was one of the great media spectacles of the age, with the couple even dropping by the White House for a heavily publicized visit with the Lincolns.

"And I'm not sure that anything we've seen today exceeds the bizarre devotion to Rudolph Valentino in the 1920s," he continues. "For that matter, the media

Reality TV Rarely Leads to Lasting Fame

Reality TV shows have introduced the viewing public to instant celebrities like "The Bachelorette" lovebirds Trista and Ryan, "The Apprentice" villain Omarosa and "Survivor" schemer Richard Hatch.

The unscripted programs have given all-too-fleeting fame to thousands of average Janes and Joes who helped provide casting directors with the many stereotypes that make up reality television, including the hypersensitive minority, the big-city neophyte, the sex siren.

"The vast majority of people on reality TV believe that it is not only going to bring a bachelor that they can marry or $1 million for surviving life on an island, but also that it's the beginning of a career that will make them celebrities," says Robert Thompson, founding director of the Center for Popular Television at Syracuse University.

But most reality alums soon learn that their celebrity has a short shelf life — six months for most, Thompson says.

"Now that we've had years to map this out — five since the first 'Survivor' in the summer of 2000 and 13 since the first 'Real World' aired in 1992 — the votes are in," Thompson says, "and the chances of making a long career in show business from a reality show are very, very small."

But there are a few exceptions. "Survivor" alumna Elizabeth Hasselbeck is now one of five hosts of "The View." And London "Real World" alum Jacinda Barrett recently had substantial roles in the films "Ladder 49" and "Bridget Jones: The Edge of Reason."

"American Idol" stars Kelly Clarkson and Clay Aiken also have found mainstream stardom, but that is largely because 'American Idol' is really a talent show, Thompson says.

But for every success story, there are hundreds of cast members who have tried and failed to extend their 15 minutes of fame.

"It's a letdown 99 percent of the time for most people," said Brian Brady, a talent booker for the casts of "Survivor," "The Apprentice" and other reality shows.[1]

"I get 10 calls a day from cast members trying to get some kind of work," Brady says. "You can hear it in their voices; they're desperate. They're trying to milk their show for anything."

Jamie Murray was 22 when he appeared as one of the roommates on the ninth season of the "Real World" in New Orleans. Now 27, Murray has spent much of the last five years using his reality experience to book college appearances, which pay about $2,000 each. He has also appeared on two MTV "Challenges," which bring back cast members from past seasons of "Real World" and "Road Rules" to compete in events like raft building and bungee jumping for plastic rings. With a little luck, he says, he won both challenges, earning $80,000 and two cars.

Murray says that was the only compensation he's received from his celebrity. "My financial situation has been less than stellar during the last few years because I've been living off the scraps of the 'Real World,' " Murray says. "All my high

today may be less celebrity-obsessed than that of 100 years ago — at least in terms of the [print] press."

Does the media's attention to celebrities lead to poor coverage of more important issues?

The performance of the American media in covering the run-up to the war in Iraq has come in for scathing criticism from press critics — and the press itself. Many media critics, including *New York Press* columnist Matt Taibbi, castigated the so-called mainstream media for failing to adequately challenge the Bush administration's rationale for going to war.[6]

And an editorial in *The New York Times* acknowledged that mistakes in the *Times'* coverage were made. "The world little noted, but at some point late last year

the American search for weapons of mass destruction in Iraq ended," the *Times* commented. "We will, however, long remember the doomsday warnings from the Bush administration about mushroom clouds and sinister aluminum tubes; the breathless reports from TV correspondents when the invasion began, speculating on when the 'smoking gun' would be unearthed; our own failures to deconstruct all the spin and faulty intelligence."[7]

There are many reasons, critics argue, why the U.S. media have failed to pay more attention to world events or even to cover important events closer to home. It is "much easier to land 'event'-oriented coverage (such as spot news, crime news, announcements or events that occur, scheduled and unscheduled," argues Tompkins, of the Poynter Institute.

school and college friends are doing big-time, corporate jobs, and I'm still making money off appearing at some bar night in Austin, Texas."

But someone is getting rich off Jamie's MTV appearances. "Viacom has a multimillion-dollar syndication deal for 'Real World,' and not a dime was thrown down to the people on the show," Murray explains. "We signed our rights away."

It's harsh, Thompson says, but potential cast members know that if they don't sign, there are thousands of others willing to do so.

Jon Murray, a co-creator of "Real World," understands that it's difficult for his cast members to have empty pockets when they are recognized on the street. "It's hard for any of us who haven't . . . gone on a reality show, to understand what it's like to be famous for being yourself, but not necessarily having a lot of money that goes with fame," Murray said.[2]

Unfortunately for most reality show stars, they rarely have skills that can take them beyond reality TV, Thompson says. "Jerri Manthey from the first 'Survivor' would love to

MTV's "Real World" is considered the first modern reality TV program. Launched in 1992, it follows the lives of seven young strangers living in a house together. Above, the show's Paris cast visits New York City.

be a big star, but she's not a great singer or a great actress. She isn't a great anything that makes you a celebrity," Thompson says.

Even for those who are great at something, reality TV is no guarantee of success. It can even hurt wannabe stars by typecasting them and showing them in a negative light, Brady says. "A lot of these people end up bartending and waitressing, and, hopefully, they're counting up nice tips because the patrons of the restaurant or bar recognize them," he says.

But without the talent to keep them in the limelight, most reality stars quickly slide into obscurity, Thompson says. "It's celebrity built on the foundation of sand, and it blows away."

— *Kate Templin*

---

[1] www.concertideas.com.

[2] Kate Aurthur, "Reality Stars Keep on Going and Going," *The New York Times*, Oct. 10, 2004, p. B22.

Taibbi is less charitable. "In the run-up to the war," he writes, "every major daily and television network in the country parroted the White House's asinine WMD claims for months on end . . . "Justice would seem to demand that a roughly equivalent amount of coverage be given to the truth, now that we know it (and we can officially call it the truth now, because even Bush admits it; previously the truth was just a gigantic, unendorsed pile of plainly obvious evidence). But that isn't the way things work in America.

"We only cover things around the clock every day for four or five straight months when it's fun," and "fun" boils down to covering celebrities at the expense of all else, Taibbi argues.[8]

On the other hand, the Annenberg School's Kaplan blames the shrinking "news hole," or the amount of space devoted to hard-news coverage. For example, the percentage of pages in news magazines dedicated to celebrities and entertainment doubled from 1980 to 2003, while coverage of national affairs dropped from 35 percent of all pages to 25 percent. (*See graph, p. 2.*)

"The smaller the hole for hard news, the less likely that people will find out what they need to know about their communities, their country and their world," Kaplan says. "Celebrity news attracts eyeballs. We can't help it. Fame is mesmerizing. The challenge for responsible media is to make the [more] important [stories] interesting."

Competing with celebrity news is a tall order, says *The Washington Post*'s Kurtz. Celebrity news is "cheap and easy to cover, easier, say, than unraveling the president's budget cuts or Social Security proposal," Kurtz

Are You Celebrity Obsessed?

A test developed by a group of American and British psychologists ranks interest in celebrities from harmless escapism to obsessive thinking that — in rare cases — may lead to delusion-driven behavior like stalking.

Answer yes or no to the following statements:

Yes No

☐ ☐ 1. I often feel compelled to learn the personal habits of my favorite celebrity.

☐ ☐ 2. I love to talk with others who admire my favorite celebrity.

☐ ☐ 3. When something happens to my favorite celebrity, I feel like it happened to me.

☐ ☐ 4. I enjoy watching, reading or listening to my favorite celebrity because it means a good time.

☐ ☐ 5. I have pictures and/or souvenirs of my favorite celebrity, which I always keep in exactly the same place.

☐ ☐ 6. When my favorite celebrity dies, I will feel like dying, too.

Celebrity Attitude Scale

If you answered "Yes" to:

Nos. 2 and 4 — Your celebrity attitudes are on the Entertainment-Social level; they are undisruptive and focused on the entertainment abilities of celebrities.

Nos. 1 and 5 — Your celebrity attitudes are on Intense-Personal level; attitudes about celebrity are more intimate and obsessive and can have a negative effect on mood and behavior.

Nos. 3 and 6 — Your celebrity attitudes are on the Borderline-Pathological level; attitudes and behaviors are dangerous, troublesome and anti-social.

Source: Lynn McCutcheon, et al., "Conceptualization and Measurement of Celebrity Worship," British Journal of Psychology, Feb. 1, 2002

that anything resembling hard news journalism is coming to an end," McCafferty says. "But I don't believe it for a minute. The last time I checked, *USA Today* and *The Washington Post* and *The Wall Street Journal* and the other usual suspects are still doing some pretty darn good hard-news stories. And my local *Fairfax Journal* is still staying on top of how local politicos are spending my tax dollars.

"I also notice that *USA Today's* "Life" section — that's supposed to be the fluffy one — devotes as many pages to health, science and other related topics as it does to Hollywood. The "Style" section [of *The Post*] still devotes 80-inch features to newsmakers, as opposed to star machinery."

While there has been a tremendous increase in time and space devoted to celebrity coverage, McCafferty says, the advent of cable and the Internet means that there is a huge appetite for all kinds of content — including hard news.

"Has there been a huge increase in celebrity-devoted magazines, cable shows and the like? Of course," says McCafferty. "There's also been a huge increase in business magazines and 24/7 financial cable shows. There are countless niches within the business-magazine industry. If you want to read about small business, you have a choice of several competing titles.

says. "It's the O. J. syndrome as a permanent feature of our journalistic culture. Martha Stewart, convicted felon, is about to get a television show. Need I say more?"

But Dennis McCafferty, who covers celebrities as senior writer for *USA Today Weekend*, says our fascination with celebrities does not mean the death of hard news.

Is journalism in trouble? McCafferty asks. "I'm sure a response of 'Yes! Mercy yes!' would come from the sanctimonious types who incessantly write letters saying

"The same with mutual funds, personal finance, venture capitalists, CEOs, and, for all I know, administrative assistants and the guys who change purified water jugs in the office everyday.

"The same massive increase in ongoing coverage is also reflected in what's available when it comes to sports, health, parenting, community, religion and every single other subject that affects our lives. Celebrity news is hardly crowding that out. There's simply more of all kinds of news, period, and that includes hard news."

Weekly magazines like *People, Us Weekly, In Touch* and *Star* reel in readers with gossip, interviews and paparazzi photographs of their favorite celebrities. Jennifer Lopez was the most featured celeb in 2004, appearing on 29 covers published by the four magazines. Jennifer Aniston, alone or with estranged husband Brad Pitt, came in second with 26 covers. The February 2005 Aniston-Pitt breakup sparked a celebrity magazine feeding frenzy, with *Us Weekly* featuring the couple on its cover for five consecutive weeks, the longest for a single news story.

Kennedy, of the *Boston Phoenix*, agrees. "I'm not so sure that the media *per se* are obsessed with celebrity," he says." Our culture is obsessed with celebrity, and the media are a reflection of that, although it's complicated, because celebrity wouldn't be possible without the media.

"So you've got a non-virtuous circle: The media cover celebrities because that's what a large swath of the public wants; and then, in response to public demand, the media end up covering celebrities even more. . . . I consider my own tastes to be fairly heavily oriented toward real news, yet even I would rather read about Ozzie Osbourne's latest stint in rehab than Social Security reform."

BACKGROUND

'Star' Gladiators

Fame and celebrity are nothing new to human civilization. As humans progressed from spending all their time hunting and gathering, those who excelled at war, sports, politics

and the arts captured the imagination, says Blake at the College of New Jersey.

"Many scholars find it useful to distinguish between fame and celebrity, connecting fame to the kind of renown people achieve for extraordinary talents or achievements, and celebrity for the kind of meretricious notoriety that is so prevalent today and so frequently criticized — the state of being known for being known," notes Blake.

"I'm inclined to see fluidity between these two terms, to see them as having differences in degree rather than kind," Blake continues. "For someone like Alexander the Great, or Caesar, fame was an important, motivating force. Ovid compared fame to a spur, propelling men to greater accomplishments.

"Being a celebrity adds a new dimension to this immortality, for it suggests that one is actively celebrated by the crowd. The original Latin meaning of celebrity is "to be thronged." Along with this comes a sense of visibility, a sense of being widely recognized and known. How frequently are you seen? How visible is your face? As one wag put it, God may be famous, but Jesus is the celebrity."

The mainstream media have long been fascinated by celebrities, as shown by this 1962 *Life* cover featuring superstars Elizabeth Taylor and Richard Burton during the filming of "Cleopatra."

The first celebrities may have been the cave dwellers who began leaving their artistic marks some 40,000 years ago, but there are no records identifying any of them. The first *known* celebrities probably were the Pharaohs, such as the first "power couple," Akhenaten and his beautiful wife Nefertiti, who lived 3,500 years ago.

The Golden Era of Greece, about five centuries before the birth of Christ, produced great thinkers like the mathematician Pythagoras and the philosopher Socrates.

Perhaps not surprisingly, one of the earliest celebrities was an athlete: Milo of Kroton, a five-time Olympic wrestling champion whose fame was at its height between 532 and 516 B.C. Hans van Wees, a lecturer in ancient Greek history at University College, London, says such athletes were accorded the same high status enjoyed by today's superstars.[9]

"They were not only widely talked about but also given red-carpet treatment," van Wees says. When they returned home, for instance, part of the city wall was demolished so they did not have to use the gates like "ordinary mortals," he says.

The athletes also won the lifelong right to free meals and would advertise their fame by commissioning hymns of praise from famous poets that would be performed in their honor — probably "the next best thing to appearing on TV," van Wees says.[10]

During the Roman Empire, other "athletes" — including the slaves who became gladiators — achieved fame. And Greeks were followed by the context of their celebrity, which speaks volumes about the political and social order of the day.

By 65 B.C., as Caesar was pitting 320 pairs of gladiators against each other in an amphitheater at one time, news of gladiators' battles spread by word of mouth.[11] Boys idolized them, often taking lessons at gladiator schools, while women were known to have affairs with them.[12]

A year later, Cleopatra, history's enduring icon of sex, beauty and political intrigue, was born. She lived for 39 years before famously committing suicide by raising an asp to her breast.

One of the first writers to win fame and celebrity was the Roman historian Tacitus (55-120 A.D.). His seminal work, *The Annals*, chronicled the nexus between fame and power in Rome.

Tacitus' description of Roman consul Caius Petronius, for instance, sounds like a precursor of the recent TV show, "Lifestyles of the Rich and Famous:" "His days he passed in sleep, his nights in the business and pleasures of life. Indolence had raised him to fame, as energy raises others, and he was reckoned not a debauchee and spendthrift, like most of those who squander their substance, but a man of refined luxury."[13]

Tacitus' nuanced examination of Petronius is just one of many instances where the writer investigated the machinations and foibles of the power players of his day. "The love of fame is the last weakness which even the wise resign," he observed.[14]

In the ensuing centuries, artists, athletes, writers, rulers, discoverers and conquerors became celebrities — until the nature of celebrity changed drastically.

Modern Celebrity

By the time William Shakespeare arrived on the scene in the mid-16th century, times were changing. England had a very

CHRONOLOGY

1880s-1930s *First modern power-generating station is invented, followed by movies, radio and television.*

1879 First radio is developed.

1910s Hollywood develops the star system . . . *Photoplay, Motion Picture Stories* and other fan magazines begin publishing, ushering in the age of celebrity worship.

1920s Silent-film comedian Fatty Arbuckle is charged with murder, becoming one of the first victims of the celebrity gossip machine . . . Gossip columnist Walter Winchell reaches more than 50 million homes with his radio show and newspaper column.

1922 First public radio broadcasting station opens in Pittsburgh.

1939 Television is introduced at the World's Fair in Flushing Meadows, N.Y.

1960s-1980s *Scientists conceive of the Internet, but mass communications is dominated by newspapers, a few television channels and radio.*

1966 Former movie star Ronald Reagan is elected governor of California. He is re-elected in 1970.

1972 U.S. computer experts unveil the ARPANET, forerunner of the Internet.

1974 *People* magazine is launched by Time Inc., paving the way for the delayed explosion, two decades later, of innumerable imitators.

1980 Reagan is elected president.

1990s *Widespread use of the Internet revolutionizes mass communication; media organizations begin filing news continuously on the Web. Traditional media are gobbled up by megacorporations.*

January 1990 Warner Communications and Time Inc. complete $14.1 billion merger, creating world's biggest media conglomerate.

1992 "Real World" debuts as the first reality TV show.

September 1993 New York Times Co. buys Affiliated Publications Inc. (*The Boston Globe*) for $1.1 billion — biggest takeover in U.S. newspaper history.

1994 Viacom buys video rental chain Blockbuster Entertainment Corp. in August for $8 billion. . . . In July Viacom buys Paramount Communications, a movie, publishing and sports company, for $10 billion.

1996 Walt Disney Co. buys Capital Cities/ABC for $19 billion in February, creating a movie, television and publishing conglomerate . . . In October Time Warner and Turner Broadcasting System complete $7.6 billion merger.

September 1999 Viacom buys CBS for $34.5 billion in the biggest media marriage ever.

2000s *Technological advancements continue to change the way people think of news. Reality TV becomes major phenomenon.*

2000 "Survivor" airs and becomes a huge hit, triggering a deluge of reality TV shows that produce hundreds of instant celebrities. . . . In the largest corporate merger in history, AOL acquires Time Warner in a stock swap valued at $166 billion.

2002 *Forbes* magazine names Jennifer Aniston the nation's top celebrity.

2003 Arnold Schwarzenegger sworn in as governor of California.

March 5, 2004 Martha Stewart is convicted of four counts of lying to investigators and obstructing justice in connection with a well-timed sale of stock.

Oct. 8, 2004 Stewart begins her five-month sentence, eluding photographers and cameramen staking out the federal prison in Alderson, W.Va.

2005 Jennifer Aniston and Brad Pitt announce they are separating on Jan. 7. . . . Michael Jackson's trial on child-molestation charges begins on Feb. 28. . . . Martha Stewart gets out of prison on March 5.

Riding Celebrity Into Politics

Twenty years ago, when "The Terminator" took moviegoers by storm, who could have predicted that the bodybuilder playing the indestructible cyborg would one day run the most populous state in the nation?

But in today's celebrity-obsessed society, no one was surprised when Arnold Schwarzenegger announced his candidacy for governor of California — and, in true celebrity fashion, did it on the "The Tonight Show With Jay Leno."

It was the latest in a growing trend toward cross-pollination between celebrity and politics. "The lines between politician and celebrity have become increasingly obscured in the past 30-40 years," says David Blake, a professor of English at the College of New Jersey, in Ewing. "I think we are only now beginning to see the consequences of that blurring."

Darrell West, director of the Taubman Center for Public Policy at Brown University and author of *Celebrity Politics*, says that when celebrities run for office, they often win by impressive margins, even though voters make fun of them initially, saying they know nothing about politics.

In fact, elected officials and celebrities need similar skills: connecting with an audience, developing a loyal fan base and cultivating the "it" factor that transcends a résumé.

Indeed, politicians have learned that showmanship and charisma can help win elections. Conversely, as the public grows increasingly skeptical of career politicians, lacking a political pedigree can be a plus. Voters are often drawn to celebrities and other political-outsiders, such as actor Clint Eastwood (former mayor of Carmel, Calif.), singer Sonny Bono (the late congressman from California) and pro

wrestler Jesse Ventura (former governor of Minnesota). In a *USA Today* poll taken shortly before California's 2003 gubernatorial election, 34 percent of likely voters said Schwarzenegger's lack of experience actually made them more likely to vote for him.[1]

"Celebrities bring a special credibility that career politicians don't have," West says. "They haven't spent their lifetimes cutting deals and doing things the public doesn't like."

Moreover, celebrity politicians operate differently than their professional peers. For instance, because celebrities are less entrenched in the political establishment, West says, they are often more likely to take risks, trust their gut instincts and support ambitious programs.

Jack Kemp, former quarterback for the Buffalo Bills, took his athletic enthusiasm with him when he went into politics, advocating enterprise zones that encouraged entrepreneurship and job creation in urban America. After serving in Congress, Kemp went on to become a Cabinet member under the first President George Bush and eventually the Republican vice presidential nominee in 1996. "Having been a quarterback, I had a quarterback mentality," he said. "In a huddle, you can't have everybody talking . . . and you're willing to throw a long ball on third and one or on fourth and one. Which I was always willing to do."[2]

Likewise, in 2004, Schwarzenegger broke with the Republican Party to endorse stem-cell research, which California voters supported last November. And when President Ronald Reagan couldn't get Congress to support

famous woman — Elizabeth I — making history, and dramatic shifts in science, religion and culture were occurring.

The Elizabethan era saw popular theater become a major source of entertainment for the masses. The clergy and scholars may have disapproved of such "corrupt" entertainment, but it turned Shakespeare into London's most celebrated playwright.

Other writers gained recognition as literacy rates improved.[15] And the more people read, the more they wanted to know about the writers.

"Celebrity, as we know the term, begins to appear in the 18th century with the increasing importance of the

public sphere," Blake says. "As people came to recognize the public as an entity separate from the government and the church, as newspapers began to turn their attention to items of public interest, a new class of people emerged as the recipients of widespread attention."

Although these individuals, were most frequently known for their exceptional skill — Alexander Pope, Jean Jacques Rousseau, Lord Byron — people were frequently interested in their private lives and their personalities, Blake says. By the middle of the 19th century, the notion of celebrity had grown to embrace well-known people in society who were glamorous or fashionable. During a trip

his economic program, he used his charisma to get grass-roots supporters to inundate Congress with letters and phone calls. Congress eventually passed the plan.

But celebrities aren't magicians, cautioned Marty Kaplan, dean of the Annenberg School for Communication at the University of Southern California, which studies the impact of entertainment on society. "Only magicians can make things like red ink disappear by waving a wand," he said after Schwarzenegger's election. "People had the mistaken impression that all our problems got solved now that we have a famous superhero in place. In truth, that was just the beginning."[3]

Celebrity politicians also can have trouble adapting to the snail-like pace of the political world, and their role within that world, West says. After his 1998 election, Ventura grew weary of the mundane realities — and the heavy responsibilities — that came with the job; he did not run for re-election in 2002.

Lawyer-turned-actor-turned-senator Fred Thompson also retired from politics in 2002 to return to acting on NBC's popular "Law and Order."

"After two years in Washington," Thompson quipped, "I often long for the realism and sincerity of Hollywood."

America probably has not seen the last of the celebrity candidate, Blake says. Actor Ben Affleck seems to be laying the groundwork for a political run, making appearances on Capitol Hill to support an increase in the minimum wage, a proposal supported by his friend Sen. Edward M. Kennedy, D-Mass. Affleck also was seen as a ubiquitous wonk during the Democratic National Convention in 2004, when he stumped for presidential candidate Sen. John Kerry.

AFP Photo/Stephen Jaffe

President George Bush greets California Gov. Arnold Schwarzenegger, the latest celebrity to combine popularity and politics.

"We're seeing more and more athletes and actors winning elections," West says. "And sometimes they even turn out to be good governors."

— Kate Templin

[1] Susan Page, "Lack of Political Resume Can Actually Boost Newcomers," *USA Today*, Sept. 29, 2003, p. 1A.

[2] *Ibid.*

[3] John Broder, "Even Celebrity Has Its Limits," *The New York Times*, July 25, 2004, p. D5.

to England, for example, American poet Ralph Waldo Emerson described meeting "celebrities of wealth and fashion."

Inventors Nikola Tesla and Thomas Edison amped up the public's newfound fascination with the private lives of the well-known. Tesla, who made alternating current usable, and Edison, whose inventions included the phonograph and moving images, gave the masses the sounds and images of the famous or the soon-to-be famous.

The increasing popularity of moving pictures in the second decade of the 20th century proved pivotal in making celebrities ubiquitous in American society.

Star System

In 1910, producer Carl Laemmle triggered the rise of the American movie-star phenomenon by creating the first movie star through a massive publicity campaign. Now forgotten, Florence Lawrence was known coast to coast as the "Vitagraph Girl."[16]

That same year, film companies began to move to the area later known as Hollywood, and director D. W. Griffith and Biograph Studios released "In Old California," the first film made in Hollywood.

In 1911 *Photoplay*, the first, true movie fan magazine debuted and gave rise to the whole idea of a celebrity

AFP Photo/Jorge Uzon

Singer Britney Spears is a popular magazine subject as much for her marriages and peccadilloes as for her talent and glamorous appearance.

culture. Soon afterward, *Motion Picture Stories*, *The Moving Picture World* and *The Motion Picture News* also offered interviews and gossipy columns about the personal lives and careers of the stars.[17]

Fascination with the private lives of public figures fueled a feverish interest in the 1921 arrest and subsequent trials of silent-film comedian Roscoe "Fatty" Arbuckle, presaging today's media frenzy over Michael Jackson. Arbuckle was charged with the rape and murder of actress Virginia Rappe during a wild party in San Francisco. Tabloids sensationalized the crime and concocted stories about Arbuckle's "bottle party." Although he was acquitted in multiple manslaughter trials, Arbuckle saw his career end, while Hollywood was forever linked in the public's view as wild and scandalous.[18]

In the wake of the Arbuckle scandal, efforts were made to police the industry, including the creation of the Hays Office, designed to clean up Hollywood through censorship and public relations.

Inevitably, the nation's fascination with Hollywood doings gave rise to Walter Winchell, the Jazz Era's most famous, and influential, gossip columnist. Each week he wrote six fast-paced columns that appeared in nearly 2,000 newspapers. In the 1930s he added Sunday radio broadcasts. With his columns and his distinctive, staccato radio voice, he reached 50 million homes.

"Feeding the public's craving for scandal and gossip, he became the most powerful — and feared — journalist of his time," wrote biographer Ralph D. Gardner. "His articles were loaded with snappy, acerbic banter. Broadcasts were slangy, narrated with machine-gun rapidity, a telegraph key clicking in the background. 'Good evening Mr. and Mrs. North and South America and all ships at sea,' his programs began, Let's go to press!' "[19]

Winchell helped foster the rise of such modern, gossipy publications as *People* magazine, launched in 1974. But it was the unveiling of a technological marvel at the 1939 World's Fair that almost single-handedly ushered in a whole new world of celebrity culture. The future rival to radio and film — television — was formally introduced to the world when the Radio Corporation of America (RCA) displayed the first TV sets for sale to the American public.[20]

Kanazawa, at the London School of Economics and Science, says TV had a profound influence on society. Kanazawa studied television's role in creating "imaginary friends" — celebrities who are seen increasingly as replacements for real friends.

"The major change in the history of celebrity worship was the invention, and then subsequent spread, of television," he says. "Before TV, the only way for people to have 'imaginary friends' was to watch a movie, or read a magazine. So the effect of the exposure to "imaginary friends" was minimal. TV changed all that. It is in your living room, you can watch it every day, and, nowadays, 24 hours a day. So we should feel a lot closer to our imaginary friends than we used to before the spread of TV."

Communications Revolution

Today's celebrity culture is largely possible because of changes in how we communicate. The printing press helped spread The Word, as well as a love of words. In 1776, it helped spread a revolution in Colonial America. A century later, the development of electrification begat the modern era of movies, radio, television and, ultimately, the Internet.

But even as recently as the late 1960s, the world of communications was a very different place from what it is today. In every major city, there were usually at least two daily newspapers, three networks and no cable or Internet. The big story of the day did not have to fight for attention with myriad other media outlets, argues David T. Z. Mindich, an associate professor of journalism and mass communication at Saint Michael's College, in Colchester, Vt.

"On Feb. 27, 1968, when [CBS anchorman] Walter Cronkite made his famous remark that 'we are mired in the stalemate of Vietnam,' he was competing against two or three other news and public-affairs shows, two movies and a couple of sitcoms — 'F-Troop' and 'I Love Lucy,'" Mindich says. "There were seven TV stations in New York City at the time, four of them devoted to news and public affairs, three to entertainment. Today, much less of the TV universe is devoted to news and public affairs, so it is much more possible to watch television all day long, and not get any news."

CURRENT SITUATION

Big Business

Brad and Jen. They are so big that movie fans know them simply by their first names. Brad Pitt and Jennifer Aniston have long been in the pantheon of celebrity newsmakers. And the amount of press coverage devoted to the recent breakup of their marriage is a textbook example of the economics of celebrity.

People, *Us Weekly* and *In Touch Weekly* took the unprecedented step of rushing out a second issue in less than a week to splash the split-up on their covers. "We're in a far more competitive environment than ever," said *People* Deputy Managing Editor Larry Hackett, explaining why his magazine could not afford to wait another week."[21]

The breakup has created a cottage industry, including the first-ever instant book from the publishers of *Us Weekly* — *Brad & Jen: The Rise and Fall of Hollywood's Golden*

> "The problem is that Americans have grown too fond of sweets, both on their tables and in their newspapers. And the new tabloids, such as the Tribune Company's *RedEye*, that are aimed at the youth market seem geared to the attention of a mayfly."
>
> — *Evan Cornog, Publisher,*
> Columbia Journalism Review

Couple. When grainy, long-lens photos of Aniston kissing friend Vince Vaughn, surfaced recently, they prompted a bidding war between upstart celebrity magazine *Life & Style* and *Us*. "Kissing Brad Goodbye?" asked a recent *Life & Style* cover, while the cover of *Us Weekly* wondered, "Dating Already?"

Indeed, during the breakup brouhaha, *Us* featured the couple on its cover for five consecutive weeks, the longest run ever for a single news story. The Feb. 7 issue, with a cover article, "How Jen Found Out," was the magazine's highest-selling issue, with 1.25 million copies. *People*, which has an exclusive first photo this week of Julia Roberts' twins on the cover, also includes a mug of Pitt with the teaser "Brad & Angelina: Their Movie Wedding!"[22]

"Anything involving hope that they might get back together or signs that either of them is moving on is fascinating to our readers and the world," says Sheryl Berk, editor-in-chief of *Life & Style*. The Johnny-come-lately among the nation's highly competitive celebrity magazines has featured Aniston on the cover four times since the split.

However, as popular as the celebrity world's pre-eminent couple have been of late, they were second-rate in 2004 — at least in terms of how many covers they graced. A recent *Daily News* tally showed that Jennifer Lopez, either alone or with Affleck or husband Marc Anthony, led all celebrities in 2004, as the dominant subject on a total of 29 covers published by *People*, *Us*, *Star* and *In Touch*. Aniston, alone or with Pitt, was a close second, on 26 covers.[23]

And there seems to be no end in sight when it comes to feeding the voracious celebrity media beast. In 2004, there were 1,006 launches of new magazines, many focused on celebrities, according to Samir Husni's *Guide to New Magazines*.[24]

AT ISSUE

Do the media devote too much attention to celebrities?

YES
James Houran, Ph.D.
Coauthor, "Celebrity Worship Scale"

Written for *The CQ Researcher*, March 2005

Unequivocally yes. Having idols and heroes is a natural part of identity development, but it's indicative of a problem when individuals shower attention and affection onto people whom they do not even know — and essentially can never know personally. This trend strongly suggests that we're a media- and entertainment-saturated culture that treats celebrities akin to religious icons. The media give celebrities a powerful pulpit and encourage the public's fascination and preoccupation with celebrities.

Undoubtedly, celebrities are more accessible than ever before due to the advent of the Internet, the myriad "real life" stories about celebrities shown on entertainment news programs and even in the mainstream press. We even have reality shows that turn normal people into stars, and these mass-produced "celebrities" are also given inordinate amounts of media attention. The media are clearly giving the public "what it wants," but in doing so the media are exacerbating the problem.

This media attention does two counterproductive things. First, it reinforces the status and prestige of celebrity in our society, even as it objectifies and trivializes celebrities themselves. People no longer need special talents or abilities to be famous — they only need to do something that gains the media's attention. Also, devoting too much space to celebrities arguably undermines the credibility or relevancy of the media outlets. The private and professional lives of celebrities are not legitimate news topics, unless their actions affect society in a meaningful way, as in the case of Arnold Schwarzenegger running for and winning political office.

What constitutes a "meaningful" story is clearly a subjective standard, but it should be a red flag when media are reporting on a person simply because of his or her celebrity. That resembles voyeurism, not journalism.

Second, overzealous coverage of celebrities decreases the psychological "distance" between fans and celebrities. This reinforces the false and unhealthy notion that the public can really come to know stars — that we can establish real, personal relationships with them. The media are often a vanguard that informs us of significant occurrences that have real implications for society.

But the news media can also act like a drug dealer, devoting far too much space to superficial stories about the rich and famous — information that has little real value but that has tremendous power to reinforce society's addiction to celebrities.

NO
Dennis McCafferty
Senior writer, USA WEEKEND magazine

Written for *The CQ Researcher*, March 2005

As a confessed newsy newspaper writer turned celebrity scribe, I hear this question quite often. Now, the response that any self-respecting journalist is supposed to give is, 'Heavens! Mercy yes!' At least that's the one that I can only imagine that 99 percent of this fine publication's readership would give.

When it comes to hard news versus celeb fluff, I've been hearing the "sky is falling" uproar for some time now. But I don't buy it for an instant.

Newspapers and other media outlets are certainly cutting budgets, along with, unfortunately, a shockingly large number of both "designated award winner" hard-news staffers as well as the incredibly undervalued grunts who deliver the nuts and bolts of day-to-day news gathering as a career calling. (By the way, I have no doubt that readers and audiences place far more value in the latter kind of coverage rather than the former. As humor columnist Dave Barry put it, those notebook-emptying newspaper series presentations should come with a warning to readers: "Caution! Journalism prize entry!"

This budgeting trend is sad to see and, unfortunately, does not appear to be reversible anytime in the near future. But that said: I'm still completely unconvinced that hard news is falling victim to celebrity coverage. The last time I checked, *USA Today, The Washington Post, The Wall Street Journal* and the other usual suspects continue to produce some pretty darn good hard-news stories.

If anything, reporters have more tools than ever to produce serious news on a daily basis, thanks to modern technology. And let's face it — most reporters early in their careers gravitate toward hard news. That's where you make a reputation. You show the older veterans that you can dig it out with the best of them, and then you move on to the (hopefully) more lucrative and less taxing "lighter" stuff.

Oh, and let's not forget sports, either. Why, Tony Kornheiser alone represents a vast media industry unto himself, with his *Washington Post* column and radio and TV gigs.

The same, massive increase in ongoing coverage is also reflected in what's available when it comes to health, parenting, community, religion and every single other subject that affects our lives. Celebrity news is hardly crowding that out. There's simply more of all kinds of news, period, and that includes hard news.

While it's more fashionable to wring our hands about the mass of celebrity news, I view it as part of a larger, expanding appetite for content in general. What's wrong with that?

Celebrities are hot, said Husni, a professor of journalism at the University of Mississippi. Hotter, even, than sex, which once was the leading subject among new magazines, especially fast, cheap, new magazines. Now sex doesn't dent the top 10 categories, Husni said. "In 1997, sex was the No. 1 category, with 110 start-ups. Last year, there were only 20 new entries [focusing on sex].

"Celebrities are becoming the sex of the 21st century," according to Husni, a self-proclaimed magazine junkie who has been tallying launches since 1985 and is known in the industry as Mr. Magazine.[25] And while that may be good news for magazines, it's bad news for broadcast television, he says.

"TV is surrendering its mass audience," Husni said. "With cable and satellite, [broadcast] TV has been converted to a narrowcast medium."[26]

In broadcast television — where the networks have seen steadily diminished ratings — the influence of celebrity culture can be summed up with the debate over the future of Ted Koppel's "Nightline," which has seemed almost sacrosanct since its launch during the 1979 Iranian hostage crisis.

"ABC News last week shot a pilot for one possible 'Nightline' replacement, a freewheeling show hosted by Washington reporter Jake Tapper and Bill Weir, the co-anchor of the weekend edition of 'Good Morning America,' two network insiders reported recently. "One of the pilot's top stories was about the Michael Jackson child-molestation trial — exactly the kind of tabloid-friendly fodder that the generally sober-minded 'Nightline' has tended to avoid."[27]

At the Associated Press Managing Editors convention in Louisville, Ky., last fall, much of the discussion was about the decline in newspaper readers. As recently as 1997, 39 percent of Americans ages 18 to 34 were reading papers regularly, writes Evan Cornog, publisher of the *Columbia Journalism Review*, but by 2001 the number had dropped to 26 percent. Similar declines have been reported in TV news viewing.[28]

Cornog says many editors pursue celebrity coverage not just because readers want it, but because they see it as a way to regain new, younger readers. Nonetheless, one of the convention sessions focused on "Celebrity Coverage — Where's the Line...And Have We Crossed It?"

"It is a common lament of newsrooms that readers often skip over the long, thoughtful series on important topics in their haste to read the latest on the Hilton sisters

Getty Images/Arnaldo Magnani

British actress Kate Beckinsale, who starred recently in the "The Aviator," greets fans in New York after appearing on the "Late Show with David Letterman" on Dec. 16, 2004.

or the specs on the best high-end cappuccino makers," Cornog writes. "Still, why not include some of that fluff? The occasional confection is fine as long as one eats a healthy, balanced diet.

"The problem is that Americans have grown too fond of sweets, both on their tables and in their newspapers. And the new tabloids, such as the Tribune Company's *RedEye*, that are aimed at the youth market seem geared to the attention of a mayfly."[29]

Science of Celebrity

Humans have a biological predisposition to celebrity interest, according to James Bailey, a research fellow at the Center for the Study of Learning at George Washington University, in Washington, D.C.

"There are two factors at work in our 'biological' reaction to celebrities," he says. "The first is the 'beauty' factor. Simply put, celebrities tend to be physically attractive, and there is a whole host of literature showing that physically attractive people are at an advantage in virtually every avenue of life."

When exposed to an attractive face, he says, the so-called pleasure centers of the brain — those associated with the release of adrenaline, epinephrine and other endorphins — "light up." That reaction to beauty, he says, has been consistent over time, even as the concept of what is beautiful has changed.

Although beauty is culturally conditioned and changes over time, there seem to be certain "golden proportions" — such as from hips to bust to shoulders, or from the eyes to the forehead, mouth and nose — that transcend both culture and time. "I guarantee that Paris Hilton's face and figure are described by a mathematical equation that could also model Mozart's music," Bailey says.

In addition, he says, the repeated exposure to celebrities' faces can also have an organic effect on the human brain. "Basically, if a person is exposed to a stimulus over and over again, that stimulus becomes familiar, and familiarity triggers those same pleasure centers," he says. "It's as if being exposed digs a neurochemical groove in the brain, that when activated, triggers a biochemical cascade that's experienced as pleasurable. That's why we like seeing things we've seen before.

"It's been postulated that there is a survival instinct behind 'liking' things that are familiar, because the familiar is safer than the unknown," he says.

But researcher Kanazawa, at the London Institute for Science and Economics, worries that the constant celebrity images bombarding our brains may be harmful.

"Celebrity interest didn't evolve; it is an exaptation," says Kanazawa, who studies evolutionary psychology. "In other words, the adaptation, the evolved trait, was our genuine interest in friends and family. When artificial images of photographs, films, TV, video and DVDs were invented, our adaptation was co-opted by these evolutionarily familiar stimuli, and our interest in celebrity was born. Now we cannot tell the difference between our 'real' friends and family, and 'imaginary' friends and family."

The result, he says, is that "We are living in an entirely evolutionarily novel, strange place, which our human brain — adapted to the conditions of the African savanna 50,000 years ago — cannot comprehend."

Studying Celebrity

Some college courses are trying to help students understand, and deal with, the reality of celebrity culture.

At Central Michigan University in Mt. Pleasant, Lorrie Lynch, an editor at *USA Weekend*, is teaching a course in celebrity journalism this fall. The advanced course is aimed at students already proficient in journalism who want to learn how to cover celebrities. They'll learn how to cover big events, like Oscar night, report on the business of entertainment, conduct a celebrity interview and write a celebrity profile.

Lynch says she will cover celebrity journalism ranging from staged events to uncovering scandal, including figuring out how celebrities' publicity operations work and meeting stars' demands without abandoning journalistic integrity.

"Rather than texts, I plan to have the students reading the news and entertainment magazines, columns on the Web and looking at entertainment-oriented TV shows so they immerse themselves in this niche of the profession and get very familiar with what's out there," Lynch says.

"We'll have weekly discussions about how big stories are handled," she says. "For example, the week of the Brad Pitt/Jennifer Aniston breakup we would talk about the timing of their announcement, how each publication played it, who had the best stuff. We might analyze the coverage looking for fairness and accuracy."

Lynch's course is among a growing number of university-level offerings focusing on celebrities and celebrity journalism, from England's University of Gloucestershire to Australia's University of Queensland, whose Centre for Critical and Cultural Studies examines celebrity culture in depth.

At the University of North Carolina in Chapel Hill, Charles Kurzman, an associate professor of sociology, is teaching a course called Celebrity Status, which examines whether celebrities constitute a "status group" in the sense described by Max Weber, a founder of modern sociology.

"It may be that celebrities usurp honor, command authority, engage in a distinctive lifestyle and pass along their status (sometimes in diminished form) to their children, just like the aristocratic elites whom Weber analyzed a century ago," Kurzman writes in his course description. "At the same time, celebrity may be unlike Weberian status in other ways."

Kurzman wants his students to understand the historical and anthropological context of celebrity status. "Celebrities appear to play a role in today's society similar in some ways to the role that the aristocracy played in earlier eras," he says. "Ordinary folks treat them with awe and exaggerated rituals of respect when they come across a celebrity in person. We peons take a bizarre interest in the mythologized details of their lives, and we willingly grant them a portion of our harvest, as it were, in the form of movie tickets, CDs, live shows and products associated — even if only contractually — with this modern aristocracy."

OUTLOOK
Diminishing Democracy?

Media pundits and social scientists, already concerned about the proliferation of celebrity, worry about the future.

"Thomas Jefferson said that democracy's strength depends on an informed electorate," says the Annenberg School's Kaplan. "Public education and quality journalism are essential channels for delivering that information. In both those enterprises, need-to-know has taken a back seat to need-to-make-dough. If that continues, the prospects for robust democracy will diminish, and the opportunities for demagoguery, which depends on mass ignorance, will increase."

Given current trends, the long march toward diminishing democracy is very likely to continue, he says. As the news hole continues to shrink, he says, traditional hard news will have to present itself as entertainment so "journalism" can become an even bigger profit center for entertainment conglomerates.

Washington Post media critic Kurtz agrees. "Given past trends," he says, "I wouldn't be surprised if there was a Celebrity Channel on TV — several of them, actually — not to mention even more magazines and Web sites devoted to the pointlessly famous."

Social scientists say rapid advances in technology will only make matters worse. "I don't think celebrity worship will ever abate," says psychologist Houran, coauthor of the Celebrity Worship Scale. "We're a media- and entertainment-saturated society, so I predict that we'll become increasingly obsessed with celebrity culture over

time as technology advances further and allows us to feed more efficiently the addiction — and false sense of connection — we have to celebrities."

Moreover, says George Washington University's Bailey, "Hollywood, advertisers and others in the selling game are gaining a greater understanding of brain functioning, which means that their persuasive attempts will be all the more effective and compelling."

In addition, he notes, "the communication media are part and parcel of modern life — the Internet, digital on-demand programming, portable entertainment — I-pods, miniature DVD players and so forth. Hence, there will be a greater probability of encountering these increasingly sophisticated and clever messages and imagery."

But not everyone worries that America's flourishing celebrity culture will hurt our democracy, or others. English Professor Blake at College of New Jersey, says the election of Arnold Schwarzenegger, one of the world's most popular celebrities, actually spread hope around the world. "In what we might think of as our peer democracies, there was comic disbelief that the Terminator had won" election as governor of California, Blake says. "This did not seem to portend well for American democracy."

But, in several developing countries, the recall process that resulted in Schwarzenegger's election signaled the openness of our democracy, he says. "Newspapers in Swaziland, Zambia and the Philippines saw in his victory a lesson for their own political situations," Blake says. "Some compared his campaign to that of their own celebrities — the pop singers, soccer players and beauty queens — who were trying to channel their fame into public service. The 'meaning' of Arnold Schwarzenegger was open to broad interpretation.

"In the next 15 years, the importance of those varying interpretations will only grow in significance."

NOTES

1. For background, see Kathy Koch, "Journalism Under Fire," *The CQ Researcher*, Dec. 25, 1998, pp. 1121-1144, and David Hatch, "Media Ownership," *The CQ Researcher*, Oct. 10, 2003, pp. 845-868.

2. Lynn McCutcheon, *et al.*, "Conceptualization and Measurement of Celebrity Worship," *British Journal of Psychology*, Feb. 1, 2002.

3. The American Society of Aesthetic Plastic Surgery, www.surgery.org.

4. McCutcheon, *op. cit.*

5. John Maltby and David Giles, "The Role of Media in Adolescent Development: Relations between Autonomy, Attachment, and Interest in Celebrities," *Personality and Individual Differences*, Vol. 36, p. 813, March 2004.

6. Matt Taibbi, "WMDUH," *New York Press*, March 9, 2005.

7. "Bulletin: No W.M.D. Found," editorial, *The New York Times*, Jan. 13, 2002, p. A34.

8. Taibbi, *op. cit.*

9. Gregory R. Crane, ed., "Perseus Digital Library Project," Tufts University, www.perseus.tufts.edu.

10. *Ibid.*

11. *Classics Technology Center*, http://ablemedia.com/ctcweb/consortium/gladiator1.html.

12. "A Brief History of Celebrity," BBC News, April 4, 2003, http://news.bbc.co.uk/1/hi/entertainment/showbiz/1777554.stm.

13. *The Annals*, translated by Alfred John Church and William Jackson Brodribb, in "The Tech Classics Archives," http://artemis.austincollege.edu/acad/hwc22/Rome/Pagans_v_Christians/Tacitus-Petronius.html.

14. GIGA quotes, www.giga-usa.com.

15. BBC News, *op. cit.*

16. *Timeline of Greatest Film Milestones and Turning Points*, www.filmsite.org.

17. *Ibid.*

18. *Ibid.*

19. Ralph D. Gardner, *The Age of Walter Winchell*, www.evesmag.com/winchell.htm.

20. *Timeline of Greatest Film Milestones, op. cit.*

21. Paul Colford, "Mags Do Double Duty on Brad-Jen Breakup," *New York Daily News*, Jan. 12, 2005, p. 54.

22. Karen Thomas, "Magazines Can't Break Up With Brad and Jen," *USA Today*, Feb. 15, 2005.

23. Paul Colford, "Hot Copy: Four Top Mags Take Circ Honors," *New York Daily News*, Jan. 29, 2005.

24. Keith Kelly, "We're Ga-Ga for Glam-a," *New York Post*, Feb. 25, 2005.

25. *Ibid.*

26. *Ibid.*

27. Scott Collins, "Signs That 'Nightline's' Days May Be Numbered," *Los Angeles Times*, Feb. 7, 2005.

28. Evan Cornog, "Let's Blame the Readers: Is it possible to do great journalism if the public does not care?" *Columbia Journalism Review*, January/February 2005, pp. 43-49.

29. *Ibid.*

BIBLIOGRAPHY

Books

Braudy, Leo, *The Frenzy of Renown: Fame and Its History, Vintage Books,* 1997.
An English professor at the University of Southern California explains the historical relationship between the famous and their audiences.

Lorenc, Z. Paul, and Trish Hall, *A Little Work: Behind the Doors of a Park Avenue Plastic Surgeon, St. Martin's Press,* 2004.
Plastic surgeon Lorenc argues that the nation's celebrity culture is spurring people to seek plastic surgery.

Orth, Maureen, *The Importance of Being Famous: Behind the Scenes of the Celebrity-Industial Complex, Henry Holt and Co.,* May 6, 2004.
Vanity Fair's veteran special correspondent describes America's evolution from a society where talent earned attention to the modern era, when the star-making machinery of the "celebrity-industrial complex" creates "a war zone of million-dollar monsters and million-dollar spin." She takes special aim at personalities — such as Tina Turner, Judy Garland, Madonna and Michael Jackson — whom she says portray themselves as victims just to hold the limelight.

Rojek, Chris, *Celebrity, Reaktion Books,* 2004.
A sociology professor at Britain's Nottingham Trent University argues that celebrity culture is an integral element in everyday life, and that — like the myths of the gods in ancient society — celebrities provide the public with role models. He also examines why the desire for celebrity can drive some people to any lengths to achieve fame or notoriety.

West, Darrell, and John Orman, *Celebrity Politics,* *Prentice Hall,* **2003.**
The director of the Taubman Center for Public Policy at Brown University (West) and a coauthor examine why celebrities like Arnold Schwarzenegger become politicians, and how politicians like President John F. Kennedy become celebrities.

Articles

Aurthur, Kate, "Reality Stars Keep on Going and Going," *The New York Times,* **Oct. 10, 2004, p. B22.**
A growing number of television-savvy men and women seem intent on pursuing careers as serial reality stars in order to extend their time as television stars and make more money from their fleeting fame.

Brooks, Carol, "What Celebrity Worship Says About Us," *USA Today,* **Sept. 14, 2004, p. 21A.**
A team of researchers found that one-third of Americans suffer from "celebrity worship syndrome," which in its most benign form manifests itself as a sense of emptiness but can progress to obsessive thinking and — in the rarest of cases — stalking.

Colford, Paul, "Four Top Mags Take Circ Honors," *New York Daily News,* **Jan. 29, 2005, p. 50.**
In Touch Weekly, the juggernaut of celebrity magazines, was among 2004's top circulation gainers.

Colford, Paul, "Mags Do Double Duty on Brad-Jen Breakup," *New York Daily News,* **Jan. 12, 2005, p. 54.**
The breakup of Brad Pitt and Jennifer Aniston ranks as the mother of all celebrity news stories.

Gardner, Ralph D., "The Age of Winchell," *Eve's Magazine,* **www.evesmag.com.**
Despite his fame, entertainment columnist Walter Winchell — the prototypical Hollywood gossip — was a foreign concept to Ralph Gardner's students at Baylor University.

Innes, John, "Middle Class Teenagers Hit By Stress," *The Scotsman,* **March 24, 2003, p. 9.**
Tom Low, senior educational psychologist for the North Lanarkshire Council, in Great Britain, believes today's youth face rising everyday demands not just to achieve academically but also to look good, attain wealth and "have it all" in a society increasingly obsessed with celebrity culture.

Kelly, Keith, "We're Ga-Ga for Glam-A," *New York Post,* **Feb. 25, 2005.**
When it comes to new magazine launches, "celebrities are becoming the sex of the 21st century," said Samir Husni, a University of Mississippi journalism professor who has been tallying launches since 1985.

Reports and Studies

"A Brief History of Celebrity," BBC News, April 4, 2004, http://news.bbc.co.uk.
This comprehensive collection of articles looks at the history of celebrity and its role in shaping our society.

"Greatest Film Milestones and Turning Points," www.filmsite.org.
A decade-by-decade brief history of film and its role in society.

McCutcheon, Lynn E., Rense Lange and James Houran, *et al.,* **"Conceptualization and Measurement of Celebrity Worship,"** *British Journal of Psychology,* **February 2002, Vol. 93, p. 67.**
The authors, all psychologists, developed a questionnaire and accompanying Celebrity Worship Scale to measure whether a person's interest in celebrities is healthy or potentially pathological.

For More Information

Center for the Study of Popular Television, S. I. Newhouse School of Public Communications, Syracuse University, Syracuse, NY 13244; (315) 443-4077; www .newhouse.syr.edu/research/poptv. Studies the role of entertainment television in shaping popular culture.

Hall's Reports, 733 Summer St., Suite 503, Stamford, CT 06901; (203) 363-0455; www.hallsreports.com. A leading provider of editorial content analysis for magazines.

Norman Lear Center, Annenberg School for Communication, University of Southern California, Los Angeles, CA 90089; (213) 821-1343; www.learcenter.org. A multidisciplinary research and public policy center "exploring implications of the convergence of entertainment, commerce and society."

Poynter Institute, 801 Third St. South, St. Petersburg, FL 33701; (888) 769-6837; www.poynter.org. A school for journalists, future journalists and teachers of journalists. The nonprofit institute owns the St. Petersburg Times as well as Congressional Quarterly and CQ Press.

Samir Husni, P.O. Box 2906, 231 Farley Hall, University, MS 38677; (662) 915-1414; www.mrmagazine.com. Husni teaches journalism at the University of Mississippi and is a nationally known expert on the magazine industry.

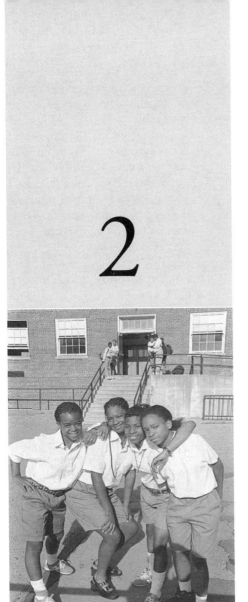

Teaching Values

*Do School-Based Programs
Violate Parents' Beliefs?*

Sarah Glazer

As 8:30 a.m. approaches, the boys and girls outside Jefferson Junior High School form separate lines at the door. They wait quietly, neat and tidy in their private school-style uniforms. At a signal from Vera White, their imposing principal, they file inside. She has a cheery "Good morning" for every child, and a personal word for many: Why were you absent yesterday? I see you're playing baseball! Why didn't you take your test yesterday? White misses nothing. Baseball caps are whisked off at her glance. Shirttail hanging out? Please go outside and tuck it in.

It's easy to see why the immaculate inner-city Washington, D.C., public school is often favorably compared with parochial schools, with their higher level of discipline.

Part of the secret of Jefferson's success is its broad character-education program. Seventh-graders take courtesy classes. Before assemblies, students are reminded about proper audience deportment. Once a week, homerooms discuss the school's values of courtesy, caring, respect and responsibility.

"I'm so proud of you this morning," math teacher Anthony Antoniswami tells his homeroom students, who have been silently doing homework. As he turns things over to the class president for a brief meeting, Antoniswami exhorts, "Let's respect our leaders. That's character."

Since White beefed up the school's character-education program in 1992, pregnancies have plummeted from 12-15 a year to one this year; thefts and fighting have dropped, too. The gains are impressive for the predominantly African-American school, where 70 percent of the students come from single-parent families.

From *CQ Researcher*,
June 21, 1996.

Students' Ethics

The Josephson Institute of Ethics conducts surveys to gather information on the behavioral ethics of young people. Here are some of the findings from its "1996 Report Card on American Integrity."

Did you cheat on an exam or quiz during the previous 12 months?

	At least once	More than once
High school	65%	47%
College	33%	21%

Have you stolen something from a parent or relative in the previous 12 months?

	At least once	More than once
High school	29%	17%
College	13%	6%

Did you take something from a store without paying for it in the previous 12 months?

	At least once	More than once
High school	39%	26%
College	17%	9%

Do you agree that "If necessary to get or keep a job, I would lie?"

	Agreed	Disagreed
High school	41%	59%
College	24%	76%

Source: "1996 Report Card on American Integrity," Joseph & Edna Josephson Institute of Ethics, 1996. Nearly 12,000 Americans over age 12 were surveyed.

Jefferson's ninth-graders score above grade level in reading, science and math on national standardized tests. The school boasts the best junior high attendance record in Washington.

Jefferson is among thousands of U.S. schools that have instituted character education in recent years. At inner-city schools, the aim is to reverse the worst signs of social decay, such as rising violence and pregnancy, and to give children the social survival skills they need to make it in mainstream society.

White, now going on 17 years as principal, introduced character education at Jefferson years ago. But in 1992, after some of her students appeared in TV news footage of a rock-throwing melee, she turned up the heat. "If I didn't do anything else," she says, "I would let them know that they have some dignity."

At the end of every day, Jefferson students file past White with their "objectives notebook" open to that night's homework assignments. Through that ritual,

White says, "We demonstrate every day that we care enough to help our children accomplish the goals they want in life."

Caring reflects the essence of White's approach. "If I had a choice between a knowledge-based teacher and one who cared about children, I would choose the caring teacher," she says. "I could retrain a teacher who really cares for children, but I can't retrain a teacher who doesn't care; that ruins a child's life."

Many suburban schools, though faced with less extreme social problems than inner-city schools, also have initiated character education, mainly to combat declining decorum and rising cheating and disrespect toward teachers.

Across the Potomac River and a world away from Jefferson Junior High, prestigious Thomas Jefferson High School for Science and Technology, in Alexandria, Va., also saw a need for character education following a controversial student prank six years ago. Two seniors cropped photos of two male faculty members to suggest homosexual activity and printed the altered picture on T-shirts they planned to sell to other students.

The students were suspended. But teachers were outraged when several students circulated a petition arguing that the pranksters' freedom of speech had been denied, and said that the feelings of the defamed faculty members were irrelevant.

"At that point we realized that the words 'right' and 'wrong' had years ago been stripped from our school vocabulary," two Jefferson High teachers wrote. "High schools in our experience had been values neutral for so long that our ability to engage students in conversations about moral issues had become rusty."[1]

The character-education movement has won endorsements from political leaders as ideologically different as President Clinton and former Education Secretary William J. Bennett, author of the *Book of Virtues* and the recent *Book of Virtues for Young People.* This month, Sens. Nancy Landon Kassebaum, R-Kan., and Bill Bradley, D-N.J., joined the White House in co-hosting the third conference on character building organized by the Communitarian Network. (*See story, p. 26.*)

So far, character-education programs have provoked relatively little controversy. About one in five schools uses character education, according to the Character Education Partnership in Alexandria, Va., a consortium of educational groups. The programs range from an occasional assembly to full-blown curricula.

By sticking to values that everyone agrees upon, like respect and honesty, character-education advocates say they can avoid controversy. But the first question skeptics raise is, "Whose values?"

More than 90 percent of the public favors public school teaching of such specific values as respect for others and hard work, according to a 1994 poll. But when pollsters asked about character education in general, 39 percent of the respondents thought that teaching values and ethical behavior should be left to parents and churches.[2]

On both the left and the right, there is concern that character education could touch on issues that bitterly divide Americans. "We believe teaching civility is a great virtue, but if included in that is, 'You must accept homosexuality as a valid alternative lifestyle,' then it's problematic," says Perry Glanzer, education policy analyst at Focus on the Family, a conservative Christian group in Colorado Springs, Colo.

Despite the potential for such conflict, Glanzer says, "We support the teaching of character in grades K-12. There are strengths and weaknesses. We see the strengths as a grass-roots response to moral decline."

At the other end of the political spectrum, the American Civil Liberties Union (ACLU) worries that character education could pose constitutional problems, though the organization cannot cite any instances of violations. "It really depends on whether you're talking about a thinly veiled way of getting religious, ethical and moral issues into the classroom that are not seen by the entire population as American civic values, or you're talking about good citizenship, fairness, tolerance — those sorts of things," says Loren Siegel, the ACLU's director of public education.

"I would be very surprised if there weren't problems," Siegel adds. "Not everyone believes a fetus is a human being or that children should always be obedient and never question authority."

"It's impossible to run a values-free school," responds A. John Martin, executive director of the Character Education Partnership. "Wouldn't these groups prefer we develop respect, rather than disrespect?" Martin is convinced the movement can dodge controversy by avoiding three lightning-rod issues: abortion, gay rights and sex education. Yet some observers believe that discussions about character inevitably will touch on stormy issues, such as tolerance for behaviors that some people consider immoral.

The current character-education movement is a self-conscious reaction against the last great wave of values education — so-called "values clarification" — which first became popular in the 1970s (*see p. 34*). The approach encouraged students to "clarify" their values in class discussions. But teachers were not supposed to offer a judgment as to whether Johnny's values were right or wrong.

"[I]t took the shallow moral relativism loose in the land and brought it into the schools," Thomas Lickona, a professor of education at the State University of New York in Cortland, wrote in *Educating for Character,* a widely read 1991 manual of the character-education movement.[3] Values clarification is "at odds with the basic premise of character education, which is that there are objective standards," Lickona says today. "One can tell the difference between right and wrong and teach it to young people."

Values clarification made generations of teachers afraid to take strong stands on basic issues of right and wrong, critics like Lickona charge. As Martin puts it, "Kids felt there was no ethical anchor there."

Values clarification developed such a bad reputation that today's movement strives to avoid the word "values"— with its connotation of mere personal preference — by discussing "character," presumably a less divisive term.

One of the political strengths of character education is the fact that it differs enormously in style and substance from school to school. But that diversity is also a weakness, because parents don't immediately know what type of character-education program a school may have. Does it fit easily into a school's atmosphere and curriculum, or does it impose an artificial layer of autocracy, with stern slogans tacked up around the school?

Task Force Enters Sex Education Debate . . .

First lady Hillary Rodham Clinton delayed her speech until 5 p.m. so that she could spend the afternoon with her daughter Chelsea. Harvard government Professor Michael J. Sandel missed the first day of the conference so that he could coach his 10-year-old son's Little League baseball game.

These last-minute schedule changes did not seem to bother the more than 300 educators, academics and parent activists from around the country who attended the character building conference co-hosted by the White House and Sens. Nancy Landon Kassebaum, R-Kan., and Bill Bradley, D-N.J., on June 6-8 in Washington, D.C.

George Washington University sociologist Amitai Etzioni, master of ceremonies for the event, mentioned these commitments as examples of the character education movement's philosophy — that children need to be taught good character through their parents' and teachers' behavior.

"Children are born without values; they're not naturally benign," he told the participants. "I have five wonderful sons. They were not born wonderful. A lot of effort went into that."

This is the third conference on character building organized by the Communitarian Network, a nonpartisan, nationwide coalition founded by Etzioni in 1993 to foster socially responsible communities. (The 1994 and 1995 conferences also were hosted by the White House.)

In addition to serving as master of ceremonies, Etzioni chaired a task force on sex education. His task force recommended that schools abandon what it called "value-free" sex education focusing on the biology of sex but also avoid moralistic programs that urge "abstinence only." Instead, the task force favored a middle ground sometimes dubbed "abstinence plus." The approach urges students to postpone sexual activity until they are mature enough to handle the consequences — including pregnancy and the risk of sexually transmitted diseases — but also provides them with information on how to use contraceptives if they do start.

The moral component of sex education is essential, Etzioni said. "We don't think sex education should take place as if it's dental hygiene."

Asked by one conference participant "whose values" would be promoted in such a curriculum, Etzioni suggested that schools focus on disturbing behaviors all parents want to prevent rather than on lists of virtues, like virginity before marriage. There is likely to be little disagreement among parents when they are asked, "Do you believe 12-year-old children should have sex? Do you believe children should have children?" Etzioni said.

An argument often raised against teaching contraception is that it will entice students into sexual activity. But the "enticement factor is so huge to begin with" in television, movies and other media, Etzioni observed, that "adding to it is almost impossible." On the other side of the moral balance, he said, are the "many lives we lose" through AIDS and teen pregnancies "when people don't have the information on how to conduct themselves."

In a paper prepared for the conference, the task force stated that sex education programs "should stress that bringing children into the world is a moral act, one that entails a set of personal and social responsibilities." The task force rejected the liberal Swedish-style approach to sex education, saying "Statements such as 'sexuality is natural and [a] healthy part of living' . . . are open to gross misinterpretation when given to children."

The recommendations on sex education were endorsed by task force members as politically divergent as conservative Neil Gilbert, professor of social welfare at the University of California-Berkeley, and liberal economist

Most supporters of character education say it should remain diverse, representing the consensus views of individual communities.

"Character education does not lend itself to didactic instruction," says Chester E. Finn Jr., a senior fellow at the conservative Hudson Institute and a supporter of the movement. "You learn it by example, by discussion." Consciously introducing character education as formal education carries the risk that it will be "fake, insincere, not very effective," says Finn, a former assistant secretary of Education in the Reagan administration.

As schools like Jefferson Junior High make apparent, character education "depends upon the quality, outlook and commitment of the teachers and administration," says Jay Winsten, associate dean at the Harvard University School of Public Health. He points out that White and her staff are routinely at school until 10 p.m. to meet with parents. "From what I've seen," he says, "Jefferson is pretty darn impressive."

… at White House Character Education Conference

Isabel Sawhill, a senior fellow at the Urban Institute in Washington, D.C., who served in the Clinton administration from 1993 to 1995. Researchers have found that the "Not yet" approach advocated by the task force is effective in reducing the percentage of students who begin having sex and in increasing birth control among those who start sexual intercourse.[1]

The character building conferences are part of a larger effort by the Communitarian Network to inject the teaching of values into all aspects of school life, including sports. The first lady strongly endorsed the movement in her remarks. Schools "must be part of a conscious effort to shape the values we expect our children to live by," Mrs. Clinton said. Noting that President Clinton had addressed last year's conference, she added, "My husband saw character education as part of a larger community vision of what our country can be."

But even character education enthusiasts see problems when it comes to specific recommendations, like those on sex education. "I think it will raise a lot of questions," said Teri J. Traaen, human resources director of the Paradise Unified School District in Phoenix, Ariz. "Some of our single parents would say, 'I never was married and that doesn't mean our family is not OK.' Others may say, 'I don't want you talking to my child about contraceptives.' The flash point will be religious differences."

Dean L. Ryerson, superintendent of the Wisconsin Rapids Public Schools, raised similar concerns. "We have a strong Roman Catholic community. [The approach to sex education advocated by the task force] would just blow us out of the water," he said. "We teach it in a very non-directional way. This is how babies are made. This is the science."

At the same time, Ryerson also said he would like to introduce more explicit discussions of good character into his school system. "I'm concerned about the amount of cheating, the casualness, the lack of self-discipline among students," he said. But, he added, teachers already feel overloaded with a host of social agendas from combating drugs to sex education. "We're just committed to death."

Picking up on that theme, an English teacher from St. Paul, Minn., asked Mrs. Clinton, "What do you say to administrators who say the real business of education is academic instruction and anything else is soft — or fluff?"

Mrs. Clinton responded like a true adherent of today's character-education movement: "It's almost impossible for me to imagine standing in front of a class of 25 to 30 kids — teaching any class — where there wasn't a chance, an opportunity nearly every day to say or do something about values and about character. Even if it's 'Jack, your being late today disadvantages everyone else.'"

Mrs. Clinton said that while many teachers are concerned about the deteriorating quality of life in schools, they also are understandably skeptical about character education, wondering if it is just the latest educational fad. "When it comes to a formal character-education curriculum," Mrs. Clinton said, "it is going to take some proof-is-in-the pudding experience for a lot of people to think it's real."

---

[1] See "Preventing Teen Pregnancy," *CQ Researcher*, May 14, 1993, pp. 409-431.

Both detractors and supporters agree that at its best, character education embodies what good teachers have always tried to do: exemplify good character by their example and their choice of great literature.

In fact, Finn says, "It's a pity" that schools have to consciously decide to include character education. "If we had a properly conceived school curriculum, we wouldn't need a dietary supplement called character," he says. "It would be built into what kids read and into classroom discourse."

As parents, educators and policy-makers debate values education, these are some of the questions being addressed:

Should schools teach values?

"The truth is, a school couldn't run a class for five minutes without some kind of moral values," Lickona says. "You need rules against fighting and cheating on tests." So unavoidable is the transmittal of values, in Lickona's

What Would You Do?

The Josephson Institute of Ethics presented students and adults with situations that called for morally challenging decisions for its "1996 Report Card on American Integrity." A majority of the high school students surveyed said they would lie or cheat to save money or get a better job.

Dilemma	Percent who said they would or probably would		
	High school	College	Adult (not in school)
You apply for a job you wanted very much. You're sure you can do it well, but unless you make some untrue statement, the job will go to someone else. Would you "enhance" your resume?	54%	43%	14%
Your 12-year-old could get into a much better school if you lived in your sister's school district. Your sister is willing to let you use her address when you enroll your child. Would you do it?	81%	75%	32%
An application for health insurance asks whether you have had previous injuries. If you admit previous injuries, premiums would go up at least $50 a month. Would you hide the fact of your previous injuries?	59%	47%	20%
You hosted a dinner party at a restaurant. When the bill comes there is a $60 error in your favor. A friend suggests you say nothing about the mistake but leave a large tip. Would you do it?	57%	51%	17%
Your 13-year-old looks 11. You could save $14 at an amusement park if you say he is under 13. Would you do it?	81%	71%	37%
You badly need a loan which you will get only if you understate what you owe. Would you do it?	52%	39%	19%
After a collision, the auto body repairman offers to fix other damages not caused by the collision in an insurance claim. Would you do it?	68%	67%	47%

Source: "1996 Report Card on American Integrity," Joseph & Edna Josephson Institute of Ethics, 1996. Nearly 12,000 Americans over age 12 were surveyed.

view, that the relevant issue is not "Should schools teach values?" but "Which values will they teach?" and "How well will they teach them?"[4]

Inevitably, teachers and principals convey values simply by how they behave and what they say in class. "What you don't teach is just as important as what you teach," maintains teacher Patricia Giegerich, who has spearheaded a new character-education effort at Annandale (Va.) High School. "And if you ignore values or ethical issues as something you can't talk about, you're

teaching nonetheless. You're saying they're not important."

What's driving the current campaign to make character education as much a part of schooling as English or math? In Lickona's words, it's a widely perceived "societywide moral decline."

In his book, Lickona cites the rising rates of youth murder and vandalism and surveys showing that half of all Americans ages 9-21 have shoplifted. Lickona's litany of social ills includes disrespect for teachers, cruelty toward fellow students, bigotry, bad language and self-destructive behavior like drug use and sexual precocity.[5]

In large part, Lickona blames the moral decline on the breakdown of the traditional family unit. He says children are no longer being socialized at home, where increasingly there is a single or divorced parent, and there is less time for the children. "Parents have not taught kids the most basic things about interacting with other people," Lickona says, "like taking turns, asking questions politely, saying please and thank you and obeying an adult."

Teachers report seeing seeds of the problem as early as pre-school. Starting in the 1970s, the prevailing educational view was that building self-esteem was the foundation for every child's success. Kindergarten teachers often had their students put together "All About Me" books that stressed what kids liked about themselves. "Now kindergarten teachers say kids are so self-centered that they need to do an 'All About You' book," Lickona reports.

At the same time, advocates point to the negative forces filling the character-education vacuum, including television and video game role models that Lickona says encourage kids to be "selfish, dishonest, sexually promiscuous and violent."

"A lot of high school students think about their rights without the corollary responsibility," Giegerich says, citing a growing tendency for students to talk back to teachers and use foul language. "We're talking about not doing homework, not getting to class on time, not being honest about attendance, not respecting other people's property."

Michael Josephson, founder of the Josephson Institute of Ethics in Marina del Rey, Calif., blames the disturbing trend on America's slide into a "no-consequences society." While nearly two-thirds of the high school students polled for the institute's annual "Report Card on American Integrity" admitted to cheating in the previous year, school records indicate that only a handful are being disciplined.[6] (*See table, p. 24.*)

It's not that young people don't know that cheating violates ethical behavior, Josephson observes. "But to an increasing degree," he says, "people are thinking they're not bound by it." Adults who lie about a child's age to get a cheaper movie ticket or who use radar detectors "have generated cynicism" among youth, he says.

But not everyone sees society going downhill. "I happen to believe that we are more moral [today] than we used to be," says ACLU Executive Director Ira Glasser. "We have far less racial prejudice and gender discrimination and intolerance for gays and discrimination against the disabled than we had when I was growing up."

To say that America is in moral decay "is an outrageous denial of the largest moral advance this or any other nation has ever experienced in so short a time," Glasser says. He stresses that the civil rights advances since the 1960s did not come about because of character education courses. Quite the contrary. "During all those years of segregation, subjugation and state-sanctioned terror," he argues, "children prayed in school and were taught virtues of politeness, civility, respect and the like."

In the new fashion for extolling virtue in political speeches, some observers perceive a politically motivated return to the Victorian inclination to blame poverty on a lack of moral rectitude.

"[A]lmost all of the virtue books emanate from the right," satirist Joe Queenan complains in a recent article on the Bennett-led boom in values publishing. Moreover, he notes, "almost all preach the same message: The liberal welfare state has created a huge underclass of urban psychopaths who are ruining life for decent people like us with their drugs, rap and guns."[7]

Yet the message has also struck home with middle-class baby-boomers, as indicated by the popularity of parenting books like William Kilpatrick's *Why Johnny Can't Tell Right from Wrong and What We Can Do About It.*

Student misbehavior was on the minds of teachers and parents from the private Sheridan School in Washington, D.C., who attended a week-long seminar on character education conducted by Lickona last summer. The Sheridan contingent traveled to Cortland to learn how to counter rowdiness in class, disrespect toward teachers and talking out of turn. They were surrounded by public school teachers discussing far more extreme social problems, like students bringing weapons to school.

"We found ourselves feeling a little silly sitting next to people talking about teen pregnancies," headmaster Hugh Riddleberger recalls.

Yet even at Sheridan, where children are overwhelmingly from well-off, professional families, Riddleberger has become increasingly concerned with students' moral sense. "I saw that kids were taking less and less responsibility for their behavior," he says, "blaming others and making excuses" when teachers reprimanded them. "I saw the teachers hardening their responses by thinking of longer lists of rules."

Families are spending less time with children, surveys show. As a result, Riddleberger says, parents are missing opportunities to deal with moral issues as they come up — recognizing a child's act of honesty explicitly and saying, "That was a decent thing to do."

"If you're not hanging around and waiting for a kid to drop a comment, you'll miss it," he says. "If it's left to a Spanish-speaking housekeeper, you're definitely going to miss it."

Since attending Lickona's seminar, Riddleberger has instituted class meetings, at which children are asked to come up with solutions to problems such as teasing. (*See story, p. 36.*) He also has begun delivering homilies on honesty and other virtues at weekly assemblies.

Can schools teach values that are acceptable to both liberals and conservatives?

Some conservatives as well as liberals fear that character education will impose a political agenda on children that runs counter to their parents' views.

While people of different political persuasions might agree on a list of common values, "you and I might define citizenship or justice differently," says Janet Parshall, a conservative syndicated radio talk show host.

Parshall cites a California grade school where the children pledged allegiance to a "mother Earth" flag to honor the environment. "What if I happen to find that against my value system because in my home we tell them . . . father God is the one who is preeminent?" Parshall asks.

She envisions a values war erupting in the classroom if a child is taught at home that homosexuality is wrong but learns in school that justice requires recognizing every kind of family equally, including lesbian marriages, as depicted in the children's book, *Heather Has Two Mommies*. In that case, Parshall says, the school's message is that " 'Your set of values, son, daughter, is up for review.' I think that flies in the face of the definition of public education and sends the message home that, 'Parent, you haven't done a good job.' "

Conservative Christian groups also question whether values can be discussed without discussing their genesis — in the case of Christians, the New Testament. "If you're going to teach the need to be self-controlled or honest, the kids will ask for justification," Glanzer says. "Will the appeals [for good behavior] mainly be based on pragmatic reasons?" If so, many Christians will find that unacceptable, he says, because "We believe the greatest motivation for goodness is belief in a transcendent and loving God."

Character-education proponents see little danger of wading into controversial political or religious areas. "In 98 percent of the cases, it won't be a matter of conflict whether you should be polite or not shove the kid at the water fountain," Lickona says. If school districts avoid such divisive issues as abortion, gay rights and sex education, some character-education advocates are convinced the movement can avoid value wars.

But Parshall is skeptical: "If you cut all of that out, basically what you're left with is good manners — which has been around since the inception of public schools."

Denis Doyle, senior fellow in education at the conservative Heritage Foundation and an advocate of character education, argues that manners and values are linked inseparably.[8] "Good manners are a lower-order form of character development," says Doyle, co-author of the 1991 book *Winning the Brain Race*. "It's the Aristotelian notion that a man becomes virtuous by behaving virtuously. Good manners are good training to become a person of character."

Because of volatile issues like homosexuality, schools must tread a delicate line between blandness and controversy when introducing character education. On the one hand, gay-rights activists see dangers in the movement's determination to avoid controversial issues. "I think the silence adds to and creates this atmosphere of neglect, harassment and non-support," says Jenie Hall, executive director of the American Friends Service Committee's Bridges Project in Philadelphia, which works with gay teenagers facing harrassment in school.

On the other hand, if schools confront controversial issues in class, "All it takes is one or two parents concerned about something, and the whole program can be blown out of the water," says James S. Leming, an expert on evaluating values-education programs who teaches at Southern Illinois University (SIU) in Carbondale. "The upshot is you ignore controversy and take these things to an inane, innocuous level."

Recently, a national parental-rights movement backed by conservative Republicans and Christian groups has put its weight behind the primacy of the home in teaching values. The movement has focused mainly on school sex education and condom distribution programs. Proposed parental-rights amendments to state constitutions have been introduced in 28 states. The model language for the bills, proposed by the pro-family group Of the People, holds that, "The right of parents to direct the upbringing and education of their children shall not be infringed."[9]

Supporters of the amendment say they are trying to give parents the right to remove their children from school events they find objectionable. The Supreme Court this year refused to hear appeals of three cases that typify the problems advocates say their amendment would address. In one case, parents sued the school district in Chelmsford, Mass., over a mandatory AIDs-awareness assembly that featured sexually explicit language.[10]

Critics of the proposed amendment, including the ACLU, say it would give parents veto power over curriculum and counseling programs. They point to the proposal as one more sign of the conflict that character education might spur between liberals and conservatives.

But leaders of the parental-rights movement say their proposed amendment does not target character education. "I'm certainly in favor of teaching character," says Jeffrey Bell, Of the People's chairman. "Most people in the parental-rights movement would say the educational elites are moving away from teaching traditional values."

Pointing to controversies over sex education, condom distribution and other such issues, civil libertarians wonder if character education will become a stalking horse for the religious right. "There are some aspects of this return [to character education] that are a cover for religious dogma," says Stephen B. Pershing, legal director of the Virginia ACLU. Noting that religious conservatives are in the majority on some Virginia school boards, he comments, "It's unlikely that tolerance will ever make it into the values curriculums of those school districts."

The Supreme Court has ruled that schools must provide "opt-out" provisions for students whose parents have religious objections to classroom material.[11] But Pershing emphasizes that the court opinions have been based narrowly on religious dissent. "One of my concerns is that an individual family with a philosophical difference of opinion will be shut out of constitutional, legal recourse," he says.

Professor Amitai Etzioni of George Washington University stresses that his vision of character education puts little emphasis on classroom content, which he agrees might incur a parent's wrath and the desire to keep the child out in the hallway. "I'm in favor of opt-out," says Etzioni, who advocates character education in his 1993 book, *The Spirit of Community*. "If classes preach something that deeply offends parents' values, they should opt out."

But when it comes to a schoolwide ambiance of civility, "You can't opt out of civil corridors," says Etzioni, who is also director of the Communitarian Network, which organized the recent White House character conference.[12] "I'm looking at character education as a set of experiences. No one should opt out of it."

Rather than merely teaching a list of virtues, Etzioni advocates fostering two basic psychological capabilities necessary for good character: control of impulses and empathy.

Yet for teachers, character instruction is "a field full of land mines," Leming cautions, even if they avoid obvious religious or political controversies. In one school, he recalls, "I saw teachers asking kids examples of being responsible. One kid said, 'My parents told me to clean up my room. I shoved everything under my bed, and that was cool.' Does the teacher point out to the kid what's wrong? Are you humiliating the kid? Quickly, this becomes difficult to teach."

Does character education make a difference in behavior?

Only one comprehensive, long-term study of character education has been completed.[13] From 1982 through 1989, the nonprofit Developmental Studies Center in Oakland, Calif., tested its Child Development Project in San Ramon, a middle-class suburban school district. The project sought to strengthen children's tendencies to be caring and responsible by creating "caring communities" within the school. Children participated in the program from kindergarten through sixth grade, and then researchers tracked their progress in seventh and eighth grade.

"We demonstrate every day that we care enough to help our children accomplish the goals they want in life."

— Vera White, Principal
Jefferson Junior High School
Washington, D.C.

The program teaches reading through literature rather than textbooks. It fosters a collaborative approach to learning and a warm, friendly classroom. A Buddies Program pairs older and younger students for reading and other activities. At family film nights, parents and children discuss a movie with a socially meaningful message.

When three local schools that used the program were compared with three schools that did not, participants were found to be more cooperative and friendly in class. They were committed not only to letting everyone voice their own opinions but also to asserting their own views, however unpopular. The children were evaluated by outside observers who did not know which of the six classrooms participated in the program.

Through the fifth grade, the researchers found no differences in the two groups' scores on standardized tests. Since San Ramon students generally score at around the 90th percentile, there is little room for improvement, the researchers observed. But starting in sixth grade, when tests required the children in the experimental group to read a short story or poem and write their own interpretation, they outscored their peers in the control group.

Eric Schaps, president of the Developmental Studies Center, which conducted the study, speculates that the higher scores were due to the program's early emphasis on literature, which "shows what it means to be a caring person, as opposed to textbooks, which are drier."

In a summary of their findings, the researchers elaborated: "[W]e think that [the project's] major impacts on achievement will ultimately be seen with measures that reflect deeper comprehension and critical thinking — because of the program's emphasis on explanation, reflection, intrinsic motivation and engagement with meaningful literature."[14]

Upon entering middle school, students from the experimental program displayed more self-esteem than their peers, participated more in extracurricular activities and continued to have superior skills at resolving conflicts, the researchers found. There was no difference between the grades of students from the experimental program and those of other district students by the time they reached middle school. But among the students who had participated in the project, better grades were received by those students whose elementary school teachers had done a superior job of carrying out the project's principles in the classroom.

In many ways, the experiment integrated two basic principles of progressive private schooling — making school interesting and friendly — with the traditional one-room schoolhouse, where children of different ages interact. For example, the Buddies Program and the use of literature to teach reading have long been staples at Sheridan and many other private schools.

"I worry about the preachy kind of character education," Schaps says. "If taught in an impersonal, punitive environment, it's not going to take. A humane, caring, stimulating school is as important as the teaching of character."

"I think this is the best example of a carefully researched program we have in the field," Leming of SIU comments. But, he adds, the greater harmoniousness found in classrooms with character education is not the goal behind the current movement. "Every one of the books on character education points to declines in society: increased drug use, promiscuity, violence, teen suicide," he observes. "Can a teacher in the classroom have an effect on those things?"

Leming is doubtful because he sees the family as the primary socializing institution. "Realistically, the children most at risk come from home environments almost beyond our understanding," he says. "If 95 percent of kids were raised in loving, caring, two-parent homes, we would not have a character-education movement."

The Child Development Project was also tested in Hayward, a poorer, more ethnically diverse school district than San Ramon. Implementation of the program was spottier in Hayward, partly because of budget problems and high teacher turnover. While Hayward pupils in the experimental program showed improved classroom behavior, the program did not affect their social development, such as concern for others and democratic values, as much as in San Ramon, researchers found.[15]

Beyond painstakingly designed studies like the Child Development Project, character education is hard to evaluate, Leming notes, because the movement embraces such a wide variety of programs. In addition, many schools already have several add-on programs aimed at preventing drug use, violence and other deviant behaviors.[16] So it's hard to know which program to credit for success.

For example, character-education advocates point to Jefferson Junior High's decline in pregnancies as a sign of success. But Principal White is quick to note that Jefferson also runs a special program for girls, Best Friends, that advocates abstinence until after high school graduation. The school also features lunchtime rap sessions held by the school nurse to steer adolescents away from early sexual activity.

Henry A. Huffman, director of the Character Education Institute at California University of Pennsylvania, has noted another difficulty in measuring how character education affects children: "[T]he real tests of character occur when no one is watching. Will a student return the extra change that a clerk mistakenly gave him?"[17]

Nevertheless, occasional success stories give the movement reason to hope. In 1989, when Principal Rudy Bernardo arrived at Allen Classical/Traditional Academy in Dayton, Ohio, an inner-city public elementary school, students were usually lined up outside his office each morning waiting to be disciplined for fighting, cursing or bringing drugs to school. By the end of the year, he had suspended 150 of the 543 students. The school rated 28th out of 33 elementary schools in the district on standardized tests.

After a year of soul-searching seminars with his faculty, Bernardo instituted a character-education program.

Each week, teachers weave one of 18 character traits throughout their class discussions.* Then, at an assembly, the "Word of the Week" is driven home through skits or speeches. When "honesty" is highlighted, for example, a problem in math class involving making change might include an ethical component: Should Johnny return the extra change to the store owner who made a mistake in his favor? In addition, academy students must perform community service, and those whose parents don't attend orientation sessions can be dismissed from the school.

Today Allen ranks No. 1 in achievement scores among the districts's 33 elementary schools. Suspensions have declined to just eight last year, and 87 percent of students turn in their homework, up from 10 percent before character education was instituted.

Perhaps the most marked change involves the faculty. Before, Bernardo says, the teachers were so worn-down that they typically reacted to their students' failings by criticizing them. A new recipe for mixing kindness with values changed that. "If we do not see the good in the student, if we focus on the bad behavior, then the student will always feel that he is a bad student," Bernardo says. "That will be instilled in his mind and heart, and the chance of his changing will be lost."

Character-education advocates point to schools like Allen Academy and Jefferson Junior High as proof that character education works. But one could argue that they merely demonstrate an obvious but often overlooked truth: that the best education involves both nurturing teachers and excitement about learning.

Character education may simply be a new name for an old-fashioned concept — good education. "In effect," Etzioni says, "all the schools you consider good schools are good because they have it."

BACKGROUND

Timeless Teachings

Since the time of Plato, philosophers have defined moral education as a primary aspect of schooling. The concept

---

*The 18 traits are respect, kindness, patience, cheerfulness, self-control, punctuality, courtesy, sportsmanship, tolerance, loyalty, responsibility, thrift, helpfulness, cleanliness, courtesy, self-reliance, citizenship and honesty.

was taken up by some of America's Founding Fathers. Thomas Jefferson's Bill for the More General Diffusion of Knowledge argued for an educational system that would educate citizens at an early age in democratic virtues like respect for the rights of the individual.[18]

Later presidents embraced the sentiment, as in Theodore Roosevelt's famous line: "To educate a man in mind and not morals is to educate a menace to society."[19]

Colonial schools were originally established to teach reading so children could study the Bible and better understand religious principles. This tradition continued into the early 20th century. By 1919, *McGuffey's Readers*, which were full of biblical stories and other moral lessons, were the nation's most widely used schoolbooks.

The prevailing approach was didactic, with teachers imparting moral wisdom to their students. But John Dewey and other educational innovators favored a "progressive" approach, which actively involved children in moral decision-making.

Some of today's champions of character education blame progressive education for moral decay in America's schools. But, says the Heritage Foundation's Doyle, "Dewey was not against rigor; he was reacting to an arch Edwardian sensibility — narrow and frequently unpleasant and dreadful for children."

'Values Clarification'

In the 1960s, moral instruction of the traditional, didactic type began to disappear from American public schools. After the Supreme Court held that school prayer and devotional Bible reading violated the Constitution, many teachers mistakenly believed that the court had prohibited moral education, according to B. Edward McClellan, a professor of education at Indiana University and author of the 1992 book *Schools and the Shaping of Character*.[20]

During this period, cultural "relativism" took hold with a new generation of young people, popularized by the expression "Do your own thing" and the indignant objection, "That's a value judgment!"

"Values clarification," a new approach to character education, reflected that relativism. "Clarification" made its debut in 1966 with the book *Values and Teaching* by New York University education Professor Louis Raths. He advised teachers to help children clarify their values

by acting as moderators of discussions, rather than as moralizers.

The approach took off with the 1972 publication of a values-clarification handbook.[21] More than 600,000 copies were sold, placing it in practically every school in the country. The teacher, the authors explained, was to be presented as "just another person with values (and often with values confusion) of his own."[22]

In one typical handbook exercise, the teacher asks students to raise their hands in answer to the question, "How many of you think there are times when cheating is justified?" In practice, writes Lickona, "teachers often weren't sure what to do after students had clarified their values."[23]

"It was a very optimistic view that, given enough discussion, students would come to realize on their own why cheating was not a good way to go," handbook co-author Howard Kirschenbaum says today.

The values-clarification approach and a competing approach developed by Harvard psychologist Lawrence Kohlberg often posed dilemmas pitting one moral value against another. Michael Bocian, a staffer at the Communitarian Network who organized the recent White House conference on character building, cites the "Alligator Island" dilemma. In this classic classroom exercise, the only way students can reach a loved one stranded on a desert island is to pay an armed boatman $1,000. But the students don't have the money. The lesson is, apparently, that stealing the boat is justified for the greater good of helping someone they care about.

In a critique of values clarification, Bocian wrote: "In the 'Alligator Island' example, the students had to decide between respect for other people's property and care for a loved one. But in the vast majority of human experiences, respecting other people's property is unquestionably the right thing to do."[24]

The philosophical roots of values clarification lay in the "Human Potential" movement, started in the 1960s and '70s by two prominent American psychologists, Carl Rogers and Abraham Maslow. They believed that human beings were basically good and needed to have their true selves liberated. "The basic nature of the human being, when functioning freely, is constructive and trustworthy," Rogers said.[25]

Encounter groups and experiments in group dynamics at the Esalen Institute in California were among the

CHRONOLOGY

1700s-1800s *Moral education with strong Christian overtones is an integral part of American public schooling.*

1836 First publication of *McGuffey's Readers*, emphasizing patriotism, parental respect and adherence to Christianity.

1900s *"Progressive" educator John Dewey challenges the use of moral tales to teach character.*

1918 A National Education Association report, Cardinal Principles of Secondary Education, endorses "progressive" principle of learning by doing rather than memorizing moral lessons.

1960s *American society increasingly views ethics as a matter of personal choice, and moral instruction wanes in public schools.*

1962 Supreme Court rules in *Engel v. Vitale* that daily prayer in New York State public schools violates constitutional separation between church and state. Many teachers respond by avoiding discussions of morals.

1966 "Values clarification" debuts as a teaching method with the publication of *Values and Teaching* by New York University Professor Louis Raths. He urges teachers to help students "clarify" their values without being judgmental.

1970s-1980s *Initial popularity of "values clarification" fades amid growing youth violence, prompting several public school systems to begin teaching positive values.*

1972 *Values Clarification,* a handbook for teachers, sells more than 600,000 copies.

September 1975 Alan L. Lockwood of the University of Wisconsin writes in *Teachers College Record* that "values clarification embodies ethical relativism as its moral point of view."

1977 Warsaw, Ind., drops values clarification in schools and burns the textbook.

1978 Lawrence Kohlberg endorses the presentation of moral dilemmas to students as a teaching method in *Essays on Moral Development, Vol. 1: The Philosophy of Moral Development.*

1982 In Baltimore, Md., a countywide values-education program is based on 24 common moral values from the Constitution and the Bill of Rights.

1982 Supreme Court declares, in *Board of Education, Island Trees Union Free School District, No. 26 v. Pico,* that "local school boards must be permitted to establish and apply their curriculum in such a way as to transmit community values."

1990s *Educators organize to promote character education in American schools and gain a presidential endorsement.*

July 1992 Josephson Institute of Ethics convenes educators to draft a statement endorsing character education. "Aspen Declaration" becomes the basis of the Character Counts! Coalition.

Feb. 5, 1993 Educational organizations form Character Education Partnership to promote character education.

1994 Congress authorizes funds to encourage the development of character education under the Elementary and Secondary Education Act.

July 1994 White House hosts the first of three conferences on character building organized by the Communitarian Network.

Nov. 1, 1995 An Alabama law requiring at least 10 minutes a day of values education takes effect.

Jan. 23, 1996 President Clinton endorses character education in his State of the Union address.

June 6-8, 1996 White House co-hosts the third conference on character building.

---

most famous offshoots of this movement. In many ways, writes Kilpatrick, the movement inherited the doctrine of "natural goodness" propounded by 18th-century philosopher Jean Jacques Rousseau. In *Emile,* Rousseau developed the idea that children are naturally good if not corrupted by society.

Giving Lessons in Character ...

Fifth-grade teacher Kathleen Higley keeps a folder at her desk where her students can place notes about their personal crises. Usually they're along the lines of "Someone called me a wimp."

Mrs. Higley often takes these notes as her cue to call class meetings, where students discuss moral issues like teasing and cruelty. The meetings are a new wrinkle at the private K-8 Sheridan School, in Washington, D.C., which is experimenting with teaching values.

Headmaster Hugh Riddleberger got the idea from a workshop last summer held by Thomas Lickona, an education professor at the State University of New York in Cortland and a leader in the movement to introduce "character education" into schools. Riddleberger says his interest in character education grew out of the increasing tendency of Sheridan students to make excuses for their misbehavior — whether it involved doing a sloppy job on homework, making demeaning remarks about a friend or calling answers out of turn. The class meetings, he says, are designed to teach the children to take greater responsibility for their actions.

One morning recently, Mrs. Higley decided to convene a class meeting after one of her students — we'll call her "Anna" — came to her in tears. The problem had started a few days before when a science teacher asked the students to list occupations that required electricity. As a joke, Anna passed a note to a girlfriend: "Does a hooker need electricity?" The friend passed the message on, and soon it had made the rounds of the class. Like the children's party game "telephone," the message changed into a hurtful rumor — "Anna wants to be a hooker" — as it went from student to student. Anna found herself being teased by other children. One boy told Anna another boy in the class would pay her for her services.

Anna responded by criticizing another girl she believed responsible for spreading the hurtful message. The girls in the class began to divide into camps behind the two girls.

Kathleen Higley's fifth-grade class

Mrs. Higley assembles the fifth-grade girls in the library and explains the problem. "Can you see how it went from a joke to meanness to outright sexual harassment?" she asks.

Each girl insists she is not to blame. "I only told three people," the girl whom Anna has accused says indignantly.

Mrs. Higley makes it clear she is not looking for people to blame. In keeping with the character-education philosophy, she tries to place responsibility for recognizing the problem — and solving it — on the students. "This has happened," she says. "It's reached the boys, and now they're

Kirschenbaum, an adjunct professor of education at the State University of New York in Brockport, now says the movement made a "mistake" in advocating, exclusively, that students should determine their own values.

"I'm sorry to say we denigrated the direct teaching of traditional civic values," he says. "We took it for granted and assumed it happened. History clearly shows that was a very bad assumption," he adds, citing the same litany of social ills that concern character-education proponents: youth violence, teen pregnancy and rising levels of cheating. Learning values "doesn't just happen unless society consciously works at teaching those values," he says.

Kirschenbaum and his colleagues feared direct indoctrination because they equated it with the propagandistic campaigns of pre-World II fascism. Looking back at the Nazi period, they believed the greatest good came from

... One Teacher's Experience

saying very inappropriate things to Anna. How can we make it better?"

Several girls pick up on the sexual-harassment theme, saying the boys in the class have started to make embarrassing comments to them about women's bodies. "When someone says a rude joke to me," advises Mrs. Higley, "I say, 'I don't find it very funny.'"

The emotional breaking point comes when Anna tearfully faces her rival. "You made me feel really bad yesterday," she says. "People came up and started teasing me."

The other girls are anxious to see the two contenders make peace. At Mrs. Higley's urging, the two girls hug, followed by a spontaneous group hug.

But when Mrs. Higley brings the boys into the meeting, they become indignant at the charge of "sexual harassment" and demand a meeting of their own. In the boys' meeting, most of them complain that the girls frequently taunt them but never seem to get into trouble. "We have feelings, too," says one boy. But, he adds, "We're not supposed to cry."

Mrs. Higley's meetings point up the difficult task teachers face when it comes to labeling behavior and deciding what moral lesson to draw from a troubling incident.

Though character-education proponents insist they can teach morals in schools without bumping up against controversy, experts asked about this case gave differing advice. Kevin Ryan, director of the Center for the Advancement of Ethics and Character at Boston University, sees Mrs. Higley's use of the term "sexual harassment" as unnecessarily political. "Why didn't she just call it unkindness?" he asks.[1]

Lickona says he would have discussed sexual harassment under the larger umbrella of disrespect. "I think there's a consensus that sexual harassment is wrong and that schools ought to prevent it," he adds. Both Ryan and Lickona agree

that any real-life problem is more complicated than it may appear at first.

Mrs. Higley's view is pragmatic. Afterward, she expresses confidence that the meetings achieved their primary goal — killing the rumor and ending the teasing. The class meeting, rather than preaching to the children, "gives them ownership" of the problem, she says. "I'm no psychologist, but it's like group therapy — or a family meeting."

"I like the class meetings," says Kathi Sullivan, former president of the Sheridan Parents Association. "I think it sets up a forum where kids are comfortable discussing things they're not normally comfortable discussing." Sullivan hopes the children will apply the conflict-resolution skills they have learned from class meetings to settling their tiffs on the playground.

As for changes in the school, Sullivan says, "One thing I've noticed this year is kids now hold doors open for me." She attributes that to a week that the school devoted to teaching courtesy.

In addition, fewer children are being sent out into the hall this year for misbehaving in class, teachers have told Sullivan. She believes that's due to a new rule the students helped develop: Do nothing in the classroom to prevent others from learning. "I think the teachers have better control over the classroom," she says. "Part of it is by getting the kids to buy into this rule. When the kids are acting up, the teachers say, 'Are you making this a classroom where everyone can learn?'"

Headmaster Riddleberger favors class meetings for teaching values because they deal with real-life problems. And with students ages 12 to 14, "You won't get far by preaching to them about honesty," he says. "It's only our actions that speak to them."

---

[1] For background on sexual harassment in schools, see "Education and Gender," *The CQ Researcher*, June 3, 1994, pp. 494-495.

individuals who had resisted the cultures that perpetuated atrocities. Values clarification "came as a desire to help people stand up for their own values even against peer and society pressure," Kirschenbaum says.

The same concern about state-imposed values is occasionally voiced today. Responding to a generally positive *New York Times Magazine* article on character education, one skeptical reader wrote, "Do you suppose that if [Nazi]

Germany had had character education ... it would have encouraged children to fight Nazism or to support it?"[26]

Results Questioned

Though enormously popular in the 1970s, values clarification dropped off the radar screen of public schools in the 1980s. "As the political climate changed, parents would go to school boards and say, 'You mean my kids can go through values clarification and still believe in cheating and lying?'" recalls Leming. "School boards dropped it like a hot potato."

In addition, researchers were finding, for the most part, that values clarification wasn't having the hoped-for effect on students. After reviewing nearly 100 studies, Leming found fewer than 20 percent that produced the desired outcomes, such as clearer student thinking about their values or lower rates of youth crimes. "In the over-whelming number of cases, there was no change in any of those things," Leming says.

Though values clarification waned as a formal approach, it left an enduring impression on teachers. Kirschenbaum agrees with critics that "it made teachers more timid" about telling students directly that certain values were wrong. Kevin Ryan, a professor of education and director of the Center for the Advancement of Ethics and Character at Boston University, says schools of education still give teachers the message that when it comes to moral education, "They should stay away from it and stick to the information-dispensing role."

Partly, this grew out of a psychological stress on the importance of self-esteem. "There was a great deal of fear that any criticism would damage a child's self-esteem," says Lickona. "So in discussion, a teacher would never want a child to think their judgment about something was wrong."

Today, Kirschenbaum says he finds himself directly inculcating values in his young daughter, contrary to his old approach. Just as a conservative is often described as a liberal who has been mugged, he says, "A character educator is a values clarifier who's had children."

What should a parent say to a child who remarks, for example, that radar detectors are a good way to avoid speeding tickets? "In the old days," Kirschenbaum says, "we would have confined our remarks to, 'Do you think the world would be a better place if everyone did that?'" Today, he says, as a parent, "I would feel comfortable adding my own opinion: 'I don't think the world would be a good place.'"

CURRENT SITUATION

Methods Debated

At least a dozen states have recently incorporated character education into their school curricula, either through legislative mandates, regulation or pilot projects, according to the Character Education Partnership. In the most explicit mandate to date, Alabama passed legislation requiring 10 minutes of each school day to be dedicated to teaching character.

Many other states have had moral-education mandates in their curricula for decades, legacies of an earlier era, but have not implemented them, according to Martin of the Character Education Partnership.

In his Jan. 23 State of the Union address, President Clinton said, "I challenge all our schools to teach character education, to teach good values and good citizenship." Under the Elementary and Secondary Education Act of 1994, $1 million a year was authorized for grants to states to develop character-education curricula and pay for teacher training. Last year, state education departments in California, Iowa, New Mexico and Utah received grants.

While some schools have introduced separate ethics classes, most proponents of values education say character should be taught through behavior and literature. Teachers are urged to seize opportunities to point out moral lessons as they arise in classroom crises or in the material they are teaching.

Early this year, Boston University's Ryan issued a manifesto, signed by more than 30 other educators, criticizing some existing approaches to character education and defining how it should be taught. "The public support character education has won is threatened [by] several skimpy or mislabeled programs being marketed as character education," Ryan charged in a statement.[27]

Ryan has also singled out the Josephson Institute's popular "Character Counts" program for criticism. Five hundred communities have adopted its principles, according to the institute, including school districts in Toledo, Ohio, Albuquerque, N.M., and Dallas, Texas. Last month, the institute released a 60-minute children's videotape about values, starring 30 popular TV characters from Barney the dinosaur to dashing Tom Selleck.

The character-education movement is being trivialized, Ryan says, by the production of coffee cups, pens and T-shirts with the "Character Counts" slogan. "I think that characterizes their approach," he says. "It's in no way a return to teaching about virtue and helping children acquire virtue."

In its videotape and educational materials, "Character Counts" stresses six "pillars of character":

AT ISSUE

Should schools teach moral values?

YES
Thomas Lickona
Professor of Education, State University of New York, Cortland, and author of Educating for Character

From "The Return Of Character Education,"
Educational Leadership, November 1993

In the 1990s, we are seeing the beginnings of a new character-education movement, one which restores "good character" to its historical place as the central desirable outcome of the school's moral enterprise. No one knows yet how broad or deep this movement is.... But something significant is afoot....

In the face of a deteriorating social fabric, what must character education do to develop good character in the young? First, it must have an adequate theory of what good character is, one which gives schools a clear idea of their goals. Character must be broadly conceived to encompass the cognitive, affective and behavioral aspects of morality. Good character consists of knowing the good, desiring the good and doing the good. Schools must help children understand the core values, adopt or commit to them and then act upon them in their own lives.

The cognitive side of character includes at least six specific moral qualities: awareness of the moral dimensions of the situation at hand, knowing moral values and what they require of us in concrete cases, perspective-taking, moral reasoning, thoughtful decision-making and moral self-knowledge.... People can be very smart about matters of right and wrong, however, and still choose the wrong. Moral education that is merely intellectual misses the crucial emotional side of character, which serves as the bridge between judgment and action....

At times, we know what we should do, feel strongly that we should do it, yet still fail to translate moral judgment and feeling into effective moral behavior. Moral action, the third part of character, draws upon three additional moral qualities: competence (skills such as listening, communicating and cooperating), will (which mobilizes our judgment and energy) and moral habit (a reliable inner disposition to respond to situations in a morally good way).

Once we have a comprehensive concept of character, we need a comprehensive approach to developing it. This approach tells schools to look at themselves through a moral lens and consider how virtually everything that goes on there affects the values and character of students. Then, plan how to use all phases of classroom and school life as deliberate tools of character development....

As we close out our turbulent century and ready our schools for the next, educating for character is a moral imperative if we care about the future of our society and our children.

NO
David R. Carlin Jr.
Associate Professor of Social Sciences, Rhode Island Community College, Warwick, Rhode Island

From "Teaching Values In School," *Commonweal,* Feb. 9, 1996

There is much talk nowadays about the need for public schools to teach moral values. In a society that, for 30 years, has been drifting downriver toward the Niagara of moral anarchy, there is no doubt about it: Somebody needs to teach moral values to the young. But can the public schools do it? I doubt it.

Leaving aside a number of other difficulties, let's focus on the vexed question of whose values will be taught. Will the schools teach liberal or conservative values? Values of self-expression or self-control? Values rooted in religion or in secularism? Values of individual autonomy or of community?

Now there happens to be a standard way of trying to meet this difficulty. It is argued (by Bill Bennett, for one) that, no matter what our moral disagreements, all Americans share many important values....

Everyone agrees that fairness, honesty, courage and respect for others are good qualities, while unfairness, dishonesty, cowardice and disrespect are bad.... So let schools teach a broad range of noncontroversial values while maintaining a prudent silence about the narrow range of controversial questions.

Note well, we are told, that this sensible policy does not mean that children will learn nothing about controversial matters. Far from it. Parents, churches and other non-school agencies of socialization will be quite free to give instruction on such issues.... As an abstract proposal, this seems reasonable. But as usual, the devil is in the details....

[C]onsider fairness. We all believe in fairness as an abstract principle, but what does it mean in practice? What does it mean, for instance, when applied to divorce? Or when applied to social policy questions, like affirmative action, food stamps, Medicare? Or when applied to abortion and euthanasia?...

Maybe the schools will respond to this dilemma by saying: "Our fairness curriculum will teach kids not to cut in line and not to steal one another's pencils; but we'll take no stands on divisive questions like divorce, social policy, abortion and euthanasia." But this is tantamount to teaching that fairness applies to small matters only, not to big matters. Once again, what's the point?...

Schools can be effective moral teachers when they represent communities that are morally homogeneous. The trouble is, American society is no longer a morally homogeneous community.

trustworthiness, respect, responsibility, fairness, caring and citizenship.

Ryan has no problem with the list — in fact he helped develop it. But Ryan and others active in the movement are concerned that proliferating character-education programs will not move beyond preaching and posters. Ryan, for example, stresses the importance of high-quality teacher-training seminars, such as those conducted at Boston University, where teachers discuss works by great philosophers like Aristotle.

"My fear is schools will adopt programs under the delusion they represent a quick fix to social problems they've identified," says Huffman, another signer of the manifesto. When Huffman consults with school boards interested in adopting character education, he tells them not to get involved unless they are prepared to carry out the values they espouse in every aspect of the school's management. That includes being respectful to employees across the bargaining table and training janitors to be kind to a crying child they encounter in a hallway.

Huffman also advocates that community meetings be held with parents and residents to come up with the consensus values the community will stress. "One of the challenges of character education is to get everyone to become morally reflective," Huffman says.

But Josephson, a former law professor, argues that it's "an academics' myth that if you have a debate you will get everyone to agree. If you believe that, you have to do it every year, every moment you change the city council or the school board."

Josephson says the intention behind the six pillars of character was to develop a list of widely accepted ethical values that would attract the backing of a broad political spectrum. This would "provide the political permission to unleash the talents of teachers," he says.

The program has won the endorsement of conservatives like former Education Secretary Bennett and liberals like Marian Wright Edelman, who heads the Children's Defense Fund. Josephson has also attracted a bipartisan coalition of 10 members of Congress who support Character Counts, led by Sens. Pete V. Domenici, R-N.M., and Sam Nunn, D-Ga. "That political success is what has made it so popular," Josephson says.

In addition to the cups and T-shirts that "celebrate" character, the program provides serious curriculum material and teacher-training seminars, Josephson adds. "We

bring in three to four trainers and train 35 people for 3 $1/2$ days." The program typically costs $20,000 to $30,000 for a community, and about 20 percent of the attendees receive scholarships from the institute.

Responding to his critics, Josephson says: "Why did [Dallas and Toledo] decide this was a significant program? Here are people who decided that they wanted to do this for their entire school systems. None of the other programs can show this broad-based support."

OUTLOOK
Potential Pitfalls

The debate over how character and values should be taught is sure to continue. Some of the more sophisticated observers of the movement are wary of a return to the dull moralizing of the 19th century. In a review of Bennett's *The Book of Virtues for Young People*, children's author Katherine Paterson noted that much of its poetry is "primly Victorian." She questioned whether teaching virtues is simply a matter of "getting rules down."

"We all know," Paterson wrote, "how literature has made us grow in understanding and compassion, but the learning was our own choice — not something imposed by someone else, but something we gained by entering emotionally, intellectually and spiritually into the heart of a great writer. Shouldn't we trust our children to do the same?"[28]

Meanwhile, the ACLU and conservative Christian groups are taking a wait-and-see attitude toward character education. Olivia Turner, executive director of the Alabama ACLU, views Alabama's 10-minute-a-day character-education mandate as a legislative capitulation to the religious right. While she knows of no instances in which religion is being taught under the mandate, she notes that Christian groups have offered religious textbooks to impoverished school districts. The mandate "creates a window of opportunity for the religious right," she says.

Glanzer of Focus on the Family commends the character-education movement for its grass-roots orientation and emphasis on seeking a consensus of parents' common values. "It has not been a top-down movement, for the most part," he says.

Neither the right nor the left can cite objectionable programs in character education today. But both sides frequently bring up sex education and parental rights as potential pitfalls. Community-service requirements, a component of some character-education programs, also have come under fire from parents. In Mamaroneck, N.Y., the parents of Daniel Immediato sued Rye Neck High School, arguing that its community-service requirement would "destroy any moral value in serving others" because of its involuntary nature.[29]

On Jan. 2, 1996, the 2nd U.S. Circuit Court of Appeals upheld Mamaroneck's community-service requirement; the Immediato family has appealed to the Supreme Court.[30]

Some observers believe the character-education movement will succeed as long as it sticks to an individual school-by-school approach and does not evolve into mandates. At the federal and state level, they note, efforts to establish mandatory educational standards have become mired in controversy over graduation requirements that students meet character-related objectives, such as working cooperatively with people of diverse lifestyles or obeying authority.[31]

"A lot of parents think schools should stick to teaching knowledge, not get into behavioral aspects of kids," Finn says. "It's a lot more palatable to have school 'A' say we'll teach virtues the Bennett way, school 'B' say we'll teach it the Ten Commandments way and school 'C' say we'll teach tolerance and equality."

Martin of the character-education movement is quick to declare that, "We're not necessarily wanting state mandates. Often, when a mandate comes down there's resistance to what people think is going on in schools."

To the extent that parents like what they see of the character-education movement, they will vote with their feet. Jefferson Junior High and Allen Academy boast long waiting lists. But both schools are doing so much besides character education — from providing mentors to improving academics — that it may be hard to write a formula easily followed for success.

NOTES

1. Carolyn Gecan and Bernadette Mulholland-Glaze, "The Teacher's Place in the Formation of Students'
Character," *Journal of Education*, Vol. 175, No. 2, 1993, p. 46.

2. Gallup/Phi Delta Kappa poll cited in *Character Education in U.S. Schools: The New Consensus,* Character Education Partnership, February 1996, p. 21.

3. Thomas Lickona, *Educating for Character* (1992 ed.), p. 11.

4. *Ibid.*, p. 21.

5. *Ibid.*, pp. 13-19.

6. Joseph & Edna Josephson Institute of Ethics, "1996 Report Card on American Integrity," 1996.

7. Joe Queenan, "Good as Gold," *George,* April/May 1996, p. 108.

8. See David T. Kearns and Denis Doyle, *Winning the Brain Race* (1991).

9. See Peter Applebome, "Array of Opponents Battle Over 'Parental Rights' Bills," *The New York Times,* May 1, 1996, p. A1.

10. *Ibid.*

11. The Supreme Court's *Wisconsin v. Yoder* (1972) ruling created an exception to a state compulsory school-attendance law. Several Amish families had argued that requiring their children to attend high school violated their religious beliefs. See "Parents and Schools," *The CQ Researcher,* Jan. 20, 1995, pp. 49-72.

12. The Communitarian Network is a nonpartisan, nationwide coalition devoted to fostering socially responsible communities.

13. Developmental Studies Center, *The Child Development Project: Summary of Findings in Two Initial Districts and the First Phase of an Expansion to Six Additional Districts Nationally,* August 1994.

14. *Ibid.,* p. 20.

15. *Ibid.,* p. 31.

16. For background, see "Preventing Juvenile Crime," *The CQ Researcher,* March 15, 1996, pp. 217-240.

17. Henry A. Huffman, "The Unavoidable Mission of Character Education," *The School Administrator,* September 1995, p. 14.

18. Cited in William J. Bennett, "Parents, Schools and Values," *Network News & Views,* January/February 1996, p. 59, and Lickona, *op. cit.,* p. 6. *Network News & Views* is published by the Education

Excellence Network, a project of the Hudson Institute.

19. Quoted in Boston University press release accompanying *Character Education Manifesto,* April 3, 1996.

20. The Supreme Court's series of rulings on school prayer and Bible reading began with its 1962 decision in *Engel v. Vitale,* declaring the daily prayer in New York State public schools a violation of the constitutional separation between church and state. See "Religion in Schools," *The CQ Researcher,* Feb. 18, 1994, pp. 145-168.

21. Sidney Simon, Leland W. Howe and Howard Kirschenbaum, *Values Clarification: A Handbook of Practical Strategies for Teachers and Students* (1972).

22. Cited in Stephen Bates, "A Textbook of Virtues," *The New York Times,* Education Life section, Jan. 8, 1995, p. 18.

23. Lickona, *op. cit.,* p. 11.

24. Michael Bocian, "A Communitarian Approach to Character Education," *Basic Education,* March 1996, p. 8.

25. Cited in Kilpatrick, *op. cit.,* p. 105.

26. Letter from Marc Desmond, "Letters" column, *New York Times Magazine,* May 21, 1995, in response to Roger Rosenblatt, "Teaching Johnny to Be Good," *The New York Times Magazine,* April 30, 1995.

27. Boston University press release accompanying *Character Education Manifesto,* April 3, 1996.

28. Katherine Paterson, "Family Values," *The New York Times Book Review,* Oct. 15, 1995, p. 32.

29. See Chester E. Finn Jr. and Gregg Vanourek, "Charity Begins at School," *Commentary,* October 1995.

30. The Institute for Justice, a libertarian group in Washington, D.C., filed the Immediato family's appeal.

31. See "Education Standards," *The CQ Researcher,* March 11, 1994, pp. 217-240.

BIBLIOGRAPHY

Books

Etzioni, Amitai, *The Spirit of Community,* Crown Publishers, 1993.

George Washington University Professor Etzioni, founder of the Communitarian Network, describes his vision of moral education. Distancing himself from an "authoritarian" approach to discipline, he advocates helping students to develop psychological "muscles" like self-discipline, then imparting values through positive experiences at the school.

Fine, Melinda, *Habits of Mind: Struggling over Values in America's Classrooms,* Jossey-Bass, 1995.

Fine describes "Facing History and Ourselves," a middle school curriculum on the Holocaust that draws moral lessons on tolerance and racism. Fine, a supporter of the program, chronicles attacks by conservatives on the program and gives a historical overview of conservative-liberal struggles over values education in public schools.

Kilpatrick, William K., *Why Johnny Can't Tell Right from Wrong and What We Can Do About It,* Touchstone, 1993.

Boston College Professor Kilpatrick decries American education's values-neutral turn in an extended attack on the moral relativism of the 1960s and its philosophical roots. His conservative viewpoint comes through in his passing criticisms of progressive educator John Dewey and Mark Twain's *Tom Sawyer* as contributors to the moral decline of American children.

Lickona, Thomas, *Educating for Character,* Bantam Books, 1992.

Lickona, a professor of education at the State University of New York in Cortland, is a widely recognized authority in the character-education movement. In this book, orginally published in 1991, he describes how character education works, with numerous anecdotes from schools.

Articles

Bates, Stephen, "Morality 101," *The New York Times,* Education Life, Jan. 4, 1995, p. 16.

This is an excellent summary of the current debate over teaching values in American public schools.

Etzioni, Amitai, "Who's to Say What's Right or Wrong?" *Washington Post Education Review,* April 2, 1995.

In a review of *Habits of Mind,* Etzioni says the conservative criticism of the Holocaust curriculum "Facing History and Ourselves" is fair: It depicts America as a society with no core values other than procedural ones. But he maintains it is possible to teach values without pressing either leftist or right-wing opinions.

Gecan, Carolyn, and Bernadette Mulholland-Glaze, "The Teacher's Place in the Formation of Students' Character," *Journal of Education,* **Vol. 175, No. 2, 1993, pp. 45-58.**
In a special issue of this Boston University journal devoted to ethics in education, two teachers describe how they helped start a character education program at Thomas Jefferson High School for Science and Technology in Alexandria, Va.

Henry, Tamara, "Growing Debate Centers on Who Teaches Values," *USA Today,* **March 20, 1996, p. 1.**
The parental-rights movement, led by parents who object to the values taught by schools in sex education, condom distribution and community service programs, indicates potential controversies that could plague the character-education movement.

Lawton, Millicent, "Values Education: A Moral Obligation or Dilemma?" *Education Week,* **May 17, 1995, p. 1.**
Lawton describes the national movement among schools to introduce character education and some of the criticisms that have been raised.

Ryan, Kevin, "Character and Coffee Mugs," *Education Week,* **May 17, 1995, p. 48.**
Boston University education Professor Kevin Ryan, a proponent of character education, expresses concern that the movement is becoming a "cluttered bandwagon," emphasizing slogans more than substance.

Rosenblatt, Roger, "Teaching Johnny to be Good," *The New York Times Magazine,* **April 30, 1995, p. 36.**
Rosenblatt profiles character-education proponent Thomas Lickona and examines his ideas in the context of the current movement to teach values in schools.

Reports and Studies

Character Education Partnership, *Character Education in U.S. Schools: The New Consensus,* **February 1996.**
This is a useful report on developments from 1993 to 1995 in government and schools in support of character education. The partnership is a coalition of education groups that promotes character education.

Joseph & Edna Josephson Institute of Ethics, "1996 Report Card on American Integrity," 1996.
This annual survey shows high rates of lying and cheating among high school and college students and is often cited by advocates of character education.

For More Information

American Civil Liberties Union, 132 West 43rd St., New York, N.Y. 10038; (212) 944-9800. The ACLU has not taken an official position on character education in schools, but some of its affiliates express concern that the movement could provide an entrée for the religious right into schools.

Character Education Partnership, 809 Franklin St., Alexandria, Va. 22314-4105; (703) 739-9515. This consortium of national education associations promotes character education in the nation's schools and provides information on school programs.

Josephson Institute of Ethics, 4640 Admiralty Way, Suite 1000, Marina del Rey, Calif. 90292-6610; (310) 306-1868. This nonprofit group established the Character Counts! Coalition, a popular program promoting character education in schools and community groups.

Focus on the Family, 8605 Explorer Dr., Colorado Springs, Colo. 80920; (719) 531-3400. This Christian group dedicated to "the preservation of the home" produces moral-education materials for schools and endorses a conservative Christian perspective on character education.

3

Cyber Socializing

Are Internet Sites Like MySpace Potentially Dangerous?

Marcia Clemmitt

Katherine Lester, 17, and her father leave the courthouse in Caro, Mich., on June 29, 2006, after prosecutors decided not to treat her as a runaway for flying to the Middle East to meet a man she had met on the Internet. As cyber socializing grows, so do fears that the Internet exposes the vulnerable — especially the young — to sexual predators.

Getty Images/Bill Pugliano

From *CQ Researcher*, July 28, 2006

Last year, Eddie Kenney and Matt Coenen were kicked off the Loyola University swim team after officials at the Chicago school found they belonged to a group that posted disparaging remarks about their coaches on the Internet social-networking site Facebook.[1]

Like many people who post profiles and photos and exchange messages on cyber-networking sites like Facebook, MySpace, Xanga and Bebo, the students were shocked to find that university officials, not just their friends, were checking out the site. But Facebook, whose 8 million members are high-school or college students, alumni or faculty and staff members, is considered slightly less risky than sites with open membership rolls.

Nevertheless, said Kenney, who has since transferred to Purdue, "Facebook is dangerous right now. I've learned my lesson. You're supposed to have fun with this Facebook thing, but you need to be careful."[2]

But others see greater dangers than a lack of privacy lurking on social-networking Web sites. Last month, a 14-year-old Texas girl and her mother filed a $30 million lawsuit against MySpace, claiming the girl had been sexually assaulted by a 19-year-old man she met on the site. The man allegedly contacted the girl through her MySpace site in April, posing as a high-school senior. After a series of e-mails and phone calls, they arranged a date, when the alleged assault occurred.[3]

MySpace shares blame for the incident, the lawsuit argues, because users aren't required to verify their age, and security measures intended to prevent contacts with children under age 16 are "utterly ineffective."[4]

Teens Feel Safe on MySpace

More than 80 percent of teens believe the cyber-networking site MySpace is safe. Teens spend an average of two hours a day, five days a week on the site. However, 83 percent of parents of MySpace users worry about online sexual predators.

What Teens Say About MySpace	What Parents of MySpace Users Say
• Typically visit 2 hours a day, 5 days a week	• 38% have not seen their teen's MySpace page
• 7-9% have been approached for a sexual liaison	• 43% don't know how often their teens are on MySpace
• 20% feel MySpace negatively affects school, job, family and friends	• 50% allow their teen to have a computer in the bedroom
• 83% believe MySpace is safe	• 62% have never talked to their teen about MySpace
• 70% would be comfortable showing their parents their MySpace page	• 83% worry about sexual predators on MySpace
• 35% are concerned about sexual predators on MySpace	• 75% worry MySpace fosters social isolation
• 15% are concerned that MySpace fosters social isolation	• 81% worry about their teen meeting online friends in person
• 36% are concerned about meeting online friends in person	• 63% believe there are "quite a few" sexual predators on MySpace
• 46% believe there are "some, but not too many," sexual predators on MySpace	

Source: Larry D. Rosen, "Adolescents in MySpace: Identity Formation, Friendship and Sexual Predators," California State University, Dominguez Hills, June 2006

Though computer networking was developed in the late 1960s to allow scientists to access remote computers for research, its users — initially just academic researchers — quickly saw its possibilities as a socializing tool. As early as 1973, for example, 75 percent of electronic traffic was e-mail, much of it purely social in nature.

Over the past 15 years, a slew of new Internet applications — from chat rooms to instant messaging and, most recently, social-networking Web sites — have made online socializing easier than ever. By the end of 2004, for example, about 70 million adults logged onto the Internet every day in the United States alone — up from 52 million four years earlier — and 63 percent of American adults were Internet users. Teens were logging on at even higher rates — 87 percent of those ages 12 to 17.[5]

Today, a great deal of online activity remains social in nature. For example, in a survey by the nonprofit Pew Internet and American Life Project, 34 percent of the people who said the Internet played an important role in a major decision they'd made said they had received advice and support from other people online.[6] And 84 percent of Internet users belong to a group or organization with an online presence; more than half joined only after they got Internet access. Members of online groups also say the Internet brings them into more contact with people outside their social class or their racial or age group.[7]

But as Internet socializing grows, so do fears that the practice exposes the vulnerable — especially young people — to sexual predators.[8] Some also worry that networking sites create added peer pressure for teens to engage in risky behavior, such as posting risqué pictures of themselves.

In the cyber social world, there has always been the possibility that the friendly stranger chatting about mountain biking or a favorite rock band is not who he says he is. Older socializing technologies, such as Internet discussion boards and chat rooms, allow users to converse about favorite topics, from quilting to astrophysics. Participants generally use screen names — pseudonyms — and conversation centers on the forum topic, often with a minimum of personal information exchanged.

But social-networking sites have greatly increased Internet users' ability to discover other users' full personal information. For instance, newer social-networking sites utilize a personal profile — usually with photos and detailed descriptions of the person's likes and dislikes — as well as the names of friends with whom the person e-mails or instant messages. The page owner also can post comments and message on friends' pages.

Thus, while most Internet social networkers use pseudonyms, the wealth of information on their pages — plus information gleaned by reading their friends' pages — allows strangers to learn far more about a user than they could about someone posting a comment in a traditional cyber chat room.[9]

Moreover, today most cyber social-network users are between 12 and 25 years old. The largest networking site, MySpace, had more than 51 million unique U.S. visitors in May and boasts about 86 million members.[10] Traffic on the site jumped 367 percent between April 2005 and April 2006 while overall traffic on the top 10 social-networking sites grew by 47 percent.[11] Similarly, adult-oriented online dating sites are also attracting tens of millions of users. (*See sidebar, p. 54.*)

By now, most people — including teens — know it's risky to post personal information such as last name and phone number on the Internet, says Michelle Collins, director of the exploited child unit at the National Center for Missing and Exploited Children (NCMEC). But on today's social-networking sites, "You're only as safe as your friends are," says Collins. For example, a teenage girl may think she's playing it safe by not naming her school on MySpace, "but if she has four friends who all reveal the name of their school, then anyone who reads their pages can surmise" that she also goes there and could potentially track her down.

Last spring, concern about child predators spurred the Suburban Caucus — a new group in the House of Representatives — to introduce the Deleting Online Predators Act (DOPA). Building on the 2000 Children's Internet Protection Act (CIPA), DOPA would require schools and libraries to block young people's access to Internet sites through which strangers can contact them.[12]

Young Adults Are Most Likely to Date Online

About one-in-10 Internet users — or 16 million people — visited an online dating site in 2005. Those ages 18-29 were most likely to have used an online dating site.

Online Daters
(% of Internet users who have visited a dating site)

All Internet Users	11%
Sex	
Men	12%
Women	9
Race/Ethnicity	
White	10%
Black	13
Hispanic	14
Location	
Urban	13%
Suburban	10
Rural	9
Age	
18-29	18%
30-49	11
50-64	6
65+	3
Education Level	
Less Than High School	14%
High School Grad	10
Some College	11
College+	10

Source: Pew Internet and American Life Project, March 5, 2006

The bill's purpose is to shield children from being approached by strangers when using the Internet away from home, an aim that's important to suburban families, according to sponsor Rep. Michael Fitzpatrick, R-Pa. "One-in-five children has been approached sexually on the Internet," he told a House subcommittee on June 10. "Child predation on the Internet is a growing problem."[13]

But many wonder if DOPA addresses the right problem. For one thing, the "one-in-five" figure is "more complicated than is being implied," given today's cultural norms, says Tim Lordan, executive director of the nonprofit Internet Education Foundation. As one teenage girl admitted, " 'Dad, if I wasn't getting sexually solicited by my peers, I would be doing something wrong,' " says Lordan.

"The media coverage of predators on MySpace implies that 1) all youth are at risk of being stalked and molested because of MySpace; and 2) prohibiting youth from participating in MySpace will stop predators from attacking kids," said Danah Boyd, a University of California, Berkeley, doctoral student who studies how teens use online technology. "Both are misleading; neither is true. . . . Statistically speaking, kids are more at risk at a church picnic or a Boy Scout outing than . . . when they go to MySpace."[14]

In fact, the NCMEC report from which the one-in-five figure originates shows that 76 percent of online sexual solicitations "came from fellow children," and 96 percent of the adult solicitations came from adults 18 to 25, said Boyd. "Wanted and unwanted solicitations are both included. In other words, if an 18-year-old asks out a 17-year-old and both consent, this would still be seen as a sexual solicitation."[15]

Survival Tips for Online Socializing

In the long run, the Internet is good for teens, bolstering their social development, creativity and even writing skills. Dangers do exist, however, mainly caused by teenagers not understanding how easily strangers can access posted information.

There's clear evidence that writing and creating art, music and videos on the social Internet is building literacy and creative skills in today's teens, says Northwestern University Professor of Communications Studies Justine Cassell.

In the early 1990s, educators were concerned about seriously declining interest in writing by American students. But today, "we have striking evidence that kids are willing to write, when they weren't before," a change that many analysts attribute to the popularity of e-mail, instant messaging and blogging, says Cassell. Today's teens even show sophisticated understanding of literary niceties such as tailoring one's writing style to suit the audience. "They don't use emoticons [symbols] with parents," for example, because they "understand that's a dialect," she says.

Benefits aside, however, dangers and misunderstandings also exist, exacerbated by the fact that kids have raced ahead of many adults in their use of Internet socializing tools, says John Carosella, vice president for content at the Internet security company Blue Coat. "We are the first generation of Internet parents, and we need to learn how our job has changed," Carosella says.

Here are experts' tips for handling the online social world:

- Parents must learn how to use the technology, says Cassell. "At the very least, IM (instant message) your kids." Parents who IM "report much less fear about the technology and more happiness because their kids keep in touch," she says.
- Parents should play with Internet search engines "to learn how easily they turn up information and then share that knowledge as they talk with their children about Internet privacy," she says.
- Privacy rules between parents and kids can't remain the same in an Internet world, says Kaveri Subrahmanyam, an associate professor of psychology at California State University, Los Angeles. "If it's a kid's diary, you don't look at it." But diaries are different from MySpace pages, "because nobody else is looking at them," Subrahmanyam says.

  When it comes to publicly posted information that strangers can access, "You do need to know what they do," she says. "You can say to your child, 'I don't need to know the content of the IM, but I do need to know whom you're sending it to.'"

Cyberspace presents no more danger than the real world, says Michigan State University Professor of Psychology Linda Jackson. "But there are dangers," she says, "such as the ease with which you can give away your information."

Unwary users giving out private information — perhaps permanently, since so much Internet content is archived — is the chief new danger posed by social networking, say many analysts.

Most teenagers who post MySpace pages "seem to have the sense that nobody is watching" except their closest friends, says Tamyra Pierce, an assistant professor of mass communication at California State University, Fresno, who is studying social-networking use among high-school students. For example, "a boy who posted about banging a mailbox last night" apparently was unaware that the posting exposed him to vandalism charges, she says.

"Teenage girls would be petrified if you read their diary, yet they are now posting online stuff that is much more personal," said Jeffrey Cole, director of the University of Southern California's Center for the Digital Future. "Clearly kids need guidance."[16]

Given the popularity of online socializing, it would be impossible to ban it, says Kaveri Subrahmanyam, an associate professor of psychology at California State University, Los Angeles. "The Internet is here to stay. If you ban it, they'll find a way to get around the ban. It will become a cat-and-mouse game," she says. "We need to teach kids how to keep safe."

Unfortunately, the new dangers posed by sites like MySpace "have not been integrated into the society's knowledge base," said Kevin Farnham, author of a book of safety tips for social-networking. "Common-sense teaching is [not] automatically passed from parents to child."[17]

- Age matters when it comes to teens understanding Internet privacy issues, according to Zheng Yan, an education professor at the State University of New York at Albany. Only children ages 12 to 13 or older can grasp the Internet's "social complexity," such as the large number of strangers who can access information posted on Web sites.[1]

- Web sites vary widely in how much public and private access they allow to posted material, and it's important to think about this when posting, says Alex Welch, founder and CEO of the photo-sharing Web site Photobucket. On Photobucket, the photo albums of people under age 18 are automatically kept private.

  However, even if they weren't, posting photos on Photobucket would be less risky than posting the same pictures on a social-networking site like MySpace, says Welch. That's because on MySpace photos are linked to additional personal information that may pique strangers' interest and provide clues to help them contact posters.

- It's also important to consider the future, including how employers or college admissions officers might view your online postings, says Henry Jenkins, director of the comparative media studies program at the Massachusetts Institute of Technology. Much of what appears on the Internet today ends up archived somewhere and can be retrieved tomorrow, he says. "Kids don't recognize the permanence of what they put up there."

- Adults will probably learn a lot about the Internet from their children, and they should be open to that, says Jenkins. Parents "need to recognize that some unfamiliar experiences look scarier from the outside than they are. Take time to understand what you're seeing."

  Talking about teens' MySpace pages can open the door to family discussions of important, sometimes touchy issues, like contemporary fashion, media images and ideals, Jenkins says. "Ask your kid how they choose to represent themselves" on their MySpace pages "and why."

- Teens — and adults — who socialize on the Web should remember that, "when it comes to the rules for getting to know people," the Internet "parallels our world perfectly," says Patricia Handschiegel, founder and CEO of StyleDiary.net, a social-networking site focused on fashion.

  Often, young people have "a false sense that you can't be tracked" by people they correspond with online, Handschiegel says. "If you want to correspond, fine, but take your time getting to know people. Watch for cues" to ulterior motives, "such as somebody pushing too fast to know you."

---

[1] Bruce Bower, "Growing Up Online," *Science News*, June 17, 2006, p. 376, www.sciencenews.org/articles/20060617/bob9.asp.

But predators are not the only ones reading social-networking sites. Some employers routinely scan pages posted by job candidates, with potentially disastrous results, says Matthew Smith, a professor of communications at Ohio's Wittenberg University. "You may put up a birthday-party picture of yourself in your underwear, thinking you are showing how carefree you are," says Smith. "But a future employer may see it and decide you're irresponsible."

And the federal government may be next. The National Security Agency (NSA) is funding "research into the mass harvesting of the information that people post about themselves on social networks," according to Britain's *New Scientist* magazine.[18]

"You should always assume anything you write online is stapled to your résumé," said Jon Callas, chief security officer at PGP, a maker of encryption software.[19]

As lawmakers, parents and Internet companies confront new Internet security issues, here are some of the questions being discussed.

Is cyberspace more dangerous than real space?

Some legislators and worried parents warn that the online world increases opportunities for sexual predators to reach victims. Likewise, some women won't use online dating services because they fear they might meet unsavory characters. But defenders of online socializing argue that real-life encounters pose just as much risk of unwanted sexual advances or of being bullied or defrauded.

"There's a child-abuse epidemic that we don't even know about on the Internet, which is how it stays so invisible, when a 45-year-old man engages in sexually explicit dialogue with a 12-year-old girl," says John

Shortly after graduating from Hobart and William Smith Colleges in 2004, Jessica Ostrowski joined the popular college social-networking site Facebook. She lives in Boston but stays connected to friends nationwide through the site, which has 8 million members.

Carosella, vice president for content control at Blue Coat Systems, a maker of security software and other tools for online communication.

"When it happens in real life you know it's happening," but on the Internet, even the girl herself may not know it's happening initially, he says. "This is why it can go on with such facility, and it's so easy to escalate because nobody can overhear it."

Internet communication is riskier than real-life meetings, according to Carosella, because many of the cues humans use to size up other people — such as gestures and tone of voice — are missing. "Most of our evolutionary clues [to risk] are stripped out" of online encounters, he says.

In addition, online acquaintances can do harm — especially to young people — even without offline meetings, he says. The Internet "is the most powerful tool yet invented" capable of bringing "every aspect of human behavior from the most sublime to the most debauched and depraved" right into our own homes, he says. And it's burgeoned so fast that "we've had no time to figure out how to deal with it."

Carosella also suggests that — while there is not enough research yet to prove it — easy access to online porn may prime some people to become aggressive or engage in sexual predation. "Lots of data suggest that pornography is a very significant factor in the emergence of criminal, aggressive behavior," he says. "You're whipping around the most powerful urge that people have. And the Internet has made pornography more accessible, more private, more extreme."

Moreover, peer pressure among teens may encourage risky sexual expression on cyber-networking sites, says Pierce of California State University. "The more stuff they have that is graphic and shocking, like links to porn and photos in risqué poses, the more friends they have on their lists," she says of recent studies of MySpace and Xanga and their use by more than 300 local high-school students. "Most who have upward of 500 friends have links to porn sites and graphic stuff."

Looking at social-networking sites can sometimes expose teenagers to porn, even when they don't seek it out, says Pierce. For example, in a group of 50 to 60 sites she recently examined, 10 contained automatic links that unexpectedly shift the viewer to a pornography site.

It "concerns me that our youth may be exposed to pornography" at the inadvertent click of a mouse, she says. "I don't know if individuals in the porn industry are creating 'fake' MySpace sites and then autolinking them or if the young persons themselves are getting involved. I just know that what appeared to be an innocent, young person's site automatically turned into porn" when the site user's photo was clicked.

Collins of the National Center for Missing and Exploited Children says the Internet provides a "target-rich environment" for pedophiles. "Twenty years ago, people had to go to a park or a soccer game" to meet children and teens, she says. And the Internet allows predators to learn more about a potential victim, which may help him build a relationship.

There's little disagreement that dangers can lurk on the Internet. However, many analysts note that, even as Internet socializing has burgeoned, sex crimes have been decreasing, and data do not indicate an increase in sexual predation due to teens going online. "There is no evidence that the online world is more dangerous," says Justine Cassell, a professor of communications studies at Northwestern University.

In fact, predatory crimes against girls have declined during the past decade, even as the Internet was bringing more young people online, she says. All national data sets show that from 1994 to 2004, single-offender crimes, including assaults, by men against 12-to-19-year-old girls decreased, demonstrating that the online world has not put young people in more danger, says Cassell.

The number of sexual-abuse cases substantiated by child-protection agencies "dropped a remarkable 40 percent between 1992 and 2000," and evidence shows it was a "true decline," not a change in reporting methodology, wrote sociology Professor David Finkelhor and Assistant Professor of Psychology Lisa M. Jones, of the University of New Hampshire's Crimes Against Children Research Center.[20]

Meanwhile, says Collins, tips about online enticement of young people reported to the NCMEC have risen from 50 reports a week in 1998 to about 230 a week. But does that mean there are more online incidents today? Probably not, says Collins. "More people know where to report it today," thanks to the center's education campaigns and social-networking companies' putting links to NCMEC's tipline on their sites. In addition, "more people, more kids, are wired today than ever before," Collins notes.

Even teenagers' fascination with online porn may be something of a passing fancy, not a long-term, negative behavior change, says Michigan State's Jackson, who recently completed research that tracked teens' Internet usage. "We went everywhere they went for 16 months," she says. "Porn sites were popular for the first three months, then it really died down."

As for the dangers of online dating, Eric Straus, CEO of Cupid.com, a popular dating site, rates the dangers of online and offline dating as about equal. Meeting potential romantic partners online is "a numbers game," and the Web lets you search for geographic proximity or mutual interests across a wide variety of potential dates. But "you shouldn't be fooled into feeling safe" from deception in either environment, he says. "If you talk to 100 people online, some are going to be unsavory. If you meet them in a bar, and they say they're not married, you shouldn't believe that either, and if you give them your phone number, that's dangerous."

New technology always spurs panic, says Paul DiMaggio, a sociology professor at Princeton. "MySpace is generating the same fear reaction that films and vaudeville got," says DiMaggio. "And all new technology and media generate more hysteria than threat."

"A lot of the behavior on sites like MySpace has been going on in teen hangouts for generations," wrote Anne Collier, editor of *NetFamilyNews*.[21] Dangerous behavior existed, "but parents weren't privy to it."

"One child being molested by an online predator is too many and has to be addressed," says Lordan, of the Internet Education Foundation. "Nevertheless, statistically, most molestations are family and acquaintance molestations, and the chance of your child being dragged through the computer screen by a predator is low."

Furthermore, "a lot of the stories that you hear" about teens running off with adults they met on MySpace, for example, "appear to be kids who would get in trouble in some other way if it weren't for the Internet," he says. "Once you start digging down, these don't appear to be typical families."

MySpace Users Are Most Loyal

Two-thirds of the visitors to MySpace return each month — more than any other social-networking site. MSN Groups caters to those with special interests, such as computers, cars, music, movies or sports.

Top Five Socializing Sites
(based on retention rate)

Site	Retention Rate (%)*
MySpace	67.0
MSN Groups	57.6
Facebook	51.7
Xanga.com	48.9
MSN Spaces	47.3

* Based on the number of March 2006 visitors who returned to the site in April

Source: Nielsen/NetRatings

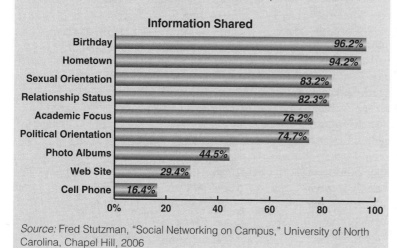

Freshmen Share Many Personal Details

Most University of North Carolina freshmen reveal their birthday and hometown on social-networking sites. But only 16 percent give out more detailed personal information, like cell phone numbers.

Information Shared

Birthday	96.2%
Hometown	94.2%
Sexual Orientation	83.2%
Relationship Status	82.3%
Academic Focus	76.2%
Political Orientation	74.7%
Photo Albums	44.5%
Web Site	29.4%
Cell Phone	16.4%

Source: Fred Stutzman, "Social Networking on Campus," University of North Carolina, Chapel Hill, 2006

All youngsters are not at equal risk of being victimized, says Collins. "Some kids are using the Internet to fill a void, and these kids are going to be more susceptible," she says.

Should schools and public libraries block access to social-networking sites?

Rep. Fitzpatrick and other members of the Suburban Caucus have introduced legislation that would require public schools and libraries to bar access to social-networking sites and chat rooms as well as to pornography sites.

For adults, online social-networking sites "are fairly benign," but "for children they open the door to many dangers," including online bullying and exposure to child predators that have turned the Internet into a virtual hunting ground for children," Fitzpatrick said on the House floor May 9.[22]

"There are thousands of online predators who are trying to contact our kids using powerful engines like MySpace.com," said Rep. Mark Kirk, R-Ill.[23]

Blocking Internet social tools, at least temporarily, is a valid response, says Carosella of Blue Coat Systems. While educational responses are vital in the long run, "we haven't invented [that education] yet."

Furthermore, "while we are developing our educational response, predators are building and improving their game plans. It's an arms race," says Carosella. Blocking access in schools and libraries "is absolutely an answer," because supervision currently "is completely inadequate."

But critics of such proposals say blocking access is nearly impossible without inadvertently blocking valuable portions of the Internet. "It's brain-dead to say you should stop people from using some technology," says John Palfrey, clinical professor of law and executive director of Harvard Law School's Berkman Center for Internet and Society, pointing out that the technology isn't going away.

Boyd, of the University of California, pointed out that because technology plays a major role in the business world, it is unwise to deprive students of it in schools and libraries. "The law is so broadly defined that it would limit access to any commercial site that allows users to create a profile and communicate with strangers." While it ostensibly targets MySpace, as written it would block many other sites, she said, including blogging tools, mailing lists, video and pod-cast sites, photo-sharing sites and educational sites like NeoPets, where kids create virtual pets and participate in educational games.[24]

Moreover, she continued, many technology companies now are using social software, such as features that help users find information, get recommendations and share ideas. "This would all be restricted," she wrote.[25]

Lynn Bradley, director of the American Library Association's government relations office, agrees. Fitzpatrick's proposal "is like using a water hose to brush your teeth." The legislation could also affect distance-learning programs, which use many technologies that the legislators want blocked. "Rural schools have increasingly started to rely on distance learning to supplement their curricula," said Bradley. "It would appear to us on reading this bill that [such programs] would be swept up in the blockage."[26]

CHRONOLOGY

1960s-1980s *As the Internet develops, its academic and technical users begin socializing online.*

1967 The term "six degrees of separation" enters the lexicon after Yale University psychologist Stanley Milgram claims to experimentally validate the theory that any two people on Earth are connected by an average of six intermediate contacts.

1972 E-mail is invented and quickly becomes the most widely used tool on the first wide-scale computer network, ARPANET.

1975 First ARPANET mailing lists link people with shared interests.

1979 First Usenet newsgroups are established.

1982 Carnegie Mellon University computer-science Professor Scott Fahlman invents the first emoticon, the smiley :-)

1988 Internet Relay Chat is invented, allowing computer users to exchange real-time messages with a group.

1990s *People flock to the Internet, mostly drawn by socializing tools like e-mail and chat rooms. Worries grow about sexual predators contacting teens through online socializing, but sex crimes against teenagers decrease overall.*

1990 John Guare's play "Six Degrees of Separation" captures the public imagination with its portrayal of the power of social networks.

1994 Netscape's Mosaic Web browser is offered free on the company's Web site, helping draw millions of non-technical users online.

1996 AOL introduces the Buddy List, which alerts users when friends are online and ready to IM (instant message) each other.

1997 Sixdegrees.com becomes one of the first Internet social sites to link users to friends of friends, through three "degrees of separation." . . . AOL introduces AIM, allowing users for the first time to IM non-AOL users.

1998 eHarmony.com, the first online dating site to require users to complete a personality-matching test, is founded by evangelical Christian Neil Clark Warren.

2000s *Social-networking sites become teenagers' socializing tool of choice, while rock bands and other performers begin using the sites to build and strengthen their fan bases. Lawmakers become concerned that cyber social networking makes it easier for sexual predators to find teens.*

2000 President Bill Clinton signs the Children's Internet Protection Act (CIPA), requiring federally subsidized public and school libraries to use filtering software to block children's access to pornography.

2001 Columbia University Professor Duncan Watts uses forwarded e-mails to confirm Milgram's six-degrees-of-separation finding.

2003 MySpace is founded. . . . The U.S. Supreme Court upholds CIPA's requirement for public libraries to block Internet sites. . . . Vermont Gov. Howard Dean becomes a leading contender for the 2004 Democratic presidential nomination by using the social-networking site Meetup to spur grass-roots organization.

2005 News Corp., a global company headed by Australian media mogul Rupert Murdoch, buys MySpace for $580 million. . . . Several state legislatures consider bills requiring online dating sites to conduct background checks of users.

2006 NBC's television newsmagazine "Dateline" films sting operations around the country in which adults pose as teens in Internet chat rooms and arrange to meet sexual predators. . . . The Deleting Online Predators Act is introduced in Congress to require federally subsidized schools and libraries to block children's access to social-networking sites, chat rooms and other socializing technology. . . . Social software continues to spread with Internet companies including AOL, Netscape, Google and Yahoo jumping on board. . . . Americans spend $521 million on online dating, making it one of the biggest income generators on the World Wide Web.

Do You Take This Online Stranger . . . ?

Soon after moving to Washington, D.C., from upstate New York to attend community college, Cait Lynch signed up with the popular Internet dating site Match.com. "I'd just moved to the area and I didn't really know anyone. I was interested in meeting new people," she says. Moreover, her mother had met her husband of eight years online. "I figured if it went well for her, I should give it a shot."

After a few months, Lynch came across a promising profile. Paul Schnetlage was getting his master's degree at Johns Hopkins University, and they shared an interest in cafes, movies, reading and art. Moreover, they were the same age — 22 — and both were in school full time while working.

After e-mailing for a month, they met for coffee, and Paul admitted he had "met several really strange girls" before meeting her. For her part, Lynch says she "didn't take anything seriously at all" on the site. Nonetheless, their relationship slowly blossomed, and last March, nearly two years after meeting, they married. Without Match.com, "I don't think I ever would have met someone like Paul," says Lynch, who works for a real estate developer. Her husband is a software designer.

Launched in 1995 as the first online dating service, Match.com now has some 15 million paid members. It claims that more than 300 marriages or engagements occur between members or former members each month and that 400,000 people find the person they are seeking each year. The "vast majority aren't marriages," says spokesperson Kristin Kelly, "but a lot of people are telling us they found a great relationship."

Today, Internet sites devoted to matchmaking constitute one of the top online income generators, with the U.S. market valued at $521 million.[1] Nearly one-third of American adults know someone who has used a dating Web site, according to the Pew Internet and American Life Project.[2]

But after steady increases in membership, the 25 most-visited online dating sites showed a 4 percent decline in American visitors in the past year compared to a 4 percent increase in the total U.S. Internet audience.[3]

"There is a natural limit to the number of people who want to participate in this industry, and they are getting to that number," explained Jupiter Research analyst Nate Elliot."[4]

And Mary Madden, a Pew senior research specialist, points out the Internet is still very low on the list of ways people meet their significant others. "We asked all Internet users who are married or in a committed relationship if they met their partner online or offline, and only 3 percent said they met them online," she says.

There are clearly cases where online dating does foster meaningful relationships, says Barbara Dafoe Whitehead, co-director of Rutgers University's National Marriage Project. "But to what degree it's a very reliable source of finding a partner is the open question," she says.

Stan Woll, a psychology professor at California State University, Fullerton, says Internet dating sites are good at introducing people to large numbers of potential mates they might not otherwise meet. However, he finds that the number of possibilities tends to make members overlook people they would ordinarily enjoy. "There is an array of other people, and you keep wanting to go on and find somebody better or closer," argues Woll.

Andrea Baker, a sociology professor at Ohio University and author of the 2005 book *Double Click: Romance and Commitment of Online Couples*, argues that online dating sites are better than chance meetings at helping members find partners with common interests.

Henry Jenkins, director of comparative media studies at the Massachusetts Institute of Technology (MIT), says it would be unwise to keep teachers and librarians away from social-networking technology because "they are the best people to teach people how to use it constructively and safely." And, while Fitzpatrick's bill contains loopholes for educational uses, history suggests that fear and uncertainty would likely stop many schools from using those loopholes, he adds.

Blocking access at public schools and libraries would also worsen the already troubling "digital divide," says Jenkins.[27] Children in wealthier families could still access school-blocked Internet sites from home, but students without Internet at home "will be shut out," he says.

Policymakers who propose blocking as a "silver bullet" don't know how important online socializing is to teenagers, says the Internet Education Foundation's

But online dating has its drawbacks. Whitehead says she often hears criticism about untruthful member profiles. Lynch says members she met online often embellished their résumés to enhance their appeal, such as the man who said he liked chess but didn't know the rules when they sat down to play. And the Pew project found that 66 percent of Internet users think cyber dating is dangerous because its puts personal information online.[5]

Online-dating experts have found that people are turning to the Internet to find love for a number of different social and cultural reasons. For instance, busy work schedules make it harder for people to find compatible partners in the offline world, Baker says.

The Internet also allows busy singles to meet people who live across town whom they would not normally bump into, contends Kelly. It can "cut across boundaries that used to limit people's opportunities to meet people, like geography," adds Whitehead.

Compared to a generation ago, people are much more likely to live in a city where they did not grow up, Kelly points out. Because people's lives are no longer dictated by geography, traditional "friendship networks are down," says Bernardo Carducci, a professor of psychology at Indiana University Southeast. He argues that people are lonely and don't have a local friendship network that can produce dates.

And Whitehead and Kelly agree that singles are also turning to the Internet to find a mate because they are not finding them in the old places, such as college or high school.

While online-dating sites do provide more conveniences, Madden says the majority of people she surveyed did not find online dating more efficient than offline dating. Users "still have the complex challenges of negotiating relationships [including] norms and social skills that are different from face-to-face communication."

Some, like Carducci, describe online dating as just one more tool in the dating arsenal. Others, including

Yahoo Drew Most Visitors

Each month Internet dating sites entice millions of hopeful visitors searching for the perfect mate.

Top 10 Dating Sites

Site	Unique Visitors (in thousands)
Yahoo! Personals	6,052
Match.com Sites	3,893
MarketRange Inc.	2,676
Spark Networks	2,638
Mate1.com	2,354
True.com	2,093
eHarmony.com	1,796
Love@AOL	1,516
Zencon Technologies Dating Sites	1,091
Lovehappens.com	976

Source: comScore Media Metrix, a division of comScore Networks, Inc., January 2006

Whitehead, think it is a big tool that is going to have a lasting force in society.

Kelly predicts that "some of the stigma about online dating is a generation gap" that will disappear over time.

— *Melissa J. Hipolit*

---

[1] Ginanne Brownell, "The Five-Year Itch," *Newsweek*, Feb. 27, 2006.

[2] Pew Internet and American Life Project, "Online Dating," March 5, 2006.

[3] comScore Media Metrix, January 2006.

[4] Brownell, *op. cit.*

[5] Pew Internet and American Life Project, *op. cit.*

Lordan. "Kids feel quite attached to their own space," whether on MySpace or another site, and are "incredibly worried" about it being taken away, he says. If sites are blocked, "they'll go underground."

The bill should be called the "Send Social Networking Sites Off Shore Act," says Lordan, because there are social-networking sites all over the world and "stories of kids creating these sites themselves." If that happens, "we can't even help them."

Moreover, there are serious technical barriers, such as age verification, he says. Many blocking proposals would depend on age verification of site users, either to block out older users who might prey on kids or to keep younger kids from using the sites. But "age verification is a problem even in the real world," where kids use fake ID cards to buy alcohol and cigarettes, says Loran, and the problem is "worse in the online world."

Does Internet social networking foster good relationships?

For nearly as long as the Internet has existed, people have used it to stay in touch with old friends and meet new ones.[28] For just as long, skeptics have argued that online relationships are less rich, real and reliable than real-world interactions and that Internet socializing actually isolates people. Over the years, studies have found evidence bolstering both sides of the question.

No online community can support human bonding the way real-world communities do, said Clifford Stoll in his 1996 best-seller *Silicon Snake Oil: Second Thoughts on the Information Superhighway.* "What's missing from this ersatz neighborhood? A feeling of permanence, a sense of location, a warmth from the local history. Gone is the very essence of a neighborhood: friendly relations and a sense of being in it together."[29]

Some studies have found that Internet usage pulls people away from their real-world friends and family and isolates the user. Research by Stanford University's Institute for the Quantitative Study of Society (SIQSS), for example, finds that Internet use directly relates to social isolation. Based on survey findings, the Stanford researchers say that for every hour a person spends online, their face-to-face time with family and friends decreases by 23.5 minutes.[30]

Furthermore, "face-to-face interaction with close [friends and family] qualitatively differs from interactions in the virtual world online and is more important to one's psychological wellness," according to Lu Zheng, a Stanford doctoral student in sociology.[31]

Another SIQSS analysis found that "the more time people spend using the Internet the more they lose contact with their social environment." Internet users, for example, spend much less time talking on the phone to friends and family, according to the study.[32]

"E-mail is a way to stay in touch, but you can't share a coffee or a beer with somebody on e-mail or give them a hug," said SIQSS Director Norman Nie. "The Internet could be the ultimate isolating technology that further reduces our participation in communities."[33]

But other analysts insist that Internet socializing strengthens online and offline relationships. What looks to some like a "fading away" of social life as Internet usage increases is actually just a shift to new communication modes that strengthen many people's social ties, says a recent report from the Pew Internet and American Life Project and the University of Toronto.[34]

"The traditional human orientation to neighborhood- and village-based groups is moving toward communities . . . oriented around geographically dispersed social networks," the Pew report said. "The Internet and e-mail play an important role in maintaining these . . . networks" and "fit seamlessly" into people's social lives. As a result, the study concluded, Americans today "are probably more in contact with members of their communities and social networks than before."

For example, the analysts found that those who e-mail 80-100 percent of their closest friends and family weekly also speak regularly by phone with 25 percent more of their closest associates than those who e-mail less. For so-called second-tier (or "significant") social ties, the effects were even more pronounced: Those who weekly e-mail 80-100 percent of their second-tier contacts also regularly telephone twice as many of those contacts as do people who do not e-mail their friends.[35] Furthermore, 31 percent of those surveyed said their Internet use had increased their number of second-tier contacts — while only 2 percent said it had decreased them.[36]

A Canadian study of a housing development found similar results. University of Toronto researchers found that residents with high-speed connections had more informal, friendly contact with neighbors than residents who were not on the Internet.[37] Residents with broadband service knew the names of 25 neighbors, on average, compared to non-wired residents, who knew eight. Wired residents made 50 percent more visits to neighbors' homes than non-wired people, and their visits were more widely scattered around the housing development, according to the paper.[38]

And in a new and interesting wrinkle, the Internet may be helping to keep the lines of communication open between people on both sides of the recent Middle East violence. Since the cross-border shelling between Israel and Lebanon erupted earlier this month, Internet message boards, discussion forums and blogs have exploded with posts from Israeli and Lebanese nationals commenting on the fighting. "The fact that the citizens of two warring countries are maintaining a dialogue while a war is going on cannot be ignored," said Lisa Goldman, a Canadian-born freelance journalist who blogs from Tel Aviv and who points out that Internet discussions and socializing between Israelis and Lebanese predated the current conflict.[39]

Another significant new phenomenon among adolescents equipped with cell phones, instant messaging and social-networking pages is "tele-cocooning," says MIT's Jenkins. Coined by Mimi Ito, a research scientist at the University of Southern California's Annenberg Center for Communication, tele-cocooning refers to "carrying your friends around with you, using technology to be literally in contact with them all the time."

Online socializing also helps kids who would otherwise have a hard time finding friendships, says Jenkins. "Kids who may be outcasts, or pariahs, or have interests that nobody in their school shares now can go online and meet kids all around the country who like the same comic books, music and sports," Jenkins says. "I watched my son go through that."

In addition, while online socializing creates opportunities for deception or misunderstanding, meeting people online first can sometimes help avoid superficial judgments. The Internet allows disadvantaged or physically different people to socialize without suffering the instantaneous negative judgment that may happen in person.

"If people consider you unattractive, for example — you have moles, a big nose — . . . you can get negative reactions in public," says Wittenberg's Smith. "Online, I can talk about my love of pets or my work with my church. I can put down what is lovable about me and bypass whatever I think is keeping me from getting fulfilled relationships" offline.

Internet dating also saves time, says Mark Brooks, editor of *Online Personals Watch.* "The great thing about Internet dating is being able to find the right people," something that is harder in the real world, where there's no search function, he says. By meeting someone online first, "I don't need to go out with somebody who smokes or who has kids. You can ask the difficult questions up front."

Social-networking sites also draw criticism for their emphasis on having a lengthy list of friends, as sites like MySpace tend to do. Encouraging the how-many-friends-do-my-friends-have game sets up bad behaviors, according to technology entrepreneur Christopher Allen. "It is not the number of connections but the quality" that counts, a fact obscured by concentration on long friend lists, he said.[40]

Social networking is based on a faulty view of friendship — the "premise that . . . if A is a friend of B, and B is a friend of C, then A can be a friend of C, too," said the University of California's Boyd. "Just because you're friends with somebody doesn't mean their friends are similar" to the friends you would choose.[41]

A social-networking site can also be used as "an electronic bathroom wall," potentially increasing the reach of school bullying, says NCMEC's Collins. "There's capacity for large-scale humiliation," Collins says, citing an incident in which a girl created a page laden with pornography and put an ex-friend's name on it.

But Boyd isn't so sure. "Bullying, sexual teasing and other peer-to-peer harassment are rampant among teenagers" because they are "tools through which youth learn to make meaning of popularity, social status and cultural norms," she said. It is unclear whether online embarrassment is any more damaging than offline humiliation, she added, conceding the Internet "may help spread rumors faster."[42]

BACKGROUND

Wired Love

Virtually all communications technology, no matter why it was developed, has quickly become a socializing tool, with teenagers usually leading the way. And — from the telegraph to MySpace — parents have always worried what kind of trouble teens may get into with the new technology.

Northwestern's Cassell recounts a newspaper story entitled "Wired Love," which describes how a father followed his 16-year-old daughter to a tryst she'd arranged with a man she had met online and was arrested after threatening to kill the man and the girl. The father had bought his daughter the new technology — a telegraph — according to the 1886 story in the magazine *Electrical World.*

"At first people thought the telegraph would be good for girls," who might land jobs as telegraph operators, says Cassell. "But in the 1880s, an attempt was made to legislate who could be a telegraph operator because people worried that girls would contact men through the device," she says.

What's Next for the Social Net?

The mega social-networking site MySpace boasts more than 30 million visitors a day. But with the bulk of its users teenagers, Internet industry observers say the MySpace phenomenon may have peaked and that teens are ready to move on to the next big thing.

But social-networking technology — from people searches to video sharing — is here to stay. And Internet entrepreneurs are scrambling to develop new ways to give social-networking sites more staying power and attract a broader — and older — audience.

"Could it be that MySpace peaked this past April?" mused Scott Karp, a technology and publishing analyst. "When a fad becomes overhyped, teens will eventually retreat," and MySpace daily traffic began falling off somewhat in late spring, Karp said.[1]

Some observers seconded Karp's observation. "MySpace is hot now, but teen audiences are the most fickle market ever invented," wrote Mathew Ingram, an online business writer for the Toronto *Globe and Mail*. "MySpace has gone (or is becoming) mainstream, and mainstream is the kiss of death."[2]

Media mogul Rupert Murdoch, whose News Corp. bought MySpace last year for $580 million, has been reticent about how he plans to make money on the free site. But the man who once described newspapers' paid classified advertising as "rivers of gold," confounded the industry recently when he announced that he will offer free classified advertising on MySpace.[3]

Social-networking sites are really about communicating with people you know or specifically want to know, not publication to strangers, according to Dalton Caldwell, founder and CEO of the social-networking site imeem.

com. "You don't want to see strangers' home movies," Caldwell says.

Some people in the social-networking business think of the sites as "destinations," cool places to hang out and spend time, says Caldwell. But that's a recipe for a site whose popularity wanes fairly quickly, Caldwell believes. "The site'll be like a night club. It'll be cool for a while, but then it'll fade."

But a site envisioned as a collection of popular, useful, top-of-the-line communication tools won't likely suffer that fate, because "you're plugging into something that people always do." IM — instant messaging — for example, "is not a fad. It hasn't changed since its inception" because it's "a cool tool that people continue to want to use," Caldwell says.

A big draw of social networking will always be connection, especially for the older-than-teenage crowd that entrepreneurs hope to lure in bigger numbers, says Mark Brooks, editor of *Online Personals Watch*. Facebook — a site that caters to college students — has an advantage for the long run in that regard, says Brooks. Providing a connecting point for school pals that will help them avoid losing touch with old friends, Facebook "hits people at their point of pain," Brooks says.

The fact that the site provides a venue to keep in touch with old school mates will eventually "drive Facebook beyond MySpace," Brooks predicts.

To draw older, hopefully more permanent users than MySpace's teens, some entrepreneurs are developing sites where people can join groups discussing topics of interest rather than hanging out and posting personal profiles. Such sites are similar to traditional Internet social venues like Yahoo Groups, whose thousands of discussion groups are

In the early 1900s, similar worries and proposals for legislation arose about the telephone, she says. The anti-telephone "rhetoric was identical to anti-MySpace rhetoric today," stressing "fears that girls were at risk," she says.

Yet socializing has always been a top human need, as the history of communications technology has shown, says Wittenberg's Smith. When new technologies change how people communicate, people always use them for social interactions — rather than just for business or educational use, he says.

The telephone, for example, was first marketed as a tool to speed up workplace transactions, says Smith. When the fledgling phone industry found that people were getting

on and talking for a long time, they "were horrified," he says. For example, for efficiency's sake a single party line was usually assigned to several families to share. "But people wanted to get on and yak," clogging up the lines for their neighbors, he says.

Internet socializing, which began almost as soon as computer networks were established in the late 1960s, also took network developers by surprise. By 1973, e-mail — much of it purely social in nature — made up 75 percent of traffic on ARPANET, the computer network designed by the Department of Defense to allow researchers to exchange data and access remote computing capability.[43]

focused on individual topics like hobbies or alternative sexual lifestyles.

Unlike traditional discussion groups, however, the new "social media" groups capitalize on social networking's ability to connect an individual not only to friends but also to friends of friends, says Tom Gerace, founder and CEO of Gather.com, a social-networking site aimed at adults.

Gather's approach is to encourage its users — envisioned as the sort of folks who are regular listeners to National Public Radio, for example — to post thoughts on subjects ranging from politics to recipes. Those postings constitute a new "social media," movie reviews and political rants composed not by professional journalists but by any interested Internet user. Sites like Gather then use social networking's "friends of friends" structure to help people link up to the posting they'll be most interested in, because they interest others in their circle of contacts, says Gerace.

Gather users will also provide "social filtering" of the media they create, providing links and reviews that will bring the best socially created content to the top of the heap, Gerace says. The site currently pays members a modest fee if content they create gets high marks from fellow users. And "eventually some writers will earn a living" by creating top-ranked Gather content, Gerace says.

Content is also king at Buzznet, a social-networking site on which users share writings, photo, music and video celebrating popular culture, especially music. "A classic social network is all about your profile, but on Buzznet our emphasis is 100 percent focused on what you're producing," says co-founder Anthony Batt.

Buzznet "is the upside-down version of MySpace," says Batt. It's a catalog of people. We're about . . . what people are interested in. You connect with an interest, then you meet the people." Buzznet founders plan to build on its success to found other communities linked by different interests, potentially attracting different age groups, says Batt.

But Buzznet is not the replacement for MySpace that its founders expected it to be, says Batt. "People tend to have three stops. They hang out on MySpace, write a bit at LiveJournal" — a blogging and social-networking site — "and then spend time with us."

Online dating services will ultimately succumb to the social-networking boom, predicts Markus Frind, creator and owner of the Canada-based dating site Plentyoffish. com. Unlike most other dating sites, Plentyoffish is free to users. Frind's operation is "half social networking," because, unlike most other dating sites, Plentyoffish doesn't attempt proactive matchmaking. Instead, like MySpace, Plentyoffish simply allows members to post profiles, converse and arrange outside meetings — one-on-one dates or multi-member parties — on their own.

Contrary to what MySpace and Facebook members claim, the social-networking sites "are all about sex," says Frind. "People go there to hook up," and they're succeeding at it, he says. Eventually, that success will pull most people away from online dating services — which charge — to social-networking sites, which are free. "The hot girls are only on the social-networking sites now, not on dating sites. It's socially unacceptable for hot girls to say they're looking for dates," Frind says. "Eventually, all the guys will follow them there."

---

[1] Scott Karp, "Has the MySpace Downturn Begun?" *Publishing 2.0*, May 25, 2006, http://publishing2.com.

[2] Quoted in *ibid.*

[3] Murdoch was interviewed on "The Charlie Rose Show," on July 20, 2006. Also see "Murdoch predicts demise of classified ads," *Financial Times*, Nov. 24, 2005.

By 1975, ARPANET users had developed the first Internet communities, mailing lists through which people could send messages on topics of interest to a whole list of others who shared their passions. Some lists were work-related, but lists linking science-fiction fans and wine-tasting enthusiasts were among the most popular.

In 1978, computer scientists developed Usenet, a network intended to allow Internet users to exchange technical information about the Unix computer operating system. Again, to the surprise of its developers, Usenet almost immediately became a tool for long-distance socializing. Usenet's "originators underestimated the hunger of people for meaningful communication," wrote Internet historians Michael and Ronda Hauben, pointing out that the possibility of "grass-roots connection of people" around the globe is what attracted users.[44]

Although other online groups joined Usenet, until the early 1990s most users were technical people and academics. In the early 1990s, however, America Online and Netscape began making it easy for non-technical users to move online. AOL, in particular, popularized chat rooms, where people can exchange messages in real time. The live nature of chat room discussions raised new fears among parents because users were generally anonymous. Some chat rooms were eventually found to be heavily trafficked by pedophiles. By 1997, AOL had 14,000 chat

Flight instructor Harold Spector, 67, is arrested last April in Marshfield, Mass., where he'd flown in his private plane to meet what he thought was a 15-year-old girl he'd been "talking" to in an Internet chat room. Spector was charged with attempted statutory rape and attempting to entice a minor under age 16 for sex after the "girl" turned out to be two police officers.

rooms, which accounted for about a third of the time AOL members spent online.[45]

Later, AOL introduced instant messaging (IM) — real-time e-mail discussions with one person — and buddy lists, which alert users when friends are online and available to receive IMs.

"Community is the Velcro that keeps people there," said AOL President Theodore Leonsis.[46]

Six Degrees

In the 1960s, Yale University psychologist Stanley Milgram tested and apparently verified the theory that any two people on Earth are connected to each other by an average of six intermediate contacts.[47]

Scientific doubt remains about the validity of Milgram's so-called "six-degrees-of-separation" theory, but the idea has intrigued the popular imagination. In the late 1990s, Internet developers recognized that the Internet provides tools to seek and contact others in the farther circles of connection. For example, those who post personal Web pages invite friends to link their personal pages to theirs, then their friends link to their friends, creating a chain of social linkages.

In 1997, the pioneering Web site Sixdegrees.com allowed users to send and post messages viewable by their first, second and third-degree contacts.[48] Founder Andrew

Weinreich, a lawyer-turned-entrepreneur, explained social networking's usefulness in business and social terms.

"Say you're coming out of college and you want to be a lawyer in Dallas. You ask, 'Who knows an environmental lawyer in Dallas?' You want advice. We give you a shot at that." Less serious queries also pay off, said Weinreich. "You can get a movie review from Siskel and Ebert, but wouldn't you rather hear it from friends you trust?"[49]

The first really big-name social-networking site, Friendster, was launched in 2002, and the blockbuster site, MySpace, quickly followed, in 2003. While other sites also attract visitors in the millions, MySpace struck gold with a concept that, for the first time, brought teen Internet users together at a single spot.

"Curiosity about other people" drove social networking's initial fast growth, says Brooks, of *Online Personals Watch.* "But to make a site grow you need to hit something more powerful than curiosity," something that attracts key people or "connectors" — the "socialites . . . or loudmouths . . . those who run into hundreds of other people all the time," Brooks says.

MySpace seized upon music as a tool to reach the young "connectors." The site contacted music promoters and got band members to engage with the popular people on the site, said Berkeley's Boyd. The bands then created their own pages on MySpace, giving musicians an opportunity to link to multiple kids' pages. "Eventually, other young people followed the young people that followed the music," she says.[50]

And follow they have. With more than 80 million members, MySpace was growing by about 250,000 members a day in early 2006.[51] Exact rankings among the world's top Web sites shift daily, but MySpace averages the sixth-highest number of daily visitors, alongside English-language sites Yahoo, Microsoft Network, Google and eBay, and several Chinese-language sites. That translates to more than 30 million users visiting MySpace daily.[52]

Following right behind have been advertisers, with everybody who wants to contact young people putting up a MySpace page. Even the Marine Corps began collecting "friends" in early 2006. Some 12,000 people now link their pages to the Corps' page, and at last count 430 people had contacted recruiters through the site.[53]

Techno Kids

What has really brought MySpace to public attention, however, is not its sheer numbers, but its demographics.

Teens and young adults make up the overwhelming majority of users, triggering fear among parents, law-enforcement agencies and some legislators that the site may offer sexual predators easier access to young victims or encourage adolescents to engage in unhealthy behavior, such as posting sexually suggestive photos of themselves.

"The dangers our children are exposed to by these sites are clear and compelling," said Rep. Fitzpatrick.[54]

These worries aren't new and didn't start with social networking, says MIT's Jenkins. "Children and young people have always been early adopters of technology," he says, noting that the Boy Scouts were early users of radio, and in the 19th century children used toy printing presses to create magazines and newspapers.

But the speed with which new technologies appear on the scene today, combined with teens' propensity to quickly embrace new technologies, makes it especially difficult for parents, lawmakers and technology companies to figure out how to respond, says Blue Coat's Carosella. "The social behaviors that involve the Internet are not going to go away." In fact, "the kids . . . are inventing these behaviors."

"Kids are always one or two steps ahead" of older generations, says California State University's Subrahmanyam. "Chat, IM, social networking have all developed as teen-heavy technology," she says. "That makes sense, because figuring out sex and their own place in the social order . . . makes talking with peers very important."

But while teens have long discussed sex and relationships via instant messaging, adolescent interchanges on sites like MySpace "can now be seen by others," she adds.

Background Checks

While most of the uproar over potential dangers in online socializing concerns teenagers and children, some fear that online dating sites may also make it easier for sexual predators to reach adult victims.

Many Americans Know Online Daters

Nearly one-third of American adults know at least one person who has used an online dating Web site.

Percent (of U.S. adults)	No. of People	Who Know Someone Who Has . . .
31%	63 million	used a dating Web site
26	53 million	dated a person they met on a dating site
15	30 million	been in a long-term relationship or married someone they met online

Source: Pew Internet and American Life Project, "Online Dating," March 5, 2006

In the past year and a half, legislators in several states, including California, Florida, Illinois, Michigan, Ohio, Texas and Virginia, proposed requiring online dating sites to conduct criminal background checks of all prospective members or prominently inform users that they do not conduct such checks.[55]

The bills were suggested by Herb Vest, founder and CEO of the True.com dating service, which checks the criminal and marital backgrounds of its members. "The primary motivation is to protect people from criminal predation online," said Vest. "I can't imagine anyone with a hatful of brains being against that."[56]

Many online daters think dating sites already take such precautions, said Republican Florida state Rep. Kevin Ambler, who sponsored a similar bill in Florida last year after hearing that 20 percent of survey respondents thought background checks were already required on dating sites. "Many online daters have a false sense of security," he said.[57]

But some Internet companies and dating sites say the bills aren't needed or would create a false sense of security. "It would be just as easy to argue that True.com should be required to post labels on each page, saying, 'Warning. True.com's background searches will not identify criminals using fake names,' " said Kristin Kelly, a spokeswoman for the Match.com dating site.[58]

True.com has contracted with Rapsheets.com, a private firm trying to build a national database of criminal convictions, according to the Internet Alliance, an advocacy group whose members include the dating sites Match.com and eHarmony.com, as well as other Internet companies such as eBay, AOL and Yahoo. But mechanisms for tracking criminal convictions are state-based, the group

AT ISSUE

Should Congress require schools and public libraries to block social-networking Web sites?

YES — Rep. Michael Fitzpatrick, R-Pa.
Sponsor, Deleting Online Predators Act

From remarks on House floor, May 9, 2006

My most important job is my role as a father of six children. In a world that moves and changes at a dizzying pace, being a father gets harder all the time. Technology is one of the key concerns I have as a parent, specifically the Internet and the sites my kids visit, register with and use on a daily basis.

One of the most interesting and worrying developments of late has been the growth in what are called "social-networking sites." Sites like MySpace, Friendster and Facebook have literally exploded in popularity in just a few short years.

For adults, these sites are fairly benign. For children, they open the door to many dangers, including online bullying and exposure to child predators that have turned the Internet into a virtual hunting ground for children. The dangers our children are exposed to by these sites are clear and compelling. MySpace, which is self-regulated, has removed an estimated 200,000 objectionable profiles since it started in 2003.

This is why I introduced the Deleting Online Predators Act as part of the Suburban Caucus agenda. Parents have the ability to screen their children's Internet access at home, but this protection ends when their child leaves for school or the library. The Deleting Online Predators Act requires schools and libraries to implement technology to protect children from accessing commercial networking sites like MySpace.com, and chat rooms, which allow children to be preyed upon by individuals seeking to do harm to our children.

Additionally, the legislation would require the Federal Trade Commission [FTC] to design and publish a unique Web site to serve as a clearinghouse and resource for parents, teachers and children for information on the dangers of surfing the Internet. The Web site would include detailed information about commercial networking sites like MySpace. The FTC would also be responsible for issuing consumer alerts to parents, teachers, school officials and others regarding the potential dangers of Internet child predators and others and their ability to contact children through MySpace.com and other social-networking sites.

In addition, the bill would require the Federal Communication Commission to establish an advisory board to review and report commercial social-networking sites like MySpace.com and chat rooms that have been shown to allow sexual predators easy access to personal information of, and contact with, our nation's children.

NO — Sen. Edward M. Kennedy, D-Mass.
Chairman, Senate Health, Henry Jenkins Director, Comparative Media Studies program, Massachusetts Institute of Technology

From interview posted online by the MIT News Office, accessed July 2006

As a society, we are at a moment of transition when the most important social relationships may no longer be restricted to those we conduct face-to-face with people in our own immediate surroundings. We are learning how to interact across multiple communities and negotiate with diverse norms. These networking skills are increasingly important to all aspects of our lives.

Just as youth in a hunting society play with bows and arrows, youth in an information society play with information and social networks. Rather than shutting kids off from social-network tools, we should be teaching them how to exploit their potential and mitigate their risks.

Much of the current policy debate around MySpace assumes that the activities there are at best frivolous and at worst dangerous to the teens who participate. Yet a growing number of teachers around the country are discovering that these technologies have real pedagogical value.

Teachers are beginning to use blogs for knowledge-sharing in schools; they use mailing lists to communicate expectations about homework with students and parents. They are discovering that students take their assignments more seriously and write better if they are producing work that will reach a larger public rather than simply sit on the teacher's desk. Teachers are linking together classrooms around the country and around the world, getting kids from different cultural backgrounds to share aspects of their everyday experience.

Many of these activities would be threatened by the proposed federal legislation, which would restrict access to these sites via public schools or library terminals. In theory, the bill would allow schools to disable these filters for use in educationally specified contexts, yet, in practice, teachers who wanted to exploit the educational benefits of these tools would face increased scrutiny and pressure to discontinue these practices.

Teens who lack access to the Internet at home would be cut off from their extended sphere of social contacts.

Wouldn't we be better off having teens engage with MySpace in the context of supervision from knowledgeable and informed adults? Historically, we taught children what to do when a stranger telephoned them when their parents are away; surely, we should be helping to teach them how to manage the presentation of their selves in digital spaces.

points out, and some states decline to participate in national databases. So it is impossible for Rapsheets to have complete information, they said.[59]

Internet Alliance also argues that new laws aren't needed because unregulated dating services — such as newspaper ads and singles hotlines — have run "smoothly for years without legislative interference," while "providing even less information [than] a typical online profile."

So far, the bills have gone nowhere. The Michigan House passed a bill, but it later died in the state Senate. In Florida, bills were approved in committee last year but did not advance. A California bill that would have fined online dating services $250 for each day they don't conduct background checks was introduced but later pulled from consideration.

CURRENT SITUATION

Big Brother

Nowadays, teenagers aren't the only ones hanging out on MySpace.com. Law-enforcement officials now are increasingly staking out the site, looking to head off crimes. Some high-profile arrests in MySpace-related cases have raised concerns about social networking similar to worries that arose in the 1990s about chat rooms.

For example, a 39-year-old Pennsylvania man faces federal charges that he molested a 14-year-old Connecticut girl he met through her MySpace page. The girl had listed her age as 18. In another Connecticut case, a 22-year-old man traveled from New Jersey to visit an 11-year-old girl, whom he molested in her home while her parents slept.[60]

Besides monitoring for sexual predation, law-enforcement officials worry that teenagers may use MySpace to plot violence or vandalism. For example, a 15-year-old New Jersey girl was charged with harassment when school officials found an apparent "hit list" on her MySpace page. In Denver, a 16-year-old boy was arrested after allegedly posting photos of himself holding handguns on MySpace.[61] And in Riverton, Kan., five high-school boys were arrested in April after school officials found a message on one boy's MySpace page apparently threatening a Columbine-style shooting. Law-enforcement officers later found weapons and documents related to a plot in a student's bedroom and in school lockers.[62]

Meanwhile, the NBC program "Dateline" recently highlighted potential online dangers to children from adults. In a series of programs, "Dateline" photographed men who had arranged to meet what they thought were young teenagers but were actually adult members of an activist group. While the encounters took place in chat rooms, not on social-networking sites, the shows raised further alarms.

In response, Rep. Fitzpatrick introduced his Deleting Online Predators Act. It would expand the anti-pornography Children's Internet Protection Act (CIPA) by requiring schools and libraries to prohibit access to any commercial social-networking site or chat room through which minors could access sexual material or be subject to sexual advances.

"This is a new and evolving problem" that requires amendments to CIPA, since social networking didn't exist when that law was written, said Michael Conalle, Fitzpatrick's chief of staff.[63]

But the American Library Association said the bill is so broadly written that it would block not only education that would teach kids to go online safely but also "a wide array of other important applications and technologies."[64]

Lordan of the Internet Education Foundation said evidence suggests that more teens are abused in their own homes and neighborhoods than online. "We could end up diverting resources" to attack online predation "when the main need is really elsewhere," he says.

Although the House Energy and Commerce Committee has held subcommittee hearings on Fitzpatrick's bill, and it has been discussed on the House floor, discussions so far have focused heavily on child pornography on the Internet. No action on the legislation has yet been scheduled, and no bills have been introduced in the Senate.

Safety First

State attorneys general, parents, entrepreneurs and social-networking companies recently have launched safety initiatives for online socializing.

Connecticut Attorney General Richard Blumenthal, for example, has asked social-networking companies to implement tougher measures to block teenagers' access to pornography and rid the sites of sexual predators. Voluntary efforts, once the state and businesses agree on what steps should be taken, would "avoid the costs and time required for any sort of legal action," he said.[65]

In April, MySpace hired Hemanshu Migam — Microsoft's former director of consumer security and child safety and a former federal prosecutor of online child-exploitation cases — to manage its safety, privacy and customer-education programs. The company also partnered with the National Center for Missing and Exploited Children and the Advertising Council to post public-service announcements about online safety on TV, MySpace and other Internet sites.[66]

While acknowledging the company's first steps, Blumenthal said he had urged MySpace to adopt other, "more significant, specific measures," such as tougher age-verification efforts and free software for parents to block MySpace from home computers.[67]

Alarms about social networking also are drawing interest from parents' groups and some entrepreneurs. In Utah, for example, the state parent-teacher association is creating materials to teach parents how to make their children safe online. It will also recommend filtering and blocking software to parents in collaboration with Blue Coat, the Internet security company.[68]

Other Internet-technology developers also are offering help. Sales have tripled in the last three years for programs like eBlaster, Content Protect, IM Einstein and Safe Eyes, which allow parents to monitor their kids' e-mails, instant messages and online chats in real time from a separate computer — such as while the parent is at work.[69] Software developer Alex Strand, for instance, has established MySpacewatch.com, where users can sign up to monitor changes — such as new photos, additional listed friends — on a MySpace Web page for free.

"I started it as . . . a way for parents to check out what their kids are doing," said Strand.[70]

OUTLOOK

Here to Stay

In the future, online communications and social networking will become even more deeply rooted in our lives, say most analysts.

That makes it imperative to learn as much as possible about how online activities affect people, says Carosella, of Blue Coat Systems. "There's a critical role for mental health and social scientists," he says. "We should be doing studies on why there are so many sexual predators out there. Where did they all come from? Is there a vicious cycle between easy access to pornography online and the emergence of online predators?"

Future generations will make even more use of social networking, predicts Diane Danielson, who created www.DWCFaces.com, a social-networking site for businesswomen. "We will see Generation Y bringing their social networks into the workspace," she says, referring to the 20-25-year-old age group. "They will also remain connected to more people from their high schools and colleges" thanks to the persistent presence on social-networking sites of links to Web sites like Classmates.com. "In a transient society, a social network Web page might be your most consistent address."

Social networking may also transform some political campaigns into more grass-roots affairs, says Zephyr Teachout, a professor of constitutional law at Vermont Law School, who directed Internet organizing for Howard Dean's 2004 presidential campaign. Dean encouraged people around the country to communicate on their own via social-network software, allowing local and individual momentum to drive many activities. "We discovered that the human need to be political is important" and, if tapped, increases participation, she says. "But to do it you need a candidate willing to devolve power."

When media mogul Rupert Murdoch's News Corp. bought MySpace last year for $580 million, some speculated that the conservative Murdoch might use the site to influence the politics of the site's young users, perhaps by pushing Republican-slanted commentary on the site during the 2008 presidential election cycle. Many analysts doubt whether such an effort could succeed, though, since social-networking users have notoriously fled sites when owners have tried to exercise regulatory clout.

But Murdoch says he spent $1.5 billion in the past year to buy MySpace and other online companies to empower people to create their own content. "Technology is shifting power away from the editors, the publishers, the establishment, the media elite. Now it's the people who are taking control," he said.[71]

As Internet technology draws more people to publish their personal information in cyberspace, a new set of ethics is needed for presenting oneself, says Bill Holsinger-Robinson, chief operating officer of Spout.com, a social-networking site focused on film. "I see it as harking back to earlier, simpler times — a town square where people can gather."

When town squares were common, he says, "there was a certain sense of responsibility on how we presented ourselves. We've lost that. Now we have to reinvent it for a new generation."

NOTES

1. Erik Brady and Daniel Libit, "Alarms Sound Over Athlete's Facebook Time," *USA Today*, March 8, 2006.

2. Quoted in *ibid.*

3. Clair Osborn, "Teen, Mom Sue MySpace.com for $30 Million," *Austin American-Statesman*, June 20, 2006.

4. Quoted in *ibid.*

5. "Internet: The Mainstreaming of Online Life," Pew Internet and American Life Project, www.pewinternet .org.

6. John Horrigan and Lee Rainie, "The Internet's Growing Role in Life's Major Moments," Pew Internet and American Life Project, April 19, 2006.

7. "Internet: The Mainstreaming of Online Life," *op. cit.*

8. For background, see Brian Hansen, "Cyber-predators," *CQ Researcher*, March 1, 2002, p. 169-192.

9. For background, see David Masci, "Internet Privacy," *CQ Researcher*, Nov. 6, 2998, pp. 953-976.

10. "Social Networking Sites Continue to Attract Record Numbers as MySpace.com Surpasses 50 Million U.S. Visitors in May," PRNewswire, comScore Networks, Inc., June 15, 2006.

11. Marshall Kirkpatrick, "Top 10 Social Networking Sites See 47 Percent Growth," the socialsoftwareweblog, May 17, 2006, http://socialsofware.weblogsinc.com.

12. For background, see Kenneth Jost, "Libraries and the Internet," *CQ Researcher*, June 1, 2001, pp. 465-488.

13. Michael Fitzpatrick, testimony before House Energy and Commerce Subcommittee on Oversight and Investigations, June 10, 2006.

14. Henry Jenkins and Danah Boyd, "Discussion: MySpace and Deleting Online Predators Act," interview published online by Massachusetts Institute of Technology News Office, May 24, 2006, www .danah.org/papers/MySpaceDOPA.html.

15. *Ibid.*

16. Quoted in Anne Chappel Belden, "Kids' Tech Toys: High-Tech Communication Tools," Parenthood. com, http://parenthood.com.

17. Kevin Farnham, reply to "Friendster Lost Steam. Is MySpace Just a Fad?" Corante blog, March 21, 2006, http://man.corante.com.

18. Paul Marks, "Pentagon Sets Its Sights on Social Networking in Washington," *NewScientist.com*, June 9, 2006, www.newscientist.com.

19. Quoted in *ibid.*

20. David Finkelhor and Lisa M. Jones, "Explanations for the Decline in Sexual Abuse Cases," *Juvenile Justice Bulletin*, U.S. Department of Justice, Office of Juvenile Justice and Delinquency Prevention, January 2004, www.ojp.usdoj/gov/ojjdp.

21. Quoted in Larry Magid, "Plug In, or Pull the Plug," Staysafe.org for Parents, www.staysafeonline.org.

22. *Congressional Record*, House, May 9, 2006, p. H2311.

23. *Ibid*, p. H2315.

24. Jenkins and Boyd, *op. cit.*

25. *Ibid.*

26. Quoted in Robert Brumfield, "Bill Calls for MySpace Age Limit," *eSchoolNews online*, May 16, 2006, www.eschoolnews.com.

27. For background, see Kathy Koch, "The Digital Divide," *CQ Researcher*, Jan. 28, 2000, pp. 41-64.

28. For background, see Marcia Clemmitt, "Controlling the Internet," *CQ Researcher*, May 12, 2006, pp. 409-432.

29. Clifford Stoll, *Silicon Snake Oil: Second Thoughts on the Information Highway* (1996), p. 43.

30. Killeen Hanson, "Study Links Internet, Social Contact," *The Stanford Daily Online Edition*, Feb. 28, 2005.

31. Quoted in *ibid.*

32. Norman H. Nie and Lutz Erbring, "Internet and Society: A Preliminary Report," Stanford Institute for the Quantitative Study of Society, Feb. 17, 2000.

33. Quoted in *ibid.*

34. Jeffrey Boase, John B. Horrigan, Barry Wellman and Lee Rainie, "The Strength of Internet Ties," Pew Internet and American Life Project, Jan. 25, 2006.

35. *Ibid.*

36. *Ibid.*

37. Barry Wellman, Jeffrey Boase and Wenhong Chen, "The Networked Nature of Community: Online and Offline," *IT & Society*, summer 2002, pp. 151-165, www.itandsociety.org.

38. *Ibid.*

39. Sheera Claire Frenkel, "Israelis and Lebanese Are Still Talking — on the Net," *The Jerusalem Post*, July 21, 2006.

40. Christopher Allen, "My Advice to Social Networking Services," Life With Alacrity blog, Feb. 3, 2004, www.lifewithalacrity.com.

41. Quoted in Michael Erard, "Decoding the New Cues in Online Society," *The New York Times*, Nov. 27, 2003.

42. Quoted in Jenkins and Boyd, *op. cit.*

43. For background, see *The Internet's Coming of Age* (2000).

44. Michael Hauben and Ronda Hauben, "The Social Forces Behind the Development of Usenet," First Monday, www.firstmonday.org.

45. "Internet Communities," *Business Week Archives*, May 5, 1997, www.businessweek.com.

46. Quoted in *ibid.*

47. For background, see Judith Donath and Danah Boyd, "Public Displays of Connection," *BT Technology Journal*, October 2004, p. 71.

48. Doug Bedell, "Meeting Your New Best Friends," *The Dallas Morning News*, Oct. 27, 1998.

49. Quoted in *ibid.*

50. Danah Boyd, "Friendster Lost Steam. Is MySpace Just a Fad?" March 21, 2006, www.danah.org.

51. Dawn Kawamoto and Greg Sandoval, "MySpace Growth Continues Amid Criticism," ZDNet News, March 31, 2006, http://news.zdnet.com.

52. MySpace traffic details from www.alexa.com.

53. Audrey McAvoy, "Marines Trolling MySpace.com for Recruits," *Chicago Sun-Times*, July 25, 2006.

54. *Congressional Record, op. cit.*, p. H2310.

55. For background, see Javad Heydary, "Regulation of Online Dating Services Sparks Controversy," *E-Commerce Times*, March 3, 2005, www.ecommerce times.com.

56. Quoted in "Online Dating Background Checks?" CNNMoney.com, April 25, 2006, http://cnnmoney.com.

57. Quoted in *ibid.*

58. Quoted in Declan McCullagh, "True Love With a Criminal Background Check," C/Net News.com, Feb. 28, 2005, http://news.com.com.

59. "Online Dating," white paper from Internet Alliance, www.internetalliance.org.

60. "Twenty Youths Suspended in MySpace Case," The Associated Press, March 3, 2006.

61. "MySpace in the News," *Bergen* [New Jersey] *Daily Record*, May 14, 2006, www.dailyrecord.com.

62. "Charges Mulled in Alleged School Shooting Plot," The Associated Press, April 23, 2006. For background, see Kathy Koch, "School Violence," *CQ Researcher*, Oct. 9, 1998, pp. 881-904.

63. Quoted in Declan McCullagh, "Congress Targets Social Network Sites," C/Net News.com, May 11, 2006, http://news.com.com.

64. Michael Gorman, "ALA Opposes 'Deleting Online Predators Act,'" May 15, 2006, statement, American Library Association.

65. Quoted in "Making MySpace Safe for Kids," Newsmaker Q&A, *Business Week online*, March 6, 2006, www.businessweek.com.

66. Maria Newman, "MySpace.com Hires Official to Oversee Users' Safety," *The New York Times*, April 12, 2006.

67. Quoted in *ibid.*

68. "Utah PTA and Blue Coat Systems Join Forces to Drive Greater Internet Safety, Community Action," Blue Coat Systems press release, June 20, 2006, www.bluecoat.com.

69. See Ned Potter, "Watching Your Kids Online," ABCNews.com, July 24, 2006.

70. Quoted in Stefanie Olsen, "Keeping an Eye on MySpace," C/Net News.com, June 29, 2006, http://news.com.com.

71. Spencer Reiss, "His Space," *Wired*, July 14, 2006, www.wired.com.

BIBLIOGRAPHY

Books

Farnham, Kevin M., and Dale G. Farnham, *MySpace Safety: 51 Tips for Teens and Parents, How To Primers*, 2006.
Parents of a veteran teenage MySpace user explain what their experiences have taught them about how social networking works and what safety rules families should follow.

Rheingold, Howard, *The Virtual Community: Homesteading on the Electronic Frontier*, The MIT Press, 2000.
A technology analyst and longtime participant in Internet communities recounts the history of the social Internet and his two decades of personal experience with it.

Articles

Armental, Maria, "Site Started by New Jersey Teens Is Growing Fast," *Asbury Park Press*, May 16, 2006, www.app.com.
MyYearbook.com, a social-networking site created by two New Jersey high-school students in 2005, has membership rolls that are growing by about 40 percent a month. The brother and sister team who created it became millionaires before high-school graduation earlier this year when an investor paid $1.5 million for a 10-percent stake in the operation.

Bower, Bruce, "Growing Up Online," *Science News Online*, June 17, 2006, www.sciencenews.org.
Research psychologists dissect the appeal of Internet social-networking software to teenagers and describe the findings in several recent studies of how teenagers behave online.

Brumfield, Robert, "Bill Calls for MySpace Age Limit," *eSchoolNews online*, May 16, 2006.
Educators discuss the possible effects on schools and students of a congressional proposal to require schools and public libraries to block social software.

Hof, Robert D., "Internet Communities," *Business Week*, May 5, 1997, *Business Week Archives*, www.businessweek.com.
As women, teenagers and non-technical users flock online, Internet use turns away from pre-created content toward socializing and Internet communities sharing common interests.

Koppelman, Alex, "MySpace or OurSpace?" *Salon.com*, June 8, 2006, www.salon.com.
A growing number of school administrators and law-enforcement officials regularly monitor MySpace and other social-networking sites in search of evidence of rule-breaking and criminal activity. The practice is raising awareness among teens that spaces they believed were private are not and also is raising questions about how far schools can go in policing students' out-of-school activities.

Leonard, Andrew, "You Are Who You Know," *Salon.com*, June 15, 2004.
Social-networking entrepreneurs and Internet analysts explain why Internet technology led to development of social software and why they believe social networks matter, online and off.

Marks, Paul, "Pentagon Sets Its Sights on Social Networking Websites," *New Scientist*, June 9, 2006, NewScientist.com.
The National Security Agency is researching the possibility of mining social-networking sites for data to assemble extensive personal profiles of individuals and their social circles to help sniff out terrorist plots.

Reiss, Spencer, "His Space," *Wired*, July 14, 2006, www.wired.com.
Since media mogul Rupert Murdoch paid $580 million to buy MySpace in 2005, technology and business analysts have argued about whether Murdoch can make the investment pay by selling advertising and/or using the site as a distribution channel for music and video content.

Studies and Reports

Boase, Jeffrey, John B. Horrigan, Barry Wellman and Lee Rainie, *The Strength of Internet Ties, Pew Internet and American Life Project*, January 2006.
Statistics show that the Internet and e-mail aid users in maintaining social networks and providing pathways to support and advice in difficult times, according to University of Toronto and Pew researchers.

Finkelhor, David, Kimberly J. Mitchell and Janis Wolak, *Online Victimization: A Report on the Nation's Youth, Crimes Against Children Research Center, University of New Hampshire*, June 2000.

In cooperation with the National Center for Missing and Exploited Children and the U.S. Department of Justice Office of Juvenile Justice and Delinquency Prevention, University of New Hampshire researchers detail national statistics on online crimes against children and teens, online sexual solicitation of young Internet users and responses to Internet sexual solicitations by young people, their families and law enforcement.

Lenhart, Amanda, and Mary Madden, *Teen Content Creators and Consumers, Pew Internet and American Life Project,* **November 2005.**
Creating and sharing content through blogs, videos and other sites is an integral part of online social life for today's teens.

For More Information

Apophenia: Making Connections Where None Previously Existed, www.zephoria.org/thoughts/. News, data and commentary on the social Internet by social-media researcher Danah Boyd.

Berkman Center for Internet and Society, Harvard Law School, Baker House, 1587 Massachusetts Ave., Cambridge, MA 02138; (617) 495-7547; http://cyber.law.harvard.edu. A research program investigating legal, technical and social developments in cyberspace.

Center for the Digital Future, University of Southern California Annenberg School, 300 South Grand Ave., Suite 3950, Los Angeles, CA 90071; (213) 437-4433; www.digitalcenter.org. A research program investigating the Internet's effects on individuals and society.

Crimes Against Children Research Center, University of New Hampshire, 20 College Rd., #126 Horton Social Science Center, Durham, NH 03824; (603) 862-1888; www.unh.edu/ccrc/. Studies criminal victimization of young people and how to prevent it, including online.

Internet Education Foundation, 1634 I St., N.W., Suite 1107, Washington, DC 20006; (202) 638-4370; www.neted.org. Educates the public and lawmakers about the Internet as a tool of democracy and communications; the industry-supported GetNetWise project (www.getnetwise.org) disseminates information on safety and security.

Many2Many, http://many.corante.com/. Academics and technical experts provide information and commentary on social networking in this blog.

National Center for Missing and Exploited Children, 699 Prince St., Alexandria, VA 22314-3175; (703) 274-3900; www.missingkids.com. Provides education and services on child exploitation to families and professionals; maintains a CyberTipline and other resources for reporting and combating exploitation via the Internet.

Online Personals Watch, http://onlinepersonalswatch.typepad.com. A blogger and industry veteran provides news and commentary on social networking and online dating.

Pew Internet and American Life Project, 1615 L St., N.W., Suite 700, Washington, DC 20036; (202) 419-4500; www.pewinternet.org. Provides data and analysis on Internet usage and its effects on American society.

Progress and Freedom Foundation, 1444 I St., N.W., Suite 500, Washington, DC 20005; (202) 289-8928; www.pff.org. A free-market-oriented think tank that examines public policy related to the Internet.

Closing Guantánamo

Can Obama Close the Detention Camp Within One Year?

Kenneth Jost

President Barack Obama signs an executive order on Jan. 22 — his second full day in office — to close the U.S. prison at Guantánamo Bay within a year. Human-rights advocates and lawmakers on both sides of the political aisle agree the controversial facility should be closed. But finding countries willing to take the 241 detainees remains a problem, and Republicans are warning they will oppose efforts to house prisoners in the United States.

From *CQ Researcher*, February 27, 2009.

Mohammed Jawad has spent more than a quarter of his young life in the prison at Guantánamo Bay, Cuba, for an offense he says he didn't commit.

The government says the Afghani teenager threw a grenade at a U.S. military jeep in Kabul in 2002, wounding two American soldiers and their Afghan interpreter.

Jawad, who was 16 or 17 at the time, claims he was working to clear land mines when the attack occurred and that another youth was responsible. Jawad says he confessed under coercion while in custody in Afghanistan and again at the U.S. detention camp at Guantánamo Bay, Cuba, only after more than a year of abusive interrogation.

Then, in a pair of rulings in October and November, an Army judge threw out the confessions that the prosecution had said were central to the case. Col. Stephen Henley ruled that Jawad had confessed the first time only after Afghan soldiers threatened to kill him and his family. The statements made in Guantánamo, Henley said, were also coerced.[1]

With a case so badly handled, Jawad would seem to be an obvious candidate for release from the controversial prison camp that President George W. Bush ordered to be established in 2002 for "enemy combatants" captured in the Afghanistan war or elsewhere. In its final week in office, however, the Bush administration on Jan. 13 urged the review panel that acts as an appeals court for the military commission system at Guantánamo to reverse the rulings in Jawad's case and allow the prosecution to go forward.

After taking office only a week later, President Obama signed an executive order for a review of all pending Guantánamo cases

America's Controversial Prison in the Caribbean

The U.S. Naval Station at Guantánamo Bay has come a long way since its days as a coaling station. The U.S. acquired the 45-mile-square base on Cuba's southeastern tip in 1903 after helping the island, a former Spanish colony, gain its independence during the Spanish-American war. Today the base houses about 8,400 U.S. personnel — including about 2,200 military and civilian personnel at the detention facilities — and 241 prisoners.

the order. "The message this sends to the world could not be clearer: The United States is ready to reclaim its role as a nation committed to human rights and the rule of law."

Even some former Bush administration officials agree the time has come to close the facility. John Bellinger, who was legal adviser at the State Department and National Security Council during the Bush administration, says he has "very strongly" supported closing this "albatross around our necks."

"The benefits of Guantánamo have been outweighed by the legacy costs of Guantánamo, and that has been true for some time," says Charles "Cully" Stimson, former assistant secretary of Defense for detainee affairs and now a senior legal fellow at the conservative Heritage Foundation. The facility has taken "a moral toll" on the U.S. image at home and abroad, he says.

Robert Chesney, a respected national security expert now a visiting professor at the University of Texas Law School in Austin, says Guantánamo reflects a broader failure of policy on how to deal with

and the closure of the facility — which now houses 241 prisoners — within one year. In Jawad's case, however, the new administration returned to the military review panel to ask for a 120-day delay before it rules on Jawad's case. The panel granted the request, over the objections of Jawad's lawyers.

Obama's action — on his second full day in office — moved toward fulfilling his repeated campaign pledge to close Guantánamo, known as "Gitmo." Human rights advocates, who have strongly criticized Guantánamo and the legal rules the Bush administration established for enemy combatant cases, are applauding Obama's move.

"Today is the beginning of the end of this sorry chapter in our nation's history," Elisa Massimino, executive director and CEO of Human Rights First, said after Obama signed

suspected terrorists captured both within and outside the United States since the Sept. 11, 2001, attacks on the World Trade Center and Pentagon by the Islamic terrorist group al Qaeda.

"For more than seven years, we've struggled to define a counterterrorism policy that is effective, that is politically sustainable and simultaneously reflects our core values as Americans," Chesney remarked as he opened a panel discussion at the school on Feb. 3. "We have not yet succeeded in doing this."[2]

Despite indications of editorial and public support for Obama's action, Republicans are raising questions and apparently setting the stage to criticize the closure if suspected terrorists are transferred to facilities within the United States. "Most families neither want nor need hundreds of terrorists seeking to kill Americans in their communities," House GOP Whip Eric Cantor of Virginia said in a statement issued the same day.

Another critic, however, notes that Obama's executive order did nothing other than promise a review of case files and set a goal of closing the facility. "He hasn't really done much," says Andrew McCarthy, legal editor of *National Review* and chairman of the Center for Law and Counterterrorism at the Foundation for the Defense of Democracies.

McCarthy has backhanded praise for the interim nature of Obama's move, which he says contrasts with candidate Obama's "demagogic, overheated rhetoric" during the campaign. "What he has obviously found is that there are very difficult issues, very complex issues that have to be worked through with respect to the detainees," McCarthy says.

"It was unfortunate that he and people who are like-minded were critical of Guantánamo," McCarthy adds, "when in point of fact if you didn't have Guantánamo, you would need to have something like it, whether it was inside the United States or outside." (*See sidebar, p. 80.*)

In establishing Guantánamo, President Bush said the camp would be used to house "the worst of the worst" suspected terrorists. But the national security and human-rights camps diverge on how to regard the 779 prisoners who have been held at Guantánamo over its seven-year history, some 540 of whom have been released. Among the detainees still being held, Bellinger predicts the Obama administration will find "a lot of bad people left, or at least many in the gray area."

"We've known all along that not everyone held at Guantánamo had any business being held there at all," counters Sharon Bradford Franklin, senior counsel with the Constitution Project, an advocacy group that seeks to find consensus on constitutional issues.

Outside government, the most extensive study of the Guantánamo detainees appears to have been conducted by Benjamin Wittes, a senior fellow at the Brookings Institution think tank and author of a highly regarded new book on counterterrorism policies, *Law and the Long War*. Wittes writes that his examination of the information publicly available on the detainees indicated many of them had incriminating ties to al Qaeda. An updated compilation by Wittes available on the Brookings Web site, however, shows that only a small

Former detainee Said Ali al-Shihri's return to terrorist activity in Yemen underscores the complications in carrying out Obama's decision to close down Guantánamo, *The New York Times* said.

AP Photo

fraction of the prisoners still being held are considered major al Qaeda leaders.[3] (*See chart, p. 75.*)

Paradoxically, Obama's pledge to close Guantánamo appears to be contributing to an increase in tensions in the prison camp. A special Defense Department review team that spent almost two weeks at Guantánamo concluded in February that detainees are being treated humanely in compliance with the provisions of the Geneva Conventions regarding wartime captives. But the report noted a nearly sixfold increase in disciplinary incidents by detainees since September 2008 and tied the increase in part to detainees' "uncertainty and anxiety about the future."[4]

Along with Guantánamo, the Obama administration inherits an array of legal proceedings. Obama's order froze proceedings in the military tribunal system, which thus far has secured three convictions: Ali Hamza Ahmad Suliman al-Bahlul, Osama bin Laden's alleged media secretary, was found guilty of 35 counts relating to support of terrorism and sentenced to life in prison; Salim Ahmed Hamdan, bin Laden's former driver, was convicted on reduced charges; and David Hicks, the so-called Australian Taliban, pleaded guilty to one count of providing material support of terrorism. He and Hamdan were essentially sentenced to time served and have since been released. But pending cases in federal courts up to and including the Supreme Court are continuing.

The high court is scheduled to hear a case on April 22 testing whether the government can hold without trial a Qatari native — Ali Saleh Kahlah al-Marri — as an enemy combatant after he was arrested while lawfully residing in the United States on a student visa. In a closely divided ruling, the federal appeals court in Richmond, Va., said yes — but with more judicial scrutiny than proposed by the Bush administration. In one of its first moves, the Obama administration asked for and was granted an extension of time to file the government's brief with the high court.

In a separate case, the federal appeals court for the District of Columbia is reconsidering whether former Guantánamo detainees can sue government officials for alleged torture and religious discrimination. The Supreme Court sent the case back to the appeals court in December to consider the impact of the justices' decision in June that Guantánamo detainees can use federal habeas corpus to challenge their confinement.[5]

The pending cases are forcing the Obama administration to make policy decisions sooner than the one-year timetable outlined for closing Guantánamo, according to Chesney. "There will not be as much new time as the administration would like," Chesney said at the panel discussion. "The litigation calendar will force them to take positions much faster than that."

As the administration's review continues, here are some of the major questions being debated:

Should the government continue repatriating Guantánamo detainees to other countries?

The day after President Obama signed the executive order on closing Guantánamo, a front-page *New York Times* story stated that one of the detainees already released from the camp, Said Ali al-Shihri, had returned to terrorist activity as the head of al Qaeda's Yemeni branch. The newspaper said that Al Shihri's role — allegedly announced in an Internet statement and confirmed by U.S. counterintelligence officials — "underscores the complications" in carrying out Obama's decision.[6]

In the seven years it has taken to complete legal proceedings against three Guantánamo detainees, the Bush administration released nearly 540 other prisoners, most of them to their home countries after what former State Department legal adviser Bellinger describes as "arduous negotiations." In the administration's last week in office, the Defense Department claimed that 18 of those released have been confirmed as "returning to the fight," and another 43 are suspected of having done so.[7]

The Defense Intelligence Agency report has no specifics and — like earlier DIA compilations on the subject — is widely questioned. "I don't want to deride that as an urban myth, but I don't have a very high level of confidence in the claims," says Eugene Fidell, president of the National Institute of Military Justice, who teaches military law at Yale Law School. Nevertheless, the total figure of 61 — representing 11 percent of those released — is cited by national security hawks as evidence that the emphasis on reducing the Guantánamo population has been and still is mistaken.

"Thus far, it's shown itself to be a terrible idea," says McCarthy, a former federal prosecutor. "To the extent that we're trying to shovel people into other countries, all that does is to empty out Gitmo, but it doesn't make the problem any better. It makes the problem in many ways worse."

Sending Guantánamo detainees to third countries, however, is the first — and principal — step in the blueprint for closing the facility that human-rights advocates issued while the presidential campaign was under way. Under the plans outlined by Human Rights First and the Center for Strategic and International Studies (CSIS), detainees who could not be tried in the United States might be transferred to their home countries or third countries for prosecution. Those not suspected of criminal activity should be repatriated or — if they faced the likelihood of torture — sent to third countries.[8]

The report by a CSIS working group, questions the claimed number of detainees who have returned to terrorist activities, but acknowledges the "security risks" in the policies. "There are risks associated with keeping Guantánamo open and there are risks with closing Guantánamo," says report author Sarah Mendelson, director of the center's human-rights and security initiative. "We came to the conclusion that the cost of keeping Guantánamo open is far greater than the cost of closing it."

The blueprint for closing the camp may rest, in part, on unrealistic premises, however. Prosecution, repatriation and resettlement have all proved to be elusive goals. "Virtually no country has been able to detain or prosecute the people we've returned to them," Bellinger says. As one example, he says there are no domestic laws making it a crime to travel to Afghanistan, where many of the detainees were taken into custody.

As for repatriation, the Saudi government is credited with operating a model rehabilitation program, but in early February it listed as terrorism suspects 11 former Guantánamo prisoners who went through the program. Yemen, the home country of the largest number of remaining detainees, has no rehabilitation program whatsoever.[9]

In any event, McCarthy mocks the whole concept of rehabilitation. "You can't really seriously think that you're going to send these people to Saudi Arabia, which is the cradle of Wahhabism, and through re-education camp and then they're not going to be a jihadist any more," he says. "It's just a silly idea."

Resettlement presents its own difficulties, Bellinger says. "Most of these countries don't want these people back," he says. "They view them as troublemakers." And third countries — notably, in Europe — are reluctant to admit detainees that the U.S. government has publicly labeled as dangerous terrorists. Matthew Waxman, a Columbia Law School professor who served in the Bush administration in the then new position of assistant secretary of Defense for detainee affairs, says the Obama administration will face "a very difficult road" in persuading third countries to admit released detainees unless the United States itself is willing to resettle some of them.

Mendelson predicts that European governments will be more willing to work with the Obama administration — given its commitment to closing Guantánamo — than they were while Bush was in office. The CSIS report also outlines steps to strengthen law enforcement, detention facilities and reintegration programs in other countries where detainees are sent after release. Still, Mendelson writes, "We cannot guarantee nor will we pretend that the risk of releasing or transferring detainees is zero."

Waxman agrees. "All options to close Guantánamo carry some risks," he says.

Should Guantánamo detainees be prosecuted in civilian courts?

FBI agents arrested Jose Padilla as he arrived at Chicago's O'Hare International Airport on May 8, 2002, on suspicion of planning to plant so-called dirty bombs at sites in the United States. The Bush administration held the Brooklyn-born Padilla as an enemy combatant for more than three years. But because of his U.S. citizenship, Padilla was held not in Guantánamo but at the U.S. Naval Brig in Charleston, S.C.

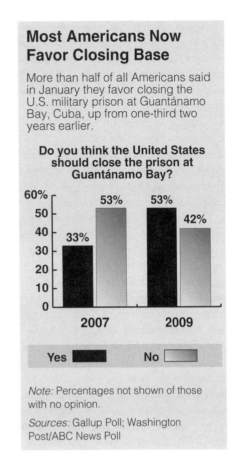

Most Americans Now Favor Closing Base

More than half of all Americans said in January they favor closing the U.S. military prison at Guantánamo Bay, Cuba, up from one-third two years earlier.

Do you think the United States should close the prison at Guantánamo Bay?

2007: Yes 33%, No 53%
2009: Yes 53%, No 42%

Yes ■   No □

Note: Percentages not shown of those with no opinion.

Sources: Gallup Poll; Washington Post/ABC News Poll

Fearing an adverse ruling in Padilla's habeas corpus challenge to his confinement, the government decided early in 2006 to indict Padilla and try him in a civilian criminal court. The strategy paid off on Aug. 16, 2007, when a federal jury in Miami convicted Padilla of conspiracy and material support of terrorism — charges that led to the 17-year prison sentence he is now serving.[10]

Critics of Guantánamo — and its military tribunals — point to the Padilla trial and scores of others since 2001 as evidence that civilian courts are up to the task of prosecuting suspected terrorists. The U.S. criminal justice system "has proven an effective venue for prosecuting terrorist suspects," Mendelson writes in the CSIS report, "especially when compared with the military commissions." The report counts 107 jihadist terrorist cases — some with multiple defendants — tried in civilian courts since 2001 with 145 convictions.[11]

"We recommend very firmly that prosecutions should be handled by Article III courts," says the Constitution Project's Franklin, referring to the constitutional provisions establishing the federal judiciary. Military law may be applicable for "actual combatants captured on the battlefield," she acknowledges, but all others "can be and should be prosecuted in civilian courts."

National security hawks like McCarthy and a range of other experts, including some Guantánamo critics, counter by pointing to a host of practical difficulties in prosecuting enemy combatants in civilian courts. Speaking at the Texas law school panel, Wittes, a former editorial writer on legal issues for *The Washington Post*, outlined several reasons why such prosecutions "might not be viable" for many Guantánamo detainees. Evidence against some may be tainted by coercion or torture, unavailable because classified as secret or inadmissible because of mundane courtroom issues, such as proving chain of custody or the like. And in many cases the quantity and quality of the evidence may simply be insufficient, Wittes says, to meet the beyond-a-reasonable-doubt standard applicable in criminal trials.

McCarthy, one of the prosecutors in the 1995 conspiracy conviction of Omar Abdel-Rahman, the so-called blind sheik implicated in the 1993 World Trade Center bombing, says terrorism defendants' rights to discover prosecution evidence present a problem even in a successful case. "The discovery rules and the trial process itself [are] an intelligence gold mine for the terrorist organization at large," he says. "And there is no real way to prevent that from happening if you're going to have a trial that deserves the name of a trial."

McCarthy and Wittes both favor creation of what is being called a national security court to handle terrorism cases. In general, proponents of such courts envision a specialized federal civilian court applying the substantive and procedural law of military tribunals. A national security court would be "a better fit," McCarthy says, than either the regular civilian justice system or the military commissions operating at Guantánamo. (*See "At Issue," p. 86.*)

The military tribunals, in fact, appear not to have performed to anyone's satisfaction. CSIS's Mendelson calls them "ineffective and inefficient." From a somewhat different perspective, ex-Pentagon official Stimson says he pronounced the system "dead" more than a year ago. Despite his service in the Defense Department,

Stimson favors federal court trials for the more complex Guantánamo cases because federal prosecutors are likely to be more experienced than most military lawyers.

Human-rights advocates generally minimize the difficulties claimed by the critics of civilian trials. The CSIS report notes, for example, that the federal statute against material support of terrorism does not require "heavy evidentiary burdens" and permits long prison sentences. In any event, the report adds, the use of civilian trials "denies terrorist suspects the symbolic value of special, extrajudicial treatment."

Columbia Law School's Waxman expects the Obama administration to look at criminal prosecutions as "the preferred option" for detainees who are not released to other countries, and he supports that stance. "One of the lessons of the Bush years is that legitimacy is not only important from a rule of law perspective but also from a strategic perspective," he says. "The United States has a strategic interest in promoting certain rule-of-law principles and in demonstrating the legal durability and legitimacy of its counterterrorism policies to garner additional international cooperation."

Should some Guantánamo detainees be held indefinitely without a military or civilian trial?

Four times, the Bush administration went before the Supreme Court to claim the power to hold suspected enemy combatants for indefinite periods with no more than a limited right to challenge their detentions and no more than the most limited judicial review. And four times the Supreme Court said no.

In the first of the rulings, the court in 2004 said U.S. citizens held as enemy combatants were entitled to some hearing before "a neutral decisionmaker." On the same day, the court issued the first of a series of three decisions on the foreigners held at Guantánamo. The series culminated in the 2008 ruling that guaranteed the detainees the right to challenge their incarceration through federal habeas corpus proceedings.[12]

The expansive claim to hold wartime captives indefinitely was one of the primary criticisms that human-rights advocates made against the Bush administration's Guantánamo policies. "The sheer fact of using the geographic location of Guantánamo is not . . . the source of the problem," explains Franklin of the Constitution Project. "The source of the problem was the notion of the prior administration that this created a law-free zone, a legal black hole, and that they could be exempt from any of the principles of the U.S. Constitution, international law or the Geneva Conventions" dealing with wartime captives.

Now, two of President Obama's principal appointees on detention and interrogation policies are signaling that this administration, too, will claim some power to hold suspected terrorists captured in war-like circumstances for indefinite periods. "There's going to be a group of prisoners that, very frankly, are going to have to be held in detainment for a long time," CIA Director-designate Leon Panetta told the Senate Intelligence Committee during his confirmation hearing on Feb. 5.

In his confirmation hearing earlier, Attorney General-designate Eric Holder had endorsed the power to hold an enemy combatant "for the duration of the conflict," but only with judicial review at least once a year to determine whether the prisoner is dangerous. "That kind of review has to be a part of what we do," Holder told the Senate Judiciary Committee on Jan. 16.

Human-rights advocates are concerned. "If we go down the road of a detention-without-charge regime, we will ultimately be moving Guantánamo to the United States rather than closing it," says Mendelson at CSIS.

Few Leaders Among Guantánamo Detainees

The 241 prisoners being held at the U.S. Naval Station at Guantánamo Bay include 36 alleged al Qaeda or Taliban leaders and 199 fighters or operatives, according to an independent examination of the records (top). Among the detainees who have responded to the military's allegations against them, nearly two-thirds have admitted some affiliation with terrorist organizations; the remainder deny any association with al Qaeda or the Taliban (bottom).

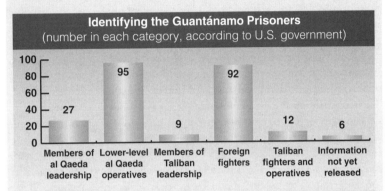

Identifying the Guantánamo Prisoners
(number in each category, according to U.S. government)

Members of al Qaeda leadership	Lower-level al Qaeda operatives	Members of Taliban leadership	Foreign fighters	Taliban fighters and operatives	Information not yet released
27	95	9	92	12	6

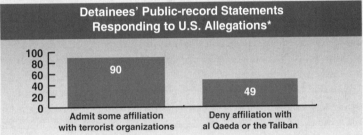

Detainees' Public-record Statements Responding to U.S. Allegations*

Admit some affiliation with terrorist organizations	Deny affiliation with al Qaeda or the Taliban
90	49

* In addition to the relevant statements from 139 detainees, 80 detainees made no statements or made statements that do not materially bear on the military's allegations against them.

Source: Benjamin Wittes and Zaahira Wyne, *The Current Detainee Population of Guantánamo: An Empirical Study,* Brookings Institution, January 2009, www.brookings.edu/reports/2008/1216_detainees_wittes.aspx

"The Obama administration made great gains when it announced the plan to close Guantánamo," Jennifer Daskal, senior counterterrorism counsel for Human Rights Watch, told the *Los Angeles Times*. "Many of those gains will be undercut if the administration is perceived as merely transferring the system of indefinite detention to U.S. soil."[13]

Security-minded experts, however, say the power to hold enemy combatants without trial is both well established and essential. "In our wars, we have held millions of prisoners of war, and they've never had legal proceedings," says McCarthy. He labels as "preposterous" the human-rights groups' stance that the government cannot hold enemy combatants indefinitely.

Unlike the Bush administration, however, defenders of the detention power are calling for Congress and the courts to be involved in establishing rules for holding enemy combatants and reviewing their confinement. "There are some number of people that we're going to have to continue to detain outside the criminal justice system," the Brookings Institution's Wittes said during the Texas law school panel. He calls for Congress to authorize detention and for any detentions to be reviewed by federal courts.

Former Bush administration officials Bellinger and Stimson also say the other branches of government need to be involved. Bellinger wants Congress to authorize detentions for specified periods of time after a given date. Stimson says detentions should be "subjected to periodic review — a very robust review — with heavy lawyer and court involvement."

Military-law expert Fidell, however, opposes any legislation authorizing preventive detention. "I don't think it can be done consistent with our legal system," he says. The Constitution Project's Franklin agrees. "We should be taking people into custody only if we can make a probable-cause showing that they have committed a terrorism offense," she says.

Stimson believes only a "very few" detainees may be too dangerous to release but impossible to try. Over time, he says, there will be a growing presumption against continued detention. He even predicts that the Obama administration might decide that it is not worthwhile to continue to hold this small category of detainees and will find a way to transfer them.

University of Texas law Professor Chesney, however, is not surprised the administration is defending the detention power for now. "The Obama administration is not going to completely forswear the power to hold clandestine, non-state actors," he says. Like others, however, he calls for judicial review to ensure proper treatment and adequate basis for continued detention.

"The executive order leaves open a range of options of how [the administration] is going to detain terrorism suspects in the future, and that's a good thing," says Columbia law Professor Waxman. "It would be an error to close off options."

BACKGROUND

'The Least Worst Place'

With the nation still reeling from the 9/11 terrorist attacks, President Bush made the fateful decision in fall 2001 that suspected enemy combatants captured in Afghanistan and suspected terrorists apprehended elsewhere would be held at the U.S. Naval Station at Guantánamo Bay, Cuba. The decision came to be criticized on two related grounds. The detainees were held out of sight and largely incommunicado, and the administration argued against any judicial review over the detainees. More than two years after the first detainees arrived at Gitmo, the Supreme Court dealt the administration's legal strategy a setback by ruling in June 2004 that federal courts had jurisdiction to hear habeas corpus challenges by the detainees.

Congress set the stage for the detentions by passing, at Bush's request, a resolution authorizing the use of force against "those nations, organizations and persons" responsible for the 9/11 attacks. Bush used the resolution to launch a war against Afghanistan's Taliban government, which he said had harbored the al Qaeda terrorist organization. He then signed an executive order on Nov. 13 authorizing the Defense Department to detain al Qaeda members or anyone else responsible for terrorist activity against the United States and to try the detainees, if at all, before military tribunals.

After consulting with the Justice Department, Defense Secretary Donald Rumsfeld decided to hold the detainees at Guantánamo, which he described on Dec. 27 as "the least worst place" for confinement. The Justice Department's Office of Legal Counsel supported the decision, predicting that federal courts likely would rule detainees at Guantánamo had no access to federal courts to challenge their confinement.

Initially, the decision appeared to be drawing tentative acceptance, with Guantánamo viewed as a logical place to hold suspected terrorists — a secure facility removed from areas of conflict. Military authorities welcomed reporters to the base to tout the speed in constructing the jail cells of the so-called Camp X-Ray.

Criticism began, however, within days of the first prisoners' arrival on Jan. 11, 2002. The International Red Cross faulted the administration for permitting release of pictures of the detainees as they arrived. Amnesty International criticized holding the detainees incommunicado. Criticism increased as Bush announced — and reconfirmed over opposition from Secretary of State Colin Powell — that the detainees would not be regarded as covered by the provisions of the Geneva Conventions. Several European governments said the detainees should be granted prisoner-of-war status.

The controversy moved into the courts on Feb. 19, when lawyers from the Center for Constitutional Rights filed a habeas corpus petition challenging the detentions on behalf of two Britons, Shafiq Rasul and Javaid Iqbal, and David Hicks, an Australian. The petition claimed the prisoners were being held in "indefinite" and "unreviewable" detention in violation of international law and the U.S. Constitution. "The government has just gone too far here," said the center's president, Michael Ratner.

Legal experts questioned that day cast doubt on the prisoners' chances. "The case has a gut appeal," George Washington University law professor Mary Cheh remarked. "But not everything that is deeply troubling is unconstitutional."[14] The experts' doubts appeared to be well-founded when U.S. District Judge Colleen Kollar-Kotelly rejected the challenge in July, ruling that the detainees had no constitutional rights because Guantánamo was not formally part of the United States.

Meanwhile, the administration was making the momentous decision behind the scenes to authorize "enhanced interrogation" techniques against some of the Guantánamo detainees as well as some of the suspected terrorists being held at then-secret CIA prisons in Europe. In a critical step, the Justice Department's Office of Legal Counsel issued an advisory memorandum on Aug. 1, 2002, advising the CIA that a specific list of interrogation techniques would not constitute torture because they were not "specifically intended to inflict severe pain and physical suffering."

Armed with that opinion, military intelligence agents sought and obtained permission to use techniques such as sleep deprivation against some Guantánamo detainees. FBI agents disagreed on the legality and the effectiveness of the techniques, but Rumsfeld signed off on some of the methods in a memo dated Dec. 2. He rescinded the approval for some of the techniques in April 2003. Rasul and Iqbal contend in pending lawsuits that they continued to be mistreated up until their eventual release in March 2004.

In the habeas corpus case, the Guantánamo detainees lost a second round with a ruling in March 2003 by the U.S. Court of Appeals for the District of Columbia Circuit. The appellate judges upheld the lower court's ruling that the Guantánamo detainees could not use habeas corpus to challenge their confinement. In November the Supreme Court agreed to rule on the issue

in a pair of consolidated cases — *Rasul v. Bush* and a second case on behalf of a dozen Kuwaitis.

The move set up the first definitive test of the administration's legal strategy and ended seven months later with a limited but decisive setback. By a 6-3 vote, the court ruled on June 28, 2004, that federal courts had jurisdiction to hear the habeas corpus cases. The ruling left all other issues for future cases. Dissenting justices said the ruling was bad law and bad policy. But Ratner hailed the decision. "This is a major victory for the rule of law," he said.

'A Law-Free Zone'?

The Bush administration dug in its heels after the Supreme Court's habeas corpus decision by twice persuading Congress to approve legislation to keep the Guantánamo detainees out of federal court. Instead, the administration planned to try the detainees before specially created military "commissions" with limited substantive and procedural rights. Legal challenges slowed the commissions, however, and eventually led to two additional Supreme Court rulings rejecting the administration's strategy. At the same time, the administration worked to transfer Guantánamo detainees to their home countries but encountered reluctance and resistance as other countries came to view the facility as emblematic of a policy of lawless mistreatment of wartime captives.

The administration fashioned its Guantánamo policies against the backdrop of a worldwide controversy over mistreatment of enemy combatants symbolized by the documented abuse of Iraqi prisoners at the U.S.-run jail at Abu Ghraib, outside Baghdad. Images broadcast around the world beginning in May 2004 touched off demands at home and abroad for stronger rules to require humane treatment of all U.S.-held prisoners, including the Guantánamo detainees. But the administration also wanted to cut off habeas corpus petitions.

The result was the Detainee Treatment Act, passed in December 2005, which prohibited inhumane treatment of prisoners and limited interrogation techniques used by military intelligence agents to those approved in the *U.S. Army Field Manual.* But the act also sought to bar federal courts from hearing habeas corpus petitions by Guantánamo detainees, including cases already pending.

Meanwhile, the military commission proceedings were being challenged by one of the Guantánamo prisoners, Salim Ahmed Hamdan, a Yemeni-born Muslim accused of

CHRONOLOGY

2001-Present *U.S. Naval Station at Guantánamo Bay, Cuba, is used to hold enemy combatants after 9/11; Bush administration loses effort to block federal court challenges; Obama administration promises to close prison camp by Jan. 22, 2010.*

2001 Terrorists attack the United States (Sept. 11). . . . Congress authorizes use of military force against countries, organizations or individuals responsible for attacks (Sept. 18). . . . U.S. forces capture hundreds of suspected "enemy combatants" in Afghanistan war; President George W. Bush signs executive order authorizing military detention (Nov. 13). . . . Guantánamo Bay is chosen for detention facility; Justice Department predicts detainees cannot use habeas corpus to challenge their confinement (Dec. 28).

2002 First detainees brought to Guantánamo (Jan. 11); detentions prompt immediate controversy and legal challenges (January-February); challenges rejected in several courts. . . . Justice Department memo narrows definition of "torture," approves some "enhanced" interrogation techniques (Aug. 1). . . . Pentagon officials sign off on enhanced interrogation (November/December).

2003 Federal appeals court bars habeas corpus by Guantánamo detainees (March 11). . . . Defense Secretary Donald Rumsfeld approves new interrogation rules, including 24 of 35 recommended enhanced techniques (April 16). . . . Supreme Court agrees to rule on habeas corpus issue (Nov. 10).

2004 Four Britons released to United Kingdom, with no charges; later file civil suits for damages (March 9). . . . Images of inmate-abuse at Abu Ghraib prison in Iraq broadcast worldwide (May). . . . Supreme Court upholds federal court jurisdiction over detainees' habeas corpus petitions (June 28).

2005 Detainee Treatment Act (DTA) limits interrogation techniques, bars habeas corpus for detainees (December).

2006 Supreme Court allows pending habeas corpus cases despite DTA; rules military commissions violate U.S. military law, Geneva Conventions (June 29). . . . President

Bush announces transfer of 14 "high-value" prisoners to Guantánamo (Sept. 6). . . . Congress passes, Bush signs Military Commissions Act; law prohibits all pending, future habeas corpus actions by detainees, allows limited review of military commissions by appeals court (October).

2007 Military Commissions Act upheld by D.C. Circuit court in *Boumediene* case (Feb. 20). . . . Australian David Hicks pleads guilty to supporting terrorism; sent to Australia to finish serving sentence (March). . . . Supreme Court decides to review *Boumediene* decision; hears arguments (Dec. 5).

2008 Democrat Barack Obama, Republican John McCain cinch nominations for president, promise to close Guantánamo (spring). . . . Supreme Court rules, 5-4, Military Commissions Act unconstitutional; ruling guarantees habeas corpus rights for detainees (June 12). . . . Salim Ahmed Hamdan, driver for al Qaeda leader Osama bin Laden, convicted on reduced charges, given 5-1/2-year sentence (Aug. 6-7); released to Yemen (Nov. 25). . . . Federal judge orders 17 Chinese Muslims held at Guantánamo brought to U.S. for release (Oct. 7). . . . Obama elected, begins transition (November-December). . . . Five of six detainees in *Boumediene* case ordered released by federal judge for lack of evidence (Nov. 21).

2009 Military judge says would-be hijacker Mohammed al Qahtani was "tortured" while at Guantánamo (Jan. 14). . . . Obama sworn in; denies "false choice" between liberty, security (Jan. 20). . . . Obama signs executive order to close Guantánamo within one year; orders review of detainee cases; limits use of "enhanced interrogation" techniques (Jan. 22). . . . Administration reaffirms opposition to habeas corpus for prisoners held at Bagram Air Force Base in Afghanistan (Feb. 20). . . . First Guantánamo detainee released under Obama administration; upon arrival in Britain, Binyam Mohamed describes "medieval" torture during detention in Morocco (Feb. 22-23). . . . Pentagon review team says Guantánamo meets Geneva Conventions standards; human rights groups disagree (Feb. 23). . . . Defense Department review team says Guantánamo conditions comply with Geneva Conventions; rights groups disagree (Feb. 23).

being the driver for the al Qaeda leader, Osama bin Laden. Hamdan had been captured in Afghanistan, brought to Guantánamo in January 2002 and designated as eligible for trial in July 2003. He then filed a habeas corpus petition challenging the military commissions as inconsistent with the Uniform Code of Military Justice (UCMJ) and the procedural provisions of the Geneva Conventions.

After conflicting rulings from lower courts, the Supreme Court ruled, 5-3, in Hamdan's favor in June 2006. Preliminarily, the court held that the Detainee Treatment Act did not eliminate habeas corpus cases filed before the law was enacted. The court went on to hold that President Bush had failed to show any need to depart from the UCMJ's regular procedures for military trial. The majority also said the military commissions did not comply with the Geneva Conventions.

The administration pushed Congress to respond quickly by passing a new law, the Military Commissions Act, which revised procedures for the military tribunals and explicitly barred all habeas corpus petitions from Guantánamo. To help move the bill, Bush announced on Sept. 6, 2006, that he was transferring 14 "high-value" terror suspects to Guantánamo from the secret CIA prisons. The transferred prisoners included Khalid Shaikh Mohammed, accused of masterminding the 9/11 attacks and eventually identified as one of three prisoners subjected to waterboarding during interrogation. With midterm congressional election campaigns under way, Bush's decision to move high-profile al Qaeda suspects to Guantánamo spurred Congress to act on a supposed Guantánamo fix before the elections. Congress completed action on the bill at the end of the month; Bush signed it into law on Oct. 17.

The revamped military commissions finally produced the system's first conviction in March 2007, when the Australian Hicks pleaded guilty to a newly codified charge of providing material support to terrorism. Hicks, a convert to Islam, had been turned over to U.S. forces by the U.S.-backed Northern Alliance in Afghanistan in November 2001. A military panel sentenced Hicks to seven years' imprisonment on the charge, but under the plea deal he was returned to Australia to serve only an additional nine months. After his release, Hicks acknowledged that he trained at al Qaeda camps and fought for the Taliban, but he denied any hostile actions against United States.

Hicks' guilty plea came not long after the D.C. Circuit in February had given the Bush administration a legal victory by upholding the Military Commissions Act, including its bar to habeas corpus petitions by the Guantánamo prisoners. In June, however, the Supreme Court agreed to hear the appeals by the detainees. The lead case, *Boumediene v. Bush*, was brought on behalf of six Algerians arrested in Bosnia on suspicion of plotting to bomb the U.S. Embassy in Sarajevo. In dramatic arguments in December, Solicitor General Paul Clement contended that the act gave the detainees sufficient judicial review to substitute for habeas corpus.

Representing the detainees, former Solicitor General Seth Waxman insisted the government's argument amounted to treating Guantánamo as a "law-free zone." In June 2008, the court sided narrowly but decisively with the detainees against the government. Writing for a five-vote majority, Justice Anthony M. Kennedy said the detainees were entitled to habeas corpus because the United States exercised "de facto sovereignty" over Guantánamo.

'An Enormous Failure'

Political and legal events combined in 2008 to create broader support for closing Guantánamo. The Bush administration resisted the conclusion, however, even as both major-party presidential candidates endorsed the step, and the government's stance in Guantánamo cases met a host of legal difficulties in court. Obama's election — after he had strongly criticized Bush policies in a range of areas — seemed to portend quick action on Guantánamo. But the executive order that Obama signed on his second full day in the White House only started a process and gave the administration a full year to complete the closure of the camp.

By the time of the Supreme Court decision recognizing habeas corpus rights for Guantánamo prisoners, Obama and McCain had cinched their respective nominations. Both had already endorsed closing Guantánamo, but they differed on the court ruling. Obama hailed the decision as a victory for the rule of law, while McCain called it "one of the worst decisions" in the country's history. President Bush said he agreed with the dissenters in the case, but promised to comply. Behind the scenes, Bush's advisers considered closing the camp but rejected the idea — despite Bush's professed desire to do so — because of what the advisers concluded were unacceptable political and legal risks.[15]

From Imperial Outpost to Post-9/11 'Gulag'

Inside the prison at Guantánamo Bay.

The prison camp that became a reviled symbol of the Bush administration's "war on terror" welcomes visitors — in person or online — to witness what is officially described as "safe, humane, legal and transparent care and custody of detained enemy combatants."

The online tour of the detention camp at the U.S. Naval Station, Guantánamo Bay, Cuba (www.jtfgtmo.southcom.mil) depicts a state-of-the-art correctional facility with a 12,000-volume library, exercise equipment and recreational areas.

The pictures contrast sharply with the indelible images first shown to the world after suspected terrorists from Afghanistan and elsewhere began arriving in January 2002: prisoners in orange jumpsuits forced to crouch near chain-link fences or being led around in shackles by burly guards.

The 320 open-air, steel-mesh cages of Camp X-Ray were in use only until April 2002, but a federal judge has ordered them preserved as evidence in litigation challenging the conditions of confinement. Today, most of the 241 detainees are housed in three facilities built later as Guantánamo — widely known as "Gitmo" — was transformed from short-term expedient to long-term policy.

There are three main camps, according to an online primer provided by *The Miami Herald* (www.miamiherald.com/Guantánamo). Camp 4 resembles a traditional prisoner-of-war camp with 10-cot bunkhouses and common eating area, communal showers and athletic facilities. Camp 5 is a maximum-security facility with single-prisoner cells controlled by a centralized locking system and closed-circuit surveillance. Camp 6 was converted from minimum to maximum security after a fight between guards and detainees in Camp 4 in May 2006. The facility has single-occupancy cells where prisoners are locked up 22 hours a day.

In addition, about 15 men are believed to be housed in super-secret Camp 7, which was built for the "high-level" former CIA captives that Bush ordered transferred to Guantánamo in September 2006. The existence of the camp was not confirmed until December 2007, and its exact location remains shrouded. Among the prisoners believed to be there is Khalid Sheikh Mohammed, charged along with four others with masterminding the Sept. 11, 2001, terrorist attacks on the United States.

A separate facility, dubbed Camp Justice, was built during the Bush administration to hold trials of detainees before the specially created military commissions. President Obama suspended those proceedings for 120 days pending an inter-agency review of detainees' case files by Attorney General Eric Holder.[1]

The naval base itself houses about 8,400 U.S. military and civilian personnel — including 2,200 at the detention facilities. It occupies a 45-square-mile tract — about three-fourths of the size of the District of Columbia — near the southeastern tip of Cuba. The United States acquired the base under a lease with Cuba in 1903 after helping the former Spanish colony gain its independence as part of the Spanish-American War.[2]

U.S. rights to the base — then used as a coaling station — were reaffirmed in a 1934 treaty that calls for the United States to pay Cuba about $4,000 per year. Termination of the lease requires consent of both governments. The United States has continued to pay the lease amount, but since the Cuban Revolution of the late 1950s the government of Fidel Castro has refused to cash the checks.

The base made news in 1958, when 29 sailors and Marines were kidnapped by Cuban rebels and held for 22 days. During the Cuban Missile Crisis four years later, civilian and military personnel and families were evacuated. Two years later, Castro's government cut off water and supply routes to the base. Since then, the base has had its own power and water sources.

Guantánamo faded from the news and receded in strategic significance until the 1980s and '90s, when it became a holding facility for Cuban and Haitian refugees fleeing to the United States. Under Presidents George H. W. Bush and Bill Clinton, the Coast Guard picked up the refugees on rafts on the high seas and held them in ramshackle facilities on the base to avoid bringing them to U.S. soil.

The Clinton administration lost a legal battle over the refugee issue in June 1993, when a federal judge ordered that 150 HIV-positive Haitian refugees who qualified for political asylum could not be excluded from the United States because of their health status. Later that month, however, the Supreme Court ruled the government could return Haitian refugees captured at sea to their home country without giving them the chance to apply for asylum.

Guantánamo receded from the news for the rest of the decade but emerged in fall 2001 when the Defense and Justice departments decided it was the best place to bring enemy combatants captured in the Afghanistan war and suspected terrorists rounded up elsewhere. A legal opinion by John Yoo, then a deputy in the Justice Department's Office of Legal Counsel and now a law professor in California, forecast — wrongly — that the Supreme Court would not allow Guantánamo prisoners to challenge their detention in court because the base was not on U.S. soil.

A guard talks to a Guantánamo detainee inside the open yard at Camp 4, the medium-security detention center that resembles a military prisoner of war camp.

forward, we were not engaged in any practices that would be considered cruel, inhuman or degrading at all," he says.

Stimson and others note the many stresses for the young guards at the camp, ranging from periodic hunger strikes to urine- and feces-filled baggies thrown in a guard's face. The Guantánamo Web site notes without comment the different conditions of confinement for "compliant" and "noncompliant" prisoners. But the site also says that all prisoners "regardless of compliancy" are furnished a Koran, prayer mat, prayer beads and cap. The Muslim prisoners are aided in their religious observances by arrows in each cell pointing to Mecca and by prison schedules that take account of the observance of daily prayers.

In a new book detailing the initial history of the detention camp, Karen Greenberg, executive director of the Center on Law and Security at New York University School of Law, says the first commander, Marine Brig. Gen. Michael Lehnert, welcomed visits by the International Committee of the Red Cross (ICRC) and sought to comply with the Geneva Conventions even though President Bush had said they did not apply. Lehnert was eased out within a few months, however, after Defense Secretary Donald Rumsfeld ordered more stringent treatment of the prisoners to aid interrogation.[3]

For more than a year, official policy approved by Rumsfeld sanctioned "enhanced interrogation" techniques at the camp such as sleep deprivation, "stress positions" and forced nakedness. Rumsfeld withdrew approval for many of the challenged techniques in a memorandum signed in April 2003. Criticism of the camp continued, however. In November 2004 a leaked ICRC report described some discipline for prisoners as "tantamount to torture." In May 2005, the human-rights group Amnesty International labeled Guantánamo "the gulag of our time."

Through the years, some international visitors have dissented from this image of the camp. After visiting in 2007, Brookings scholar Wittes described the camp as "coolly professional."[4] Charles "Cully" Stimson, an assistant secretary of Defense who helped oversee the facility from January 2006 to February 2007, says interrogation complied with the *U.S. Army Field Manual.* "During my tenure and going

As part of the Obama administration's decision to close Guantánamo within one year, Defense Secretary Robert Gates ordered a review to determine whether the camp complies with all provisions of the Geneva Conventions on humane treatment of wartime captives. The report, released on Feb. 23, found no violations but recommended some changes to improve conditions. Human-rights groups faulted the report and called for broader changes.

Meanwhile, White House Counsel Gregory Craig and Attorney General Holder visited the prison in separate day visits on Feb. 19 and 23, respectively. No reporters accompanied either official, and neither had immediate comments after returning to Washington.

---

[1] For a description, see Jeffrey Toobin, "Camp Justice," *The New Yorker*, April 14, 2008, pp. 32-38.

[2] Some background drawn from U.S. Navy, "History of Guantánamo Bay," undated (www.cnic.navy.mil/Guantanamo/index.htm). For a detailed history, the site recommends M.E. Murphy, *The History of Guantánamo Bay 1494-1964*, published in two volumes in 1953 and 1964.

[3] Karen Greenberg, *The Least Worst Place: Guantánamo's First 100 Days* (2009). For a summary, see Karen Greenberg, "When Gitmo Was (Relatively) Good," *The Washington Post*, Jan. 25, 2009, p. B1.

[4] Benjamin Wittes, *Law and the Long War: The Future of Justice in the Age of Terror* (2008), p. 73.

Meanwhile, the administration was heading toward a serious embarrassment in the first-ever military commission trial: the prosecution of Salim Hamdan, the detainee most closely identified with challenging the Guantánamo regime. When Hamdan's trial opened on July 22, the prosecution depicted his role as bin Laden's former driver as vital to al Qaeda's war against the United States. Defense lawyers instead painted him as a poor Muslim in need of a job. After a two-week trial, the military jury sided mostly with the defense. The panel acquitted Hamdan of the most serious charges and convicted him on Aug. 6 only of material support of terrorism. The next day, the jury sentenced Hamdan to an unexpectedly short 5-1/2 years; the judge said he would credit Hamdan with the 61 months he had already served.

With economic issues dominating the presidential campaign, Obama and McCain devoted scant attention to Guantánamo during their parties' national conventions in late August and early September, respectively. Obama had laid out his position earlier, on June 18, where he called Guantánamo "an enormous failure" and "legal black hole" that had weakened support abroad for U.S. anti-terrorism policies. In his acceptance speech, Obama made only a less specific promise to "restore our moral standing." Accepting the GOP nomination a week later, McCain thanked Bush for keeping the country safe after 9/11 and left any criticism of Bush policies unspoken.

In the fall, the administration sustained two more blows to its Guantánamo policies when two federal judges in Washington — for the first time — ordered the release of detainees. On Oct. 7, Judge Ricardo Urbina directed the government to free 17 Chinese Muslims, known as Uighurs, who had been held at Guantánamo since they were rounded up in Afghanistan in fall 2001. The Uighurs, ethnic Turkic Muslim separatists from western China, contended they had been seeking refuge from the Chinese government. The administration initially described them as terrorists, but by 2008 acknowledged they were not enemy combatants. The dissident Uighurs remained at Guantánamo, however, because they could not be sent back to China, which classifies Uighurs as terrorists, and no other country would take them. The government won an appeals court stay of Urbina's ruling quickly, and later — after Obama took office — a reversal of the decision. (*See sidebar, p. 84.*)

The next month, Judge Richard Leon ruled the government had no legal basis for holding five of the six detainees from the *Boumediene* case. Speaking from the bench — with the prisoners listening by teleconference from Guantánamo — Leon noted that the government had dropped charges relating to the alleged bombing plot but still contended the group planned to go to Afghanistan to fight against U.S. forces. Leon said the government's case relied "exclusively" on classified evidence from one unnamed source whose reliability and credibility could not be adequately evaluated. He called the evidence "too thin a reed" to justify continued imprisonment and ordered five of the six released "forthwith." Leon found sufficient evidence, however, to justify the charge that Bensayah Belkacem had aided transportation logistics for al Qaeda members.

With Obama preparing to take office, the government suffered one more setback: On Jan. 14 Leon ordered the release of a Chadian-born detainee, Mohammed al Gharani, who had been captured at age 14 and accused by other detainees of having lived in al Qaeda guest houses. The evidence was inconsistent and unverified, Leon said.

Eight days later, Obama signed the executive order promising a new review of all the Guantánamo case files and closure of the facility within a year. McCain was among those who endorsed the president's decision.[16]

CURRENT SITUATION
Detainee Cases Reviewed

A Justice Department task force is beginning close review of case files on the 241 remaining Guantánamo detainees following visits to the prison camp by two top Obama administration officials and the first release of a detainee since Obama took office.

White House Counsel Gregory Craig and Attorney General Holder made separate day trips to Guantánamo in late February, the first visits to the facility for each. No reporters accompanied Craig or Holder on the trips. Craig was accompanied on his Feb. 18 trip by Jeh C. Johnson, the Defense Department's general counsel. Holder took half a dozen close aides with him for a similar trip on Feb. 23. A Justice Department spokesman told reporters in Washington that Holder was to discuss case histories of specific detainees and tour detention facilities and Camp Justice, the courtroom complex built for the military commissions.

Among those accompanying Holder was Matthew Olsen, a veteran Justice Department official whom Holder named on Feb. 20 to head an interagency task force charged with assembling information on each of the remaining detainees and recommending proper dispositions of their cases. Olsen had been a Justice Department prosecutor for nearly 10 years before being named in September 2006 as deputy assistant attorney general for the then newly established National Security Division.

Olsen's task force will be dealing with case files that one former military prosecutor has described as being in "a state of disarray." Darrel Vandeveld, a former lieutenant colonel in the Army Reserve, made the critical statement in January, four months after he had resigned as a prosecutor in Guantánamo for what he said were reasons of conscience.[17]

Vandeveld, a senior deputy attorney general in Pennsylvania in civilian life, told *The Washington Post* that case files were disorganized, information scattered between different databases and physical evidence stored in unknown locations or in some instances missing. Military officials denied Vandeveld's accusations, *The Post* said. The newspaper quoted Col. Lawrence Morris, chief military prosecutor, as saying that Vandeveld had not raised concerns with him and also suggesting that Vandeveld had resigned after being passed over for a promotion.

In a second story, however, *The Post* quoted ex-Defense official Stimson as saying that while at the Pentagon he had persistent problems compiling information on individual detainees. The newspaper also noted references in Justice Department filings in habeas corpus cases to the unexpected difficulties the government faced in assembling case files on individual detainees.

The officials' trips came as the administration was completing preparations for the release of one of the highest-profile Guantánamo detainees: Binyam Mohamed, an Ethiopian-born British citizen who had been accused of planning to detonate "dirty bombs" in the United States. Mohamed claimed that after being held in Afghanistan and Pakistan, he was transferred to Morocco for 18 months and tortured there before being brought to Guantánamo.

Mohamed was flown from Guantánamo on Feb. 22 and arrived in England the next day. As part of the release, Mohamed reportedly agreed to a lifetime prohibition against travel to the United States. *The New York Times* reported that the British government told U.S. officials that, under British and European human-rights laws, it could not impose other travel or surveillance restrictions on Mohamed.[18]

The Justice Department announced Mohamed's departure in a press release instead of the Defense Department, as had been the practice under the Bush administration. The Justice Department has said that an additional 57 detainees have been approved for transfer or release, but are awaiting agreements with third countries. That number includes the 17 Chinese Muslims and three others who have won habeas corpus cases but are not yet released.

Obama administration officials are counting on increased cooperation between other countries, including U.S. allies in Europe, to help empty Guantánamo before Obama's one-year deadline for closing the facility. At least three countries — Spain, Estonia and Latvia — have signaled a willingness to accept released detainees, but Italy says it won't because no Italians are being held there. "I can absolutely rule out that the closing of Guantánamo will have any consequences for Italy," Gianfranco Fini, the speaker of Italy's Chamber of Deputies and a close ally of Prime Minister Silvo Berlusconi, was quoted as telling House Speaker Nancy Pelosi, D-Calif., on Feb. 16 during a visit by the U.S. lawmaker.[19]

The Guantánamo developments come against a backdrop of concern among some human-rights and civil liberties advocates about the direction of Obama administration policies on national security issues. The American Civil Liberties Union, for example, criticized the administration after Justice Department lawyers in February reaffirmed before a federal appeals court the invocation of the state secrets privilege to try to block the trial of a suit by former prisoners attacking the practice of "rendition" of detainees to other countries.[20]

Meanwhile, the administration is giving no encouragement to proposals on Capitol Hill for an in-depth investigation of Bush administration detention and interrogation policies. Senate Judiciary Committee Chairman Patrick J. Leahy, D-Vt., is proposing a "truth commission" to look at interrogation and detention,

What Can Be Done With the Uighurs?

Even the U.S. wants to release 17 Chinese Muslims.

Seventeen Chinese Muslims held at the Guantánamo prison camp since 2002 deny that they are "enemy combatants" against the United States. The government agrees and wants to release the men, members of the Uighur Muslim community in western China.

When a federal judge last fall ordered that the Uighurs be brought to the United States to be released, however, the Bush administration appealed the decision. And last week the federal appeals court for the District of Columbia Circuit agreed that federal courts cannot order a foreigner admitted into the United States — a ruling that leaves the puzzle for the Obama administration to try to solve.[1]

The Uighurs are members of a Turkic ethnic group considered by the Chinese government to be separatist terrorists. Before the Sept. 11, 2001, terrorist attacks on the United States, the dissident Uighurs had been receiving firearms training at a camp run by the Eastern Turkistan Islamic Group near Tora Bora, Afghanistan — the same area where al Qaeda training camps are found. They fled to Pakistan after U.S. air strikes destroyed their camp but were captured, turned over to U.S. forces and brought to Guantánamo.

Initially, the government depicted the Uighurs as enemy combatants because of alleged connections between

Seventeen Chinese Muslims, or Uighurs, held since 2002 have been ordered released from Guantánamo. Five other Uighurs, including the four above, were recently released to Albania.

the Turkistan group and al Qaeda or Afghanistan's Taliban government. But the U.S. Court of Appeals for the District of Columbia, ruling on a habeas corpus case brought by one of the Uighurs, said the government had not produced enough evidence to support the accusation.[2]

The Bush administration bowed to the ruling and stepped up efforts to release the men to third countries. The Uighurs cannot be returned to their home country because they contend — and the U.S. government does not dispute — that they could face arrest, torture or execution in China. But the

among other topics. But Obama gave the proposal no support when questioned at his first prime-time news conference. "Generally speaking, I'm more interested in looking forward than I am in looking back," Obama said on Feb. 9.

Torture Suits Stymied

A civil suit by four Britons released from Guantánamo in 2004 after two years' confinement could result in the first detailed courtroom airing of allegations of torture and abusive treatment of detainees at the U.S. prison camp. But — barring an unlikely shift by the Obama administration — the case will come to trial only if a federal appeals court decision dismissing the suit is reversed either by that court or by the Supreme Court.

The suit is one of several cases seeking to air former detainees' allegations of torture that have been stymied because of legal or diplomatic hurdles. The roadblocks are persisting even after the Pentagon's top judge in the Guantánamo detainee cases in January confirmed allegations of torture used against a Saudi national identified as a would-be 9/11 hijacker. And so far the Obama administration has shown no signs of easing barriers to former detainees seeking compensation in civil courts for mistreatment while prisoners at Guantánamo or elsewhere during the Bush administration.

The four British Muslims all claim they were rounded up by mistake during the Afghanistan war in fall 2001 and subjected to abusive interrogation amounting to torture at Guantánamo before essentially being cleared and

government's six-year-long depiction of the Uighurs as dangerous terrorists has left other countries reluctant to accept them.

In October, U.S. District Judge Ricardo Urbina moved to resolve the dilemma by ordering the Uighurs to be released into the United States. After questioning the government's claim that the Uighurs could be dangerous if admitted into the country, Urbina ruled on Oct. 7 that their continued detention was unlawful. "Separation of powers concerns do not trump . . . the unalienable right of liberty," he said.[3]

The government immediately asked for and obtained a stay to Urbina's ruling pending an appeal. The three-judge panel's Feb. 18 ruling on the appeal backed the government's position that Urbina had exceeded his authority.

"It is not within the province of any court, unless expressly authorized by law, to review the determination of the political branch of the government to exclude a given alien," Senior Circuit Judge A. Raymond Randolph wrote for a two-judge majority. The third judge, Judith Rogers, disagreed with the legal ruling but said Urbina had acted prematurely because the Uighurs had never sought admission to the United States.

The case was argued before the appellate panel on Nov. 24, while the Bush administration was still in office. With the case pending, lawyers for the Uighurs wrote to Obama administration officials on Jan. 23 — the day after President Obama signed an executive order promising to close Guantánamo within one year — urging that the men be immediately released.

A Washington, D.C.-based association of Uighurs offered to help the prisoners establish residences in the United States. "We have people offering them places to stay, English training, employment," said Nury Turkel, a past president of the Uyghur American Association. "We don't want anyone to think they will be a burden on society."[4]

Lawyers for the Uighurs said they would continue their efforts to free the men, but one said the appeals court decision limits the impact of the Supreme Court's decision in June 2008 guaranteeing Guantánamo detainees the right to habeas corpus. "You win and still can't get out," Susan Baker Manning told *The Washington Post*. The administration had no immediate comment on the decision.[5]

---

[1] The decision is *Kiyemba v. Obama*, 08-5424, U.S. Court of Appeals for the District of Columbia Circuit, Feb. 18, 2009, http://pacer.cadc.uscourts.gov/common/opinions/200902/08-5424-1165428.pdf. For coverage, see Lyle Denniston, "Uighurs Barred From U.S.," SCOTUSBlog, Feb. 18, 2009, www.scotusblog.com/wp/uighurs-barred-from-us/#more-8725. Background drawn from court opinion and ongoing coverage on SCOTUSBlog.

[2] The case is *Parhart v. Gates*, 532 F.3d 834 (D.C. Cir. 2008). For coverage, see William Glaberson, "Evidence Faulted in Detainee Case," *The New York Times*, July 1, 2008, p. A1.

[3] For coverage, see Ben Winograd, "Judge Orders Uighurs to U.S.; Government Appeals," SCOTUSBlog, Oct. 7, 2008. The story links to a transcript of the Oct. 7 hearing before Urbina.

[4] Quoted in Steve Hendrix, "D.C. Area Families Are Ready to Receive Uighur Detainees," *The Washington Post*, Oct. 8, 2008, p. A8. The association uses a different spelling of Uighur.

[5] Quoted in Del Quentin Wilber and Carrie Johnson, "Court Blocks Release of 17 Uighurs Into U.S.," *The Washington Post*, Feb. 19, 2009, p. A4.

---

released after diplomatic pressure from the British government. Three of the men — Shafiq Rasul, Asif Iqbal and Rhuhel Ahmed — say they were aiding humanitarian relief efforts in Afghanistan when they were captured by forces aligned with the notorious Uzbek warlord Rashim Dotsum and turned over to U.S. forces for a bounty. The fourth, Jamal al-Harith, was taken into custody when U.S. forces took over a Taliban jail where he was being held on suspicion of being a British spy.

In their civil suit filed in federal court in Washington in 2004 after their release, the men claim that they were subjected at Guantánamo to beatings, solitary confinement, exposure to extreme heat and cold, threats of attack from unmuzzled dogs, nudity and sleep deprivation. In addition to those claimed constitutional violations, the suit claims that alleged interference with their religious beliefs violated the federal Religious Freedom Restoration Act.

Without addressing the allegations, the government won a ruling from the U.S. Court of Appeals for the District of Columbia Circuit in January 2008 dismissing the suit on legal grounds. The three-judge panel rejected the constitutional claims because the plaintiffs were held outside U.S. territory. As an alternative basis for dismissal, the court said the military officials named as defendants were entitled to qualified immunity from suit.[21]

The Supreme Court in December ordered the appeals court to reconsider the decision in light of its June 12 ruling permitting Guantánamo detainees to bring habeas corpus actions. The appeals court has ordered a new

Should Congress create a national security court for enemy combatant cases?

YES

Andrew C. McCarthy
Legal-Affairs Editor,
National Review

Written for *CQ Researcher*, Feb. 20, 2009

It has been a relief to see President Obama retreat from the irresponsible rhetoric of his campaign regarding various security measures that have protected the nation from a reprise of the Sept. 11 attacks. The president now explicitly recognizes that there are numerous terrorists who threaten the United States but cannot be tried in the civilian courts — his preferred forum. The answer is a special national security court.

As we learned in the 1990s, the federal courts are more than adequate in providing due process for jihadists hell-bent on killing Americans. All of the terrorists indicted were convicted. Nevertheless, due process for our enemies, while not unimportant, can never be our primary aim, not if government is to tend to its first responsibility — the security of the governed.

Between the 1993 bombing of the World Trade Center and its destruction on 9/11, radical Islam became bigger and bolder. American targets were repeatedly attacked — including Khobar Towers (19 U.S. Air Force members killed), the U.S. embassies in eastern Africa (over 200 killed) and the *USS Cole* (17 U.S. sailors killed). Yet, because of the high burdens and elaborate protections of the criminal justice system — a system designed to protect Americans — only 29 terrorists were successfully prosecuted in the eight-year period when prosecution in federal court was our nation's principal counterterrorism strategy.

The effect of this weak response was to encourage more attacks. Indeed, Osama bin Laden himself has been under charges by the Justice Department since June 1998 but has killed thousands of Americans in the ensuing decade — adding counts to the indictment does not seem to deter him much. Ditto Khalid Shaikh Mohammed, who had also been under indictment for years while he planned the 9/11 atrocities. It told our enemies that we could be attacked with virtual impunity.

We have not suffered another attack since 9/11 primarily because we moved in late 2001 to a law-of-war paradigm, which permits terrorists — enemies in war, not just defendants in a case — to be detained without trial, until the conclusion of hostilities. That philosophy, coupled with a comprehensive counter-terrorism strategy that does not unduly rely on criminal prosecutions, has helped us prevent terrorist attacks from happening, rather than contenting ourselves with prosecuting a handful of jihadists after innocents have been slaughtered.

If we don't want 9/11 results, we can't go back to a 9/10 mentality.

NO

Edward L. Dowd Jr.
Former U.S. Attorney, Eastern District of Missouri
Earl Silbert
Former U.S. Attorney, District of Columbia

Written for *CQ Researcher*, Feb. 20, 2009

As former federal prosecutors, we have a deep understanding and appreciation for the enormity of the crimes that terrorists commit. We strongly support the severe punishment of convicted terrorists. However, we should not create national security courts to handle these prosecutions.

For over 230 years, federal courts have protected our fundamental constitutional rights while overseeing the prosecution and punishment of criminals, including terrorists. Indeed, over the past 20 years, more than 120 terrorism-related cases were prosecuted without jeopardizing our national security. This well-tested system is responsible for the convictions of Timothy McVeigh and Terry Nichols (Oklahoma City bombers), Ramzi Yousef and Sheikh Abdel Rahman (1993 World Trade Center bombers), Zacarias Moussaoui (member of al Qaeda who was involved in 9/11), and many others.

National security court proposals, by lessening due-process standards, threaten to undermine the constitutional rights safeguarded by our existing criminal justice system. Moreover, by depriving suspects of basic constitutional rights, any convictions by national security courts would be subject to challenge.

The argument that terrorist suspects require a special "terrorist court" with fewer rights undermines the presumption of innocence at the heart of the American judicial system. We do not yet know who among the detainees are guilty of acts of terrorism and who might be innocent. While we share the goal of convicting those who commit terrorist crimes, we cannot support a separate and unequal criminal justice system that does not protect basic constitutional rights. Nor should we adopt national security courts to oversee a legalized system of indefinite preventive detention without trial for terrorist suspects. Detaining individuals indefinitely without charge simply because we "believe" they are dangerous would violate both our Constitution and fundamental American values.

We join with the Constitution Project's bipartisan Liberty and Security committee in urging that our traditional federal courts continue to be the venue for prosecutions for terrorism offenses.

To do otherwise would allow our ideals and rights to be destroyed by the very terrorists we are seeking to convict. As we undertake the critical task of closing detention facilities and prosecuting detainees for crimes of international terrorism, we should reject this dangerous proposal.

round of briefs to be filed in March, but Eric Lewis, the private lawyer representing the men, is pessimistic about getting a ruling from a panel that he describes as "not sympathetic." He says he will appeal an unfavorable ruling to the Supreme Court.

"Civil accountability is the one mechanism of accountability that's out there," says Lewis, a Washington attorney handling the case on a pro bono basis. "There's been a fair amount of confirmation [of mistreatment] that's come in, essentially through statements made, books written, but no judicial accountability."

In another high-profile Guantánamo-related case, the Obama administration in February followed the Bush administration's stance in invoking a "state secrets" privilege to block a civil suit by five current or former detainees over the Bush administration's practice of "extraordinary rendition," or sending suspected terrorists to other countries, where they allege they were tortured. The plaintiffs are seeking civil damages from a private airline for its alleged role in transporting them in cooperation with the CIA.

The Bush administration won a lower court ruling to dismiss the case on the ground that a trial would inevitably disclose state secrets. When the case was argued on Feb. 9 before the federal appeals court in San Francisco, Justice Department lawyers reaffirmed that position and said under questioning the stance had been "thoroughly vetted with appropriate officials" in the new administration.

Two of the five plaintiffs were eventually taken to Guantánamo. One was released in 2008; the other — Binyham Mohamed — was released on Feb. 22. The British government says it has evidence Mohamed was tortured while in Moroccan custody, but blocked its release after the Bush administration threatened to review intelligence sharing arrangements with Britain if the material was disclosed. After a British court reluctantly bowed to that decision, the White House issued a statement thanking the British government "for its continued commitment to protect sensitive national security information."[22]

Allegations of torture at Guantánamo gained new currency after the Defense Department judge overseeing the military commissions system confirmed that she blocked the prosecution of Mohammed al-Qahtani in May 2008 because she was convinced he had been tortured. Susan Crawford, who has the title of convening authority of the military commissions, made the statement in an interview with *The Washington Post's* Bob Woodward published in January.[23]

Qahtani is alleged to have planned to join the 9/11 hijackings but was denied entry into the United States. He was captured in Afghanistan, transported to Guantánamo and interrogated over 50 days from November 2002 to January 2003.

In the interview, Crawford details "abusive" techniques that included prolonged interrogation and forced nudity that had "a medical impact" on him. "His treatment met the legal definition of torture," Crawford is quoted as saying. Military prosecutors attempted to file new charges without using statements made during the interrogation, but Crawford said in the interview that she would not allow the case to proceed.

OUTLOOK
Looking for Closure

With the Guantánamo prison camp now slated to be closed, the Pentagon is making available on its Web site the most complete picture of conditions at the facility the government has ever published. The 85-page report by the review team appointed by Defense Secretary Gates in January details everything from the detainees' bedding, clothing and food and water to religious practices, health care and access to lawyers and others.

Despite finding the facility in compliance with humane-treatment requirements of the Geneva Conventions, the report recommends a number of steps "consistent with the approach of Chain of Command to continually enhance conditions of detainment." As examples, the team — headed by Adm. Patrick Walsh, vice chief of naval operations — recommends increasing detainees' opportunities for socialization, improving trust between health providers and detainees and video recording all interrogations.

At some length, the report describes the procedures for force-feeding hunger strikers and concludes the practices comply with international law standards. But the report fails to note — except in a letter from the American Civil Liberties Union attached as an appendix — that some 30 detainees, more than 10 percent of the population, are now on hunger strikes to protest conditions at the camp. Two prisoners, the ACLU says, have been force-fed through their noses since August 2005.

Human-rights groups and lawyers for the detainees rejected the report's conclusions. Susan Havens, a New York City lawyer who has been visiting Guantánamo since 2004, told *The New York Times* that conditions "are worse than they have ever been." The ACLU pronounced the conditions in violation of domestic and international law, and along with Amnesty International and Human Rights First called for a host of specific changes plus monitoring by independent human-rights groups.[24]

As the dispute illustrates, the Obama administration is not yet satisfying the groups that waged seven years of legal and political warfare against the Bush administration's policies on detention and interrogation. Whether or not President Obama succeeds in closing the Guantánamo prison camp by Jan. 22, 2010, the Guantánamo story — in all its ramifications — seems likely to continue, perhaps for years to come.

The administration's increased transparency regarding Guantánamo is apt to result in increased news coverage as detainees are transferred or released to other countries or brought to the United States for trial or detention. Many of the detainees will themselves seek out coverage. When he arrived in Britain this week, ex-detainee Binyam Mohamed issued a statement through the human-rights group Reprieve: "I am not asking for vengeance, only that the truth should be made known, so that nobody in the future should have to endure what I have endured."[25]

Other detainees are less likely to seek attention, but critics of the administration probably will scrutinize the background and biographies of prisoners as they are released and watch for any evidence that any of them turn to anti-U.S. activities. "The Republican Party or at least parts of it are ready, willing and able to jump if some person who is released creates some havoc," says military-law expert Fidell.

Court cases are certain to drag on, repeatedly giving the administration hard choices to adopt or repudiate legal stances the government took under President Bush. The administration may be able to skirt one high-profile case: the habeas corpus appeal by Ali Saleh Kahlah al-Marri, the Qatari arrested as an al Qaeda sleeper agent while in the United States on a student visa. Chesney, the Texas law professor, and other observers speculate

that the government could avert the April 27 arguments at the Supreme Court by indicting him and prosecuting him in a civilian criminal court. Civil cases seeking damages for past conduct, however, are less susceptible to being sidestepped.

The possible transfer of any of the prisoners to U.S. facilities is already stirring opposition from lawmakers or other officials in communities that might be affected. Possible detention facilities in the United States include the U.S. Disciplinary Barracks at Fort Leavenworth, Kansas, the military's only maximum-security prison; Camp Pendleton in California; the Charleston Naval Brig in South Carolina, and the federal Supermax prison in Florence, Colo.

Lawmakers from all four states are raising objections. Both Kansas senators — Republicans Sam Brownback and Pat Roberts — have introduced legislation along with Missouri Republican. Sen. Christopher "Kit" Bond to require a 90-day study before any transfer. Rep. Henry Brown, R-S.C., has a similar bill for his state. Rep. Duncan Hunter, R-Calif., wants to prohibit use of federal funds to transfer detainees to Camp Pendleton, which is near his San Diego-area district. And members of Colorado's congressional delegation had earlier argued that a civilian prison is unsuitable for military purposes.[26]

Meanwhile, the war in Afghanistan could further increase the number of prisoners at Bagram Air Base — and the number of legal challenges. "Afghanistan is still a physical location of actual counter-insurgency where the war is heating up not cooling down," Chesney says.

Any congressional moves to investigate Bush administration policies will also serve to prolong the story and help spotlight Obama policies as well. In addition, legislative proposals to regulate terrorism-related detention, interrogation and surveillance — including preventive detention — could force the administration's hands on some policy areas. But, says Brookings scholar Wittes, "Congress since the war on terror has never been the lead actor and it will not be."

In a somewhat surprising comment, White House counsel Craig left open the possibility of administration support for preventive detention. "It's possible but hard to imagine Barack Obama as the first president of the United States to introduce a preventive-detention law," Craig told the *New Yorker's* Jane Mayer.[27]

Facing innumerable economic issues, Congress is showing no interest so far in revisiting the detention and interrogation issues that sharply divided Democrats and Republicans over the past seven years. But Craig is making clear that the White House understands the administration's actions will be closely watched.

"We don't own the problem — it was created by the previous administration," Craig said in the interview. "But we'll be held accountable for how we handle this."

NOTES

1. For coverage, see Lyle Denniston, "Jawad Torture Case Put on Hold," SCOTUSBlog, Feb. 4, 2009, www.scotusblog.com/wp/?s=jawad.

2. A Webcast of the panel discussion, "The Post-Guantánamo Era: A Dialogue on the Law and Policy of Detention and Counterterrorism," is available at www.utexas.edu/law/news/2009/020309_web cast_post_Guantánamo.html. Other speakers included Bellinger; Stephen Vladeck, a professor at American University College of Law in Washington; and Benjamin Wittes of the Brookings Institution. Quotes in this report from Bellinger and Wittes are from the panel discussion. For background on counterterrorism policies since 9/11, see these *CQ Researcher* reports: Peter Katel, "Homeland Security," Feb. 13, 2009, pp. 129-152; Peter Katel and Kenneth Jost, "Treatment of Detainees," Aug. 25, 2006, pp. 673-696; Peter Katel, "Global Jihad," Oct. 14, 2005, pp. 857-880; Kenneth Jost, "Re-examining 9/11," June 4, 2004, pp. 493-516; Mary H. Cooper, "Hating America," Nov. 23, 2001, pp. 969-992 and David Masci and Kenneth Jost, "War on Terrorism," Oct. 12, 2001, pp. 817-848. See also these *CQ Global Researcher* reports: Robert Kiener, "Crisis in Pakistan," December 2008, pp. 321-348; Sarah Glazer, "Radical Islam in Europe," November 2007, pp. 265-294; and Seth Stern, "Torture Debate," September 2007, pp. 211-236.

3. Benjamin Wittes, *Law and the Long War: The Future of Justice in the Age of Terror* (2008), pp. 72-102; Benjamin Wittes and Zaahira Wyne, "The Current Detainee Population of Guantánamo," Brookings Institution, Dec. 16, 2008 (periodically updated), www.brookings.edu/reports/2008/1216_detainees_wittes.aspx.

4. Department of Defense, "Review Of Department Compliance With President's Executive Order On Detainee Conditions Of Confinement," February 2009, p. 5, App. 18, www.defenselink.mil/pubs/pdfs/REVIEW_OF_DEPARTMENT_COMPLIANCE_WITH_PRESIDENTS_EXECUTIVE_ORDER_ON_DETAINEE_CONDITIONS_OF_CONFINEMENTa.pdf.

5. The case is *Boumediene v. Bush*, 553 U.S. — (June 12, 2008). For an account, see Kenneth Jost, "Guantánamo Detainees Entitled to Habeas Corpus," *Supreme Court Yearbook 2007-2008*.

6. Robert F. Worth, "Freed by U.S., Saudi Becomes a Qaeda Chief," *The New York Times*, Jan. 23, 2009, p. A1.

7. For coverage, see David Morgan, "Pentagon: 61 ex-Guantánamo detainees return to terrorism," Reuters, Jan. 13, 2009.

8. Human Rights First, "How to Close Guantánamo: Blueprint for the Next Administration," August 2008 (updated November 2008), www.human-rightsfirst.org/pdf/080818-USLS-gitmo-blueprint.pdf; Center for Strategic and International Studies, "Closing Guantánamo: From Bumper Sticker to Blueprint," September 2008, www.csis.org/hrs/gtmoreport. See also Human Rights Watch, "Fighting Terrorism Fairly and Effectively," Nov. 16, 2008, www.hrw.org/en/reports/2008/11/16/fighting-terrorism-fairly-and-effectively.

9. Robert F. Worth, "Saudis Issue List of 85 Terrorism Suspects," *The New York Times*, Feb. 4, 2009, p. A5.

10. For coverage, see Abby Goodnough and Scott Shane, "Padilla Is Guilty on All Charges in Terror Trial," *The New York Times*, Aug. 17, 2007, p. A1; Adam Liptak, "A New Model of Terror Trial," *The New York Times*, Aug. 18, 2007, p. A1.

11. CSIS Report, *op. cit.*, pp. 15-16.

12. The cases are *Hamdi v. Rumsfeld*, 542 U.S. 507 (2004); *Rasul v. Bush*, 542 U.S. 466 (2004); *Hamdan v. Rumsfeld*, 548 U.S. 557 (2006); and Boumediene, *op. cit.* For accounts, see respective editions of Kenneth Jost, *Supreme Court Yearbook*, CQ Press.

13. Quoted in Julian E. Barnes, "Review of Guantánamo Detainees Begins," *Los Angeles Times*, Feb. 14, 2009, p. A11.

14. Ratner quoted in Philip Shenon, "Suit to Be Filed on Behalf of 3 Captives," *The New York Times*, Feb. 19, 2002, p. A5; Cheh quoted in Naftali Bendavid, "U.S. illegally holding 3 detainees in Cuba, suit claims; Legal experts say families' lawyers face uphill battles," *Chicago Tribune*, Feb. 20, 2002, p. 3.

15. See Steven Lee Myers, "Bush Decides to Keep Guantánamo Open," *The New York Times*, Oct. 21, 2008, p. A16.

16. Executive Order: Review and Disposition of Individuals Detained at the Guantánamo Bay Naval Base and Closure of Detention Facilities, www.whitehouse.gov/the_press_office/Closure_Of_Guantanamo_Detention_Facilities/, Jan. 22, 2009.

17. See Peter Finn, "Evidence in Terror Cases Said to Be in Chaos," *The Washington Post*, Jan. 14, 2009, p. A8. Additional quotes and background from a follow-up story by Karen De Young and Peter Finn, "Guantánamo Case Files in Disarray," *ibid.*, Jan. 25, 2009, p. A5.

18. See Raymond Bonner, "Detainee to Return to Britain, as Efforts to Prove Torture Claims Continue," *The New York Times*, Feb. 23, 2009, p. A5.

19. "Officials says Italy will not take Gitmo inmates," The Associated Press, Feb. 16, 2009.

20. The U.S. case pending before the Ninth Circuit is *Mohamed v. Jeppesen Dataplan*, Inc., 08-5693. For coverage, see Maura Dolan and Carol J. Williams, "Court urged to deny rendition trial," *Los Angeles Times*, Feb. 10, 2009, p. A10. See also Glenn Greenwald, "Binyam Mohamed, war crimes investigations, and American exceptionalism," *Salon.com*, Feb. 19, 2009.

21. The decision is *Rasul v. Myers*, 06-5209, D.C. Circuit, Jan. 11, 2008, http://pacer.cadc.uscourts.gov/docs/common/opinions/200801/06-5209a.pdf.

22. For a critical account before Mohamed's release, see Glenn Greenwald, "Binyam Mohamed, war crimes investigations, and American exceptionalism," *Salon.com*, Feb. 19, 2009.

23. Bob Woodward, "Detainee Tortured, Says U.S. Official," *The Washington Post*, Jan. 14, 2009, p. A1.

24. Havens quoted in William Glaberson, "Administration Draws Fire for Report on Guantánamo," *The New York Times*, Feb. 24, 2009, p. A13. The ACLU, Amnesty International and Human Rights First letters are included as appendices to the Pentagon report, *op. cit.*

25. Reprieve-UK represents about 30 Guantánamo detainees; Mohamed's statement is available on its Web site: www.reprieve.org.uk/Press_Statement_of_Binyam_Mohamed.htm.

26. Suzanne Gamboa, "Lawmakers: Guantánamo detainees should 'Keep Out,' " The Associated Press, Feb. 2, 2009.

27. Jane Mayer, "The Hard Cases," *The New Yorker*, Feb. 23, 2009, p. 41.

BIBLIOGRAPHY

Books

Cole, David, *Justice at War: The Men and Ideas That Shaped America's War on Terror,* **New York Review Books, 2008.**

A professor at Georgetown University Law Center critically examines the roles played by, among others, Vice President Dick Cheney, attorneys general John Ashcroft and Alberto Gonzales and Justice Department lawyer John Yoo in the formation of the Bush administration's legal policies in the war on terror. Includes chapter notes.

Greenberg, Karen, *The Least Worst Place: Guantánamo's First 100 Days,* **Oxford University Press, 2009.**

This early history of the prison camp at Guantánamo Bay depicts the supplanting of a military commander's liberal policies by more stringent conditions and treatment as ordered by Defense Secretary Donald Rumsfeld. Greenberg is executive director of the Center on Law and Security, New York University School of Law. Includes notes, six-page bibliography.

Marguiles, Joseph, *Guantánamo and the Abuse of Presidential Power,* **Simon & Schuster, 2006.**

This critical account is by one of the lawyers in the Supreme Court case that opened the door to habeas corpus challenges by Guantánamo detainees. Includes notes.

Mayer, Jane, *The Dark Side: The Inside Story of How the War on Terror Turned into a War on American Ideals, Doubleday*, 2008.
A writer for *The New Yorker* provides a detailed, critical account of the Bush administration's policies on detention, interrogation and surveillance. Includes notes, nine-page bibliography.

Wittes, Benjamin, *Law and the Long War: The Future of Justice in the Age of Terror, Penguin*, 2008.
A legal scholar at the Brookings Institution argues in this influential, ideology-crossing book for new bodies of law — to be crafted by Congress and the executive — dealing with detention, interrogation, trial and surveillance in the new national security environment in "the age of terror."

Worthington, Andy, *The Guantánamo Files: The Stories of the 774 Detainees in America's Illegal Prison, Pluto Press*, 2007, www.andyworthington.co.uk/.
An avowedly leftist British journalist gives detailed accounts of the experiences of prisoners held at Guantánamo and at Bagram Air Base in Afghanistan, relating disturbing allegations of mistreatment and intimidation. Includes detailed notes. Worthington updates his coverage on his Web site: www.andyworthington.co.uk/. For first-person accounts by former detainees, see Moazzam Begg with Victoria Brittain, *Enemy Combatant: My Imprisonment at Guantánamo, Bagram, and Kandahar* (New Press, 2006) and Murat Kurnaz with Helmut Kuhn, *Five Years of My Life: An Innocent Man at Guantánamo* (Palgrave/Macmillan, 2008).

Articles

Chandrasekaran, Rajiv, "From Captive to Suicide Bomber," *The Washington Post*, Feb. 22, 2009, p. A1; "A 'Ticking Time Bomb' Goes Off," *ibid.*, Feb. 23, 2009, p. A1.
The two-part story traces the story of Abdallah al-Ajmi from his capture in Afghanistan and nearly four-year imprisonment at Guantánamo through his release to his native Kuwait and his death as a "suicide bomber" in Iraq in an attack on an Iraqi outpost that killed 13 Iraqi soldiers.

Toobin, Jeffrey, "Camp Justice," *The New Yorker*, April 14, 2008, p. 32.
The CNN legal affairs correspondent provides a close look at the court facilities at Guantánamo — built for military commission proceedings that President Obama suspended as part of his review of detainees' cases and his plan to close the prison camp by 2010.

Reports and Studies

Garcia, Michael John, *et al.*, "Closing the Guantánamo Detention Center: Legal Issues," *Congressional Research Service*, Jan. 22, 2009, http://assets.opencrs .com/rpts/R40139_20090122.pdf.
The 37-page, carefully annotated report thoroughly covers the legal background and current legal issues relating to the closing of the Guantánamo detention center.

Prieto, Daniel B., "War About Terror: Civil Liberties and National Security After 9/11," *Council on Foreign Relations,* February 2009, www.cfr.org/publication/ 18373/.
The 116-page "working paper" by an adjunct fellow at the Council on Foreign Relations comprehensively examines civil liberties issues in regard to post-9/11 national security policies. The working paper is based on work by a task force composed of more than two dozen members that — according to the council's president — "was unable to agree on a set of meaningful conclusions" on the issues discussed.

Wittes, Benjamin, and Zaahira Wyne, "The Current Detainee Population of Guantánamo," *Brookings Institute*, Dec. 16, 2008 (periodically updated), www .brookings.edu/reports/2008/1216_detainees_wittes .aspx.
The site provides the most up-to-date information on the Guantánamo detainees.

On the Web

The Miami Herald has provided comprehensive coverage of Guantánamo and compiled much of that coverage on a continuously updated section of its Web site: www.miamiherald.com/Guantánamo/.

For More Information

Center for Constitutional Rights, 666 Broadway, 7th Floor, New York, NY 10012; (212) 614-6464; www .ccrjustice.org. Dedicated to advancing and protecting rights outlined in the Constitution and Universal Declaration of Human Rights.

Center for Strategic and International Studies, 1800 K St., N.W., Washington, DC 20006; (202) 887-0200; www .csis.org. Provides insights and policy solutions to decision-makers in government, international institutions and the private sector.

Constitution Project, 1200 18th St., N.W., Suite 1000, Washington, DC 20036; (202) 580-6920; www .constitutionproject.org. Bipartisan organization working towards consensus on controversial legal and constitutional issues.

Council on Foreign Relations, 58 E. 68th St., New York, NY 10065; (212) 434-9400; www.cfr.org. Think tank promoting a better understanding of foreign policy choices facing the United States and the world.

Foundation for Defense of Democracies, P.O. Box 33249, Washington, DC 20033; (202) 207-0190; www .defenddemocracy.org. Nonpartisan policy institute dedicated to promoting pluralism and defending democratic values.

Human Rights First, 333 Seventh Ave., 13th Floor, New York, NY 10001; (212) 845-5200; www.humanrights first.org. Protects refugees fleeing prosecution through advocacy efforts at national and international levels.

Human Rights Watch, 350 Fifth Ave., 34th Floor, New York, NY 10118; (212) 290-4700; www.hrw.org. Advocates for the rights of Guantánamo detainees.

National Institute of Military Justice, 4801 Massachusetts Ave., N.W., Washington, DC 20016; (202) 274-4322; www .nimj.org. Works to advance fair administration of military justice and improve public understanding of the military justice system.

Middle-Class Squeeze

Is More Government Aid Needed?

Thomas J. Billitteri

5

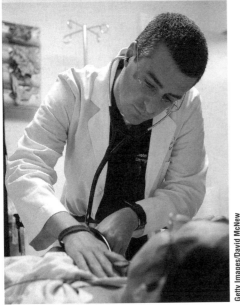

Getty Images/David McNew

Affordable health care for all Americans is a key element of the budget recently announced by President Barack Obama, along with other policies aimed squarely at helping the middle class. Nearly half of home foreclosures in 2006 were caused, at least partly, by financial issues stemming from a medical problem, according to the advocacy group Families USA. Above, emergency room physician Jason Greenspan cares for a patient in Panorama City, Calif.

From *CQ Researcher*, March 6, 2009.

Cindy Dreeszen, 41, and her husband may have seemed like unlikely visitors to the Interfaith food pantry last month in affluent Morris County, N.J., 25 miles from New York City. Both have steady jobs and a combined income of about $55,000 a year. But with "the cost of everything going up and up" and a second baby due, the couple was looking for free groceries.

"I didn't think we'd even be allowed to come here," Ms. Dreeszen told *The New York Times.* "This is totally something that I never expected to happen, to have to resort to this."[1]

Countless middle-class Americans are thinking similar thoughts these days as they ponder their suddenly fragile futures.

Millions of families who once enjoyed the American dream of upward mobility and financial security are sliding rapidly down the economic ladder — some into poverty. Many are losing their homes along with their jobs, and telling their children to rethink college.[2] And while today's economic crisis has made life for middle-class households worse, the problems aren't new. Pressure on the middle-class has been building for years and is likely to persist long after the current recession — now 14 months old — is over.

The middle class "is in crisis and decline," says sociologist Kevin Leicht, director of the Institute for Inequality Studies at the University of Iowa.

"Between wages that have been stagnant [in inflation-adjusted terms] since the middle of the 1970s and government policies that are weighted exclusively in the direction of the wealthy, the only thing that has been holding up most of the American middle class is access to cheap and easy credit."

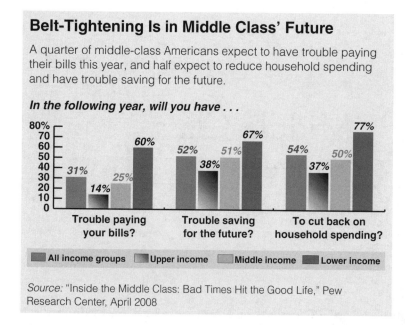

Belt-Tightening Is in Middle Class' Future

A quarter of middle-class Americans expect to have trouble paying their bills this year, and half expect to reduce household spending and have trouble saving for the future.

In the following year, will you have . . .

Trouble paying your bills?
- All income groups: 31%
- Upper income: 14%
- Middle income: 25%
- Lower income: 60%

Trouble saving for the future?
- All income groups: 52%
- Upper income: 38%
- Middle income: 51%
- Lower income: 67%

To cut back on household spending?
- All income groups: 54%
- Upper income: 37%
- Middle income: 50%
- Lower income: 77%

Legend: All income groups | Upper income | Middle income | Lower income

Source: "Inside the Middle Class: Bad Times Hit the Good Life," Pew Research Center, April 2008

No official definition of the "middle class" exists. (*See sidebar, p. 104.*) But most Americans — except perhaps the very richest and poorest — consider themselves in that broad category, a fact not lost on Washington policy makers.

Indeed, President Barack Obama announced a 10-year budget on Feb. 28 that takes direct aim at the challenges facing America's middle class and the growing concentration of wealth at the top of the income scale.[3] Key elements of the plan include shifting more costs to the wealthiest Americans and overhauling health care to make it more affordable.[4]

In further recognition of the importance of the middle-class, Obama has named Vice President Joseph R. Biden to chair a new White House Task Force on Middle Class Working Families. It will examine everything from access to college and child- and elder-care issues to business development and the role of labor unions in the economy.[5]

"Talking about the middle class is the closest that American politicians and maybe Americans are willing to go to emphasize the fact that we have growing inequality in this country," says Jacob Hacker, a political scientist at the University of California, Berkeley, and a leading social-policy expert. "A very small proportion of the population is getting fabulously rich, and the rest of Americans are getting modestly richer or not much richer at all."

What's at stake goes far beyond economics and family finances, though, experts say. "A large middle class, especially one that is politically active, tends to be a kind of anchor that keeps your country from swinging back and forth," says sociologist Teresa Sullivan, provost and executive vice president for academic affairs at the University of Michigan and co-author of *The Fragile Middle Class: Americans in Debt.* What's more, she says, "there are typical values that middle-class families acquire and pass on to their children," and those values "tend to be very good for democracy."

Right now, though, the middle class is under threat.

In a study of middle-class households, Demos, a liberal think tank in New York, estimated that 4 million families lost their financial security between 2000 and 2006, raising the total to 23 million. Driving the increase, Demos said, were declines in financial assets, then-rising housing costs and a growing lack of health insurance.[6]

"In America the middle class has been a lifestyle, a certain way of life," says Jennifer Wheary, a co-author of the study. "It's been about being able to have a very moderate existence where you could do things like save for your retirement, put your kids through school, get sick and not worry about getting basic care. And those kinds of things are really imperiled right now."

In another study, the Pew Research Center found this year that "fewer Americans now than at any time in the past half-century believe they're moving forward in life."[7]

Among the findings:

• Nearly two-thirds of Americans said their standard of living was higher than that of their parents at the same age, but more than half said they'd either made no progress in life over the past five years or had fallen backward.

• Median household income rose 41 percent since 1970, but upper-income households outperformed those in the middle tier in both income gains and wealth accumulation. The median net worth of upper-income families rose 123 percent from 1983 to 2004, compared with 29 percent for middle-income families.

• Almost eight in 10 respondents said it was more difficult now for those in the middle class to maintain their standard of living compared with five years ago. In 1986, 65 percent felt that way.

Lane Kenworthy, a sociology and political science professor at the University of Arizona who studies income inequality and poverty, says "the key thing that's happened" to the middle class over the past three decades "is slow income growth compared to general economic growth." Moreover, Kenworthy says a bigger and bigger portion of economic growth has accrued to the wealthiest 1 percent, whether the measure is basic wages or total compensation, which includes the value of employee-sponsored and government benefits.

Even the economic boom leading up to today's recession has proved illusory, new Federal Reserve data show. While median household net worth — assets minus debt — rose nearly 18 percent in the three years ending in late 2007, the increase vanished amid last year's drastic declines in home and stock prices, according to the Fed's triennial "Survey of Consumer Finances." "Adjusting for those declines, Fed officials estimated that the median family was 3.2 percent poorer as of October 2008 than it was at the end of 2004," *The New York Times* noted.[8]

A hallmark of middle-class insecurity reflects what Hacker calls "the great risk shift" — the notion that government and business have transferred the burden of providing affordable health care, income security and retirement saving onto the shoulders of working Americans, leaving them financially stretched and vulnerable to economic catastrophe.

"Over the last generation, we have witnessed a massive transfer of economic risk from broad structures of insurance, including those sponsored by the corporate

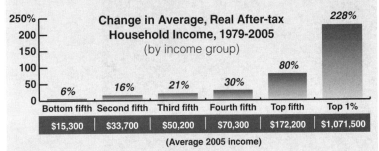

Income Gap Getting Wider

The gap between the wealthiest Americans and everybody else grew to its widest point since at least 1979.* The top 1 percent of households received 70 times as much in average after-tax income as the bottom one-fifth and 21 times as much as the middle one-fifth — in both cases the widest gaps on record. From 1979-2005, the top 1 percent saw its income rise 228 percent compared to a rise of only 21 percent for the middle one-fifth of Americans.

Change in Average, Real After-tax Household Income, 1979-2005
(by income group)

	Bottom fifth	Second fifth	Third fifth	Fourth fifth	Top fifth	Top 1%
Change	6%	16%	21%	30%	80%	228%
Average 2005 income	$15,300	$33,700	$50,200	$70,300	$172,200	$1,071,500

* Data go back only to 1979.

Source: Arloc Sherman, "Income Inequality Hits Record Levels, New CBO Data Show," Center on Budget and Policy Priorities, December 2007

sector as well as by government, onto the fragile balance sheets of American families," Hacker wrote. "This transformation . . . is the defining feature of the contemporary American economy — as important as the shift from agriculture to industry a century ago."[9]

The challenge of solving the problems facing the American middle class will confront policy makers for years to come. Some experts say the key is growth in good jobs — those with good pay, good benefits and good, secure futures. Others argue that solving the nation's health-care crisis is the paramount issue.

One thing is certain, experts say: Leaving the fate of the American middle class to chance is not an option.

"We're believers in hard work, and we're increasingly in a situation where the difference between whether or not a middle-class family prospers comes down to luck, says Amelia Warren Tyagi, co-author of *The Two-Income Trap: Why Middle-Class Mothers and Fathers Are Going Broke.* "And that's an idea that makes us really uncomfortable."

Here are some of the questions that policy makers and average Americans are asking about the middle class:

Getty Images/William Thomas Cain

Vice President Joseph Biden, chair of the new White House Task Force on Middle Class Working Families, listens to a presentation on creating "green" jobs at the University of Pennsylvania on Feb. 27, 2009. President Obama directed the panel to examine issues such as access to college, business development and the role of labor unions in the economy.

Is a stable middle class a thing of the past?

First lady Michelle Obama remembers what some call the good old days of middle-class security.

"I am always amazed," she told a gathering, "at how different things are now for working women and families than when I was growing up. . . . When I was growing up, my father — as you know, a blue-collar worker — was able to go to work and earn enough money to support a family of four, while my mom stayed home with me and my brother. But today, living with one income, like we did, just doesn't cut it. People can't do it — particularly if it's a shift-worker's salary like my father's."[10]

Brookings Institution researchers noted in 2007 that two-thirds of American adults had higher family incomes than their parents did in the late 1960s and early '70s, but a third were worse off. Moreover, they pointed out,

the intergenerational gains largely stemmed from dual paychecks in families.[11]

"Men's earnings have grown little, if at all," while those of women "have risen along with their greater involvement in the work world," they said. "So, yes, today's families are better off than their own parents were. . . . But they are also working more and struggling with the greater time pressures of juggling work and family responsibilities."[12]

At the same time, many economists say the earnings of middle-class working families have not kept pace with gains by the wealthy. They point to Congressional Budget Office data showing that from 1979 through 2005, the average after-tax income of the top 1 percent rose 228 percent, compared with 21 percent for the middle fifth of the population. For the poorest fifth, the increase during the 25-year period was just 6 percent.[13]

Emmanuel Saez, an economist at the University of California, Berkeley, concluded last year that those in the top 1 percent of income distribution captured roughly half of the overall economic growth from 1993 to 2006, and almost three-fourths of income growth in the 2002-2006 period.[14]

"It's the very top of the economic ladder that's pulled away from the rest," Berkeley political scientist Hacker says. "Depending on which source you look at, it's the top 1 percent, or the top one-half of 1 percent, or the top one-tenth of 1 percent that's really received the lion's share of the gain in our economy overall. . . . It would be one thing if we saw middle-class Americans hold onto or even expand their wealth and economic security. But they're more in debt and less secure than they were 20 years ago."

The reasons the middle class is running in place or falling behind can be elusive, though. Kenworthy of the University of Arizona cites a litany of factors — technological changes in the workplace, globalization of trade and the outsourcing of jobs overseas, declining influence of labor unions, slow growth in the proportion of workers with at least a high-school diploma and a stagnant minimum wage — that have helped dampen the economic progress of the middle class. But, he says, social scientists and economists don't have a "good handle on which matter most."[15]

John Schmitt, senior economist at the Center for Economic and Policy Research, a liberal think tank, disputes the notion that technology and globalization are immutable forces that have, in themselves, hurt the middle class. "We've had technological growth at a rapid pace in the United States from the early 1800s," he says, and after World War II the country saw "massive technological innovation," including the introduction of computers.

Those "were huge, potentially disruptive innovations, but we had a social structure that had a lot of protections and guarantees for workers," including "a decent minimum wage, significant union representation" and a strong regulatory framework.

"The real story is that we've made a lot of decisions about economic policy that have had the effect of shifting the playing field toward employers and away from workers at a whole lot of levels," Schmitt says.

As that shift occurred, job security has suffered, many economists say.

In a recent study, Schmitt found that the share of "good jobs" — those paying at least $17 per hour and offering health insurance and a pension — declined 2.3 percentage points in the 2000-2006 business cycle, far more than in comparable periods in the 1980s and '90s. A "sharp deterioration" in employer-provided health plans was a "driving force" in the decline of good jobs, which was most pronounced among male workers, he found.[16]

Meanwhile, career employment — employment with a single employer from middle age to retirement — is no longer the norm, according to researchers at Boston College. Only half of full-time workers ages 58 to 62 are still with the same employer for whom they worked at age 50, they found.[17]

And manufacturing — long a bedrock of middle-class lifestyles — has shrunk from about a third of non-farm employment to only 10 percent since 1950.[18]

Still, interpretations of income and other economic data can vary widely among economists, depending on their political viewpoint. While not diminishing the severe pressures many in the middle class are feeling right now, some conservative economists have a more optimistic view of the jobs issue and long-term middle-class gains in general.

In a study last year, James Sherk, Bradley Fellow in Labor Policy at the Heritage Foundation, challenged the notions "that the era of good jobs is slipping away" and that workers' benefits are disappearing.[19]

"Throughout the economy, jobs paying high wages in fields requiring more education are more available today than they were a generation ago, while low-wage, low-skill jobs are decreasing," he wrote. And, he added, "employer-provided health insurance and pensions are as available now as they were in the mid-1990s. Worker pension plans have improved significantly, with most employers shifting to defined-contribution pensions that provide workers with more money for retirement and do not penalize them for switching jobs."

In an interview, Sherk said that while many middle-class families are struggling today, over the long term they have not, on average, fallen behind overall growth in the economy. Average earnings have risen in step with productivity, he said.

But others are not sanguine about the status of the middle class, long-term or otherwise.

"For quite some time, we've had a sizable minority of the middle class under enormous strain and on the verge of crisis, and since the recent meltdown the proportion of middle-class families in crisis increased exponentially," says Tyagi, who co-authored *The Two-Income Trap* with her mother, Harvard law Professor Elizabeth Warren, chair of a congressional panel overseeing last fall's $700 billion financial bailout. "Many families teetering on the uncomfortable edge have been pushed over."

Is overconsumption at the root of the middle class' problems?

In a recent article about the collapse of the Florida real estate market, *New Yorker* writer George Packer quotes a woman in Cape Coral who, with her husband, had built a home on modest incomes, borrowed against its value, spent some of the money on vacations and cruises, and then faced foreclosure after her husband was laid off.

"I'm not saying what we did was perfect," the woman said. "We spent our money and didn't save it. But we had it, and we didn't see that this was going to happen."[20]

Such vignettes are commonplace these days as the economy plummets and home foreclosures soar. So, too, is the view that many middle-class consumers brought trouble to their own doorsteps by overconsuming and failing to save.

Thomas H. Naylor, a professor emeritus of economics at Duke University and co-author of *Affluenza: The All-Consuming Epidemic*, says the vulnerability of the

Being Middle Class Takes More Income

The minimum income needed for a three-person household to be considered in the middle class was about 40 percent higher in 2006 than in 1969.

Economic Definition of Middle-class Household of Three
(in constant January 2008 dollars)

Year	Income
1969	$31,755 to $63,509
1979	$37,356 to $74,712
1989	$41,386 to $82,771
1999	$45,920 to $91,841
2006	$44,620 to $89,241

Source: "Inside the Middle Class: Bad Times Hit the Good Life," Pew Research Center, April 2008

middle class has been "enhanced by [its] behavior." He blames both consumer excess and the influence of advertising and media.

"On the one hand, consumers have done it to themselves. They've made choices to spend the money," Naylor says. "On the other hand, they've had lots of encouragement and stimulation from corporate America. The big guns are aimed at them, and it's very difficult to resist the temptation."

Pointing to the Federal Reserve's recent "Survey of Consumer Finances," Nobel laureate and *New York Times* economic columnist Paul Krugman wrote that the fact that "the net worth of the average American household, adjusted for inflation, is lower now than it was in 2001" should, at one level, "come as no surprise.

"For most of the last decade America was a nation of borrowers and spenders, not savers. The personal savings rate dropped from 9 percent in the 1980s to 5 percent in the 1990s, to just 0.6 percent from 2005 to 2007, and

household debt grew much faster than personal income. Why should we have expected net worth to go up?"

But, Krugman went on to say, until recently Americans thought they were getting wealthier, basing their belief on statements saying their homes and stock portfolios were appreciating faster than the growth of their debts.[21]

In fact, many economists say the picture of consumer behavior and household savings is far more complex than simple theories of overconsumption suggest.

President Obama weighed in at a press conference in early February, saying, "I don't think it's accurate to say that consumer spending got us into this mess." But he added that "our savings rate has declined, and this economy has been driven by consumer spending for a very long time. And that's not going to be sustainable."

Schmitt, of the Center for Economic and Policy Research, contends that what has hurt the middle class the most are steep cost increases of necessities, not spending on luxuries. "There's a lot of argument about overconsumption, but my argument is that consumption of basic necessities is not subject to big price savings," he says. "Housing, education, health care — those are much more expensive than they used to be. That's where people are feeling the pinch."

Housing prices doubled between the mid-1990s and 2007.[22] Average tuition, fees and room-and-board charges at private four-year institutions have more than doubled since 1978-79, to $34,132.[23] And growth in national health expenditures has outpaced gross national product (GNP) growth every year at least since the late 1990s.[24]

One study found that among adults earning $40,000 to $60,000, the proportion of adults spending 10 percent or more of their income on health care doubled between 2001 and 2007, from 18 percent to 36 percent.[25]

"Health care is the epicenter of economic security in the United States today," says Hacker, the University of California political scientist. "It's not the only thing impinging on families finances, but it's one of the areas where the need is greatest."

Economist Robert H. Frank, author of *Falling Behind: How Rising Inequality Harms the Middle Class*, argues that as the wealthiest Americans have acquired bigger and more expensive houses and luxury possessions, their behavior has raised the bar for middle-class consumers, leading them to spend more and more of their incomes on bigger houses and upscale goods.

While some of the spending may be frivolous, he says, many consumers have felt compelled to keep up with rising economic and cultural standards — and often for practical reasons: Bigger, more expensive homes typically are in neighborhoods with the best schools, and upscale clothing has become the norm for those who want to dress for success.

"There are people you could say have brought this on themselves," Frank says of the troubles middle-class families are now facing. "If you've charged a bunch of credit cards to the max [for things] that aren't really essential, is that your fault? You bet. But most of it I don't think is. You need a decent suit to go for a job interview. You can buy the cheap suit, but you won't get the call-back. You can break the rules at any turn, but there's a price for that."

In their book on two-income middle-class families, Tyagi and Warren attacked the "rock-solid" myth that "middle-class families are rushing headlong into financial ruin because they are squandering too much money on Red Lobster, Gucci and trips to the Bahamas."[26]

In fact, they wrote, after studying consumer bankruptcy data and other sources, "Today, after an average two-income family makes its house payments, car payments, insurance payments and child-care payments, they have less money left over, even though they have a second, full-time earner in the workplace," than an average single-earner family did in the early 1970s.[27]

One-paycheck households headed by women are among the most vulnerable. In an analysis of 2004 Federal Reserve Board data, the Consumer Federation of America found that the 31 million women who head households had median household income of $22,592, compared with $43,130 for all households. And women on their own had a median net worth of less than $33,000 compared with about $93,000 for all households.[28]

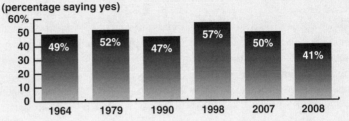

Fewer Americans Say They Are Better Off

The percentage of Americans who said they were better off in 2008 than they were five years earlier dropped to 41 percent in 2008, the lowest confidence level since 1964.

Are you better off now than you were five years ago?

(percentage saying yes)

Source: "Inside the Middle Class: Bad Times Hit the Good Life," Pew Research Center, April 2008

Are aggressive new government programs needed to bolster the middle class?

Last year, former Republican Rep. Ernest Istook of Oklahoma criticized then-presidential candidates Hillary Clinton and Obama for arguing that "America is a place where the middle class is repressed" by rising income inequality, stagnating wages, soaring medical and college costs and other woes.

"For both candidates," wrote Istook, a Heritage Foundation fellow, "the answer to all these problems is a rush of new government programs." He pointed to Heritage Foundation studies arguing that wage-growth data have been understated and that the poor are doing better than they were 14 years earlier.

"Convincing Americans that they need government to do all these things," he wrote, "hinges on convincing them that they are victims in need of rescue. . . . It's not enough for America's left to show sympathy for victims of real tragedies like 9/11 or Katrina. Now they must elevate every challenge into a crisis, provoking a sense of desperation that more and bigger government is the answer."[29]

Yet that is not how many policy advocates view the question of government help for the middle class. The pressures weighing on working families — heightened by the current economic crisis — are so great, they argue, that bold government action is needed to keep working Americans from further economic harm.

"We talk about the big financial institutions as too big to fail," says University of California political scientist Hacker. "But most Americans have until recently been apparently viewed as too small to save."

Without policy changes, including ones that make education and health care more affordable and help people build assets, "instability is going to stay," argues Wheary of Demos.

Yet, while the needs of the middle class are a favorite rhetorical device for politicians, they often disagree about the best way to advance those interests. This year's $787 billion stimulus package, which emerged from a cauldron of partisan bickering, is a case in point.

President Obama, speaking to employees of Caterpillar Inc. in February, said the stimulus plan is "about giving people a way to make a living, support their families and live out their dreams. Americans aren't looking for a handout. They just want to work."[30] But Rep. John A. Boehner of Ohio, a key Republican opponent of the president's recovery plan, said it "will do little to create jobs, and will do more harm than good to middle-class families and our economy."[31]

An overhaul of health-care policy is a key priority for many policy experts. Families USA, an advocacy group supporting affordable health care, pointed to research showing that nearly half of home foreclosures in 2006 were caused, at least partly, by financial issues growing out of a medical problem.[32]

Also key, many liberal policy analysts say, is solving what they see as a growing pension crisis, made more perilous for middle-class workers by the Wall Street crash. (*See sidebar, p. 106.*) Rep. George Miller, D-Calif., chairman of the House Education and Labor Committee, says private retirement-savings vehicles like 401(k) plans "have become little more than a high-stakes crap shoot. If you didn't take your retirement savings out of the market before the crash, you are likely to take years to recoup your losses, if at all."[33]

And crucial to the future of the middle class, many experts say, are sound policies for job creation and retention.

"The major policy change we need is to decide that good steady jobs with good wages are a family value," says Leicht of the University of Iowa. "It's good jobs at good wages that last — that's the Rosetta Stone."

Leicht says "our entire system of consumption is built around the idea that you accumulate a lot of debts when you're young, then you get a steady job and your income steadily rises and you gradually pay off your debt as you age." But nowadays, he says, the average job lasts only four to five years. "If you're constantly starting over, you never get out of the hole."

Leicht wants to see a 25 percent break on corporate taxes for businesses that create "high-quality jobs" — ones lasting at least five years and paying at least 30 percent above the median income of a family of four, which in 2007 was $75,675, according to the U.S. Census Bureau.

Kenworthy, the University of Arizona sociologist, advocates temporary "wage insurance" that would "prop up your earnings for a little while if you lost your job and took a new one that paid considerably less."

Not counting the current economic crisis, Kenworthy says, "there really isn't a problem in the United States with long-term unemployment. Most people are able to get a job within six months." Even so, he adds, such jobs often come "at a lower salary."

BACKGROUND

Evolving Concept

During the 2008 presidential campaign, the Rev. Rick Warren, pastor of giant Saddleback Church in Lake Forest, Calif., asked Democrat Obama and Republican John McCain to define "rich."

Obama said that "if you are making $150,000 a year or less as a family, then you're middle class, or you may be poor. But $150 [thousand] down you're basically middle class." He added, though, that "obviously, it depends on [the] region and where you're living." McCain answered the question another way, saying — perhaps with tongue in cheek — that as a definition of rich, "if you're just talking about income, how about $5 million?"[34]

Besides helping to open a window on the candidates' views and personalities, the exchange underscored how highly subjective social and economic class can be.

That's nothing new. For centuries, the concept of a "middle class" has been evolving.

"The middle class first came into existence in early modern Europe as a new social class for which the economic basis was financial rather than feudal — the system in which the nobility owned land and others (serfs, peons) worked it," according to Andrew Hoberek, an associate professor of English at the University of Missouri, Columbia, and author of *The Twilight of the Middle Class: Post World War II American Fiction and White-Collar Work.*[35]

In the United States, the term "middle class" didn't start showing up until the 1830s or 1840s, says Jennifer L. Goloboy, an independent scholar.[36] But years earlier, she says, a segment of the population began to embrace values that would come to define the American middle class, including diligence, frugality, self-restraint and optimism.

"The early republic was such an aspirational time, and it was disproportionately young," Goloboy says. "These young people came to the cities hoping for the best, and they clung to ideas of how they would make it. That's sort of the root of middle-class values. They believed that if they held to these values they were middle class, even if they were not necessarily successful yet."

As the American economy matured in the 20th century, industrialization both nurtured and threatened the nation's budding middle class. Pioneering automaker Henry Ford helped nurture it by paying high wages and encouraging mass consumption of his cars. But the gap between rich and poor remained wide, and industrialization made life precarious for the working class when jobs disappeared.

"The paramount evil in the workingman's life is irregularity of employment," Supreme Court Justice Louis D. Brandeis wrote in 1911.[37] Historian David Kennedy noted that Brandeis' view "was echoed in Robert and Helen Merrell Lynd's classic study *Middletown* a decade later, when they cited 'irregularity of employment' as the major factor that defined the difference between the life trajectories of the working class and the middle class."[38]

During the Great Depression of the 1930s, unemployment soared to 25 percent, and many Americans fell from middle-class stability into destitution. But from the ashes of the Depression came President Franklin D. Roosevelt's New Deal program, which *New York Times* columnist Krugman says created the modern middle class.[39]

"Income inequality declined drastically from the late 1930s to the mid-1940s, with the rich losing ground while working Americans saw unprecedented gains," he wrote.[40]

Consumerism at Its Finest

Some economists say the higher cost of necessities like health care, rather than spending on luxury items like big-screen TVs or new cars, has hit consumers hardest. Moreover, Americans' personal savings rate from 2005 to 2007 was just 0.6 percent — down from 9 percent in the 1980s — with household debt growing faster than personal income.

The New Deal "made America a middle-class society," Krugman wrote this year in *Rolling Stone* magazine. "Under FDR, America went through what labor historians call the Great Compression, a dramatic rise in wages for ordinary workers that greatly reduced income inequality. Before the Great Compression, America was a society of rich and poor; afterward it was a society in which most people, rightly, considered themselves middle class."[41]

After World War II, the U.S. economy blossomed, aided by the GI Bill, which helped millions of former service members buy homes and get college educations. In 1946, construction began on Levittown, one of a series of massive housing developments that became national models of middle-class suburbia.

The postwar boom helped spawn the contemporary notion of the American Dream — a home, a car or two (or three), a good job, paid vacation and a comfortable suburban lifestyle. By 1960, median family income was 30 percent higher in purchasing power than a decade earlier, and more than 60 percent of homes were owner-occupied, compared with 44 percent just before World War II.[42]

Downward Slide

But many economists say the good times began to wane in the 1970s, and for a variety of reasons that can be difficult to untangle. The shift away from manufacturing toward a service economy helped erode middle-class security, as did the increasingly competitive nature of globalization, many economists say. Some also cite the declining power of unions. In 1979, 27 percent of employed wage-and-salary workers in the United States were covered by a collective bargaining agreement, but that figure has steadily declined over the years. It stood at less than 14 percent in 2008.[43]

In remarks tied to formation of his middle-class task force, Obama said, "I do not view the labor movement as part of the problem; to me it's part of the solution. We need to level the playing field for workers and the unions that represent their interest, because we know that you cannot have a strong middle class without a strong labor movement."[44]

Hacker, the University of California political scientist, says that "employers at one time were encouraged by unions, the federal tax code and their own competitive instincts to provide very strong guaranteed benefits to many of their workers in the form of defined-benefit pension plans [and] good health insurance coverage."

But, he says, "over the last generation the work force has changed, and the competitive environment in which employers have operated changed in ways that have made it much less attractive for many employers to provide such

benefits. There used to be a kind of implicit long-term contract in many workplaces, enforced in part by unions, that is no longer there. So it's much more of a free-agent economic culture, which means that it's good for some workers but imposes a lot more risk on all of them."

Many conservatives disagree, though, on the role of unions in helping the middle class. "Numerous studies have shown that unions are not the answer to increasing prosperity for American workers or the economy," the U.S. Chamber of Commerce stated in a paper on the issue. It added: "Organized labor's claims that unionization is a ticket to the middle class cannot be squared with data showing that increased unionization decreases competitiveness and leads to slower job growth."[45]

Besides the issue of union influence, critics often cite Reagan-era economic policies, which included cuts in tax rates for those in upper-income brackets, as contributing to inequality and hurting the middle class.

The criticism is not universal. George Viksnins, a professor emeritus of economics at Georgetown University, argues that so-called Reaganomics was a plus for the middle class. "Perhaps the most significant positive aspect of the Reaganomics program of lower taxes and regulatory reforms is the tremendous increase in employment," he wrote.[46] In an interview, he said that "lowering marginal tax rates held out a lot of hope for young members of the middle class that they might get to keep some of the income" they earned "and didn't need to work quite as hard in sheltering it."

But others see the Reagan years differently. "Yes, there was a boom in the mid-1980s, as the economy recovered from a severe recession," Krugman, the Nobel economist and *Times* columnist, wrote. "But while the rich got much richer, there was little sustained economic improvement for most Americans. By the late 1980s, middle-class incomes were barely higher than they had been a decade before — and the poverty rate had actually risen."[47]

The University of Iowa's Leicht is highly critical of another legacy of the 1980s: deregulation of the banking industry, which he says set the stage for a massive increase in easy credit. The explosion in consumer lending that began in the 1980s helped millions of working Americans buy homes and cars, Leicht acknowledges, but he says the credit binge has come back to haunt the middle class now as home-foreclosure rates and personal bankruptcies soar.

C H R O N O L O G Y

1800-1929 *Industrial age shifts employment from farm to factory, setting stage for rise of middle class.*

October 1929 Stock market crash marks end of a speculative bubble on Wall Street.

1930-1970 *Great Depression sends unemployment soaring, President Roosevelt crafts New Deal social and economic legislation and postwar boom spurs growth of middle class.*

1933 Unemployment rate reaches 25 percent; Congress passes flood of New Deal legislation.

1935 President Franklin D. Roosevelt signs Social Security Act into law.

1939 Food Stamp program starts.

1944 Roosevelt signs Servicemen's Readjustment Act, or GI Bill, into law; by 1952, the law backed nearly 2.4 million home loans for World War II veterans, and by 1956 nearly 8 million vets had participated in education or training programs.

1946 Construction starts on New York's Levittown, one of three low-cost post-World War II residential communities that would come to define middle-class suburbia.

1960 Median family income is 30 percent higher in purchasing power than a decade earlier, and more than 60 percent of homes are owner-occupied, compared with 44 percent just before World War II.

1970-1995 *Oil shocks, inflation, foreign competition, and other changes mark tougher era for middle-class Americans.*

1979 U.S. manufacturing employment peaks at 21.4 million workers.

1981 President Ronald Reagan fires 11,000 striking members of the Professional Air Traffic Controllers Organization, helping to weaken the power of organized labor; Reagan persuades Congress to pass largest tax cuts in U.S. history.

1981-82 Severe recession rocks U.S. economy, sending the unemployment rate to 10.8 percent, the highest since the Great Depression.

Oct. 19, 1987 Dow Jones Industrial Average loses 23 percent of its value.

1996-Present *Home ownership peaks, and consumer spending soars, but good times end as home values plummet, financial institutions collapse and nation sinks into recession.*

1996 Congress ends 60-year welfare entitlement program, imposing work requirements and putting time limits on cash benefits.

1997 Federal minimum wage raised to $5.15 an hour.

2000 Federal poverty rate falls to 11.3 percent, lowest since 1974.

2001-2006 Housing prices in many cities double, and home-equity loans help lead to soaring consumer spending.

2004 Home-ownership rate peaks at 69 percent.

2008 Federal minimum wage rises to $6.55 an hour; it is set to increase to $7.25 effective July 24, 2009. . . . U.S. seizes Fannie Mae and Freddie Mac, Lehman Brothers files for bankruptcy and Washington Mutual collapses in biggest bank failure in history. . . . President George W. Bush signs $700 billion financial rescue bill but recession deepens.

2009 President Barack Obama announces budget seeking to aid middle class and forms Middle Class Task Force headed by Vice President Joseph Biden; first meeting focuses on "green jobs." . . . Federal unemployment rate rises to 7.6 percent in January (12.6 percent for African-Americans and 9.7 for Hispanics). . . . Labor Department says employers took 2,227 "mass layoff actions" in January, resulting in nearly 238,000 job cuts; from December 2007 through January 2009, mass layoff events totaled more than 25,700. . . . Claims for unemployment benefits exceed 5 million for first time in history. . . . Home foreclosures are reported on 274,399 U.S. properties in January, up 18 percent from January 2008.

What Does 'Middle Class' Really Mean?

Does the definition include income? Number of cars in the garage?

At his first White House press conference, President Barack Obama promised tax relief for "working and middle-class families." But what, exactly, does it mean to be in the "middle class"?

No official definition exists. Politicians, journalists and pundits freely use the term, often without attaching a precise meaning to it. And in opinion polls, most Americans — uncomfortable defining themselves as "rich" or "poor" — place themselves in the category of the middle class, even if their incomes reflect the outer limits of wealth or poverty.

In a report last year, the Pew Research Center noted that the term "middle class" is both "universally familiar" and "devilishly difficult to pin down."

"It is both a social and economic construct, and because these domains don't always align, its borders are fuzzy," Pew said. "Is a $30,000-a-year resident in brain surgery lower class? Is a $100,000-a-year plumber upper middle class?"

In a national survey of more than 2,400 American adults, Pew asked people to define themselves. It found that 53 percent said they were middle class. But, Pew said, "behind the reassuring simplicity of this number lies a nest of anomalies."

For example, it said, 41 percent of adults with annual household incomes of $100,000 or more said they were middle class, as did 46 percent of those with household incomes below $40,000. And of those in between, roughly a third said they were not middle class.

"If being middle income isn't the sole determinant of being middle class, what else is?" Pew added. "Wealth? Debt? Homeownership? Consumption? Marital status? Age? Race and ethnicity? Education? Occupation? Values?"[1]

Christian Weller, an associate professor of public policy at the University of Massachusetts, Boston, and a fellow at the liberal Center for American Progress, says that often, people count the number of cars in a garage or the square footage of a house to judge another person's economic standing. But, he says, "that's not really how people perceive and define middle class. . . . One part of middle class is an aspirational definition: 'I'll be able to send my kids to college, I'll be able to create a better future for my children, and do I have a secure lifestyle right now?'

"That goes beyond just simply having a good job," he says. "That means, do you have health insurance coverage, do you have enough savings, do you own your own home, do you

have retirement savings?" And, Weller adds, "By all those measures middle-class security has been eroding substantially."

Many economists look at the concept of a middle class through the lens of household-income data gathered by the federal government. Median household income was $50,233 in 2007, the latest year for which data are available.[2] That was the midpoint in the distribution, with half of households having more income and half less.[3]

The government also separates household income into five "quintiles," from lowest to highest. Some might consider "middle class" to mean only the third quintile — the one in the very middle — with incomes between $39,101 and $62,000. But many economists consider that view to be too cramped. Some count the third and fourth quintiles, with an upper limit of $100,000 in household income in 2007. Among the broadest definitions of middle class is one encompassing the three income quintiles in the middle, from $20,292 to $100,000.

Of course, using household income to measure the middle class has its own problems. For example, a family might seem solidly middle class based on its income, but parents may be toiling at two jobs each to raise their income level into the middle tier of the distribution tables. They might make good incomes but lack health insurance, putting them and their children at risk of a catastrophic financial collapse. Or they may live in a high-cost region of the country, where a supposed middle-class income of around $50,000 or $60,000 a year simply can't cover the bills.

One thing is certain, say those who have studied the American middle class: Its survival is crucial to the nation's future.

"It is the heart of the country, it's the heart of our democracy, it's the heart of our economy, it's the heart of our population," says Amelia Warren Tyagi, co-author of *The Two-Income Trap: Why Middle-Class Mothers and Fathers Are Going Broke*. "So while it may not be easy to define with precision, it's extremely important."

---

[1] Paul Taylor, et al., "Inside the Middle Class: Bad Times Hit the Good Life," Pew Research Center, April 9, 2008, p. 3, http://pewsocialtrends.org/assets/pdf/MC-Middle-class-report.pdf.

[2] U.S. Department of Commerce, Bureau of the Census, "Historical Income Tables — Households," www.census.gov/hhes/www/income/histinc/h05.html.

[3] In 2007, the United States had about 116,783,000 households.

"Starting in about the mid-1980s, we decided as a nation, through a number of mechanisms, that being loaned money was a perfect substitute for being paid it as long as you could buy things that represented middle-class status like houses and cars," Leicht says.

Impact of Globalization

Like the impact of so-called supply-side Reaganomics, the effects of globalization and trade policy are often hotly debated. While some argue they have, on balance, helped the U.S. economy, others say they have undermined middle-class security. (*See "At Issue," p. 109.*)

In his 2006 book *War on the Middle Class*, CNN anchor Lou Dobbs wrote "[i]n their free-trade fervor, Republicans and Democrats alike, most economists, certainly corporate leaders, and business columnists assure us that

Middle Class Enjoys Some of 'Life's Goodies'

More than two-thirds of middle-class Americans enjoy at least three of "life's goodies," such as high-speed Internet and more than one vehicle, according to the Pew Research Center. But half as many middle class as wealthy Americans have vacation homes, household help and children in private school.

Percentage of Americans who have. . .

Item	All incomes	Upper income	Middle income	Lower income
Cable or satellite service	70%	80%	71%	62%
Two or more cars	70	83	72	57
High-speed Internet	66	80	67	50
High-definition or flat screen TV	42	59	42	28
Young child in private school	15	31	14	6
Paid household help	16	36	13	7
A vacation home	10	19	9	4

Source: "Inside the Middle Class: Bad Times Hit the Good Life," Pew Research Center, April 2008

the loss of millions of jobs to other countries is the inevitable result of a modern global economy. The result, they promise us, will be a higher standard of living for everyone in America — and especially for the rest of the planet."

But Dobbs went on to say that millions of U.S. manufacturing jobs already had vanished and that many more jobs — including millions of white-collar service positions — were expected to do so in coming years, with the information-technology industry leading the way. "The free-trade-at-any-price enthusiasts once promised us that all those millions of people who lost their positions in manufacturing would find even better ones in the tech industry. But today no one is saying which industry will be the source of replacement for those jobs lost to outsourcing."[48]

C. Fred Bergsten, director of the Peterson Institute for International Economics, appearing on the PBS show "The NewsHour with Jim Lehrer," said studies by his organization have shown that the U.S. economy is $1 trillion a year richer as a result of globalization during the past 50 years.

Nonetheless, Bergsten said "there are losers . . . , costs . . . [and] downsides" to globalization and that the United States "has done a very poor job" in dealing with

those problems. "You lose your health care when you lose your job. Unemployment insurance is miserably inadequate. Trade-adjustment assistance works, but it doesn't even cover [service] workers who get outsourced, and it's inadequate."

But Thea Lee, policy director and chief international economist at the AFL-CIO, who also appeared on the PBS program, was more critical of globalization than Bergsten. "We've had the wrong kind of globalization," she said. "It's been a corporate-dominated globalization, which has not really served working people here or our trading partners very well. . . . We've seen this long-term, decades-long stagnation of wages and growth of wage inequality in the United States even as we've been in a period of tremendous economic growth, productivity growth, technological improvements and increase in globalization."[49]

However one may interpret the economic history of recent decades, few observers would disagree that the middle class is now caught in the greatest economic downdraft in generations.

"We've really had an erosion of economic security and economic opportunity," and it occurred "very rapidly" after 2001, says Christian Weller, an associate professor of public policy at the University of Massachusetts,

Economic Meltdown Batters Retirement Plans

Reform proposals call for limiting risk to workers.

The economy may look bleak for millions of middle-class Americans, but for those in or near retirement, it's downright scary.

Experts say the steep downturns in stock and real estate values, along with soaring layoffs among older workers, have left millions worrying that they won't have enough income to see them through their golden years. And the crash has underscored what critics see as the weaknesses of 401(k) accounts — tax-advantaged plans that require employees to assume the primary responsibility for building and managing their retirement nest eggs.

"The collapse of the housing bubble, coupled with the plunge in the stock market, has exposed the gross inadequacy of our system of retirement income," Dean Baker, co-director of the Center for Economic and Policy Research, a liberal think tank in Washington, told a House committee in February.[1]

At the same hearing, Alicia H. Munnell, director of the Center for Retirement Research at Boston College, said the center's National Retirement Risk Index, which projects the share of households that will not be able to maintain their living standard in retirement, jumped from 31 percent in 1983 to 44 percent in 2006 and rises to 61 percent when health-care expenses are factored in.

Munnell said that in the two years following the stock market's peak on Oct. 9, 2007, the market value of assets in 401(k) retirement plans and Individual Retirement Accounts fell roughly 30 percent. For people ages 55 to 64, she said, median holdings in 401(k) plans went from a modest $60,000 or so in 2007 to $42,000 at the end of 2008.[2]

Critics have long warned of serious faults in the nation's private system of retirement savings. The number of so-called defined-benefit plans, which provide for guaranteed pensions, has been shrinking, while defined-contribution plans like 401(k)s have risen from supplemental savings vehicles in the early 1980s to what they are now: the main or sole retirement plan for most American workers covered by an employer-sponsored retirement plan.[3]

Jacob S. Hacker, a political scientist at the University of California in Berkeley, said the historical "three-legged stool" of retirement security — Social Security, private pensions and personal savings — is now precarious.

"The central issue for retirement security is . . . the risk," he told a congressional hearing last fall. "Retirement wealth has not only failed to rise for millions of families; it has also grown more risky, as the nation has shifted more of the responsibility for retirement planning from employers and government onto workers and their families.[4]

Several proposals have surfaced for revamping the retirement system, some bolder than others.

Teresa Ghilarducci, a professor at the New School for Social Research in New York, wants Congress to establish "Guaranteed Retirement Accounts," in which all workers not enrolled in an equivalent or better defined-benefit pension plan would participate. A contribution equal to 5 percent of each worker's earnings would go into an account each year, with the cost shared equally between worker and employer. A $600 federal tax credit would offset employees' contributions.

Money in the accounts would be managed by the federal government and earn a guaranteed 3 percent rate of return, adjusted for inflation. When a worker retired, the account would convert to an annuity that provides income until death, though a small portion could be taken in a lump sum at retirement. Those who died before retirement could leave only half their accounts to heirs; those who died after retiring could leave half the final balance minus benefits received.[5]

Boston, and a fellow at the liberal Center for American Progress.

After a "five-year window" of employment and wage growth during the late 1990s, Weller says, pressure on the middle class began accelerating in 2001. "There are different explanations, but one is . . . that after the 2001 recession [corporate] profits recovered much faster than in previous recessions, to much higher levels, and corporations were unchecked. They could engage in outsourcing and all these other techniques to boost their short-term profits, but obviously to the detriment of employees. I think what we ended up with was very slow employment growth, flat or declining wages and declining benefit coverage."

The plan has drawn criticism. Paul Schott Stevens, president and CEO of the Investment Company Institute, which represents the mutual-fund industry, called it "a non-starter."[6] Jan Jacobson, senior counsel for retirement policy at the American Benefits Council, said, "We believe the current employer-sponsored system is a good one that should be built on."[7]

But Ghilarducci told the AARP Bulletin Today that "people just want a guaranteed return for their retirement. The essential feature of my proposal is that people and employers would be relieved of being tied to the financial market."[8]

Hacker advocates an approach called "universal 401(k)" plans. The plans would be available to all workers, regardless of whether their employer offered a traditional retirement plan. All benefits would remain in the same account throughout a worker's life, and money could be withdrawn before retirement only at a steep penalty, as is the case with today's 401(k) plans. The plans would be shielded against excessive investments in company stock, and the default investment option would be a low-cost index fund that has a mix of stocks and bonds. Over time, the mix would change automatically to limit risk as a worker aged.

At age 65, government would turn a worker's account into a lifetime annuity that guarantees a flow of retirement income, unless the worker explicitly requested otherwise and showed he or she had enough assets to withstand market turmoil.

Employers would be encouraged to match workers' contributions to the plans, and government could give special tax breaks to companies offering better matches for lower-paid workers.[9]

Teresa Ghilarducci, a professor at the New School for Social Research, says Congress should establish "Guaranteed Retirement Accounts" for workers not enrolled in similar pension plans.

teresaghilarducci.org

Says Hacker, "We have to move toward a system in which there is a second tier of pension plans that is private but which provides key protections that were once provided by defined-benefit pension plans."

[1] "Strengthening Worker Retirement Security," testimony before House Committee on Education and Labor, Feb. 24, 2009, http://edlabor.house.gov/ documents/111/pdf/testimony/20090224DeanBakertestimony.pdf.

[2] "The Financial Crisis and Restoring Retirement Security," testimony before House Committee on Education and Labor, Feb. 24, 2009, http://edlabor.house.gov/documents/111/pdf/testimony/20090224AliciaMunnellTestimony.pdf.

[3] Ibid. For background, see Alan Greenblatt, "Pension Crisis," *CQ Researcher*, Feb. 17, 2006, pp. 145-168, and Alan Greenblatt, "Aging Baby Boomers," *CQ Researcher*, Oct. 19, 2007, pp. 865-888.

[4] "The Impact of the Financial Crisis on Workers' Retirement Security," testimony before House Committee on Education and Labor field hearing, San Francisco, Oct. 22, 2008.

[5] For a detailed explanation, see, Teresa Ghilarducci, "Guaranteed Retirement Accounts: Toward retirement income security," Economic Policy Institute, Briefing Paper No. 204, Nov. 20, 2007, www.sharedprosperity.org/bp204/bp204.pdf.

[6] Stevens and Jacobson are quoted in Doug Halonen, "401(k) plans could be facing total revamp," *Financial Week*, Oct. 29, 2008.

[7] *Ibid.*

[8] Quoted in Carole Fleck, "401(k) Plans: Too Risky for Retirement Security?" AARP Bulletin Today, Dec. 17, 2008, http://bulletin.aarp.org/yourmoney/retirement/articles/401_k_plans_too_risky_for_retirement_security_.html.

[9] See Jacob S. Hacker, The Great Risk Shift (2006), pp. 185-187. See also Testimony before House Committee on Education and Labor, Oct. 22, 2008, op. cit.

And overlaid on all of that, Weller says, was the unprecedented boom in housing.

Even before the housing bubble burst, though, the middle class was on shaky ground, as Weller noted in an article early last year. In 2004, fewer than a third of families had accumulated enough wealth to equal three months of income, he found. And that was counting all financial assets, including retirement savings, minus debt.[50]

"For quite some time," says *Two-Income Trap* co-author Tyagi, "we've had a sizable minority of the middle class under enormous strain and on the verge of crisis, and since the recent meltdown the proportion of middle-class families in crisis increased exponentially.

"Many families teetering on the uncomfortable edge have been pushed over. I really see the [home] foreclosure crisis as front and center in this. We can't overestimate how important home ownership is to the middle class is, and what a crisis losing a home is."

CURRENT SITUATION

Narrowing the Gap

Joel Kotkin, a presidential fellow at Chapman University in Orange, Calif., and author of *The City: A Global History,* wrote recently that "over the coming decades, class will likely constitute the major dividing line in our society — and the greatest threat to America's historic aspirations."[51]

With the gap between rich and poor growing and even a college degree no assurance of upward mobility, Kotkin wrote, President Obama's "greatest challenge . . . will be to change this trajectory for Americans under 30, who supported him by two to one. The promise that 'anyone' can reach the highest levels of society is the basis of both our historic optimism and the stability of our political system. Yet even before the recession, growing income inequality was undermining Americans' optimism about the future."

Obama's legislative agenda, along with his middle-class task force, aims to narrow the class gap. But the deep recession, along with a partisan divide on Capitol Hill, could make some of his key goals difficult and costly to reach.

In announcing his budget, Obama did not hesitate to draw class distinctions between "the wealthiest few" and the "middle class" made up of "responsible men and women who are working harder than ever, worrying about their jobs and struggling to raise their families." He acknowledged that his political opponents are "gearing up for a fight" against his budget plan, which includes tax cuts for all but the richest Americans, universally available health-care coverage and other policies aimed squarely at the middle class. Yet, he said, "The system we have now might work for the powerful and well-connected interests that have run Washington for far too long, but I don't. I work for the American people."[52]

Republicans also are invoking middle-class concerns in expressing their opposition to Obama's budget. Delivering the GOP response to Obama's weekly address, Sen. Richard Burr, R-N.C., said the budget would require the typical American family to pay $52,000 in interest alone over the next decade.[53]

"Like a family that finds itself choking under the weight of credit-card balances and finance charges," said Burr, "the federal government is quickly obligating the American people to a similar fate.

The stimulus package signed by the president in February includes payroll-tax breaks for low- and moderate-income households and an expanded tax credit for higher-education expenses. But costly overhauls of health and retirement policies remain on the table.

Douglas W. Elmendorf, director of the Congressional Budget Office, told a Senate budget panel in February that without changes in health-insurance policy, an estimated 54 million people under age 65 will lack medical insurance by 2019, compared with 45 million this year. The projection "largely reflects the expectation that health-care costs and health-insurance premiums will continue to rise faster than people's income."[54]

Meanwhile, the abrupt collapse of the global financial markets has decimated middle-class retirement accounts. Between June 30 and September 30 of 2008, retirement assets fell 5.9 percent, from $16.9 trillion to $15.9 trillion, according to the latest tally by the Investment Company Institute, which represents the mutual-fund industry.[55]

In announcing his middle-class task force, Obama said his administration would be "absolutely committed to the future of America's middle class and working families. They will be front and center every day in our work in the White House."[56]

The group includes the secretaries of Labor, Health and Human Services, Education and Commerce, plus the heads of the National Economic Council, Office of Management and Budget, Domestic Policy Council and Council of Economic Advisors.[57]

According to the White House, the task force will aim to:

- Expand opportunities for education and lifelong training;
- Improve work and family balance;
- Restore labor standards, including workplace safety;
- Help to protect middle-class and working-family incomes, and
- Protect retirement security.

AT ISSUE

Has U.S. trade and globalization policy hurt the middle class?

YES

Thea Lee
Policy Director, AFL-CIO

Written for *CQ Researcher*, March 2, 2009

The middle class is not a single entity — nor is trade and globalization policy. The clothes we wear, the food we eat, the air we breathe, the jobs we have, the places we choose to live — all are affected by trade and globalization policy, but in many different ways.

I would argue, nonetheless, that U.S. trade and globalization policy has failed the middle class in numerous ways. It has eroded living standards for a large majority of American workers, undermined our social, environmental, consumer safety and public health protections, exacerbated our unsustainable international indebtedness, weakened our national security and compromised our ability to innovate and prosper in the future.

Most significant, especially during this global downturn, the negative impact of globalization on American wages should be a top concern — both for policy makers and for business. Economists may disagree about the magnitude of the effect, but few would dispute that globalization has contributed to the decades-long stagnation of real wages for American workers.

The Economic Policy Institute's L. Josh Bivens finds that the costs of globalization to a full-time median-wage earner in 2006 totaled approximately $1,400, and about $2,500 for a two-earner household. It only makes intuitive sense that if the point of globalization is to increase U.S. access to vast pools of less-skilled, less-protected labor, wages at home will be reduced — particularly for those workers without a college degree. And this impact will only grow in future years, as trade in services expands. We won't be able to rebuild our real economy and the middle class if we can't figure out how to use trade, tax, currency and national investment policies to reward efficient production at home — not send it offshore.

That is not to say, however, that trade and globalization in themselves are inherently pernicious. U.S. globalization policies in recent decades prioritized the interests of mobile, multinational corporations over domestic manufacturers, workers, farmers and communities. At the same time, they undermined prospects for equitable, sustainable and democratic development in our trading partners.

If we are going to move forward together in the future, we need to acknowledge that our current policies have not always delivered on their potential or their promise — particularly for middle-class workers. If new trade and globalization initiatives are to gain any political momentum, we will need deep reform in current policies.

NO

C. Fred Bergsten
Director, Peterson Institute for International Economics

Written for *CQ Researcher*, March 2, 2009

The backlash in the United States against globalization is understandable but misplaced. Despite widespread and legitimate concerns about worsening income distribution, wage stagnation and job insecurity, all serious economics studies show that globalization is only a modest cause of these problems. In the aggregate, globalization is a major plus for the U.S. economy and especially for the middle class.

An in-depth study by our nonpartisan institute demonstrates that the U.S. economy is $1 trillion per year richer as a result of global trade integration over the last half-century, or almost $10,000 per household. These gains accrue from cheaper imports, more high-paying export jobs and faster productivity growth. The American economy could gain another $500 billion annually if we could lift the remaining barriers to the international flow of goods and services.

Of course, any dynamic economic change, like technology advances and better corporate management, affects some people adversely. The negative impact of globalization totals about $50 billion a year due to job displacement and long-term income reductions. This is not an insignificant number, but the benefit-to-cost ratio from globalization is still a healthy 20-to-1.

The United States could not stop globalization even if it wanted to. But it must expand the social safety net for those displaced while making sure that our workers and firms can compete in a globalized world.

The Obama administration and the new Congress have already begun to shore up these safety nets through the fiscal stimulus package. Unemployment insurance has been substantially liberalized. Sweeping reform of the health care system has begun. Most important, Trade Adjustment Assistance has been dramatically expanded to cover all trade-impacted workers and communities.

We must also remember that globalization has lifted billions of the poorest citizens out of poverty. No country has ever achieved sustained modernization without integrating into the world economy, with China and India only the latest examples. The flip side is that products and services from these countries greatly improve the purchasing power and an array of consumer choices for the American middle class.

Fears of globalization have expanded during the current worldwide downturn. But strong export performance kept our economy growing through most of last year, and global cooperation is now necessary to ignite the needed recovery.

Getty Images/Ethan Miller

Soaring home foreclosures and job losses are battering middle class families. Home prices doubled between the mid-1990s and 2007, prompting many families to borrow against the higher values and take out cash for vacations and other expenditures. When values began plummeting, families with job losses and limited or no savings found themselves underwater.

The group's first meeting, on Feb. 27, focused on so-called green jobs.

Jared Bernstein, Vice President Biden's chief economist and a task force member, told *The Christian Science Monitor* that the group "has a different target" than the recently enacted $787 billion economic-stimulus plan, which includes huge government outlays with a goal of creating millions of jobs. "It's less about job quantity than job quality," Bernstein told the *Monitor* in an e-mail. "Its goal is to make sure that once the economy begins to expand again, middle-class families will reap their fair share of the growth, something that hasn't happened in recent years."[58]

Biden expressed a similar sentiment in an op-ed piece in *USA Today.* "Once this economy starts growing again, we need to make sure the benefits of that growth reach the people responsible for it. We can't stand by and watch as that narrow sliver of the top of the income scale wins a bigger piece of the pie — while everyone else gets a smaller and smaller slice," he wrote.[59]

In late January, as he pushed Congress to pass the stimulus plan, Obama said that not only would the task force focus on the middle class but that "we're not forgetting the poor. They are going to be front and center because they, too, share our American Dream."

Cash-Strapped States

Cash-strapped state governments are on the front lines of dealing with the swelling ranks of the nation's poor. States are struggling to handle a rising number of Americans in need of welfare assistance as the economy weakens — some of them middle-class households pushed over the financial edge by job losses and home foreclosures.

Despite the economic collapse, 18 states reduced their welfare rolls last year, and the number of people nationally receiving cash assistance was at or near the lowest point in more than four decades, a *New York Times* analysis of state data found.[60]

Michigan, with one of the nation's highest unemployment rates, reduced its welfare rolls 13 percent, and Rhode Island cut its by 17 percent, the *Times* said.

"Of the 12 states where joblessness grew most rapidly," the *Times* said, "eight reduced or kept constant the number of people receiving Temporary Assistance for Needy Families, the main cash welfare program for families with children. Nationally, for the 12 months ending October 2008, the rolls inched up a fraction of 1 percent."

While the recession has devastated households across the demographic spectrum, it has been especially hard on minorities. The overall unemployment rate in January stood at 7.6 percent, but it was 12.6 percent for African-Americans and 9.7 percent for Hispanics. The rate for whites was 6.9 percent. What's more, unemployment among minorities has been rising faster than for whites.[61]

The crash of the auto industry, among the most spectacular aspects of the past year's economic crisis, has devastated African-Americans.

About 118,000 African-Americans worked in the auto industry in November 2008, down from 137,000 in December 2007, the start of the recession, according to researchers at the Economic Policy Institute, a liberal think tank.[62]

"One of the engines of the black middle class has been the auto sector," Schmitt, told *USA Today* in January. In the late 1970s, "one of every 50 African-Americans in the U.S. was working in the auto sector. These jobs were the best jobs. Particularly for African-Americans who had migrated from the South, these were the culmination of a long, upward trajectory of economic mobility."[63]

For those living in high-cost urban areas — whether black, Hispanic or white — the strain of maintaining a middle-class standard of living is especially acute. According to 2008 survey data by the Pew research organization, more than a fourth of those who defined themselves as middle class who lived in high-cost areas said they had just enough money for basic expenses, or not even that much, compared with 16 percent living in low-cost metropolitan areas.[64]

In New York City, the nation's biggest urban area, people earning the median area income in the third quarter of 2008 could afford only about 11 percent of the homes in the metro area — the lowest proportion in the country — according to the Center for an Urban Future, a Manhattan think tank. To be in the middle class in Manhattan, according to the center's analysis, a person would need to make $123,322 a year, compared with $72,772 in Boston, $63,421 in Chicago and $50,000 in Houston.[65]

"New York has long been a city that has groomed a middle class, but that's a more arduous job today," said Jonathan Bowles, the center's director and a co-author of the report. "There's a tremendous amount of positives about the city, yet so many middle-class families seem to be stretched to their limits."[66]

OUTLOOK

Silver Lining?

No cloud is darker over the middle class than the deepening recession. "Everything points to this being at least three years of a weak economy," Nobel economist Krugman told a conference in February sponsored by several liberal groups.[67]

The economic crisis, he said, is "out of control," and "there's no reason to think there's any spontaneous mechanism for recovery. . . . My deep concern is not simply that it will be a very deep slide but that it will become entrenched."

Still, Krugman said, "If there's any silver lining to [the crisis], it's reopening the debate about the role of public policy in the economy."

Many liberals argue that policy changes in such areas as health-care coverage and higher-education benefits offer avenues for lifting middle-class families out of the economic mire and that getting medical costs under control is a key to the nation's long-term fiscal health. But many conservatives oppose more government spending. Advancing major reforms amid partisan bickering and a budget deficit inflated by bailouts, recession and war will be difficult.

University of Arizona political scientist Kenworthy says focusing policy changes on people living below the poverty level could be "easier to sell politically" and would still benefit those in higher income brackets.

"For example, think about the minimum wage," he says. Raising it "has an effect further up the wage distribution. The same with the earned-income tax credit," a refundable credit for low and moderate working people and families. "If it's made more generous, it has effects a bit further up. The same with health care."

As Washington grapples with potential policy changes, the plunging economy is forcing many middle-class consumers to live within their means. Many see that as a good thing.

"I certainly think some of the entitlements we have come to expect, like two homes, a brand-new car every couple of years, college education for all the kids, a yacht or two, expensive vacations — some of this will need to be reoriented," says Georgetown University's Viksnins. "Some reallocation of people's priorities is really necessary."

Yet, long-term optimism hasn't vanished amid the current economic gloom. "We will rebuild, we will recover and the United States of America will emerge stronger than before," Obama declared in an address on Feb. 24 to a joint session of Congress.

Viksnins says he is "utterly hopeful" about the future of the middle class.

And Sherk, the Heritage Foundation labor-policy fellow, says that "as long as your skills are valuable, you're going to find a job that pays you roughly at your productivity.

"For people in jobs disappearing from the economy, it's going to mean a substantial downward adjustment in standard of living," Sherk says. But "those in the middle class who have some college education, or have gone to a community college or have skills, broadly speaking, most will wind up on their feet again."

NOTES

1. Julie Bosman, "Newly Poor Swell Lines at Food Banks," *The New York Times*, Feb. 20, 2009, www.nytimes .com/2009/02/20/nyregion/20food.html?scp=1&sq= newly%20poor&st=cse.

2. For background, see the following *CQ Researcher* reports: Marcia Clemmitt, "Public Works Projects," Feb. 20, 2009, pp. 153-176; Kenneth Jost, "Financial Crisis," May 9, 2008, pp. 409-432; Thomas J. Billitteri, "Financial Bailout," Oct. 24, 2008, pp. 865-888; Marcia Clemmitt, "Mortgage Crisis," Nov. 2, 2007, pp. 913-936, and Barbara Mantel, "Consumer Debt," March 2, 2007, pp. 193-216.

3. For background see Kenneth Jost, "The Obama Presidency," *CQ Researcher*, Jan. 30, 2009, pp. 73-104.

4. Jackie Calmes, "Obama, Breaking 'From a Troubled Past,' Seeks a Budget to Reshape U.S. Priorities," *The New York Times*, Feb. 27, 2009, p. A1. For background, see the following *CQ Researcher* reports: Marcia Clemmitt, "Rising Health Costs," April 7, 2006, pp. 289-312, and Marcia Clemmitt, "Universal Coverage," March 30, 2007, pp. 265-288.

5. For background, see the following *CQ Researcher* reports: Thomas J. Billitteri, "Domestic Poverty," Sept. 7, 2007, pp. 721-744; Alan Greenblatt, "Upward Mobility," April 29, 2005, pp. 369-392; and Mary H. Cooper, "Income Equality," April 17, 1998, pp. 337-360.

6. Demos and Institute on Assets & Social Policy at Brandeis University, *From Middle to Shaky Ground: The Economic Decline of America's Middle Class, 2000-2006* (2008).

7. Paul Taylor, *et al.*, "Inside the Middle Class: Bad Times Hit the Good Life," Pew Research Center, April 9, 2008, http://pewsocialtrends.org/assets/pdf/ MC-Middle-class-report. pdf, p. 5.

8. Edmund L. Andrews, "Fed Calls Gain in Family Wealth a Mirage," Feb. 13, 2009, www.nytimes .com/2009/02/13/business/economy/13fed .html?ref=business. The study is by Brian K. Bucks, *et al.*, "Changes in U.S. Family Finances from 2004 to 2007: Evidence from the Survey of Consumer Finances," *Federal Reserve Bulletin*, Vol. 95, February 2009, www.federalreserve.gov/pubs/bulletin/2009/ pdf/scf09.pdf.

9. Jacob S. Hacker, *The Great Risk Shift* (2006), pp. 5-6.

10. Quoted in Ta-Nehisi Coates, "American Girl," *The Atlantic*, January/February 2009.

11. Julia B. Isaacs and Isabel V. Sawhill, "The Frayed American Dream," Brookings Institution, Nov. 28, 2007, www.brookings.edu/opinions/2007/1128_ econgap_isaacs.aspx.

12. *Ibid.*

13. Arloc Sherman, "Income Inequality Hits Record Levels, New CBO Data Show," Center on Budget and Policy Priorities, Dec. 14, 2007, www.cbpp .org/12-14-07inc.htm. The CBO report is "Historical Effective Federal Tax Rates: 1979 to 2005," www.cbo.gov/doc.cfm?index= 8885. Figures are inflation adjusted and are in 2005 dollars.

14. Emmanuel Saez, "Striking it Richer: The Evolution of Top Incomes in the United States," University of California, Berkeley, March 15, 2008, http://elsa .berkeley.edu/~saez/saez-UStopincomes-2006prel.pdf.

15. For background, see the following *CQ Researcher* reports: Pamela M. Prah, "Labor Unions' Future," Sept. 2, 2005, pp. 709-732; Brian Hansen, "Global Backlash," Sept. 28, 2001, pp. 761-784; Mary H. Cooper, "World Trade," June 9, 2000, pp. 497-520; Mary H. Cooper, "Exporting Jobs," Feb. 20, 2004, pp. 149-172; and the following *CQ Global Researcher* reports: Samuel Loewenberg, "Anti-Americanism," March 2007, pp. 51-74, and Ken Moritsugu, "India Rising," May 2007, pp. 101-124.

16. John Schmitt, "The Good, the Bad, and the Ugly: Job Quality in the United States over the Three Most Recent Business Cycles," Center for Economic and Policy Research, November 2007, www.cepr.net/ documents/publications/goodjobscycles.pdf.

17. Alicia H. Munnell and Steven A. Sass, "The Decline of Career Employment," Center for Retirement Research, Boston College, September 2008, http:// crr.bc.edu/images/stories/ib_8-14.pdf.

18. Richard Florida, "How the Crash Will Reshape America," *The Atlantic*, March 2009, www.theatlantic .com/doc/200903/meltdown-geography.

19. James Sherk, "A Good Job Is Not So Hard to Find," Heritage Foundation, June 17, 2008 and revised and updated Sept. 2, 2008, www.heritage.org/research/labor/cda08-04.cfm.

20. George Packer, "The Ponzi State," *The New Yorker*, Feb. 9 and 16, 2009.

21. Paul Krugman, "Decade at Bernie's," *The New York Times*, Feb. 16, 2009, www.nytimes.com/2009/02/16/opinion/16krugman.html?scp=1&sq=decade%20at%20bernie's&st=cse.

22. Federal Housing Finance Agency, "U.S. Housing Price Index Estimates 1.8 Percent Price Decline From October to November," Jan. 22, 2009, www.ofheo.gov/media/hpi/MonthlyHPI12209F.pdf.

23. College Board, "Trends in College Pricing 2008," http://professionals.collegeboard.com/profdownload/trends-in-college-pricing-2008.pdf.

24. Department of Health and Human Services, Centers for Medicare and Medicaid Services, www.cms.hhs.gov/NationalHealthExpendData/downloads/tables.pdf.

25. Sara R. Collins, *et al.*, "Losing Ground: How the Loss of Adequate Health Insurance Is Burdening Working Families: Findings from the Commonwealth Fund Biennial Health Insurance Surveys, 2001-2007," Commonwealth Fund, Aug. 20, 2008, www.commonwealthfund.org/Content/Publications/Fund-Reports/2008/Aug/Losing-Ground-How-the-Loss-of-Adequate-Health-Insurance-Is-Burdening-Working-Families-8212-Finding.aspx.

26. Elizabeth Warren and Amelia Warren Tyagi, *The Two-Income Trap* (2003), p. 19.

27. *Ibid*, pp. 51-52.

28. Press release, "'Women on Their Own' in Much Worse Financial Condition Than Other Americans," Consumer Federation of America, Dec. 2, 2008, www.consumerfed.org/pdfs/Women_America_Saves_Tele_PR_12-2-08.pdf.

29. Ernest Istook, "Land of the free and home of the victims," Heritage Foundation, Feb. 29, 2008, www.heritage.org/Press/Commentary/ed022908b.cfm.

30. "Remarks by the President to Caterpillar Employees," Feb. 12, 2009, www.whitehouse.gov.

31. Foon Rhee, "Partisan spat continues on stimulus," Political Intelligence blog, *The Boston Globe*, Feb. 17, 2009, www.boston.com/news/politics/political intelligence/2009/02/partisan_ spat_c.html.

32. Fact Sheet, "The Hidden Link: Health Costs and Family Economic Insecurity," Families USA, January 2009, www.familiesusa.org/assets/pdfs/the-hidden-link.pdf. The research cited by Families USA is by Christopher Tarver Robertson, *et al.*, "Get Sick Get Out: The Medical Causes of Home Mortgage Foreclosures," *Health Matrix Vol. 18*, 2008, pp. 65-105.

33. Reuters, "U.S. may need new retirement savings plans: lawmaker," Feb. 24, 2009, www.reuters.com/article/domesticNews/idUSTRE51N5UM20090224.

34. Lynn Sweeton, "Transcript of Obama, McCain at Saddleback Civil Forum with Pastor Rick Warren," *Chicago Sun Times*, Aug. 18, 2008, http://blogs.suntimes.com/sweet/2008/08/transcript_of_obama_mccain_at.html.

35. Quoted in Jeanna Bryner, "American Dream and Middle Class in Jeopardy," www.livescience.com, October 9, 2008, www.livescience.com/culture/081009-middle-class.html.

36. See Jennifer L. Goloboy, "The Early American Middle Class," *Journal of the Early Republic*, Vol. 25, No. 4, winter 2005.

37. Quoted in David Kennedy, *Freedom From Fear* (1999), p. 264.

38. *Ibid*.

39. For historical background, see *CQ Researcher Plus Archive* for a large body of contemporaneous coverage during the 1930s and 1940s in *Editorial Research Reports*, the precursor to the *CQ Researcher*.

40. Paul Krugman, "The Conscience of a Liberal: Introducing This Blog" *The New York Times*, Sept. 18, 2007, http://krugman.blogs.nytimes.com/2007/09/18/introducing-this-blog/.

41. Paul Krugman, "What Obama Must Do: A Letter to the New President," *Rolling Stone*, Jan. 14, 2009, www.rollingstone.com/politics/story/25456948/what_obama_must_do.

42. James T. Patterson, *Grand Expectations* (1996), p. 312.

43. Barry Hirsch, Georgia State University, and David Macpherson, Florida State University, "Union Membership, Coverage, Density, and Employment

Among All Wage and Salary Workers, 1973-2008," www.unionstats.com.

44. "Remarks by the President and the Vice President in Announcement of Labor Executive Orders and Middle-Class Working Families Task Force," Jan. 30, 2009, www.whitehouse.gov/blog_post/Todaysevent/.

45. U.S. Chamber of Commerce, "Is Unionization the Ticket to the Middle Class? The Real Economic Effects of Labor Unions," 2008, www.uschamber.com/assets/labor/unionrhetoric_econeffects.pdf.

46. George J. Viksnins, "Reaganomics after Twenty Years," www9.georgetown.edu/faculty/viksning/papers/Reaganomics.html.

47. Paul Krugman, "Debunking the Reagan Myth," *The New York Times*, Jan. 21, 2008, www.nytimes.com/2008/01/21/opinion/21krugman.html?scp=1&sq=%22Debunking%20the%20Reagan%20Myth%22&st=cse.

48. Lou Dobbs, *War on the Middle Class* (2006), p. 112.

49. Transcript, "In Bad Economy, Countries Contemplate Protectionist Measures," "The NewsHour with Jim Lehrer," Feb. 19, 2009, www.pbs.org/newshour/bb/business/jan-june 09/trade_02-19.html.

50. Christian Weller, "The Erosion of Middle-Class Economic Security After 2001," *Challenge*, Vol. 51, No. 1, January/February 2008, pp. 45-68.

51. Joel Kotkin, "The End of Upward Mobility?" *Newsweek*, Jan. 26, 2009, p. 64.

52. "Remarks of President Barack Obama, Weekly Address," Feb. 28, 2009, www.whitehouse.gov/blog/09/02/28/Keeping-Promises/.

53. "Burr delivers GOP challenge to Obama's budget," www.wral.com, Feb. 28, 2009, www.wral.com/news/local/story/4635676/.

54. Statement before the Committee on the Budget, U.S. Senate, "Expanding Health Insurance Coverage and Controlling Costs for Health Care," Feb. 10, 2009, www.cbo.gov/ftpdocs/ 99xx/doc9982/02-10-HealthVolumes_Testimony.pdf.

55. Investment Company Institute, "Retirement Assets Total $15.9 Trillion in Third Quarter," Feb. 19, 2009, www.ici.org/home/09_news _q3_retmrkt_update.html#TopOfPage.

56. Quoted in Jeff Zeleny, "Obama Announces Task Force to Assist Middle-Class Families," *The New York Times*, Dec. 22, 2008.

57. Cited at www.whitehouse.gov/blog_post/about_the_task_force_1/.

58. Mark Trumbull, "Will Obama's plans help the middle class?" *The Christian Science Monitor*, Dec. 24, 2008.

59. Joe Biden, "Time to put middle class front and center," *USA Today*, Jan. 30, 2009.

60. Jason DeParle, "Welfare Aid Isn't Growing as Economy Drops Off," *The New York Times*, Feb. 2, 2009.

61. See Bureau of Labor Statistics, "The Employment Situation: January 2009," www.bls.gov/news.release/empsit.nr0.htm.

62. Robert E. Scott and Christian Dorsey, "African Americans are especially at risk in the auto crisis," Economic Policy Institute, Snapshot, Dec. 5, 2008, www.epi.org/economic_snapshots/entry/webfeatures_snapshots_20081205/.

63. Quoted in Larry Copeland, "Auto industry's slide cuts a main route to the middle class," *USA Today*, Jan. 20, 2009, www.usatoday.com/money/autos/2009-01-20-blacks-auto-industry-dealers_N.htm.

64. D'Vera Cohn, "Pricey Neighbors, High Stress," Pew Social and Demographic Trends, May 29, 2008, www.pewsocialtrends.org/pubs/711/middle-class-blues.

65. Jonathan Bowles, *et al.*, "Reviving the City of Aspiration: A study of the challenges facing New York City's middle class," Center for an Urban Future, Feb. 2009, www.nycfuture.org/images_pdfs/pdfs/CityOfAspiration.pdf.

66. Quoted in Daniel Massey, "City faces middle-class exodus," *Crain's New York Business*, Feb. 5, 2009, www.crainsnewyork.com/article/20090205/FREE/902059930.

67. Krugman spoke at the "Thinking Big, Thinking Forward" conference in Washington on Feb. 11 sponsored by *The American Prospect*, the Institute for America's Future, Demos and the Economic Policy Institute.

BIBLIOGRAPHY

Books

Dobbs, Lou, *War on the Middle Class*, Viking, 2006.
The CNN broadcaster argues that the American government and economy are dominated by a wealthy and politically powerful elite who have exploited working Americans.

Frank, Robert H., *Falling Behind: How Rising Inequality Harms the Middle Class*, University of California Press, 2007.
The Cornell University economist argues that most income gains in recent decades have gone to people at the top, leading them to build bigger houses, which in turn has led middle-income families to spend a bigger share of their incomes on housing and curtail spending in other important areas.

Hacker, Jacob S., *The Great Risk Shift*, Oxford University Press, 2006.
A professor of political science argues that economic risk has shifted from "broad structures of insurance," including those sponsored by corporations and government, "onto the fragile balance sheets of American families."

Uchitelle, Louis, *The Disposable American*, Alfred A. Knopf, 2006.
A New York Times business journalist, writing before the current economic crises threw millions of workers out of their jobs, calls the layoff trend "a festering national crisis."

Articles

Copeland, Larry, "Auto industry's slide cuts a main route to the middle class," *USA Today*, Jan. 20, 2009, www.usatoday.com/money/autos/2009-01-20-blacks-auto-industry-dealers_N.htm?loc=interstitialskip.
The financial crisis in the auto industry "has been more devastating for African-Americans than any other community," Copeland writes.

Gallagher, John, "Slipping standard of living squeezes middle class," *Detroit Free Press*, Oct. 12, 2008, www.freep.com/article/20081012/BUSINESS07/810120483.
America's middle-class living standard "carried generations from dirt-floor cabins to manicured suburban subdivisions," Gallagher writes, but it "has sputtered and stalled."

Kotkin, Joel, "The End of Upward Mobility?" *Newsweek*, Jan. 26, 2009, www.newsweek.com/id/180041.
A presidential fellow at Chapman University writes that class, not race, "will likely constitute the major dividing line in our society."

Samuelson, Robert J., "A Darker Future For Us," *Newsweek*, Nov. 10, 2008, www.newsweek.com/id/166821/output/print.
An economic journalist argues that the central question confronting the new administration is whether the economy is at an historic inflection point, "when its past behavior is no longer a reliable guide to its future."

Weller, Christian, "The Erosion of Middle-Class Economic Security After 2001," *Challenge*, Vol. 51, No. 1, January/February 2008, pp. 45-68.
An associate professor of public policy at the University of Massachusetts, Boston, and senior fellow at the liberal Center for American Progress concludes that the gains in middle-class security of the late 1990s have been entirely eroded.

Reports and Studies

Bowles, Jonathan, Joel Kotkin and David Giles, "Reviving the City of Aspiration: A study of the challenges facing New York City's middle class," *Center for an Urban Future*, February 2009, www.nycfuture.org/images_pdfs/pdfs/CityOfAspiration.pdf.
Major changes to the nation's largest city have greatly diminished its ability to both create and retain a sizeable middle class, argues this report.

Schmitt, John, "The Good, the Bad, and the Ugly: Job Quality in the United States over the Three Most Recent Business Cycles," *Center for Economic and Policy Research*, November 2007, www.cepr.net/documents/publications/goodjobcycles.pdf.
The share of "good jobs," defined as ones paying at least $17 an hour and offering employer-provided medical insurance and a pension, deteriorated in the 2000-2006 business cycle.

Sherk, James, "A Good Job So Hard to Find," *Heritage Foundation*, June 17, 2008, www.heritage.org/research/labor/cda08-04.cfm.

Job opportunities have expanded the most in occupations with the highest wages, the conservative think tank states.

Taylor, Paul, *et al.,* **"Inside the Middle Class: Bad Times Hit the Good Life,"** *Pew Research Center,* April 2008, http://pewsocialtrends.org/assets/pdf/MC-Middle-class-report.pdf.
The report aims to present a "comprehensive portrait of the middle class" based on a national opinion survey and demographic and economic data.

Wheary, Jennifer, Thomas M. Shapiro and Tamara Draut, "By A Thread: The New Experience of America's Middle Class," *Demos and the Institute on Assets and Social Policy at Brandeis University,* 2007, www.demos.org/pubs/BaT112807 .pdf.
The report includes a "Middle Class Security Index" that portrays how well middle-class families are faring in the categories of financial assets, education, income and health care.

For More Information

Brookings Institution, 1775 Massachusetts Ave., N.W., Washington, DC 20036; (202) 797-6000; www.brookings .edu. Independent research and policy institute conducting research in economics, governance, foreign policy and development.

Center on Budget and Policy Priorities, 820 First St., N.E., Suite 510, Washington, DC 20002; (202) 408-1080; www .cbpp.org. Studies fiscal policies and public programs affecting low- and moderate-income families and individuals.

Center for Economic and Policy Research, 1611 Connecticut Ave., N.W., Suite 400, Washington, DC 20009; (202) 293-5380; www.cepr.net. Works to better inform citizens on the economic and social choices they make.

Center for Retirement Research, Boston College, 140 Commonwealth Ave., Chestnut Hill, MA 02467; (617) 552-1762; www.crr.bc.edu. Researches and provides the public and private sectors with information to better understand the issues facing an aging population.

Center for an Urban Future, 120 Wall St., 20th Floor, New York, NY, 10005; (212) 479-3341; www.nycfuture.org. Dedicated to improving New York City by targeting problems facing low-income and working-class neighborhoods.

Consumer Federation of America, 1620 I St., N.W., Suite 200, Washington, DC 20006; (202) 387-6121;

www.consumerfed.org. Advocacy and research organization promoting pro-consumer policies before Congress and other levels of government.

Dēmos, 220 Fifth Ave., 5th Floor, New York, NY 10001; (212) 633-1405; www.demos.org. Liberal think tank pursuing an equitable economy with shared prosperity and opportunity.

Heritage Foundation, 214 Massachusetts Ave., N.E., Washington, DC 20002; (202) 546-4400; www.heritage .org. Formulates and promotes public policies based on a conservative agenda.

Middle Class Task Force, 1600 Pennsylvania Ave., N.W., Washington, DC 20500; (202) 456-1414; www.white-house.gov/strongmiddleclass. Presidential task force headed by Vice President Joseph R. Biden working to raise the living standards of middle-class families.

Pew Research Center, 1615 L St., N.W., Suite 700, Washington, DC 20036; (202) 419-4300; www .pew research.org. Provides nonpartisan research and information on issues, attitudes and trends shaping the United States.

U.S. Chamber of Commerce, 1615 H St., N.W., Washington, DC 20062; (202) 659-6000; www.uschamber .com. Business federation lobbying for free enterprise before all branches of government.

6

Debating Hip-Hop

Does Gangsta Rap Harm Black Americans?

Peter Katel

The Rev. Al Sharpton leads a protest march for cleaner hip-hop lyrics on May 3, 2007, in New York City. Sharpton, who criticized shock jock Don Imus in April for using degrading hip-hop lingo to describe female basketball players, has called for equal accountability among hip-hop artists and music-industry executives.

AP Photo/Frank Franklin II

From *CQ Researcher*,
June 15, 2007.

Critics of hip-hop have much to deplore. The posturing about the "gangsta" lifestyle. The ostentatious displays of "bling" — big diamonds, big cars and big houses. And the ever-present young women who are portrayed as rappers' sexual playthings, draping themselves around star performers and shaking their booties through hundreds of near-pornographic videos. Not to mention the incessant use of the N-word.

"They put the word 'nigga' in a song, and we get up and dance to it," actor and comedian Bill Cosby fumed last year.[1]

Anger over hip-hop has been simmering for more than a decade, largely among African-Americans. Much of the concern has focused on its glorification of violence, sexual exploitation and crime. At a time when black males are increasingly endangered — six times more likely than white males to die in homicides — critics like Cosby say gangsta rappers are terrible role models for impressionable inner-city youths.[2]

Superstars Tupac Shakur and The Notorious B.I.G. (Christopher Wallace) fell to hitmen's bullets in 1996 and 1997, respectively. A protégé of hip-hop tycoon Sean "P. Diddy" Combs was sentenced to 10 years in prison for a nightclub shooting eight years ago that Combs had fled. Rapper Cam'ron (Cameron Giles) survived getting shot in 2005 in his $250,000 Lamborghini. And a bodyguard for rapper Busta Rhymes (Trevor Smith) was killed last year at a Brooklyn, N.Y., video shoot.[3]

But outside black America, the heat didn't get turned up on gangsta rap until an aging white radio shock jock, Don Imus, unleashing a word from black street vernacular, offhandedly described the

Hip-Hop Trails Rock and Country

Despite its popularity among youth, hip-hop/rap music sales in the United States only accounted for 11.4 percent of all music purchased in 2006, trailing both rock and country. Over the past nine years, hip-hop/rap sales essentially have remained constant. The figures do not include illegally downloaded computer files, which are believed to contribute to hip-hop's overwhelming popularity.

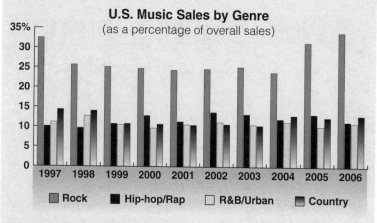

U.S. Music Sales by Genre
(as a percentage of overall sales)

■ Rock   ■ Hip-hop/Rap   ☐ R&B/Urban   ■ Country

* Yearly percentages do not total 100 due to omission of other genres.

Source: "2006 Consumer Profile," Recording Industry Association of America, 2007

Rutgers University women's basketball team as "nappy-headed hos."[4]

In the uproar that followed, Imus' show was canceled, and critics of gangsta rap intensified their demands for change. Notably, TV talk-show host and communications mogul Oprah Winfrey threw her considerable power into demonstrating that tolerance was eroding for rappers' language and lifestyle.

Winfrey broadcast two programs in which African-American writers and entertainers confronted hip-hop magnates. The critics included writer and newspaper columnist Stanley Crouch, who has been thundering for years that hip-hop performers are retailing destructive black stereotypes for the entertainment of a largely white audience.[5]

Entrepreneur Russell Simmons, whose hip-hop empire has been valued at more than $325 million, had defended rappers' artistic freedom in the past. But after Oprah's shows, he changed his tune. "We recommend that the recording and broadcast industries voluntarily remove/bleep/delete the misogynistic words 'bitch' and 'ho' and the racially offensive word 'nigger,' " Simmons and Benjamin

Chavis, co-chairmen of the advocacy group Hip-Hop Summit Action Network, said in a statement. Two major performers, Master P (Percy Miller) and Chamillionaire (Hakeem Seriki), took similar stands.[6]

The Rev. Al Sharpton — who helped force Imus off the air — supported the demand. "We plan to continue to march until those three words are gone," he said.[7]

In effect, Sharpton and his allies are trying to return hip-hop to its origins as a street art form that was largely free of violence or sexual exploitation.

Hip-hop emerged in the Bronx in the early 1970s, when the first MCs chanted their raps, and the first DJs spun their vinyl disks, for neighborhood youths who wanted to groove to the music. Hip-hop's venues weren't velvet-roped clubs, or stadiums, but apartment building community rooms, playgrounds and street corners.[8]

"B-boys" and "b-girls" — the 'b' stands for the rhythm-heavy "break" in a song — invented a whole new form of "break" dancing to the new kind of sound. And the so-called graffiti writers who considered themselves radical street artists became part of the hip-hop culture as well. But New Yorkers in general and city government officials saw the graffiti crews as vandals. In retrospect, their activities seem innocent in light of the violence that now seems inseparable from hip-hop.[9]

As they waited to see whether Oprah, Sharpton and other critics had caught a wave of public revulsion, most rappers stayed silent. But not David Banner (Levell Crump), a leading member of the sex-heavy "dirty south" school of rap. He argues that rappers are taking the fall for an entire culture's pathology.

"What does America want?" Banner asks, his voice ringing with outrage. "People go to NASCAR because they want to see somebody crash. They want to see [the movie] 'The Departed,' with people blowing each other's heads off — that's cool, that's trendy. We see what people buy.

Gangsta rap is just a reflection of America. America is sick. There's so many other things we should be complaining about, and we're talking about hip-hop."

Banner (the name is borrowed from the comic hero "The Incredible Hulk") agrees with critics that white suburban kids are rappers' biggest audience. "Truth is, there's somebody mad at rap because their son is walking around looking like a black dude. That's what the problem is."

But while Banner expresses contempt for Sharpton and other old-school black political leaders, they are voicing views that have been percolating for years, among ordinary people as well as public figures.

"There's a big correlation between rap and the breakdown of the black community in general," says Chris English, a Web site developer and fledgling rapper and record producer in Pocono Summit, Pa. "Regardless of what people say, it is affecting kids. I still love hip-hop, I still believe in artistic freedom. But I see kids who would rather be on the corner, drinking 40's [40-ounce bottles of malt liquor] and smoking [marijuana] blunts."

Anti-violence educator and filmmaker Byron Hurt explored this territory in his PBS documentary, "Hip-Hop: Beyond Beats and Rhymes." "The more I grew, and the more I learned about sexism and violence and homophobia, the more [rap] lyrics became unacceptable to me," he says in the film.[10]

Other critics within the hip-hop world draw a line between mass-marketed rap and what they consider the real thing. They point to independent-label rappers who disdain the gangsta material, as well as a handful of so-called "conscious" performers who've broken through to the mainstream with songs that reject celebrations of violence and exploitation. Among the latter is Talib Kweli, whose new single includes the line: "I'm stayin' conscious to radio playin' garbage."[11]

T.I.'s "King" Leads Rap Sales

Southern-style rapper T.I., who leads the group Pimp Squad Click — better known by the acronym P$C — led all rappers in album sales in 2006 with "King." The Notorious B.I.G.'s "Duets: The Final Chapter" was No. 5 on the list, nine years after his death.

Top-Selling Rappers, 2006
(by album sales)

Artist	Album	Label
T.I.	"King"	Grand Hustle/Atlantic/AG
Lil Wayne	"Tha Carter II"	Cash Money/Universal Motown/UMRG
Eminem	"Curtain Call: The Hits"	Shady/Aftermath/Interscope
Ludacris	"Release Therapy"	DTP/Def Jam/IDJMG
The Notorious B.I.G.	"Duets: The Final Chapter"	Bad Boy/AG
Chamillionaire	"The Sound of Revenge"	Universal Motown/UMRG
Yung Joc	"New Joc City"	Block/Bad Boy South/AG
Rick Ross	"Port of Miami"	Slip-N-Slide/Def Jam/IDJMG
Juelz Santana	"What the Game's Been Missing!"	Diplomats/Def Jam/IDJMG
Busta Rhymes	"The Big Bang"	Aftermath/Interscope

Source: Billboard, www.billboard.biz

Indeed, rappers who've based their careers on glorifying the thug life might find a censored environment impossible. "For me, don't expect me to compromise myself," rapper 50 Cent (Curtis Jackson) told a hip-hop Web site on June 1. "Fitty" has built his public persona on bullet scars from a drug-dealing past and a (now-settled) gangland-style feud with a rival that included threats of violence and at least one episode of gunplay.[12]

"If it's over then it's over," 50 Cent said. "I'll find something else to do."[13]

Mass-marketed hip-hop may not be over, but rap/hip-hop has slipped to No. 3 in nationwide music sales, accounting for 11.4 percent of a recorded-music market whose total retail value was $11.5 billion in 2006. The country category, which hip-hop had surpassed in 2000, now holds the No. 2 spot. Rock music maintained the lead it has held for years.[14]

The statistics don't reflect the irony of the hip-hop market. "Record executives, if you talk to them privately,

Getty Images/Erik S. Lesser

Gangsta rapper David Banner defends his "dirty south" rap style and misogynistic lyrics, arguing that hip-hop has been used as a scapegoat for society's problems. "Gangsta rap is just a reflection of America," he says.

will say that it's white kids who have the disposable income to buy this stuff," says hip-hop historian Yvonne Bynoe.[15] "Black kids are going to buy bootlegs or dupe songs themselves. But you need young black people to give a rap record the imprimatur of authenticity."

To hip-hop marketers, "authenticity" is a code word for thug culture and a never-ending source of some black intellectuals' contempt for mass-market hip-hop. "There was a time when Malcolm X the liberator was the patron saint of hip-hop," says Chicago-based poet Heru Ofori-Atta, speaking of the assassinated African-American leader who earlier in life was a drug dealer named Malcolm Little. "Now it's Malcolm Little the pimp, the hustler, the fool, who is given the microphone."

Indeed, when "60 Minutes" correspondent Anderson Cooper asked Cam'ron if he would tell police if a serial killer lived next door, the rapper declared, "No, I wouldn't call and tell anybody on him. But I'd probably move." After an outcry, the rapper apologized for his remarks.[16]

The retraction reflected the sudden change in tolerance for anti-social behavior. Even pioneer gangsta rapper Ice-T (Tracy Marrow), who now plays a detective on the popular TV show "Law and Order: Special Victims Unit," challenged Cam'ron. "When you and your partner are involved in a crime, and both of y'all get caught and you tell on your partner, that's snitching," Ice-T said in a video interview on the hip-hop Web site, sohh.com.

"Let's get that straight. . . . If I know somebody's in the neighborhood raping little girls — you supposed to tell the police about that sucka. That's not snitching."[17]

As criticism of hip-hop intensifies, here are some of the questions being asked:

Is hip-hop harmful to black America?

Debates over hip-hop have taken place mostly within the black community. Some of hip-hop's fiercest critics, including black intellectuals and entertainers, argue that hip-hop presents a caricature of black America that damages how young black people view themselves and how they're viewed by others.

Author and jazz critic Crouch has decried for years what he calls the ravages of hip-hop. In a 2003 column in the *New York Daily News* — one of dozens he has devoted to the theme — he lamented that ordinary African-Americans were bearing the consequences of a genre in which "thugs and freelance prostitutes have been celebrated for a number of years." The result: "Thousands upon thousands . . . have been murdered or beaten up or terrorized. After all, the celebration of thugs and thuggish behavior should not be expected to bring about any other results."[18]

Though white suburban teenagers make up the vast majority of hip-hop fans, Crouch added, the class privileges they enjoy largely shield them and their communities from the outcome. "With black teenagers . . . street behavior is defined these days as being 'authentic' and 'not trying to be white.' Those who take that seriously have been committing intellectual suicide for years by aspiring downward."[19]

Younger cultural critics, even those who partly agree with Crouch, reject such sweeping condemnations. Instead, they insist on distinguishing between mass-marketed hip-hop and what they see as its purer, original form, less tainted by the demands of the marketplace.

"I like conscious hip-hop, and the stuff that you just dance to," says Lisa Fager, a former promotional specialist at commercial radio stations and record labels, "but I don't like stuff that demeans me as a black woman, or a woman, period — or degrades my community. People don't want to be called bitches, niggas and hos." Fager co-founded an advocacy organization in the Washington area, Industry Ears, largely to call attention to the Federal Communications Commission's responsibility to regulate radio content during hours when children may be listening.[20]

Virtually the entire school of hip-hop that focuses on social criticism is kept off commercial radio, Fager contends, making a point that echoes widely in African-American intellectual circles. "The gatekeepers don't want us to be politically active," she says, referring to programming and record executives who decide what gets recorded and broadcast. "Hurricane Katrina, you don't think that was something the black community talked about? But there were no songs on the radio about it."

Rapper Banner doesn't dispute Fager on what gets played. "The labels don't want to deal with anything that creative," he says. "They don't want to develop an artist; they want the quickest thing people will buy. And as soon as one feminist gets mad, they back up."

But Banner reacts explosively to African-Americans who attack hip-hop as degrading. Hip-hop mogul Simmons' proposal to ban offensive words is "stupid," Banner says. "There was a time in history when we didn't have a choice about being called a nigger. Now that we're making money off it, it's a problem."

He adds, "If I try to change my music, I'm not going to sell my records, and that's the truth. My fans have literally come to me and they say, 'We put you on for pimp [songs]. You put anything else on, you're betraying us.'"

Other blacks argue that rappers who blame profit-motivated companies and the market are copping out. "If what you are doing is not lifting people up to a better day, then I don't think it's valuable and necessary for our people," says poet Heru, a native of Ghana, who practiced law in Florida until he took to poetry full time in 2001.

"For us, and for me, music is more than just music," says Heru. "In Africa-based communities, whether on the continent or in the Western Hemisphere, music is ceremonial and music is used for everything. So what ceremony are we introducing to people? If it's about killing random people in your community or degrading women of your community, that's not a ceremony I want to celebrate."

For others, viewing hip-hop only through the prism of race is simplistic. James McBride, a writer, jazz saxophonist and bandleader, says, "the music has demonstrated that it's bigger than race." In a recent article for *National Geographic*, McBride examined hip-hop's global reach and existential pull: "It's about identity — 'this is what I am,

Reflecting hip-hop's international popularity, 20-year-old French "b-boy" Lilou performs during the Red Bull BC One break-dancing tournament in Berlin in 2005. Break-dancing is a core element of hip-hop, along with rapping, DJ-ing and graffiti-writing.

this is what I know.' This is what every 15-year-old kid feels. This music is the music of all young people. White people loved jazz in the '40s, too."[21]

But rappers do get subjected to a higher standard of scrutiny and indignation, McBride says. "If it weren't promoted or initiated by young black men, then I think some of this debate wouldn't even be a debate," he argues. "Alice Cooper used to do disgusting things on stage" without provoking storms of condemnation. In the 1970s and '80s, Cooper's "shock-rock" stage shows depicted a woman being decapitated and Cooper being hanged.[22]

Is hip-hop a genuine artistic/social movement transcending rap and break-dancing?

Cultural critics argue that after evolving for more than 30 years, hip-hop has become an artistic and social movement extending far beyond rapping over rhythm tracks, break-dancing and spray-painting graffiti.

Like blues, jazz and rock, hip-hop has emerged from the underground to be studied in universities, examined in scholarly tomes and celebrated in museum shows. (*See sidebar, p. 122.*)[23]

"Political context, lack of resources and reappropriation . . . provide an aesthetic context from which hip-hop has sprung," writes actor, playwright and director Danny Hoch in a recent anthology of essays on hip-hop. "They have also informed — and continue to

Colleges Embrace Hip-Hop Studies

'It's worthy of scholarship'

When David Cook submitted his undergraduate thesis — "The Power of Rap" — at the University of California, Berkeley, in 1987, he didn't include sources.

"I handed it in without footnotes," he recalled recently. "I mean, [I was] talking about something I was a part of, something I knew a lot about, and [my professor] was like, 'Footnote something. There's got to be books about hip-hop.' "

But 20 years ago there were few academic works on the subject, so Cook ended up citing Davey D — the alias he still uses as a hip-hop journalist — as somebody quoted in *Bomb* magazine. He received an A on the paper.[1]

Today's students would have no trouble crafting bibliographies — or even taking college courses — on hip-hop and its origins. From "American Protest Literature from Tom Paine to Tupac" at Harvard to "Beginning Hip Hop Funkamentals" — a dance class at UCLA — colleges and universities increasingly are offering courses in tune with the current generation of students. Stanford University's The Hiphop Archive — an educational and artistic repository and research center dedicated to hip-hop culture — estimates 300 such courses are being offered nationwide, with many being taught by professors who themselves were involved in the movement's earlier stages.[2]

Historically black Howard University, in Washington, D.C., became the first college to offer a hip-hop course in 1991 and aims to become the first university to offer a hip-hop minor by 2009. "If you look at the overall impact of hip-hop on youth in the United States and abroad, you see why it's worthy of scholarship," says Howard history graduate student Joshua Kondwani Wright, who is spearheading efforts to establish the minor.

Mark Anthony Neal, an associate professor of black popular culture at Duke University who helped organize Hip-Hop Appreciation Week on campus, adds, "Pop culture, and especially hip-hop now, has a large influence on how people see the world. That demands we take it seriously. We also know that popular culture is a place where ideology is introduced."

Although the study of black culture has been a prominent fixture on college campuses since the end of the civil rights era, the study of hip-hop has been controversial.

In 2004, then-Harvard University President Lawrence H. Summers declined to offer tenure to noted hip-hop scholar Marcyliena Morgan despite unanimous support from the Department of African and African-American Studies. Many speculated the decision reflected Summers' lack of respect for hip-hop's presence in the ivory tower.

Morgan eventually accepted a tenured position at Stanford, taking The Hiphop Archive with her. Just last month, an ad hoc committee at Harvard and Interim President Derek C. Bok offered a tenured professorship to Morgan, reversing Summers' decision.[3]

But only a few within academia belittle the study of hip-hop. "It's a continuum of African-American history and African-American culture," says Wright. Moreover, some scholars are using their knowledge of the subject to propose remedies to social ills.

"We cannot ignore the sociological and economic circumstances out of which the hip-hop culture emerges," says James Peterson, assistant professor of English at Bucknell University, who teaches a course on hip-hop culture and composition. "And effectively addressing many societal problems requires an examination of the hip-hop culture." Peterson, who is also an educational consultant, has often been asked to assess the role of hip-hop in issues such as urban crime.

But will hip-hop remain a viable academic pursuit a generation from now, when society's problems are unlikely to be associated with hip-hop and today's regular fixtures on MTV and Black Entertainment Television are reduced to archives in VH1 Classic?

"If anything, it strengthens authenticity within the academy," says Peterson. "If studied from a historical perspective, hip-hop will become similar to what jazz or blues studies are now."

— Darrell Dela Rosa

---

[1] See Reyhan Harmanci, "Academic Hip-Hop? Yes, Yes Y'all," *The San Francisco Chronicle*, March 5, 2007, p. F1.

[2] For a list of courses, see www.hiphoparchive.org.

[3] See William Lee Adams, "Ivory Tower: Teachings Of Tupac," *Newsweek*, Oct. 4, 2004, p. 45; Lulu Zhou, "Hip Hop Scholar Offered Tenure," *The Harvard Crimson*, May 25, 2007, www.thecrimson.com.

inform — our artistic practice, even when the form or genre varies."[24] By "reappropriation," Hoch means "sampling," or taking snippets from other peoples' work and using them in creating something new.

A fellow essayist, playwright and actress, Eisa Davis, pushes hip-hop's boundaries even further. " 'Scarface,' 'Boiler Room,' 'Ghost Dog' — these are all hip-hop films even though there is nary a graf artist, MC, DJ or b-boy in sight," she writes. "Instead, the protagonists are dealers, schemers and samurai stylists — and through the narrative and the soundtracks comes a stance that reveals hip-hop vision."

But, Davis points out, "The hip-hop stance is not a lifestyle, it's a thinking style, and if you try to define [it], those brain waves will run away from you and set up shop on another corner."[25]

Trying to equate rap music with a samurai film could indeed overload brain circuitry. But even hip-hop artists of less sweeping vision argue that their art form should be seen as expandable. "It's a culture," says Miami-based Mecca aka Grimo (Patrick Marcelin), a Haitian-American hip-hop performer. "It expresses views and opinions; it has its own language, fashion, the way we walk — that bop, that swagger. It's a subculture that's evolved in the inner city and has become global."

He adds: "We've got to take the good with the bad. I can't tell you the pants hanging down or the cursing isn't part of the culture. But music can definitely be used as an educational tool to change these things."

Performance poetry, recited without accompaniment and known in its rap-spinoff form as "spoken word," is the clearest example of hip-hop culture extending itself beyond its turntable-spinning roots.

In its original form, hip-hop consisted of DJ-ing, or manipulating turntables, volume levels and other features of music tracks; MC-ing (rapping over an accompaniment); graffiti spray-painting and break-dancing.

"I still believe in hip-hop as its four original elements," says hip-hop historian Bynoe. "When you talk about hip-hop movies, I don't know what that means. Is it a movie about black people? A lot of times, 'hip-hop' is, like 'urban,' a euphemism to mean 'black people.' I don't think 'hip-hop' is a concrete enough definition to make it valid beyond the four elements."

Bynoe doesn't object to using hip-hop as a device to get students interested in academic subjects. And she

acknowledges that hip-hop theater may be the one valid extension of the hip-hop genre. But she cautions those involved in such projects against assuming "just because you're poor or of color that hip-hop is the only thing you're interested in." In her own case, she says, "There have been some old, dead white guys whose writing and work were relevant to me."

In May, Chicago's Museum of Contemporary Art hosted the Hip-Hop Theater Festival, which sought to introduce theater to inner-city residents who aren't part of the high-culture scene.

The festival showcased work "that has been embraced by a younger population," says Yolanda Cesta Cursach, the museum's associate director of performance programs. She says hip-hop theater is characterized by "a young attitude toward politics in the urban environment — race, class, getting a job, getting incarcerated. It's really about putting a creative face on the politics of urban life."

The performers included a women's ensemble, "We Got Issues," which presented rhymed tales of women's lives; Jerry Quickley, a performance poet whose antiwar piece, "Live from the Front," tells of his adversarial reporting experiences in Iraq; and "Still Fabulous," a story-telling and singing group that was formed at the Illinois Youth Center, a female juvenile detention center.

Rapping on a stage without music may not be a far cry from MC-ing on a dance floor. But writer-musician McBride argues that trying to extend hip-hop much further from its origins may be a fruitless exercise. For example, he says, "I don't see hip-hop in 'Spider-man 3.' Other than quick cuts, the snap-crackle-pop immediacy of in-your-face imagery — if that's what hip-hop is, then yes, it's in 'Spider-man 3.' But I see hip-hop as a deeper question about identity and purpose and sense of place."

Is hip-hop becoming a political force?

Hip-hop and politics have been entwined since the genre's birth. Musically, hip-hop may have started as good-time dance music, but the spray-painted graffiti art style that "taggers" adopted as the visual counterpart of DJ-ing and MC-ing was illegal. Its practitioners and fans celebrated graffiti "tagging" as a reclaiming of urban space by its marginalized inhabitants. "Graffiti writers had claimed a modern symbol of efficiency and progress and made it into a moving violation," writes hip-hop chronicler Jeff Chang

Getty Images/Scott Gries

Sean "P. Diddy" Combs announces his Citizen Change Campaign schedule for the 2004 election season, with Democratic political consultant James Carville. The campaign is one of several hip-hop-driven efforts to register young voters and promote political awareness.

about subway-car taggers. "Authorities took their work as a guerrilla war on civility. They were right."[26]

All of hip-hop was shaped, Chang writes in his definitive history, by the social and political upheavals of its time. These include the end of the civil rights era, the crack epidemic, the incarceration boom and globalization.

Hip-hop stars recognize the political roots of their genre and have lent their talents and checkbooks to a variety of causes. In 2003, hip-hop moguls Combs and Simmons helped organize a movement that campaigned for repeal of New York's harsh Rockefeller drug laws.[27] When the campaign failed, some politicians and activists blamed Simmons for settling for small-scale reforms.[28]

Star power may have worked better in registering young people to vote. In 2004, the Hip-Hop Summit Action Network — Simmons is co-chair and Combs is a director — launched a campaign to sign up 2 million new, young voters. The organization's other directors include the top executives of three major record labels and an ex-president of the powerful Recording Industry Association of America.[29]

Combs ran his own registration-and-turnout campaign as well, under the slogan "Vote or Die!" "When the president is running your country," he said in 2004, "he is running you, closing you out of a hospital, or taking you to war. Not just Bush. Every president."[30]

Perhaps aided by the campaign, the number of overall registrations of Americans under age 30 between the 2000 and 2004 presidential elections increased by 3.1 million, according to George Washington University's Graduate School of Political Management. Measured as a share of the general voting public, the percentage of young registered voters rose from 55 percent of the general population to 60 percent. And turnout among young voters rose by 4.3 million between the two elections — with African-Americans and Latinos accounting for more than half the surge.[31]

Youth turnout increased by at least two percentage points in the 2006 midterm elections, compared to 2002 (with Democrats reaping most of the benefits), according to the school's Young Voter Strategies project.[32]

Even so, the question remains whether hip-hop represents a marginalized subculture or a rising tide of political protest.

Some politically minded hip-hop community members argue that to focus on big-scale projects led by entertainment-industry giants is to miss the point, in part because they work with the conventional political structure. "Our country has to change, and I don't think it's going to happen through this Democratic-Republican Party medium we have," says Bakari Kitwana, co-founder of the National Hip-Hop Political Convention. "That's one party to me. Why can't we get money into education the way we've dumped money into Iraq? That would be revolutionary, and it would be hip-hop."

Kitwana, former executive editor of *The Source*, a leading hip-hop magazine, argues that the real power of hip-hop politics can be felt in community organizations that have led fights to improve schools or oppose police brutality. "What I'm talking about is off the radar," he adds. "You're not going to see it on CNN."

But Marc Lamont Hill, an assistant professor of urban education and American studies at Temple University in Philadelphia, cautions against overestimating the power of hip-hop. "If Jay-Z did a song tomorrow criticizing multinational capitalism, that's great," he says, "but even if that happens, it has to be connected to a real struggle. We need people who consume hip-hop culture to think about political conditions they can affect through their activism. If we don't think about it that way, all we're doing is making some cool music. We hip-hop scholars and academics sometimes romanticize cultural politics."

Davey D (David Cook), a radio DJ, blogger and newspaper columnist in Oakland, Calif., suggests hip-hop politics is in flux. Some influential voices in the community, for instance, challenge the usefulness of voting. The rapper Nas (Nasir Bin Olu Dara Jones) "would get on the stage and say, 'Don't vote, they're not counting our votes.' And his message gets reinforced inadvertently by grass-roots people who feel that the system is too corrupt."

The mass-market political side of hip-hop suffers from its own weaknesses, Davey D says. "Russell [Simmons] and them have raised awareness, but they didn't do voter education. There's not a lot of activism with them. I would describe activism as being politically engaged and seeing the system, or various touch points in that system, as something that needs to be dealt with day in and day out."

Still Davey D remains committed to the idea that the hip-hop community represents a political force, whether at the polls or in community matters. On the other hand, Southern-style rapper Banner sees a wave of political disillusionment spreading as a result of the 2004 Republican presidential victory.

"We had all these kids hyped, and when it didn't turn out the way it was supposed to, they're not going to vote for the next 10 years," Banner says. "A lot of people from the Republicans said, 'If it was about vote or die, then most of you people are dead.'"

So far, events haven't borne out Banner's gloomy outlook. But activists who see themselves as part of the political hip-hop community are trying to avoid the outcome he predicts.

"We rap, we utilize the art," says T. J. Crawford, 31, a community organizer whose Chicago Hip-Hop Political Action Committee registered 17,000 young people in 2004 and is now working on voter education. "The most important thing is making voting relevant, so that people understand how the system works."

BACKGROUND

Made in the Bronx

The cradle of hip-hop lies in the Bronx, a borough of New York City that has nurtured generations of middle- and working-class families, many newly arrived in the United States, including Colin L. Powell. The former chairman of the joint chiefs and Secretary of State grew up there in the 1940s and 50s, the son of Jamaican immigrants.[33]

But the engine of upward mobility that propelled Powell and thousands of other Bronxites into college and greater prosperity was sputtering by the late 1970s. New York was shedding its role as a manufacturing center, losing thousands of factory jobs in the process. At the same time, the civil rights and anti-Vietnam War movements had created a current of social rebelliousness that was still running strong.[34]

Amid this sociopolitical heavy weather a new musical style blew in from Jamaica, where many musicians and singers were leaving for jobs in Britain and the United States. The migration of performers forced disc jockeys to take over the top spots in an active song-and-dance culture. "Outfitted with powerful amplifiers and blasting stacks of homemade speakers, one only needed a selector and turntable to transform any yard," writes Chang in *Can't Stop Won't Stop: A History of the Hip-Hop Generation*.[35]

In 1967, the DJ's role expanded even further when a record-spinner happened upon a disk with the vocals missing — the result of an engineer's error. The DJ used the record on his next gig, as a rhythmic foundation for spouting rhymes. Dancers loved it, and the trend took off.

That same year, the emigrés included 12-year-old Clive Campbell, who arrived in New York with powerful memories of Jamaica's sound-system DJs. Clive's father took up DJ-ing as a sideline, buying big speakers and an amplifier to play for house parties.

By the time he reached high school, Clive was following in his father's footsteps. But he rewired the equipment to push more power to the speakers and to mix echo and other effects with his patter as DJ Kool Herc — borrowing from a popular brand of cigarettes and the name of the mythological strongman.

When his sister Cindy threw a dance party on Aug. 11, 1973, in the community room of the family's apartment building on Sedgwick Avenue, Kool Herc made his debut — and so did hip-hop. Cindy used her share of the entrance fee to buy clothes for school.

Going Global

As news of the event spread and demand for Herc's services rose, he came up with a trend-setting innovation.

CHRONOLOGY

1970s–1980s *Hip-hop is born in the Bronx and spreads throughout New York City before starting to take off elsewhere.*

Aug. 11, 1973 The first recognized hip-hop event takes place in the Bronx at a dance in an apartment building's community room.

October 1979 "Rapper's Delight," by the Sugar Hill Gang, becomes the first hip-hop record to be released.

March 1981 Afrika Bambaataa and the Jazzy Five release "Jazzy Sensation," which fuses dance-funk beats and so-called electronica sounds and helps widen hip-hop's appeal.

1984 Hip-hop powerhouse Def Jam Recordings issues its first record.

1988 Public Enemy releases "It Takes a Nation of Millions to Hold Us Back," which merges hip-hop with black nationalism.

1989 The Los Angeles-based group N.W.A. (Niggaz With Attitude), creates "gangsta rap," characterized by the song "Fuck Tha Police."

1990s *Hip-hop becomes a force in the entertainment industry, even as rap becomes increasingly identified with graphic depictions of sex, misogyny and glorification of violence and drugs; criticism mounts, as do sales.*

1992 Supreme Court lets stand an appeals court's ruling that Miami-based 2 Live Crew's "Nasty As They Want To Be" is not obscene. Bill Clinton, the Democratic presidential nominee, attacks rap performer Sister Souljah (Lisa Williamson) for comments about the Los Angeles riots, calling her remarks racist.

Feb. 23, 1994 Rap lyrics are both attacked and described as reflecting the lives of urban youths at a Senate Judiciary Committee hearing.

Sept. 13, 1996 Gangsta rapper Tupac Shakur dies after a gangland-style hit on the Las Vegas Strip. About six months later, rival rapper The Notorious B.I.G. (Christopher Wallace) dies in a similar shooting in Los Angeles.

Dec. 27, 1999 Sean Combs (then known as "Puff Daddy") is arrested for gun possession outside a New York nightclub; he is later acquitted, though a bodyguard gets prison time for having opened fire.

2000s *Music industry steadily loses steam; hip-hop magnates and stars diversify and take up politics, while "dirty south" rap, with its raunchy lyrics that many say degrade women, gains in popularity.*

December 2003 Rapper Jay-Z joins a consortium that eventually buys the NBA's New Jersey Nets.

2004 In April, students at Atlanta's Spelman College demand rapper Nelly meet with them to discuss his raunchy "Tip Drill" video, but he refuses and also cancels a charity fundraising appearance at the women's college. . . . The Hip-Hop Summit Action Network and Sean "P. Diddy" Combs each organize national voter-registration campaigns focused at African-American youth. . . . Chicago Hip-Hop Political Action Committee registers 17,000 young voters.

2005 Record and cable TV executives at a Spelman panel discussion try to explain why they market videos that portray women as sex toys.

2006 Hip-hop recordings slip to 11.4 percent of U.S. market share, from 13.3 percent in 2005.

2007 On Jan. 1, Atlanta police raid a major mixtapes production operation. . . . In April, radio shock jock Don Imus is fired for describing the Rutgers women's basketball team as "nappy-headed hos," spurring debate about hip-hoppers' use of similar language. . . . Under pressure from TV personality Oprah Winfrey and other African-American leaders, the Hip-Hop Summit Action Network calls for a ban on "bitch," "ho" and "nigger" in recordings . . . . Rappers Master P and Chamillionaire announce plans to tone down street language in the future. . . . Rapper Cam'ron is criticized after telling "60 Minutes" he wouldn't call the police even to protect neighbors from a serial killer; he later apologizes for the comment. . . . The "Hip-Hop Project" opens; the Bruce Willis-produced documentary depicts young people learning artistic expression through rap. . . . "Dirty south" rapper David Banner transports relief supplies to Hurricane Katrina-ravaged areas and organizes a fundraising concert for hurricane victims.

Other DJs were already using two turntables, so that a new song could start when the first one ended. But Herc would put the same tune on each turntable, allowing him to continue the bass-and-drums "break" section of a dance song as it was fading away on the other record. Dancers — who loved the breaks — went wild.

Other Bronx DJs soon adopted the technique, adding the Jamaican technique of rhyming over the beat. "Rapping" spread through other neighborhoods and eventually to clubs. All it took for the national spotlight to hit the localized craze was one record — the Sugar Hill Gang's "Rapper's Delight," which came out in 1979.

"Imitations popped up from Brazil to Jamaica," Chang writes. "It became the best-selling 12-inch single ever pressed."[36] The tune's first words: "I said uh hip-hop. . . ."

By 1980, even *The Washington Post* was taking notice of "the newest craze among the 14-to-21-year-olds, the record-buying majority who are putting rap records on the national charts and making money for the nightclub disc jockeys capitalizing on a bit of New York City party culture."[37]

As rap's popularity soared, its home turf was becoming a grittier, tougher place. City government was headed toward bankruptcy. Education, garbage collection and other public services were strained to their limits.

Meanwhile, the election of Ronald Reagan as president in 1980 marked a turn toward conservative policies that hit poor communities across the country. Reagan's first years in office saw a tax cut that he championed swing the country into a recession marked by unemployment and cutbacks in state and local assistance programs.[38]

Although the national economy slowly improved, many poor communities were assaulted by the arrival of a cheap, addictive stimulant — crack cocaine — and barely felt the change. The ensuing crack epidemic devastated inner cities across the country. Crack left users wanting another hit almost immediately. High demand helped make crack a big business in job-starved ghettoes, where local drug sellers served neighborhood customers as well as users driving in from the suburbs.[39]

As addiction and warfare between rival drug-trafficking gangs ravaged poor neighborhoods, law enforcement cracked down. New sentencing laws imposed tougher penalties for crack trafficking than for dealing in powdered cocaine, which didn't make an inroad in inner cities because of its relatively high price.[40]

Inevitably, popular culture, including hip-hop, began to reflect the distressing, new inner-city realities.

Sex and Violence

The lighthearted dance-party flavor that characterized early rap soon took a turn toward hard-edged social commentary with a black-nationalist spin. A key source of inspiration was a pre-hip-hop group of African-nationalist verse writers, The Last Poets, who had released a groundbreaking album of poetry recited over percussion backing in 1970. The Long Island, N.Y., group Public Enemy took the idea into the hip-hop age with an album, "It Takes a Nation of Millions to Hold Us Back," released in 1988.[41]

Public Enemy soon became engulfed in a controversy over anti-Jewish comments by Professor Griff (Richard Griffin), the group's "minister of information." After he blamed Jews for "the majority of wickedness that goes on across the globe," Griff was dropped by the group and then rehired, before it broke up for a time.

Public Enemy's founder, Chuck D (Carlton Douglas Ridenhour) became, and remains, a leading spokesman for rap as a form of social and political commentary and activism. His description of rap as "the black CNN" stands as one of the most widely circulated observations ever coined about hip-hop.

But the comment had a perhaps-unintended consequence — it tended to validate "gangsta" rap. The new hip-hop genre seemed to spin off from Public Enemy's work but departed from its message of social improvement: Violence and the drug trade were glorified; police were threatened; and women were treated purely as sex objects and prostitutes.

The Los Angeles-based group N.W.A. (Niggaz With Attitude) is considered the founder of gangsta rap, though arguably it was depicting the drug war rather than celebrating it, as the group's imitators did. In 1989, sales of N.W.A.'s second album, "Straight Outta Compton," reached gold-record status — 500,000 copies sold — within six weeks and eventually racked up 3 million sales. Its degradation of women and profanity made radio airplay impossible, so the album sold initially through a below-the-radar distribution system of mom-and-pop record stores in black communities, then through word-of-mouth among white teenagers. Once stores in white neighborhoods stocked it, "That's all it took," a record

Hip-Hop Stars Are Branching Out

Tycoons have clothing lines, TV shows

Hip-hop has come a long way from the community rooms and street corners where pioneers in the 1970s and '80s developed the arts of sampling and rapping and sold cassettes out of car trunks.

And in the view of hip-hop tycoon Russell Simmons — who has branched out into fashion, TV and jewelry — the genre still has a long way to go. "Hip-hop is not fully exploited," Simmons said.[1]

He should know. Simmons' hip-hop enterprises are worth upwards of $325 million. But music sales — traditionally hip-hop's top moneymaker — are hurting, as the music business struggles to adjust to a digital world in which consumers can easily and cheaply acquire music over the Internet.

Sales of hip-hop recordings and downloads in the United States reached about $131 million in 2006 — or 11.4 percent of the $11.5 billion U.S. music market — according to the Recording Industry Association of America (RIAA).[2] And while breakdowns by genre aren't available for the $33 billion global recorded-music market, two hip-hop albums — by 50 Cent and Eminem — were among the world's top 10 bestsellers in 2005, the most recent figures available.[3]

But music sales are plunging, both nationally and globally — by 6.2 percent in the United States and 3 percent worldwide in 2005. And sales of "physical" products, such as CDs, are down 6.7 percent, according to the International Federation of the Phonographic Industry.[4]

The diversified Simmons doesn't worry about those problems, however. The co-founder of hip-hop's first record label, Def Jam Recordings, is out of the music business now; his Rush Communications produces movies, TV shows, yoga DVDs and "urban wear."

Simmons' fellow hip-hop pioneers haven't done too shabbily either. Sean "P. Diddy" Combs — a performer as well as entrepreneur — is a designer, actor and international

fashion icon. Def Jam co-founder Rick Rubin, whose work ranges far beyond hip-hop, won his fifth producer-of-the-year Grammy in February and is considering an offer to become co-chairman of Columbia Records. And Lyor Cohen, another Def Jam alum, is chairman and CEO of U.S. Recorded Music for Warner Music Group.[5]

A handful of other rappers have also hit it big, though not quite on Simmons' level. Ice Cube (O'Shea Jackson), once a member of the much-condemned N.W.A., has diversified into acting, producing and screenplay writing for mainline Hollywood features. Jay-Z (Shawn Corey Carter) sold his Rocawear clothing line in March for $204 million, owns a chain of upscale clubs and helps promote Budweiser Select beer. He recently bought a minority share of the NBA's New Jersey Nets and is still president and CEO of Def Jam. And Lil Jon (Jonathan Smith), a rapper and producer in the Southern "crunk" style, also produces clothing and has starred in a video game and a Comedy Central cartoon show.[6]

With the push for diversification, it might appear that rappers are hedging their bets on the recording business. Rap album sales dropped 20 percent in 2006, and the top 10 best-selling albums that year included no rap records — an occurrence not seen for more than 10 years. "Hip-Hop Is Dead," declared the title of a late-2006 album by rapper Nas.[7]

No funerals are scheduled, but rap has slipped to third place (after country music) in U.S. music sales — from 13.3 percent of market share in 2005 to 11.4 percent in 2006. "It's not losing popularity," says David Banner (Levell Crump). "Downloading is up, and the music industry is going down."[8]

A top industry-watcher agrees. "I don't have any data that supports that hip-hop, particularly, is down," says Don Gorder, chair of the Music Business/Management Department at Berklee College of Music. In general, he adds, "Demand for

---

salesman said years later. Eventually, white suburban kids accounted for about 80 percent of sales.[42]

One of the album's cuts, "Fuck Tha Police," was structured as a snapshot of life on the street:

. . . Searchin' my car, lookin for the product
Thinkin' every nigga is sellin narcotics . . .
Just cuz I'm from the CPT [Compton], punk police are afraid of me

A young nigga on a warpath
And when I'm finished, it's gonna be a bloodbath
Of cops, dyin' in LA.[43]

"Gangsta, gangsta" was the title of another selection, giving the genre its name.

A pushback against gangsta rap began almost immediately, fueled by black activists and police. An assistant director of the FBI sent the record company a letter accusing the firm

music has never been greater. It's a strong market. The business models are having to adjust."

One adjustment may be dealing with "mixtapes" — CD albums with songs not yet officially released, designed to test the market and keep a performer's work before the public between formal releases. The little-known sector of the business — the name is a holdover from the old days of cassettes — came under scrutiny after an Atlanta recording studio was raided in January by police. Accompanied by RIAA officials, the police grabbed two men — along

Hip-hop entrepreneurs Sean "P. Diddy" Combs, Jay-Z and Russell Simmons (from left) have built multi-million-dollar businesses involving street clothing, jewelry, sports, television and film.

jewelry, the old hip-hop hustle is still alive and well at street level — and some mixtapes are even sold out of car trunks.

with 25,000 CDs — on charges of illegally producing CDs. (The RIAA later said the police, not the association, had instigated the raid, but industry insiders were skeptical.)[9]

The men, Tyree Simmons and Donald Cannon, turned out to be two of the country's best-known DJs and are well known for producing mixtapes.[10]

Derrick Ewan, an XM Satellite Radio DJ who goes by the name Furious Styles, says the record companies actually help get mixtapes made. "The label wants you to make tapes," Ewan says. "They give you the songs and in some cases pay you to make a mixtape. They don't care if you sell them." Ewan says he hasn't received such payments or run afoul of copyright laws because the CDs he makes showcase unsigned acts — his DJ niche.

Still, mixtapes are easily available at independent record stores and even online, and the business has its own wholesalers. For all the millions made in hip-hop clothing and

[1] See Mindy Fetterman, "Russell Simmons can't slow down," *USA Today*, May 14, 2007, p. B1.

[2] "2006 Consumer Profile," Recording Industry Association of America, undated, www.riaa.com/news/marketingdata/pdf/2006RIAAConsumerProfile.pdf.

[3] See Jeff Leeds, "Music Industry's Sales Post Their Sixth Year of Decline," *The New York Times*, April, 1, 2006, p. C2; "Digital formats continue to drive the global music market — World Sales 2005," International Federation of Phonographic Industries, March 31, 2006, www.ifpi.org/content/section_news/20060331a.html.

[4] "The Recording Industry World Sales 2001," International Federation of the Phonographic Industry, April, 2002, www.ifpi.org/content/library/worldsales2001.pdf.

[5] See Fetterman, *op. cit.*; Alana Semuels, "Rappers hear siren song of opportunity,: *Los Angeles Times*, March 12, 2007; Sia Michel, "A New Sound for Old What's-His-Name," *The New York Times*, Sept. 10, 2006, p. B67; Robert Hilburn, "The Music Industry Titans — Rick Rubin; A balance of rattle and om," *Los Angeles Times*, Feb. 11, 2007, p. F1; Charles Duhigg, "Q&A; Getting Warner Music More Upbeat," *Los Angeles Times*, Aug. 28, 2006, p. C1; "Lyor Cohen," (official biographical sketch) Warner Music Group, undated, www.wmg.com/about/biography/?id=contact400004.

[6] See Yvonne Bynoe, *Encyclopedia of Rap and Hip Hop Culture* (2006), pp. 178-181; Semuels, *op. cit*; Steve Jones, "Jay-Z is a very busy man," *USA Today*, Nov. 21, 2006, p. D5.

[7] Semuels, *op. cit.*

[8] For market-share statistics, see "2006 Consumer Profile," *op. cit.*

[9] See Samantha M. Shapiro, "Hip-Hop Outlaw (Industry Version)," *The New York Times Magazine*, Feb. 18, 2007, p. 29.

[10] See *ibid.*

of encouraging "violence against and disrespect for the law-enforcement officer.' " The letter stoked consumer interest even more.[44]

Some politically minded hip-hop radio D.J.'s, led by Davey D, in the San Francisco Bay area, organized a boycott of N.W.A., largely because of its repeated use of the N-word. But "progressive" college radio stations refused to take part in what they called censorship of black rappers,

whom college kids considered voices of the inner city. "They were fascinated with anger from the ghetto," Davey D says now. "I live in the 'hood, and I'm not fascinated by it."

Opposition to gangsta rap merged with another current that had been building since 1985. A group of political wives began decrying rock songs' lyrics, especially explicit sexual references. Tipper Gore, wife of then-Sen. Al Gore, D-Tenn., and Susan Baker, wife

Hip-Hop Pioneers

Kool Herc (top) introduced break-beat DJ-ing in the 1970s, in which the percussive instrumental "breaks" of songs are isolated and repeated during dance parties. Bronx DJ Afrika Bambaataa (middle) pioneered the use of electronic drum machines to create hip-hop beats. In the 1980s, Public Enemy (bottom) became one of the first mainstream hip-hop groups to promote social and political activism.

of then-Treasury Secretary James A. Baker III, a Republican, founded the Parents Music Resource Center to demand ratings and voluntary warnings on albums, which performers and record companies opposed during a congressional hearing. But in 1990, record companies began putting warnings on albums with language that could be considered offensive.[45] That year, with hip-hop's popularity soaring, Gore began echoing African-American concerns over gangsta rap lyrics.[46]

At roughly the same time, explicitly sexual lyrics in albums produced by Luther Campbell of Miami led to litigation that reached the U.S. Supreme Court. Campbell was an entrepreneur-turned-rapper who produced and then joined the group 2 Live Crew. Among its songs: "We Want Some Pussy" and "Dick Almighty."

A series of court battles in Florida led eventually to a 1992 ruling by the 11th U.S. Circuit Court of Appeals in Atlanta that one of the group's albums, "Nasty as They Wanna Be," was not obscene. The Supreme Court let the decision stand. The years that followed saw gangsta rappers indulging in vulgarity as enthusiastically — and profitably — as they did in violence.

The Big Time

By the end of the 1990s, rap was a billion-dollar business — much of the revenue generated by gangsta rap purveyors like California-based Death Row Records.[47]

But the boom had been started on the East Coast by Simmons, then a budding rap promoter, and Rick Rubin, a white hip-hop fan with ambitions to manage rappers and produce recordings. In 1979, they founded Def Jam Productions.

Def Jam soon branched out from promoting and producing rap shows to recording rappers. After signing an unknown rapper who became a big-time hitmaker, LL Cool J (James Todd Smith III), the fledgling Def Jam signed a distribution deal with CBS/Columbia records and went on to produce some of the biggest hip-hop acts.

The money rolled in. Simmons and his partners (Rubin had left the firm) sold Def Jam to Universal Music Group for $130 million in 1999.

Another sign of hip-hop's new status was MTV's entry into the rap world, following years in which the channel didn't show black artists' videos. In 1988, the TV show "Yo! MTV Raps," began a swing toward rap that, a year later, had the channel showing rap videos 12 hours a day. Black Entertainment Television (BET) likewise came to rely heavily on hip-hop videos.

Hip-hop's sudden new status grew out of an appeal that reached beyond the inner cities. White suburban appetites for gangsta rap grew all the more intense as some stars began living up to the name, none more than Tupac Shakur. The California rapper reached the heights of hip-hop stardom during a five-year solo career that ended in September 1996 when he died at age 25 in a hail of bullets in Las Vegas.[48]

Shakur, who combined genuine talent with an attraction to the "thug life," was seen by many as a player in an East Coast-West Coast feud between California-based Death Row Records — for which he recorded — and New York-based Bad Boy Entertainment, headed by Combs. The label's biggest star, The Notorious B.I.G., was gunned down in Los Angeles six months later. Both murders remain unsolved.[49]

Controversy

The revulsion to rap engendered by N.W.A. and 2 Live Crew intensified. In New York, columnist Crouch kept up a steady drumbeat of fierce criticism beginning in the 1990s. "Illiterates with gold and diamonds in their teeth" Crouch called hip-hop entertainers. He slammed them for, among other things, marketing a stereotype of inner-city life that both titillated white kids and damaged black inner-city kids by making the thug a hero.[50]

Throughout the '90s, a growing number of African-American critics — especially women — joined in the criticism. C. Delores Tucker, a Democratic politician from Philadelphia who had founded the National Political Congress of Black Women, spoke for "many disillusioned, middle-class, middle-aged people of color," writes hip-hop chronicler Chang.[51]

With generational and class fault lines in black America becoming exposed, Democratic presidential candidate Bill Clinton — renowned for his connection to the African-American public — harshly criticized popular rapper Sister Souljah (Lisa Williamson) for comments she'd made after the near-fatal beating of a white truck driver by black youths during the Los Angeles riots of 1992. "I mean, if black people kill black people every day, why not have a week and kill white people?" she said.[52]

She went on to indicate that she was describing rioters' state of mind rather than espousing violence. But

diverseimages/Getty Images/Chi Modu

Rapper Tupac Shakur was a major player in the feud between California-based Death Row Records — for which Shakur recorded — and New York's Bad Boy Entertainment. Shakur was killed in a Las Vegas drive-by shooting in 1996. The Notorious B.I.G., Shakur's rival and Bad Boy's most popular artist, was killed in a similar shooting in 1997 in Los Angeles.

Clinton pilloried her anyway. Speaking at a meeting of the Rev. Jesse Jackson's Rainbow Coalition, he said: "If you took the words white and black and you reversed them, you might think [former Ku Klux Klan leader] David Duke was giving that speech." Political writers showered Clinton with praise for what they saw as a courageous challenge to an outspoken member of one of the Democratic Party's major constituencies.[53]

In 1994, attacks on gangsta rap converged with growing concern about violence on TV and in video games, prompting House and Senate hearings.

"It is an unavoidable conclusion that gangsta rap is negatively influencing our youth," Tucker told the Senate

New York rapper Nas, who leans toward "conscious" hip-hop, titled his controversial 2006 album "Hip Hop Is Dead" because he thinks the genre has become too commercialized and has lost its street credibility.

Judiciary Committee on Feb. 23, 1994. "This explains why so many of our children are out of control and why we have more blacks in jail than we have in college."[54]

Others argued that Tucker and her allies were, in effect, advocating killing the messenger. "Gangsta rappers are an easy target," Michael Eric Dyson, now a professor of humanities at the University of Pennsylvania, told lawmakers. "We should be having a hearing on crime and on economic misery."[55] Dyson authored the provocative 2006 book *Is Bill Cosby Right? Or Has the Black Middle Class Lost its Mind?*

None of the hearings led to legislation. Instead, the debate continued bubbling, especially within the African-American community.

The outrage flared in 2004 at Atlanta's Spelman College. Rapper Nelly had been scheduled to visit the prestigious, historically black institution for women to encourage bone-marrow donations. But then students became aware of a Nelly song and video, "Tip Drill" — street slang for a homely woman with a good body. The video, in which David Banner appeared, featured the usual gyrating, bikini-clad women and an offensive scene in which the rapper swipes a credit card between a woman's buttocks.[56]

The students had planned to confront Nelly about the song, but he canceled his appearance. "Spelman is 10 blocks from a strip club," Nelly said. "You're not out in front of the strip club picketing."[57]

A panel discussion held the following year at the school showed that opinions on both sides remained firm. Bryan Lynch, an executive with TVT Records, a leading independent label that's home to some gangsta rappers, told the audience that record companies catered to record buyers' tastes. Panel mediator Michaela Davis, fashion editor of *Essence* magazine, snapped back: "Crack sells, too."[58]

CURRENT SITUATION

Imus Fallout

After the brief outburst of remorse that followed Imus' sudden fall, the captains of the hip-hop industry are keeping their microphones switched off. Calls and e-mails from reporters (including *CQ Researcher*) to top record and radio executives are going unanswered. A press conference that Simmons' Hip-Hop Summit Action Network had announced for a date in May to discuss language in rap songs was canceled.[59]

"I expect them to wait it out," says Don Gorder, chairman of the Music Business/Management Department at Berklee College of Music in Boston. "I would be surprised to see any strong policy statements. In pure business terms, people vote with their pocketbooks. Is it record companies' responsibility to steer them away from [gangsta rap]? I don't think they'd say that."

Some rappers made their own evaluations of how the wind was shifting — or re-examined their consciences — and acted accordingly. Master P founded Take a Stand Records to market "street music without offensive lyrics."[60]

In an open letter to the news media, he said, "Oprah Winfrey is absolutely right. We need to grow up and be responsible for our own actions. . . . Most artists' mission is to sell records. My mission is to help save and change lives."[61]

Best-selling rapper Chamillionaire spoke in similar terms. "On my new album, I don't say the word n***a," he told the hip-hop news site, AllHipHop.com. "I guarantee if I don't go out and say it in the media they're not even gonna realize that. People go back and listen to all of my old mixtapes and don't even realize that I wasn't even doing

AT ISSUE

Should hip-hop artists produce material that is socially uplifting to African-Americans?

YES

Heru Ofori-Atta
Poet, author of The Unapologetic African:
Inside The Mind of a Frontline Poet *

Written for *CQ Researcher*, June 2007

According to the American worldview, an artist should be "free" to do whatever he wants as long as he doesn't disrupt capitalism and/or white supremacy. In an intact African worldview, the village is more important than any one person. As an unapologetic African, I care about my people more than I care about hip-hop. I care about the minds of black children more than I care about my freedom of speech. I care about setting high standards for my people more than I care about capitulating to the low cultural standards set by white culture.

Any art form created by people of African descent must speak to the best of who we are and our aspirations. Art that doesn't must be thrown away. History will show that this is the correct position of a people who are still oppressed — socially, economically, politically and religiously. In fact, history judges a people based on the cultural artifacts they have left behind. In 3007, what would an archaeologist or anthropologist deduce about the African-American culture of today? The conclusion would not make our descendants proud.

I recently appeared on a panel with an artist who said if he wakes up in the morning and feels like writing about having sex with groupies and smoking crack, he would write it and feel "proud" to perform it. After all, that's what freedom of speech is about. Panelists said they didn't want to preach to their fans. I informed them that when an artist sings, raps or performs poetry bragging about shopping for acquisitions of illusory splendor, he is preaching for materialism.

The powers that be, who have always been against the progress of people of African descent in this country, have made sure the only artists to wield power are the village idiot and not the village freedom fighter and sage. The ruling elite saw the effects on the tastes, interests and values of American society when political dissenters were popular musicians in the 1960s and '70s.

As Nigerian musician and activist Fela Anikupalo Kuti said, "Music is a weapon." A weapon can be used to 1) commit suicide, 2) unjustly hurt others or 3) defend oneself, one's family and home. Black artists must ask themselves what are they using this weapon for? I proudly and unequivocally choose No. 3.

* Forthcoming in 2007.

NO

Marc Lamont Hill
*Assistant professor of urban education and
American studies, Temple University*

Written for *CQ Researcher*, June 2007

More than any other musical genre in history, hip-hop has been linked to the sociopolitical fortunes and futures of African-American and Latino youth. Critics, fans and even artists themselves argue that hip-hop practitioners should produce music that socially uplifts its constituents. Although such efforts are highly valuable, it is both unfair and unrealistic to expect all artists to produce "socially uplifting music."

To be sure, artists have always played a critical role in our collective struggles for freedom, justice and equality. From the plantation to the pulpit, black and brown creative expression has been an indispensable weapon against the most vicious forms of oppression. While hip-hop artists should not be excluded from this tradition, we must expand our understanding of "social uplift" in ways that move us beyond the explicitly political.

From its birth in the streets and parks of post-industrial urban spaces, hip-hop has been more than a reflection of our political predilections. In addition to speaking truth to power, hip-hop, like its cultural forebears, has enabled us to find joy, love and community in the midst of the most absurd circumstances. While songs about a fresh romance, a newborn baby or a fun party may not end sexism, racism or poverty, they allow us to sustain a sense of hope and possibility against the most discouraging odds. Such sensibilities are as important to social-justice struggles as political education or cultural nationalism. From this perspective, Public Enemy's exhortations to "Fight the Power" and dead prez's critiques of the school system are just as valuable as Lupe Fiasco's stories about his skateboard.

Additionally, by placing exorbitant political demands on hip-hop artists, we easily misplace our political hopes in ways that compromise our progress. Despite the importance of an incisive lyrical critique or a provocative slogan, such things cannot replace the critical work of organizing, voting, striking or marching. By overestimating the power of cultural politics, the hip-hop generation absolves itself of the responsibilities of real, on-the-ground engagements with the public and its problems.

Like all human beings, hip-hop artists have moral obligations that necessarily inform their professional and personal lives. As such, they should feel compelled not to create any music that degrades, assaults or otherwise contradicts the values and goals of our respective communities. But they should also feel empowered to produce art that reflects their own feelings, desires and beliefs about the world — uplifting or not.

all that type of stuff. I was saying n***a, but I wasn't saying the F-word or [the] B-word. I was never saying those types of things. . . . I hear that so much, and it restricts your creativity and how far it can go."[62]

Nevertheless, for members of the hip-hop world who had been sounding the alarm for years over rap lyrics, the record and radio executives bear the biggest responsibility. "Nobody is saying that porn stars need to make more substantive and dramatic films," says Industry Ears co-founder Fager. "Why are we looking at rappers to change their content? It's the gatekeeper who allows this stuff to be made and seen."

That take on the matter can be found throughout the loosely knit community of the hip-hop world's internal critics. "Certainly we've heard rap artists make [sexist] comments," says historian Bynoe, "but they don't own record companies, for the most part, or radio stations or cable companies. The decisions about what content will be aired or produced isn't up to rap artists. If tomorrow these entities decide that material is not going to be broadcast, it would end."

Whether broadcasters and cable companies are likely to be forced into that position is another question. "Congress has bigger targets in its sights," says Adam D. Thierer, senior fellow at the Progress & Freedom Foundation, a think tank funded by telecommunications and entertainment giants, including Sony Music Entertainment and Time Warner. "They are so obsessed with regulating violence on cable TV. That has sucked all the oxygen out of the room."

But that picture could change. A hip-hop equivalent of Janet Jackson's breast-baring "wardrobe malfunction" on live TV during the 2004 Super Bowl halftime show would likely put serious pressure on radio and/or radio executives. "It could lead to them being drug in front of a congressional committee to testify," Thierer says. The Jackson episode led the FCC to fine CBS $550,000. And then Congress raised fines tenfold for broadcasting "indecent" material.[63]

So far, the recent indignation over rap lyrics has the industry avoiding the congressional witness table. But the flare-up that followed Cam'ron's controversial statement on CBS' "60 Minutes" might have blazed brighter if he hadn't beat a retreat.

Days after saying that talking to police about a crime "would definitely hurt my business," Cam'ron acknowledged having refused to help in an investigation in which he

himself had been shot. "But my experience in no way justifies what I said," he said. Looking back now, I can see how those comments could be viewed as offensive, especially to those who have suffered their own personal tragedies or to those who put their lives on the line to protect our citizens from crime."[64]

Good Works

It's not just "hos and clothes." That's what an orphaned young Brooklynite tells a group of other hard-luck kids he's brought together at an alternative high school to make a hip-hop record. Chris "Kharma Kazi" Rolle's idea is to show them how to turn their experiences into material.

After considerable struggle, the kids eventually did make their record, and two New York University film students document their progress. The fledgling filmmakers assembled their material into a movie, "The Hip Hop Project," which opened in May.[65]

Both the record and the movie attracted impressive support — from Simmons, Queen Latifah and movie star Bruce Willis. "In five years, hopefully, there'll be 25 programs in 25 other cities in this country" that replicate the New York project's methods, Willis says in a promotional Web video.[66]

In effect, the movie illustrates what some hip-hop advocates have been insisting for years: that the music is a tool for social betterment, not merely a soundtrack for street crime. "If you're just seeing it on TV, you're only going to see the bad hip-hop," says rapper Blitz the Ambassador, who greets the movie as validation. "Hip-hop helps these kids cope with extreme environments they are born into," he says.

In effect, the project also supports the vision of hip-hop theater activists, who argue the art form is made to order for artistically undernourished kids who need to express themselves. "Let's not isolate the voices of these young people," says Clyde Valentín, executive director of the Hip-Hop Theater Festival, which travels from city to city. "Let's encourage them to write their voices."

A native of Brooklyn, Valentín was a break-dancer in the early days of hip-hop. Like others, he argues that using hip-hop as a vehicle of expression also combats the "criminalization" of young people, especially blacks and Latinos, by enabling them to break free of stereotypes.

Miami's Mecca aka Grimo has a contract with Miami-Dade County Public Schools to run poetry workshops, which he describes in terms similar to those that resonate in the new film. "My main purpose is to stimulate their brains," Mecca says. "It's not just putting words together. I want these kids to start making sense. Being able to recite lyrics, come up with messages — that's a beautiful thing."

Paradoxically, the self-expression that Mecca tries to encourage is aimed at combating impressions largely created by mass-marketed hip-hop. "I do believe that the hip-hop culture — rap music and visuals — has definitely gone in a direction that's detrimental to black youth," he says. "You gotta get shot, you got to be a gangsta to be cool in the streets. Hip-hop became part of a system that's implemented in urban communities to keep people down or make money off them."

To be sure, even rappers who do work the seamier side of the hip-hop street get some credit for good works.

After Hurricane Katrina devastated the Gulf Coast, Banner used his tour bus to deliver emergency supplies to southern Mississippi and organized a fund-raising concert in Atlanta starring other much-criticized rappers, including Nelly.[67]

"I am definitely critical of both of them for some of the stuff they do," says Davey D. "But David Banner — I can't write him off. I didn't go to Mississippi. He did."

OUTLOOK

New and Old

No one could have predicted that a Jamaica-born, Bronx-developed combination of turntable music and spontaneous versifying would morph into a global entertainment force. Now the people who follow hip-hop's fortunes are fairly cautious about predicting its future.

The starting point for looking toward that future is to recognize the extent of the change already wrought by hip-hop, says writer and musician McBride. "This music has changed the way all of us think about music. It's hard for those of us who grew up listening to music as melody. When rap artists pull samples from different records, there's no tonality, the center of gravity is gone, so it's hard for us listeners of Western music to lock into where the center of the song is."

With those changes in the nature of music has come a break in the chain of musical heritage. "We've had whole generations growing up without hearing the great songs of the '40s," McBride says. "Their idea of an old song is Grandmaster Flash and the Furious Five."

That severing of the link between present and past musical styles might not be as widespread as McBride argues. Some hip-hop fans haven't abandoned what might be termed old-fashioned music, the kind that you play on instruments other than a turntable. Inevitably, that practice links them with musicians of the past.

Among the hip-hop performers who put the old and new together are The Roots, a Philadelphia group, and Wyclef Jean, a Haitian-American superstar. Jean, for instance, performs on guitar, playing Santana-style solos and even picking the guitar from behind his head, and then with his mouth, techniques made famous by Jimi Hendrix. And in 2004, Queen Latifah went back even further in musical time, recording an album of standards, (including some from the '40s). This year, Mecca aka Grimo performed in Miami in May with guitar and drums behind him.[68]

Generally speaking, Mecca says, "We need to explore the art form of live music. That will jump us off into a bigger genre of music. Eventually we could be classified as world music if we take to live instruments."

Meanwhile, the debates about lyrics that have dominated discussion of hip-hop for nearly two decades strongly influence others' views of the future. "People are getting tired of gangsta hip-hop," Mecca adds. "In 10 years, you'll see another high point of good, clean, conscious hip-hop."

For others, the lessons of the recent past inspire more caution. "I had no clue in '98 that by 2007 hip-hop would have taken such a blow," says Blitz the Ambassador, describing what he calls the ill effects of Southern rap. "That whole wave of music just took over and killed the lyrics."

Still, he holds out hope that old-school hip-hop at least will survive. "I really hope we can find a way to positively progress," Blitz says. "And if it's another genre, I hope it's more positive."

English, the fledgling independent record label owner, voices the identical hope. But he adds, "I don't see that happening. The major labels haven't said anything. This is their moneymaking machine."

Meanwhile, English says, he plays non-gangsta hip-hop to his 7- and 11-year-old daughters — but exposes them to soul and disco as well. "They need to know what good music really is."

Davey D, the Oakland-based DJ and writer, sees more room for hope, largely because of his sense that gangsta rap's appeal is fading." I want it to be dead," he says. "If ratings are going down, that's great. I'm not trying to save that manifestation of hip-hop. We're very clear that there's been an enabling of people to just be immature."

But if gangsta rap dies, what will replace it? "We might not even call it hip-hop," says Davey D. But he's sure that whatever kind of music grabs hold of the public will help them better understand their world.

It will be, he says, the "soundtracks to conditions that people are living in."

NOTES

1. Quoted in Brent Jones, "Cosby calls to absent fathers," *Baltimore Sun*, Aug. 23, 2006, p. B1.

2. See "The State of Black America," The National Urban League, 2007, p. 37.

3. See Robert Hilburn and Jerry Crowe, "Rapper Tupac Shakur, 25, Dies 6 Days After Ambush," *Los Angeles Times*, Sept. 14, 1996, p. A1; Chuck Phillips, "Bad Boy II Man," *Los Angeles Times*, May 25, 1997, p. 8 [Calendar section]; Katherine E. Finkelstein, "Combs Protégé is Sentenced to 10 Years in Shooting," *The New York Times*, June 2, 2001, p. B2; Andrew Jacobs, "Security Guard Killed Outside of a Busta Rhymes Video Shoot," *The New York Times*, Feb. 6, 2001, p. B1.

4. For background, see Marcia Clemmitt, "Shock Jocks," *CQ Researcher*, June 1, 2007, pp. 481-504. See also, Teresa Wiltz and Darragh Johnson, "The Imus Test: Rap Lyrics Undergo Examination," *The Washington Post*, April 25, 2007, p. C1.

5. See "A Hip-Hop Town Hall," April 17, 2007, "The Oprah Winfrey Show," www.oprah.com/tows/slide/200704/20070417/slide_20070417_284_101.jhtml. For a sample of Crouch's views, see Bakari Kitwana, *Why White Kids Love Hip Hop* (2005), pp. 107-109.

6. Quoted in Reuters, "Hip-Hop Mogul Simmons Calls For Ban on 3 Epithets," *The Washington Post*, April 24, 2007, p. C5. See also Kelefa Sanneh, "How Don Imus' Problem Became a Referendum on Rap," *The New York Times*, April 25, 2007, p. B2; Marcus

Franklin, "Music Execs Silent as Rap Debate Rages," The Associated Press, May 11, 2007; Wiltz and Johnson, *op. cit.*

7. Franklin, *op. cit.*

8. See Jeff Chang, *Can't Stop Won't Stop: A History of the Hip-Hop Generation* (2005), pp. 67-88.

9. *Ibid.*, pp. 118-125.

10. For information on the film, see Hurt's Web site, www.bhurt.com.

11. See Talib Kweli, "Say Something," Talib Kweli Message Board, April 20, 2007, http://board.talibkweli.com/index.php?showtopic=850.

12. See "AllHipHop Direct 9 — Censorship 2007," *AllHipHop.com* (on YouTube), June 1, 2007, www.youtube.com/watch?v=YzdwISAjdos. For background on 50 Cent, see Stephanie Utrata and Tracy Connor, "The Game Fires Back at 50," *New York Daily News*, March 6, 2005, p. A4.

13. See "AllHipHop Direct 9," *ibid.*

14. "2006 Consumer Profile," Recording Industry Association of America, undated, www.riaa.com/news/marketingdata/pdf/2006RIAAConsumer Profile.pdf.

15. See Yvonne Bynoe, *Encyclopedia of Rap and Hip Hop Culture* (2006).

16. See "Stop Snitchin,' " "60 Minutes," April 22, 2007, www.cbsnews.com/stories/2007/04/19/60minutes/main2704565.shtml.

17. Interview, Ice-T, undated, www.freshflixx.com/channel/sohh-tv/index.php?bcpid=376530222&bclid=440748065&bctid=823328719.

18. See Stanley Crouch, "Hip-Hop's Thugs Hit New Low," *New York Daily News*, Aug. 11, 2003, p. 35.

19. *Ibid.*

20. See Industry Ears Web site, www.industryears.com/index.php.

21. See James McBride, "Hip-Hop Planet," *National Geographic*, April 2007, p. 100.

22. See Jim Sullivan, "When shock art goes too far," *Boston Globe*, Nov. 1, 1989, p. 75.

23. See, for example, Jeff Chang, ed., *Total Chaos: The Art and Aesthetics of Hip-Hop* (2007), which includes

Oliver Wang, "Trapped Between the Lines: The Aesthetics of Hip-Hop Journalism;" and Danny Hoch, "Toward a Hip-Hop Aesthetic: A Manifesto for the Hip-Hop Arts Movement."

24. See Hoch, *ibid.*, p. 355.

25. See Eisa Davis, "Found in Translation: The Emergence of Hip-Hop Theater," in Chang, *Total Chaos, op. cit.*, p. 72.

26. Chang, *Can't Stop Won't Stop, op. cit.*, p. 122.

27. See John J. Goldman, "Rally Protests N.Y. Drug Laws," *The New York Times*, July 5, 2003, p. A28; Marcus Franklin, "A hip-hop voting bloc," *St. Petersburg Times*, Aug. 31, 2003, p. A1.

28. See Leslie Eaton and Al Baker, "Changes Made to Drug Laws Don't Satisfy Advocates," *The New York Times*, Dec. 9, 2004, p. B1; and Dasun Allah, "Movement Hijacked by Hip-Hop?" *Village Voice* [New York], June 17, 2003, p. 24.

29. See "Hip-hop group announces voter registration drive," The Associated Press, Jan. 19, 2004; and "Hip-Hop Summit Action Network, Board of Directors," www.hsan.org/Content/main.aspx?pageid=10.

30. Quoted in Ann Gerhart, "Citizen Diddy; the Rapper-Designer is Out to Make Voting Hip," *The Washington Post*, Sept. 3, 2004, p. C1.

31. See "The 2004 Youth Vote," Center for Information & Research on Civil Learning & Engagement, University of Maryland, www.civicyouth.org/PopUps/2004_votereport_final.pdf.

32. See "New Lake-Goeas Poll Analysis Shows Iraq, Pocketbook Issues & Candidate Contact Spurred Large 2006 Youth Vote," Dec. 14, 2006, Young Voter Strategies, www.youngvoterstrategies.org/index.php?tg=articles&idx=More&topics=37&article=284.

33. See Rick Hampson, "Memory is all that's left of Powell's South Bronx," *USA Today*, Jan. 22, 2001, p. A8.

34. Except where otherwise indicated, this section is drawn from Chang, *Can't Stop Won't Stop, op. cit.*

35. *Ibid.*, p. 29. See also David Gonzalez, "Will Gentrification Spoil the Birthplace of Hip-Hop?" *The New York Times*, May 21, 2007, p. B1.

36. *Ibid.*, p. 131.

37. Quoted in Leah Y. Latimer, "Recording the Rap: Jive Talk at the Top of the Charts," *The Washington Post*, Aug. 31, 1980, p. G1.

38. For background, see Bob Benenson, "Reaganomics on Trial," *CQ Researcher*, Jan. 8, 1982, available at *CQ Researcher Plus Archives*, http://cqpress.com.

39. For background, see Mary H. Cooper, "The Business of Illegal Drugs," *CQ Researcher*, May 20, 1988, available at *CQ Researcher Plus Archives*, http://cqpress.com.

40. For background, see Kenneth Jost, "Sentencing Debates," *CQ Researcher*, Nov. 5, 2004, pp. 925-948.

41. This subsection also draws from Bynoe, *op. cit.*, and David Gates with Peter Katel, "The Importance of Being Nasty," *Newsweek*, July 2, 1990, p. 52; Linda Greenhouse, "Supreme Court Roundup," *The New York Times*, Dec. 8, 1992. p. A22.

42. Quoted in Terry McDermott, "Parental Advisory: No One Was Ready for N.W.A.'s 'Straight Outa Compton,' " *Los Angeles Times*, April 14, 2002.

43. See N.W.A., "Fuck Tha Police," www.lyricsdepot.com/n-w-a/fuck-tha-police.html.

44. See McDermott, *op. cit.*

45. See Dennis McDougal, "Music Group, ACLU Join Forces in Lyric Battle," *Los Angeles Times*, Sept. 16, 1985, Calendar Sect., p. 2; "Industry Offers Voluntary tag for Recordings," The Associated Press, May 9, 1980.

46. See Tipper Gore, "Hate, Rape and Rap," [op-ed], *The Washington Post*, Jan. 8, 1990, p. A15.

47. See Shelley Branch, "Goodbye Gangsta; Can Jimmy Iovine Make Interscope a Mainstream Success?" *Fortune*, July 7, 1997, p. 40.

48. Hilburn and Crowe, *op. cit.*

49. See Chuck Phillips, "Who Killed Tupac Shakur," *Los Angeles Times*, Sept. 6, 2002, p. A1; Chuck Phillips, "Slain Rapper's Family Keeps Pushing Suit," *Los Angeles Times*, Feb. 4, 2007, p. B11; and Phillips, May 25, 1997, *op. cit.*

50. See Stanley Crouch, "Merchants of Filth Have Worthy Foe," *New York Daily News*, April 3, 2006, p. 31.

51. Quoted in Chang, *Can't Stop, Won't Stop, op. cit.*, p. 452.

52. *Ibid.*, p. 394.

53. *Ibid.*, pp. 395-396.

54. Quoted in Linda M. Harrington, "On Capitol Hill, a Real Rap Session," *Chicago Tribune*, Feb. 24, 1994, p. A1.

55. *Ibid.*

56. See Gracie Bonds Staples and Vikki Conwell, "Spelman women dis sex-laden rap videos," *Atlanta Journal-Constitution*, April 21, 2004, p. A1.

57. Quoted in Elliott C. McLaughlin, "Spelman continues its war on hip-hop," The Associated Press, Feb. 25, 2005.

58. *Ibid.*

59. See Franklin, *op. cit.*, May 11, 2007.

60. Quoted from *AllHipHop*, in Franklin, *ibid.* See also Take a Stand Records Web site, www.takeastandrecords.com/index.html.

61. Quoted in Larry "The Blackspot" Hester, "Music News: 50 Cent Pushes Back Release Date, Master P Speaks Out," BET.com, *Music, News & Interviews*, May 25, 2007, www.bet.com/Music/News/musicnews_50_5.25.htm?wbc_purpose=Basic&WBCMODE=PresentationUnpublished.

62. See Danielle Harling and Dove, "Grammy Award-Winning Rapper Chamillionaire Profanity Free," *AllHipHop*, April 27, 2007; www.allhiphop.com/Hiphopnews/?ID=7007.

63. "Bush Signs Broadcast Decency Law," The Associated Press, June 15, 2006.

64. See "Stop Snitchin,' " *op. cit.*

65. See Mark Olsen, "More than words to tell their stories," *Los Angeles Times*, May 11, 2007, p. E6; Teresa Wiltz, "No Rhyme or Rhythm," *The Washington Post*, May 11, 2007, p. WE29.

66. See "The Hip Hop Project," film Web site, http://pressurepointfilms.com/thehiphopproject.html.

67. See Michael Brick, "Cultural Divisions Stretch to Relief Concerts," *The New York Times*, Sept. 17, 2005, p. B7.

68. See "Wyclef Shreds in Haiti," *MiamiVideo*, Dec. 1, 2006, www.brightcove.com/title.jsp?title=494388358&channel=474448254&lineup=-1; "Mecca Live From Oxygen," MiamiVideo, undated, www.brightcove.com/title.jsp?title=900691975&channel=474448254&lineup=-1; Lorraine Ali, "God Save The Queen," *Newsweek*, Oct. 4, 2004, p. 59.

BIBLIOGRAPHY

Books

Bynoe, Yvonne, *Encyclopedia of Rap and Hip Hop Culture, Greenwood Press, 2006.*
A prominent critic of some aspects of hip-hop produced this systematic look at the major players and trends.

Chang, Jeff, *Can't Stop Won't Stop: A History of the Hip-Hop Generation, Picador, St. Martin's Press, 2005.*
A leading hip-hop journalist provides a detailed but fast-moving account of hip-hop in the context of social and political developments.

Chang, Jeff, ed., *Total Chaos: The Art and Aesthetics of Hip-Hop, Basic Civitas Books, 2007.*
A variety of scholars and hip-hop creators examine the artistic significance of hip-hop in its various musical, theatrical and other incarnations.

Cobb, William Jelani, *To the Break of Dawn: A Freestyle on the Hip Hop Aesthetic, New York University Press, 2007.*
A Spelman College historian takes a close look at the messages, imagery and techniques of imaginative and inventive hip-hop songs.

Kitwana, Bakari, *Why White Kids Love Hip Hop, Basic Books, 2005.*
Race and hip-hop, one of the most complicated topics in the field, gets an insider's look by a leader of the political-activism side of the hip-hop world.

Morgan, Joan, *When Chickenheads Come Home to Roost: A Hip-Hop Feminist Breaks it Down, Touchstone, Simon & Schuster, 1999.*
A former staff writer at *Vibe,* a major hip-hop magazine, discusses hip-hop as part of an examination of male-female relations among African-Americans.

Articles

Boles, Mark A., "Breaking the 'Hip Hop' Hold: Looking Beyond the Media Hype," in "The State of Black America 2007: Portrait of the Black Males," *National Urban League*, **2007, p. 239.**
A market researcher and member of the Urban League Board of Trustees examines the role of hip-hop — especially music videos — in distorting young African-Americans' values.

Davey D, "Why commerce is killing the true spirit of hip-hop," *San Jose Mercury-News*, **March 1, 2007, p. M4.**
A veteran of the hip-hop world finds "corporate" hip-hop losing ground — and salutes that outcome — given what he calls its vulgarity and perpetuation of racial stereotypes.

Fetterman, Mindy, "Russell Simmons can't slow down," *USA Today*, **May 14, 2007, p. B1.**
One of hip-hop's biggest success stories radiates confidence to a reporter who looks at Simmons' success in branching out from record sales.

McBride, James, "Hip Hop Planet," *National Geographic*, **April, 2007, p. 100.**
An author and musician — who'd spent years ignoring hip-hop — travels throughout the country and to Africa to understand its wide appeal.

McDermott, Terry, "Parental Advisory: Explicit Lyrics," *Los Angeles Times Magazine*, **April 14, 2002, p. 12.**
In a long narrative filled with colorful characters, a *Los Angeles Times* correspondent traces gangsta rap to its very beginnings.

Michel, Sia, "A New Sound for Old What's-His-Name," *The New York Times*, **Sept. 10, 2006, Sect. 2, p. 67.**
Sean "P. Diddy" Combs gets ready to release a new album and takes a reporter on a tour of his world.

Span, Paula, "The Business of Rap is Business," *The Washington Post*, **June 4, 1995, p. G1.**
A *Washington Post* reporter takes an early look at rap stars' efforts to turn musical success into something more durable.

Williams, Clarence, "Outreach Group Tries to Foster Greater Cooperation With Police," *The Washington Post*, **May 13, 2007, p. C5.**
In a violence-plagued Washington, D.C., neighborhood, activists encourage residents to report crime to police — defying the "no-snitching" message of some rap stars.

Video

"Hip Hop: Beyond Beats and Rhymes," Byron Hurt, director-producer, *IndependentLens*, *PBS*, **2007.**
A college football star turned violence-prevention educator examines hip-hop's role in fostering stereotypes that widen the gulf between men and women.

"Stop Snitchin'," *"60 Minutes,"* Anderson Cooper, correspondent, April 19, 2007, www.cbsnews.com/ sections/i_video/main500251.shtml?channel=60Sunday.
The CBS TV newsmagazine takes on rap stars — and their record companies — who urge crime victims and witnesses not to cooperate with police.

"Wild Style," Lee Ahearn, director/writer, 1983.
A fictional film starring hip-hop fans who helped found the movement is considered a key document of its early days, with a heavy emphasis on graffiti "taggers."

For More Information

AllHipHop, www.allhiphop.com. Widely consulted site for news, music downloads and video interviews with stars and fans.

Davey D's Hip Hop Corner, http://daveyd.com. A combination news site, collection of essays about hip-hop-related topics and hip-hop history archive.

The Hiphop Archive, Department of Communication, Building 120, McClatchy Hall, 450 Serra Mall, Stanford University, Stanford, CA 94305; (650) 725-2142; http://hiphoparchive .org. Vast trove of material on multiple aspects of hip-hop.

Hip-Hop Association, P.O. Box 1181, New York, NY 10035; (212) 500-5970; http://hiphopassociation.org. Harlem-based community-development organization that uses hip-hop as a tool in educational and leadership-development projects.

Hip-Hop Summit Action Network, www.hsan.org. The organizational home of hip-hop's tycoons, who have involved themselves in voter registration and, more recently, in an effort to clean up rap lyrics.

Hip-Hop Theater Festival, 57 Thames St. #4B, Brooklyn, NY 11237; (718) 497-4240; www.hhtf.org. Organizes hip-hop events in major cities.

Industry Ears, http://industryears.com. Advocacy organization that presses the Federal Communications Commission to regulate radio stations that broadcast hip-hop songs with sexually explicit lyrics.

7

Gender Pay Gap

Are Women Paid Fairly in the Workplace?

Thomas J. Billitteri

A suit filed by Betty Dukes, right, and other female Wal-Mart employees accuses the retail giant of sex discrimination in pay, promotions and job assignments in violation of the Civil Rights Act of 1964. The case, covering perhaps 1.6 million current and former Wal-Mart employees, is the biggest class-action lawsuit against a private employer in U.S. history.

From *CQ Researcher*, March 14, 2008.

An insult to my dignity" is the way Lilly Ledbetter described it.[1] For 19 years, she worked at the Goodyear Tire plant in Gadsden, Ala., one of a handful of women among the roughly 80 people who held the same supervisory position she did. Over the years, unbeknownst to her, the company's pay-raise decisions created a growing gap between her wages and those of her male colleagues. When she left Goodyear, she was earning $3,727 a month. The lowest-paid man doing the same work got $4,286. The highest-paid male made 40 percent more than she did.[2]

Ledbetter sued in 1998, and a jury awarded her back pay and more than $3 million in damages. But in the end, she lost her case in the U.S. Supreme Court.[3]

A conservative majority led by Justice Samuel A. Alito Jr. ruled that under the nation's main anti-discrimination law she should have filed a formal complaint with the federal government within 180 days of the first time Goodyear discriminated against her in pay. Never mind, the court said, that Ledbetter didn't learn about the pay disparity for years.

"The Supreme Court said that this didn't count as illegal discrimination," she said after the ruling, "but it sure feels like discrimination when you are on the receiving end of that smaller paycheck and trying to support your family with less money than the men are getting for doing the same job."[4]

The *Ledbetter* decision has added fuel to a long-burning debate over sex discrimination in women's wages and whether new laws are needed to narrow the disparity in men's and women's pay.

141

Women Closing the Pay Gap . . . Slowly

More than 40 years after women began demanding equal rights and opportunities, they still earn 77 percent of what men earn. The pay gap has been closing, however, because women's earnings have been rising faster than men's.

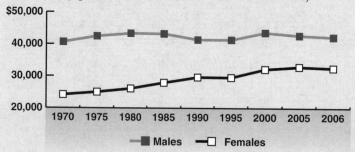

Median Annual Earnings of Full-time, Year-round Workers
(By gender, 1970-2006, in constant 2006 dollars)

Source: Carmen DeNavas-Walt, et. al., "Income, Poverty, and Health Insurance Coverage in the United States: 2006," U.S. Census Bureau, August 2007

"A significant wage gap is still with us, and that gap constitutes nothing less than an ongoing assault on women's economic freedom," declared U.S. Rep. Rosa L. DeLauro, D-Conn., at a congressional hearing on a pay-equity bill she is sponsoring, one of several proposed on Capitol Hill.

But that view is hardly universal. "Men and women generally have equal pay for equal work now — if they have the same jobs, responsibilities and skills," testified Diana Furchtgott-Roth, a senior fellow at the Hudson Institute, a conservative think tank, and former chief economist at the Labor Department in the George W. Bush administration.[5]

The wrangle over wages is playing out not just in Washington but in cities and towns across America. In the biggest sex-discrimination lawsuit in U.S. history, a group of female Wal-Mart employees has charged the retail giant with bias in pay and promotions. The case could affect perhaps 1.6 million women employees of Wal-Mart and result in billions of dollars in back pay and damages. (*See sidebar, p. 154.*)

The enormously complex gender-pay debate encompasses economics, demographics, law, social justice, culture, history and sometimes raw emotion. Few dispute that

a wage gap exists between men and women. In 2006 full-time female workers earned 81 percent of men's weekly earnings, according to the latest U.S. Labor Department data, with the wage gap broader for older workers and narrower for younger ones. Separate U.S. Census Bureau data put the gap at about 77 percent of men's median full-time, year-round earnings.[6]

The fundamental issues are why the gap exists, how much of it stems from discrimination and what should be done about it.

Some contend the disparity can largely be explained by occupational differences between women and men, variations in work experience, number of hours worked each year and other such things.

June O'Neill, an economics professor at the City University of New York's Baruch College and former director of the Congressional Budget Office in the Clinton administration, says that the most important factors affecting the pay gap stem from differences in the roles of women and men in family life. When the wages of men and women who share similar work experience and life situations are measured, the wage gap largely disappears, she says. Reasons that the earnings disparity may appear bigger in some research, she says, include the fact that many studies do not control for differences in years of work experience, the extent of part-time work and differences in training and occupational choices. O'Neill notes that Labor Department data show median weekly earnings of female part-time workers exceed those of male part-timers. She also says the wage gap has been narrowing over time as women's work experience, education and other job-related skills have been converging with those of men.

"Large amounts of discrimination? No," she says. "Individual women may experience discrimination, and it's good to have laws that deal with it," she adds. "But those cases don't change the overall picture. The vast majority of employers don't harbor prejudice against women."

Yet others argue that beneath such factors as occupation and number of hours worked lies evidence of significant discrimination — covert if not overt.

"Women do not realize the enormous price that they pay for gender wage discrimination because they do not see big bites taken out of their paychecks at any one time," Evelyn F. Murphy, president of The Wage Project, a nonprofit organization that works on eliminating the gender wage gap and author of *Getting Even: Why Women Don't Get Paid Like Men and What To Do About It*, told a congressional panel last year.[7]

In her book, she told the hearing, she wrote of employers "who had to pay women employees or former employees to settle claims of gender discrimination, or judges and juries ordered them to pay up. The behavior of these employers vividly [illustrates] the commonplace forms of today's wage discrimination: barriers to hiring and promoting qualified women; arbitrary financial penalties imposed on pregnant women; sexual harassment by bosses and co-workers; failure to pay women and men the same amount of money for doing the same jobs," and "everyday discrimination" marked by "the biases and stereotypes which influence [managers'] decisions about women."

Women's advocates point to a 2003 General Accounting Office (GAO) study concluding that while "work patterns" were key in accounting for the wage gap, the GAO could not explain all the differences in earnings between men and women. "When we account for differences between male and

Gap Widens for College Graduates

College-educated women earn only 80 percent of what their male counterparts earn a year after graduation, when both male and female employees have the same level of work experience and (usually) no child-care obligations — factors often used to explain gender pay differences. The gap widens to 69 percent by 10 years after graduation.

Gap in Average Weekly Earnings for Bachelor's Degree Recipients
(For full-time workers)

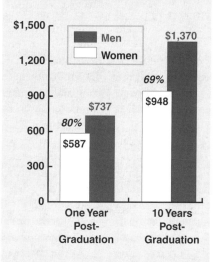

Source: "Beyond the Pay Gap," American Association of University Women, based on data from the "2003 Baccalaureate and Beyond Longitudinal Study," National Center for Education Statistics, U.S. Department of Education

female work patterns as well as other key factors, women earned, on average, 80 percent of what men earned in 2000. . . . We cannot determine whether this remaining difference is due to discrimination or other factors," the GAO report said.[8]

The study said that in the view of certain experts some women trade promotions or higher pay for job flexibility that allows them to balance work and family responsibilities.

Women's advocates point out that many women have little choice but to work in jobs that offer flexibility but pay less because they typically shoulder the bulk of family caregiving duties. And, they argue further, expectations within companies and society — typically subtle, but sometimes not — often channel women away from male-dominated jobs into female-dominated ones that pay less.

"People who argue that [wage discrimination] is small will say a lot of it is due to women's choices," such as the choice to stay home with the children, work part time or enter lower-paying fields, says Reeve Vanneman, a sociology professor at the University of Maryland, College Park, who studies gender inequality. But, he says, it's misleading to explain most of the wage gap in that way, especially when mid-career and older female workers are concerned.

"Why do women make those choices? Part of the reason is because they are discriminated against in the job. They see men getting rewarded more and promoted more than they are."

Women face unequal work not just on the job but at home, too, Vanneman says, with husbands not picking up their share.

Part of the wage gap stems from weak government enforcement, some argue. A U.S. inspector general's report stated last fall that the Equal Employment Opportunity Commission, which enforces federal employment-discrimination laws, is "challenged in accomplishing its mission" because of "a reduced workforce and an increasing backlog of pending cases." The agency has experienced a "significant loss of its workforce, mostly to attrition and buyouts . . . offered to free up resources," the report said.[9]

The news on gender discrimination in pay is not all bad. The wage gap has narrowed considerably in recent decades. For example, Labor Department data show that for 35- to 44-year-olds, the earnings ratio of women to men rose from 58 percent in 1979 to 77 percent in 2006. For 45- to 54-year-olds, it went from 57 percent to 74 percent.[10] Among the youngest workers, ages 16 to 24, only about 5 percentage points separated median weekly wages of men and women in 2006.[11]

Still, many experts say the progress of the 1980s and early '90s has slowed or stalled in recent years, with the wage gap stuck in the range of 20 to 24 percent, although it is not entirely clear why. Some argue that entrenched wage discrimination remains a major culprit.

In a study of college graduates last year, the American Association of University Women Educational Foundation found that one year out of college, women working full time earn only 80 percent as much as their male colleagues, and 10 years after graduation the gap widens to 69 percent. Even after controlling for hours worked, training and education and other factors, the portion of the pay gap that remains unexplained is 5 percent one year after graduation and 12 percent a decade afterward, the study found.[12] (*See graph, p.143.*)

"These unexplained gaps are evidence of discrimination," the study concluded.

Employer advocates challenge such conclusions, though. Michael Eastman, executive director of labor policy at the U.S. Chamber of Commerce, questions the assumption "that whatever gap is not explained must be due to discrimination. An unexplained gap is simply that — it's unexplained."

Election-year politics and the recent shift toward Democratic control of Congress — along with the Supreme Court's decision in the *Ledbetter* case — have helped to reinvigorate the pay debate. Proposed gender-pay bills have strong support from women's-rights groups and some economists, who argue that the Equal Pay Act and Title VII of the Civil Rights Act of 1964 — the main avenues for attacking wage discrimination — fall short.

Presidential contender Sen. Hillary Rodham Clinton, D-N.Y., is sponsoring the Senate version of the DeLauro bill; another presidential hopeful, Sen. Barack Obama, D-Ill., is one of the 22 co-sponsors, although he didn't sign on to it until more than a month after she introduced it. Among other things, the measure would raise penalties under the Equal Pay Act, which bars paying men and women differently for doing the same job.[13]

Obama is co-sponsoring a more controversial bill, introduced in the Senate by Sen. Tom Harkin, D-Iowa, that advocates the notion of comparable worth; the idea, generally speaking, suggests that a female-dominated occupation such as social work may merit wages that are comparable to those of a male-dominated job such as a probation officer.[14] The Harkin measure would bar wage discrimination in certain cases where the work is deemed comparable in skill, effort, responsibility and working conditions, even if the job titles or duties are different. (*See sidebar, p. 152.*)

A third effort would undo the Supreme Court's ruling in the *Ledbetter* case.[15] A bill passed the House last summer, and advocates are hoping the Senate version — sponsored by Sen. Edward M. Kennedy, D-Mass., and co-sponsored by Clinton and Obama — moves forward soon. But the Bush administration has threatened a veto, and business interests are vehemently opposed.

As the debate over wage disparities continues, these are some of the questions being discussed:

Is discrimination a major cause of the wage gap?

When economist David Neumark studied sex discrimination in restaurant hiring in the mid-1990s, he discovered something intriguing: In expensive restaurants, where waiters and waitresses can earn more than they can at low-price places, the chances of a woman getting a wait-staff job offer were 40 percentage points lower than those of a man with similar experience.[16]

The study is a telling bit of evidence that the wage gap is real and that discrimination plays a significant part in it, says Vicky Lovell, director of employment and work/life programs at the Institute for Women's Policy Research, an advocacy group in Washington. She estimates that perhaps a third of the wage gap stems from discrimination — mostly "covert" bias that occurs when people make false assumptions about the ability or career commitment of working women.

Lovell has little patience with those who say the wage gap stems from non-discriminatory reasons that simply haven't yet been identified. "That's just specious," she says. "If we can't explain why women on average get paid less, what is the alternative explanation?"

The role of discrimination lies at the heart of the pay-gap debate. Researchers fall into different camps.

Some see little evidence that bias plays a big part in the gap. When adjusted for work experience, education, time in the labor force and other variables, wages of men and women are largely comparable, they contend.

"This so-called wage gap is not necessarily due to discrimination," the Hudson Institute's Furchtgott-Roth said in congressional testimony. "Decisions about field of study, occupation and time in the work force can lead to lower compensation, both for men and women."[17]

What's more, "some jobs command more than others because people are willing to pay more for them," she said. "Many jobs are dirty and dangerous. . . . Other highly paid occupations have long, inflexible hours. . . . Women are not excluded from these or other jobs but often select professions with a more pleasant environment and potentially more flexible schedules, such as teaching and office work. Many of these jobs pay less."

Pay Gap Exists Despite Women's Choices

Those who discount the seriousness of gender pay bias often blame differences in men's and women's salaries on women's choices to study "softer" sciences or to have children. But a recent study shows that the pay gap persists even when women choose not to have children and when they choose male-dominated fields of study and occupation — such as business, engineering, mathematics and medicine. The pay gap is greatest in the biology, health and mathematics fields. Women out-earn men only in the history professions.

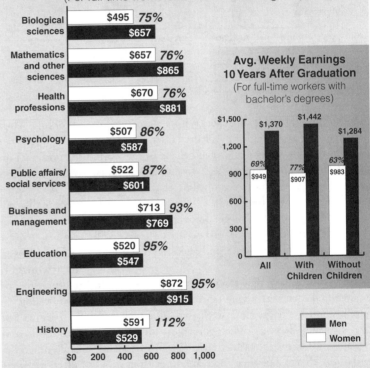

Avg. Weekly Earnings One Year After Graduation
(For full-time workers with bachelor's degrees)

Field	Women	%	Men
Biological sciences	$495	75%	$657
Mathematics and other sciences	$657	76%	$865
Health professions	$670	76%	$881
Psychology	$507	86%	$587
Public affairs/social services	$522	87%	$601
Business and management	$713	93%	$769
Education	$520	95%	$547
Engineering	$872	95%	$915
History	$591	112%	$529

Avg. Weekly Earnings 10 Years After Graduation
(For full-time workers with bachelor's degrees)

	All	With Children	Without Children
Men	$1,370	$1,442	$1,284
Women	$949 (69%)	$907 (77%)	$983 (63%)

Source: "Behind the Pay Gap," American Association of University Women, 2007

Warren Farrell, who in the 1970s served on the board of the New York City chapter of the National Organization for Women, argues in his 2005 book — *Why Men Earn More: The Startling Truth Behind the Pay Gap — and What Women Can Do About It* — that women pay an economic price by seeking careers that are

Wage Disparities Highest Among Asians

The median weekly earnings for women are lower than men's across all ethnic groups. The largest disparity is among Asians, where men earn $183 more on average per week than their female counterparts. The average difference for all groups is $143.

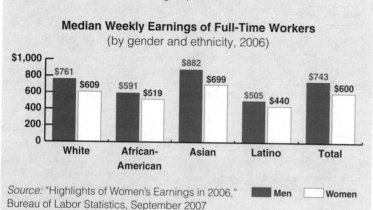

Median Weekly Earnings of Full-Time Workers
(by gender and ethnicity, 2006)

Source: "Highlights of Women's Earnings in 2006," Bureau of Labor Statistics, September 2007

■ Men □ Women

more fulfilling, flexible and safe. With a stated goal of helping women gain higher pay, Farrell offers 25 "differences in the way women and men behave in the workplace." Those who earn more, he says, work longer hours, are more willing to relocate, require less security and produce more, among other things.

O'Neill, of Baruch College, points out that women are much more likely to go into occupations that will allow them to work part time, and typically "that doesn't pay as well."

She studies data that track the work histories of women and men over a long period of time. "Women have just not worked as many weeks and hours over their lives as men," she says. "When you adjust for that, you explain most of the [pay] difference. . . . You're still left with a difference, but then there are other things that become harder to measure."

The AAUW study found that even women who make the same choices as men in terms of fields of study and occupation earn less than their male counterparts. A typical college-educated woman working full time earns $46,000 a year compared to $62,000 for college-educated male workers — a difference of $16,000.

"The pay gap between female and male college graduates cannot be fully accounted for by factors known to affect wages, such as experience (including work hours), training, education and personal characteristics," the AAUW study

says. "In this analysis the portion of the pay gap that remains unexplained after all other factors are taken into account is 5 percent one year after graduation and 12 percent 10 years after graduation. These unexplained gaps are evidence of discrimination, which remains a serious problem for women in the work force."[18]

"This research asked a basic but important question: If a woman made the same choices as a man, would she earn the same pay? The answer is no," Catherine Hill, director of research at the AAUW, told a House Committee on Education and Labor hearing last year.

Speaking more generally about pay inequity, Linda Meric, national director of 9to5, National Association of Working Women, a Milwaukee-based advocacy group, says that "when you control for all the other so-called factors" that might explain the wage gap, "there is still a gap."

"And many of those so-called factors are not independent of discrimination and stereotypes of women. One is time in the work force. If there aren't policies that allow women to get jobs and maintain and advance in employment at the same time they are meeting their responsibility in terms of family caregiving, that's not an independent factor. It's something that influences the pay gap significantly."

Heather Boushey, senior economist at the Center for Economic and Policy Research, a Washington think tank, noted that time away from the workforce strongly affects lifetime earnings. She said it is a myth that women choose lower-paying occupations because they provide the flexibility to better manage work and family. "The empirical evidence shows that mothers are actually less likely to be employed in jobs that provide them with greater flexibility."[19]

Echoing that sentiment, Beth Shulman, co-director of the Fairness Initiative on Low Wage Work, a public policy advocacy group also in Washington, says, "We have kind of an Ozzie and Harriet workplace, with a full-time worker and the wife at home," but "70 percent of women with children are in the workplace." She adds,

"Our structures haven't kept up with that. So women who are primary caregivers get punished."

Shulman, author of *The Betrayal of Work: How Low-Wage Jobs Fail 30 Million Americans*, says that while overt gender discrimination exists in the job market, an equally important contributor to the wage gap is the lack of flexibility for low-income working women with families. For example, she says, female factory employees with family responsibilities often find it difficult to accept better-paying manufacturing jobs because such jobs often require mandatory overtime.

Shulman also says that three-fourths of women in low-wage jobs don't have paid sick days. So when a child is sick or an elderly parent needs help, women may be forced to leave the workforce and then re-enter it — something that has a huge effect on wages over time.

"Low-wage workers get kind of ghettoized into these part-time jobs that have poor wages, poor benefits and less government protection," Shulman says.

In a 1998 study, Cornell University economists Francine Blau and Lawrence Kahn found that 40 percent of the pay gap is unexplained after adjusting for gender differences in experience, education, occupation and industry. Blau cautions that such an estimate is conservative, because variables such as women's choices of occupation or industry and even their education and work experience can themselves be affected by discrimination. On the other hand, she acknowledges that some of the unexplained differences may be due to unmeasured productivity characteristics that increase men's earnings relative to women's earnings.

Applying that 40 percent figure to current government wage-gap data would suggest that 8 to 9 cents of each dollar in wage disparity is unexplained, with an unknown portion of that amount caused by discrimination.

Martha Burk, who directs the Corporate Accountability Project for the National Council of Women's Organizations, a coalition of more than 200 women's groups, says some of the pay gap stems from "historical discrimination" rooted in a time when employers could legally exclude women from certain jobs and pay them less for the kinds of jobs they typically did hold, such as teaching and clerical work.

Burk, who led the fight to open the Augusta (Ga.) National Golf Club to women, says those female-dominated jobs "were systematically devalued, and that has carried through to modern times."

Are new laws needed to close the gender pay gap?

When President John F. Kennedy signed the Equal Pay Act in 1963, he called it "a first step."[20]

Over the decades, the pay gap has narrowed significantly, but the push for new laws to curb gender-pay inequity goes on, fueled in part by the view among women's advocates that progress toward wage equity has slowed or stalled in recent years.

"The best way is for corporations to behave as socially responsible corporate citizens [and] examine their wage practices," says Lovell of Women's Policy Research. "But that is not going to happen. I don't see any reason to think the private sector is going to address this issue on its own. A few will to the extent they can within their own workforces. But if corporations individually or within industry groups aren't going to make this a priority, then that's why we have a government."

Opponents of new laws have sharply different views, though.

Roger Clegg, president and general counsel of the Center for Equal Opportunity, a conservative think tank in Falls Church, Va., says some gender discrimination will always exist but that existing laws can address it. Besides, Clegg says, the amount of gender discrimination that remains in the American work force "is greatly exaggerated by the groups pushing for legislation."

Much of the support for new laws rests on the view that some jobs pay poorly because females historically have dominated them. Jocelyn Samuels, vice president for Education and Employment at the National Women's Law Center, a Washington advocacy group, told a congressional hearing last year that 95 percent of child-care workers are female while the same proportion of mechanical engineers are male.

Moreover, she said, wages in fields dominated by women "have traditionally been depressed and continue to reflect the artificially suppressed pay scales that were historically applied to so-called 'women's work.'" Maids and housecleaners — 87 percent of whom are women — make roughly $3,000 per year less than janitors and building cleaners, 72 percent of whom are men, she said. "Current law simply does not provide the tools to address this continuing devaluation of traditionally female fields."[21]

To attack that situation, some advocates back the comparable-worth theory, arguing that women should

AP Photo/Ron Edmonds

U.S. Rep. Rosa L. DeLauro, D-Conn., is sponsoring one of several pay-equity bills in Congress. Presidential contender Sen. Hillary Rodham Clinton, D-N.Y., is sponsoring the Senate version of DeLauro's bill; Sen. Barack Obama, D-Ill., is one of the 22 co-sponsors. "A significant wage gap is still with us, and that gap constitutes nothing less than an ongoing assault on women's economic freedom," DeLauro says.

be paid commensurate with men for jobs of equivalent value to a company, even if the work is different. But critics argue that such an approach violates the free-market principles of supply and demand for labor and that it could hurt both the economy and the cause of women.

"The comparable-worth approach has the government setting wages rather than the free market, and a great lesson of the 20th century is that centrally planned economies and centrally planned wage and price systems do not work," Clegg says.

Carrie Lukas, vice president for policy and economics at the Independent Women's Forum, a conservative group in Washington that backs limited government, contends that "government attempts to 'solve' the

problem of the wage gap may in fact exacerbate some of the challenges women face, particularly in balancing work and family."

In an opinion column last year, she criticized the Clinton/DeLauro bill, which calls for guidelines to help companies voluntarily "compare wages paid for different jobs . . . with the goal of eliminating unfair pay disparities between occupations traditionally dominated by men or women." Lukas wrote that the bill would "give Washington bureaucrats more power to oversee how wages are determined, which might prompt businesses to make employment options more rigid." Flexible job structures would become less common, she argued. Why, Lukas wondered, "would companies offer employees a variety of work situations and compensation packages if doing so puts them at risk of being sued?"[22]

Not only might women suffer from new laws, but so would employers, some argue. Washington lawyer Barbara Berish Brown, vice-chair of the American Bar Association's Labor and Employment Law Section, said in a hearing on the Clinton/DeLauro bill that she is "unequivocally committed" to erasing gender-pay bias, but that existing laws suffice.

"All that the proposed changes will do is encourage more employment-related litigation, which is already drowning the federal court docket, and make it much more difficult, if not impossible, for employers, particularly small businesses, to prove the legitimate, nondiscriminatory reasons that explain differences between the salaries of male and female employees," she said.[23]

But longtime activists such as Burk, author of *Cult of Power: Sex Discrimination in Corporate America and What Can Be Done About It*, say existing laws are not effective enough to stamp out wage bias. "It has always been the view of conservatives that if you pay women equally, it's going to destroy capitalism," she says. "So far capitalism has survived quite well."

Is equity possible after the Supreme Court's *Ledbetter* ruling?

After the Supreme Court ruled in the Goodyear pay-discrimination case, Eleanor Smeal, president of the Feminist Majority, urged congressional action to reverse the decision. "We cannot stand by and watch a Bush-stacked court destroy in less than a year Title VII — the bedrock of women's rights and civil rights protection in wage-discrimination cases," she said.[24]

Yet, such outrage at the Supreme Court is matched by praise from business advocates. "We think the court got it exactly right," says Eastman, the U.S. Chamber of Commerce labor policy official.

In the 5-4 ruling, the court said workers can't sue under Title VII of the Civil Rights Act, the main federal anti-discrimination law, unless they file a formal complaint with the EEOC within 180 days of a discriminatory act. And in Ledbetter's case, the clock didn't start each time a new paycheck was issued. The 180-day timeline applies whether or not the employee immediately spots the discrimination.

Critics argue that because pay decisions are seldom broadcast throughout a company, the ruling makes it difficult — if not impossible — for an employee to detect bias until it may have gone on for years. "The ruling essentially says 'tough luck' to employees who don't immediately challenge their employer's discriminatory acts, even if the discrimination continues to the present time," said Marcia Greenberg, co-president of the National Women's Law Center.[25]

"With this misguided decision, the court ignores the realities of the 21st-century workplace," Margot Dorfman, chief executive officer of the U.S. Women's Chamber of Commerce, told a congressional panel this year. "The confidential nature of employee salary information complicates workers' abilities to recognize and report discriminatory treatment."[26]

Lovell, of the Institute for Women's Policy Research, says the Ledbetter ruling "seems to reflect a complete lack of understanding of the labor market and a complete lack of concern for individuals who are at any kind of disadvantage in the labor market." Workers wouldn't necessarily know right away that they were being discriminated against, she says. When Congress passed Title VII, it "was trying to establish an avenue for people who are discriminated against to pursue their claims . . ., not trying to make it impossible."

In a strongly worded dissent to the Ledbetter ruling, Justice Ruth Bader Ginsburg noted that pay disparities often occur in small increments, evidence of bias may develop over time, and wage information is typically hidden from employees. At the end of her dissent she wrote that "the ball is in Congress' court" to correct the Supreme Court's "parsimonious reading of Title VII" in the Ledbetter decision, just as Congress dealt with a spate of earlier Supreme Court decisions with passage of the 1991 Civil Rights Act.

Business groups have stood firm in the face of such impassioned views, though.

An exchange between Eastman of the U.S. Chamber of Commerce and law professor Deborah Brake last fall on the National Public Radio show "Justice Talking" helped underscore how polarizing the Ledbetter decision has been between advocates for women and for employers.[27]

Brake, a professor at the University of Pittsburgh School of Law who once litigated sex-discrimination cases for the National Women's Law Center, said she thought it was questionable whether the ruling was even good for employers.

"What an employee is supposed to do, let's say from the moment in time that they are hired, is search around the workplace and make sure that they're not being paid less if it's a woman than her male colleagues," she said on the radio program.

"If she has the slightest inkling or suspicion that she might be paid less than her male colleagues, she'd better immediately file a pay-discrimination claim. At every raise decision she better be sniffing around to make sure that her raise wasn't less than that of her male colleagues. And if she hears that someone got a higher raise than her who was a male, to preserve her rights under [the Ledbetter ruling] she'd better immediately file an EEOC claim. I don't think that is in the best interest, long-term, of employer or employees."

Eastman, though, said Title VII "has a strong incentive for employees to file claims quickly so that matters are resolved while all the facts and evidence are fresh and in people's minds. And it is very difficult for employers to defend themselves from allegations made many, many years down the line."

Brake said it wasn't the 180-day limit that bothered her. "What I'm objecting to is a ruling that starts the clock running before any employee has enough reason or incentive to even think about filing a discrimination claim," she said.

BACKGROUND

Early Wage Gap

From the republic's beginning, women have played an integral role in American economic growth and prosperity, yet a wage gap has always been present.

C H R O N O L O G Y

1900-1940 *Women make economic gains but face discrimination.*

1914 Start of World War I marks a period of advancement in the status of women, who go to work in traditionally male jobs.

1919 Women gain the right to vote through the 19th Amendment.

1923 The Equal Rights Amendment is introduced, but it falls three states short of ratification.

1930 Half of single women are in the labor force, and the labor-participation rate among married women approaches 12 percent.

1938 Fair Labor Standards Act establishes rules for a minimum wage, overtime pay and child labor.

1940-1960 *Women make major contribution to wartime manufacturing efforts but don't gain wage equality with men.*

1942 National War Labor Board urges employers to equalize pay between men and women in defense jobs.

1945 Congress fails to approve Women's Equal Pay Act.

1955 Census Bureau begins calculating female-to-male earnings ratio.

1960-1980 *Major anti-discrimination laws helps women to fight pay bias.*

1963 Equal Pay Act bans gender pay discrimination in equal jobs.

1963 *The Feminine Mystique* by Betty Friedan challenges idea that women can find happiness only through marriage.

1964 Title VII of the Civil Rights Act bans job discrimination on the basis of race, color, religion, national origin and sex.

1965 Equal Employment Opportunity Commission founded.

1966 National Organization For Women is formed.

1973 Supreme Court's *Roe v. Wade* ruling overturns laws barring abortion, energizes the women's movement.

1979 National Committee on Pay Equity is formed.

1980-2000 *Gender pay gap continues to narrow, but progress toward wage equality shows signs of slowing in the 1990s.*

1981 Supreme Court ruling in County of *Washington v. Gunther* allows female jail guards to sue for sex discrimination but declines to authorize suits based on theory of comparable worth.

1993 Family and Medical Leave Act requires employers to grant unpaid leave for medical emergencies, birth and care of newborns and other family-related circumstances.

2001-Present *States expand laws to help working families, while several major corporations face gender-bias accusations.*

2001 Wal-Mart employees file for sex-discrimination claim against the retailer, to become the largest class-action lawsuit against a private employer in U.S. history.

2004 California grants up to six weeks partial pay for new parents.

2004 Equal Employment Opportunity Commission and Morgan Stanley announce $54 million settlement of sex-discrimination suit. . . . Wachovia Corp. agrees to pay $5.5 million in a pay-discrimination case involving more than 2,000 current and former female employees.

2007 San Francisco requires employers to provide paid sick leave to all employees, including temporary and part-time workers.

2007 In *Ledbetter v. Goodyear,* Supreme Court rules that a female worker's pay-discrimination claim was invalid because it was filed after a 180-day deadline.

During the Industrial Revolution of the 19th century, as the nation's productivity and wealth exploded, young, single women moved from farm to city and took jobs as mill workers, teachers and domestic servants.

The factory work wasn't easy, and owners exploited women and girls as cheap sources of labor. In 1830, females often worked 12 hours a day in "boarding-house mills" — factories with housing provided by mill owners. They earned perhaps $2.50 a week. "Minor infractions such as a few minutes' lateness were punished severely," historian Richard B. Morris noted, and "one-sided contracts gave them no power over conditions and no rewards for work."[28]

Still, young women flocked to manufacturing jobs in the cities. In Massachusetts, among the earliest states to industrialize, a third of all women ages 10 to 29 worked in industry in 1850, according to Harvard University economist Claudia Goldin.[29]

As demand for goods grew along with the nation's population, the wages of women working full time in manufacturing rose slowly as a percentage of men's pay. The wage gap narrowed from about 30 percent of men's earnings in 1820 to 56 percent nationwide in 1885, according to Goldin.[30]

But progress came more slowly, if at all, in ensuing years and decades.

In manufacturing, Goldin noted in a 1990 book on the economic history of American women, "The ratio of female to male wages . . . continued to rise slowly across most of the nineteenth century but reached a plateau before 1900."[31]

As the 20th century dawned, some women's advocates pushed for equal pay for equal work between the sexes. But others questioned the equal-pay idea. In 1891, the British economist Sidney Webb pointed to "the impossibility of discovering any but a very few instances in which men and women do precisely similar work, in the same place and at the same epoch."[32]

By the turn of the 20th century, women's jobs had started growing more diverse. Women found work not only in domestic service and manufacturing but also in teaching, sales and clerical positions. Still, only 21 percent of American women worked outside the home in 1900, and most left the labor force upon or right after marriage.[33]

Women seeking to move up in the business world faced huge cultural hurdles. In 1900 *Ladies' Home Journal* told its readers: "Although the statement may seem a hard one, and will unquestionably be controverted, it nevertheless is a plain, simple fact that women have shown themselves naturally incompetent to fill a great many of the business positions which they have sought to occupy. . . . The fact is that no one woman in a hundred can stand the physical strain of the keen pace which competition has forced upon every line of business today."[34]

Women's labor participation gradually rose in the early decades of the 20th century, fueled in part by World War I, which ended in 1918. By 1920, almost a quarter of all U.S. women were in the labor force, and 46 percent of single women worked.[35]

World War I advanced women's status, historian Michael McGerr noted. "Although the number of employed women grew only modestly during the 1910s, the wartime departure of men for military service opened up jobs traditionally denied to women in offices, transportation and industry. Leaving jobs as domestic servants, seamstresses and laundresses, women became clerks, telephone operators, streetcar conductors, drill press operators and munitions makers. Women's new prominence in the work force led in turn to the creation of a Women's Bureau in the Department of Labor."[36]

In 1920 women gained the right to vote with adoption of the 19th Amendment. Soon afterward, Quaker activist Alice Paul introduced the first version of today's Equal Rights Amendment. In 1982 the amendment fell three states short of ratification, and its passage remains controversial today.[37] (*See "At Issue," p. 158.*)

During the Great Depression of the 1930s, the proportion of single women who were working stayed more or less flat. But the percentage of married women who worked rose to almost 14 percent by 1940 — a jump of more than 50 percent over the 1920 rate.[38] World War II brought millions more women into the labor force, as females — characterized by the iconic image of Rosie the Riveter — took jobs in defense plants doing work traditionally performed by men.

Equal-Pay Initiatives

As women proved their mettle behind the drill press and rivet gun, advocates continued to push for equal pay. In 1942 President Franklin D. Roosevelt had the National War Labor Board urge employers to equalize wage rates between men and women "for comparable quality and quantity of work on the same or similar operations."[39]

Debating the Comparable-Worth Doctrine

Would the approach help close the gender gap?

Imagine a company whose employees include a man who supervises telephone linemen and a woman who supervises clerical employees. They oversee the same number of workers, report to the same number of bosses, work the same hours and their jobs have been deemed of equal value to the company. Should their paychecks be the same?

Should the man get extra points for having to work outside in the cold? Should the woman get extra points for having a college degree or more years of experience?

Or, as some argue, should competitive market forces and the laws of supply and demand determine how much the man and woman earn?

Such questions lie at the heart of the debate over "comparable worth." The doctrine argues that when jobs require similar levels of skill, effort, responsibility and working conditions, the pay should be the same — even if the duties are entirely different.

Advocates of comparable worth say the market historically has undervalued jobs traditionally held by women — such as social work, secretarial work and teaching — and that such inequity has been a major contributor to the gender pay gap. If comparable worth were taken into account, they argue, it would even out wage inequality between those working in jobs dominated by women and those traditionally held by men when an impartial evaluation deems the jobs are of equal value to an employer.

Advocates also say neither the Equal Pay Act of 1963 — which bars unequal pay for the same job — nor Title VII of the Civil Rights Act of 1964, which bans discrimination based on race, color, gender, religion and national origin in hiring and promotion, do what the comparable-worth doctrine would do: Root out bias against entire occupations traditionally dominated by females.[1]

Although women began entering non-traditional fields decades ago, Labor Department data show that certain occupations still are filled mostly by females. For example, in 2006, 89 percent of paralegals and legal assistants were women, while only 33 percent of lawyers were women. And only 7 percent of machinists were women, while 84 percent of special-education teachers were female.[2]

"There's a lot of [job] segregation, and the closer you look, the more segregation you find," says Philip Cohen, a sociologist at the University of North Carolina who studies gender inequality. "Under current law, it's very difficult to bring legal action successfully and say the pay gap between men and women is discrimination, because the employer can say 'they're doing different jobs.' "

But critics say comparable worth would disrupt the traditional market-based system of determining wages based on the laws of supply and demand. "You would have people moving into occupations where there was really no shortage" of workers, says June O'Neill, an economist at the City University of New York's Baruch College. "You would have gluts in some [job categories] and shortages in others."

In 2000 testimony before a congressional panel, O'Neill outlined what she saw as the dangers of adopting a

In the closing months of the war, the first bill aimed at barring gender pay discrimination came to the floor of Congress. The Women's Equal Pay Act of 1945 went nowhere, though.[40]

By 1960, more than a third of women were working, and among single, white women ages 25 to 34, the labor participation rate was a then-record 82 percent.[41] But most women continued to work in low-paying clerical, service and manufacturing jobs, and the wage gap between males and females was wide. By 1963, women made only 59 cents for every dollar in median year-round earnings paid to men.[42] Women who tried to break into so-called "men's" occupations faced huge resistance.

That year, after decades of struggle by women's advocates for federal legislation on gender pay equity, Congress passed the Equal Pay Act as an amendment to the Fair Labor Standards Act of 1938. In signing the act, President Kennedy said the law "affirms our determination that when women enter the labor force they will find equality in their pay envelopes."[43]

The measure, as finally adopted, stopped short of ensuring the elusive comparable-worth standard that

comparable-worth approach. Because there is no uniform way to rank occupations by worth, she says, such a policy would "lead to politically administered wages that would depart from a market system of wage determination." Pay in traditionally female occupations would likely rise — appointing people favorable to the comparable-worth idea "would all but guarantee that result," she said. But that higher pay would raise costs for employers, leading them to put many women out of work, she suggested. "The ironic result is that fewer workers would be employed in traditionally female jobs."

Not only that, but some employers would respond to the higher wage levels by providing fewer non-monetary benefits, such as favorable working hours, that help accommodate women with responsibilities at home, O'Neill said. "Apart from the inefficiency and inequality it would breed," she concluded, "I find comparable worth to be a truly demeaning policy for women. It conveys the message that some cannot compete in non-traditional jobs and can only be helped through the patronage of a job evaluator."

Critics also say that comparable worth would put the government into the role of setting wages for private business, an idea that is anathema to business interests.

"Who determines what is equal value?" asks Michael Eastman, executive director of labor policy at the U.S. Chamber of Commerce. "Equal value to society? Who's setting wages then? Is the government coming up with

Martha Burk directs the Corporate Accountability Project for the National Council of Women's Organizations.

guidelines? For example, are truckers equal to nurses, and who's making that comparison? We've never had the government setting private-sector wage rates like that."

Supporters of comparable worth brush off such concerns. Martha Burk, a longtime women's activist, notes that a bill proposed by Sen. Tom Harkin, D-Iowa, would require companies to disclose how they pay women and men by job categories, a practice that alone would lead to more equitable wages. "What you have is a government solution that is not telling anybody what to pay their employees," she says. It would only "increase the transparency so the company can solve its own problem if it has one."

As to the notion that comparable worth amounts to government intrusion in the private market, Burk says, "Free marketers think anything short of totally unregulated capitalism is interfering in the free market."

"It may be that markets are efficient from the point of view of employers," adds Vicky Lovell, director of employment and work/life programs at the Institute for Women's Policy Research in Washington. "But I don't think they're efficient from the point of view of workers."

---

[1] For background, see June O'Neill, "Comparable Worth," *The Concise Encyclopedia of Economics*, The Library of Economics and Liberty, www.econlib.org.

[2] "Women in the Labor Force: A Databook," U.S. Department of Labor, Report 1002, September 2007, Table 11, pp. 28-34.

women's advocates had so long sought. Instead, the bill made it illegal to discriminate in pay and benefits on the basis of sex when men and women performed the same job at the same employer.

Under the law, for example, a company couldn't pay a full-time female store clerk less per hour than a male one for doing the same job in stores located in the same city. But the law was silent on situations in which, say, the work of a female secretarial supervisor was deemed to be of comparable worth to that of a male who supervised the same company's truck drivers.

While the Equal Pay Act marked progress, it was far from an airtight guarantee of "equality in . . . pay envelopes." For example, the law initially did not cover executive, administrative or professional jobs; that exemption was lifted in 1972. Yet, one study argues that courts have interpreted the act so narrowly that white-collar female workers have had trouble winning claims through its provisions.[44]

Perhaps more significantly, the law gives companies several defenses for pay disparities: when wage differences stem from seniority or merit systems, are based on

Did Wal-Mart Favor Male Workers?

Women's suit seeks billions in damages.

Dedra Farmer, the daughter of an auto mechanic, worked in the Tire Lube Express Division of Wal-Mart Stores, the only female in her district who held a salaried manager position in that division. During her 13 years with the retail giant, she told a congressional panel last year, she saw evidence that women — herself among them — earned less than men holding the same jobs.

Farmer said she complained to Wal-Mart's CEO through e-mails, expressed her concern at a store meeting and was assured by the store manager that she'd get a response. "The response I received was a pink slip," she said.[1]

Farmer has joined a class-action lawsuit accusing Wal-Mart of sex discrimination in pay and promotions. The case, which could cover perhaps 1.6 million current and former female employees and result in billions of dollars in damages, is the biggest workplace discrimination lawsuit in the nation's history.

Filed in 2001 by Betty Dukes and five other Wal-Mart employees, the case has gone through a series of legal maneuverings, most recently in December, when a three-judge panel of the U.S. 9th Circuit Court of Appeals reaffirmed its certification as a class-action lawsuit but left the door open for Wal-Mart to ask for a rehearing on that status. If the appeals court does not reconsider the class-action designation, the company reportedly will petition the Supreme Court.[2]

The stakes in the case are high. Goldman Sachs Group last year estimated potential damages at between $1.5 billion and $3.5 billion if the retailer loses, and punitive damages could raise the figure to between $13.5 billion and $31.5 billion.[3]

The company's lawyers have asserted that a class-action suit is an inappropriate vehicle to use because Wal-Mart's employment policies are decentralized, and individual store managers and district managers make pay and promotion decisions.[4]

Theodore J. Boutrous Jr., a lawyer for Wal-Mart, has said that decisions by thousands of managers at 3,400 Wal-Mart stores during six years were "highly individualized and cannot be tried in one fell swoop in a nationwide class action."[5] He has also said the company has a "strong diversity policy and anti-discrimination policy."[6]

But Brad Seligman, executive director of the Impact Fund, a nonprofit group in Berkeley, Calif., representing the plaintiffs, said, "No amount of PR or spin is going to allow Wal-Mart to avoid facing its legacy of discrimination."[7]

A statistician hired by the plaintiffs said it took women an average of 4.38 years from the date of hire to be promoted to assistant manager, while it took men 2.86 years. Moreover, it took an average of 10.12 years for women to become managers compared with 8.64 for men.[8]

The statistician, Richard Drogin, of California State University at East Bay, also found that female managers made an average annual salary of $89,280, while men in the same position earned an average of $105,682. Female hourly workers earned 6.7 percent less than men in comparable positions.[9]

quantity or quality of production, or stem from "any other factor other than sex."

That last provision, critics say, can sometimes allow business practices that may seem gender-neutral on the surface but discriminate nonetheless.

The Equal Pay Act took effect in 1964, and that same year Congress passed Title VII of the Civil Rights Act of 1964, a broad measure that prohibits employment discrimination on the basis of race, color, religion, national origin and sex, and covers hiring, firing and promotion as well as pay. A measure called the Bennett Amendment, sponsored by Rep. Wallace F. Bennett, a Utah Republican,

sought to bring Title VII and the Equal Pay Act in line with each other.

In ensuing years, the overlap of the Equal Pay Act and Title VII created confusion but also helped to animate the battle against wage discrimination. Part of the conflict over pay equity played out in the courts in the 1970s and '80s.

Key Court Rulings

In a case that initially raised hopes for the theory of comparable worth, the U.S. Supreme Court ruled 5-4 to

Appellate Judge Andrew J. Kleinfeld has dissented in the case, arguing that certifying the suit as a class action deprived the retailer of its right to defend against individual cases alleging bias. In addition, he argued that female employees who were discriminated against would be hurt by class-action status, because women "who were fired or not promoted for good reasons" would also share in any award if Wal-Mart lost the case.[10]

Business lobbies also have urged that the class-action certification be reversed. An official of the U.S. Chamber of Commerce, which filed a "friend of the court" (*amicus curiae*) brief in the case, warned of "potentially limitless claims" against companies "with limited ability to defend against them." He added: "The potential financial exposure to an employer facing a class action of this size creates tremendous pressure to settle regardless of the case's merit."[11]

But women's advocates argue that a class-action approach is appropriate. It "provides the only practical means for most women in low-wage jobs to redress discrimination in pay because of such workers' often tenuous economic status," stated an *amicus* letter written to the appeals court on behalf of the U.S. Women's Chamber of Commerce.[12]

Added Margot Dorfman, chief executive officer of the group: "A woman with family responsibilities often isn't in a

Getty Images/Gilles Mingasson

A Wal-Mart store manager reads the store's weekly sales results to other workers. Male hourly workers at Wal-Mart earn 6.7 percent more than women in comparable positions, a pay-equity study contends.

position to quit her job or risk antagonizing her employer with a challenge to a bad workplace practice."[13]

[1] Statement of Dedra Farmer before House Committee on Education and Labor, April 24, 2007.

[2] Amy Joyce, "Wal-Mart Loses Bid to Block Group Bias Suit," *The Washington Post*, Feb. 7, 2007, p. 1D.

[3] Details of the Goldman Sachs analysis are from Steve Painter, "Judges modify sex-bias decision; Wal-Mart appeal likely to see delay," *Arkansas Democrat-Gazette*, Dec. 12, 2007.

[4] Steven Greenhouse and Constance L. Hays, "Wal-Mart Sex-Bias Suit Given Class-Action Status," *The New York Times*, June 23, 2004.

[5] Joyce, *op. cit.*

[6] Quoted in Bob Egelko, "Wal-Mart sex discrimination suit advances; Appeals court OKs class action status for 2 million women," *San Francisco Chronicle*, Feb. 7, 2007, p. B1.

[7] Joyce, *op. cit.*

[8] *Ibid.*

[9] *Ibid.*

[10] Painter, *op. cit.*

[11] "U.S. Chamber Files Brief in Wal-Mart Class Action," press release, U.S. Chamber of Commerce, Dec. 13, 2004, www.uschamber.com/press/releases/2004/december/04-159.htm.

[12] Mark E. Burton Jr., Hersh & Hersh, San Francisco, et al., letter submitted to 9th U.S. Circuit Court of Appeals, March 27, 2007, www.uswcc.org/amicus.pdf.

[13] PR Newswire, "U.S. Women's Chamber of Commerce Joins Fight in Landmark Women's Class Action Suit Against Wal-Mart," March 28, 2007.

allow female jail guards to sue for sex discrimination. The women, called "matrons," earned 30 percent less than male guards, called "deputy sheriffs."[45] The women argued that while they had fewer prisoners to guard and more clerical duties than the male guards, their work was comparable. An outside job evaluation showed that the women did 95 percent of what the men were doing, but received $200 less a month than the men.[46]

Prior to the Supreme Court's ruling, *The Washington Post* noted at the time, "the only sure grounds for a pay discrimination claim by a woman under federal law was 'unequal pay for equal work' — an allegation that she

was paid less than a man holding an identical job. The jail matrons and women's rights lawyers said that lower pay for a comparable, if not equal, job could also be the basis for a sex-discrimination charge."

Justice William J. Brennan wrote that a claim of wage discrimination under Title VII did not have to meet the equal work standards of the Equal Pay Act. Thus, noted Clare Cushman, director of publications at the Supreme Court Historical Society, "a woman employee could sue her employer for gender-based pay discrimination even if her company did not employ a man to work the same job for higher pay."[47]

Still, Cushman wrote, while the court "opened the door slightly for women working in jobs not strictly equal to their male counterparts, it also specifically declined to authorize suits based on the theory of comparable worth."

In 1985 that theory suffered a blow that continues to resonate today, partly because of the personalities who were involved. In *AFSCME v. the State of Washington*, the 9th U.S. Circuit Court of Appeals overturned a lower court's ruling ordering Washington to pay more than $800 million in back wages to some 15,000 state workers, most of them women.[48]

The case turned on the question of whether employers were required to pay men and women the same amounts for jobs of comparable worth, rather than equal wages for the same jobs. It eventually ended in a draw when the state negotiated a settlement with AFSCME (American Federation of State, County and Municipal Employees union).[49]

Judge Anthony M. Kennedy, who now sits on the U.S. Supreme Court and presumably could help decide a comparable-worth case should one arise before the justices, wrote the appellate court's decision. Kennedy wrote: "Neither law nor logic deems the free-market system a suspect enterprise." During this same period, two other personalities who now sit on the high court also expressed negative views on comparable worth. As a lawyer in the Reagan administration, John Roberts, now chief justice, described it as "a radical redistributive concept."[50] And the EEOC, then under Chairman Clarence Thomas, rejected comparable worth as a means of determining job discrimination. "We found that sole reliance on a comparison of the intrinsic value of dissimilar jobs — which command different wages in the market — does not prove a violation of Tile VII," Thomas stated.[51]

The views of Thomas and Roberts reflected the conservative policies of the Reagan administration during the 1980s. Yet despite the political tenor of that era, women made major strides toward workplace equality. From 1980 to 1992, the wage gap in median weekly earnings of full-time female wage and salary workers narrowed from 64 percent to 76 percent after adjusting for inflation. But it shrank only from 77 percent to 81 percent from 1993 — the year that Democratic President Bill Clinton took office and the Family and Medical Leave Act was enacted — to 2006.[52]

Measuring Progress

Experts debate whether and to what degree women's gains may have slowed or stopped in recent years. Some point to huge political gains in this decade, including Sen. Clinton's role in the presidential race and the rise of Rep. Nancy Pelosi, D-Calif., to speaker of the House. Others cite such evidence as a recent study showing that female corporate directors, though a small minority in boardrooms, out-earn male directors.[53]

But many scholars believe women's gains have indeed slowed.

Vanneman, the University of Maryland sociologist, has carefully charted a number of trends linked to the so-called gender revolution, and on his Web site he notes that he and several colleagues are studying the pace of women's progress.

"For much of the last quarter of the 20th century, women gradually reduced gender inequalities on many fronts," he wrote, citing such trends as women entering the labor force in growing numbers, the opening of previously male-dominated jobs to women, the narrowing wage gap, women's role in politics and a growing openness in public opinion about the participation of women in public and community life.

But, he added, "all this changed in the early to mid-1990s." A "flattening of the gender trend lines" is seen in nearly all parts of society, he added: working-class and middle-class, black, white, Asian and Hispanic, mothers with young children and those with older ones, and so on. "All groups experienced major gender setbacks during the 1990s. The breadth of this reversal suggests something fundamental has happened to the U.S. gender structure."

In an interview, Vanneman says he has no theories as to what accounts for that reversal — only hunches — as he continues to study the phenomenon. One hunch is that the flattening started happening in the 1980s but didn't show up in a big way until the 1990s. He also says he suspects the reversal in women's progress gathered momentum in the 1990s as the "culture of parenting" changed. Americans, he says, became less accepting of women trying to balance busy careers with the pressures of motherhood, a shift that has put women in more of a bind than they felt in previous periods. As a result, many women have backed away from high-paying careers and devoted more time to family, he says.

"There's been tremendous growth in expectations of what it means to be a good parent," Vanneman says.

Cornell University economist Blau agrees that progress in women's wages slowed in recent years, though she sees some evidence that the picture has brightened a bit.

One reason for the slowdown in the 1990s, she says, may have been that the increase in demand for white-collar and service workers shifted into a lower gear compared to the 1980s, when many women benefited from a surge in hiring for white-collar jobs, including ones that required computer skills, while blue-collar jobs dominated by men began to wane.

In addition, Blau says that during the eighties, as many women began to stay in the workforce even after marriage and childbirth, employers' view of the value of female workers improved. That, she says, helped narrow the wage gap at a faster pace than in earlier decades.

Blau also sees evidence that men were doing more at home in the 1980s than ever before. That trend didn't go away in the past decade, she says, but it hasn't grown much either.

CURRENT SITUATION

Prospects in Congress

As concerns over the progress of gender equity grow, women's advocates are hoping that the Democrat-controlled Congress will pass new laws this year. But proposed legislation is likely to face stiff opposition.

Reversing the Supreme Court's *Ledbetter* decision seems to have the best chance of making it through Congress. The House passed the Ledbetter Fair Pay Act last July 31 by a 225-199 vote, largely along party lines.[54] A companion bill in the Senate, called the Lilly Ledbetter Fair Pay Restoration Act, had garnered 37 co-sponsors as of early March. Momentum continued this year with a Senate hearing.

In introducing the Senate version of the bill last July, Sen. Kennedy said it "simply restores the status quo" that existed before the *Ledbetter* decision "so that victims of ongoing pay discrimination have a reasonable time to file their claims."[55]

But employer advocates such as the U.S. Chamber of Commerce dispute such descriptions. Pointing to the House version that passed last summer, chamber officials said it would broaden existing law to apply to unintentional as well as intentional discrimination and would lead to an "explosion of litigation second-guessing legitimate employment and personnel decisions."[56]

The Bush administration has threatened a veto, saying last year that if the House bill came to the president, "his senior advisers would recommend that he veto" it.[57]

The measure would "impede justice" by allowing employees to sue over pay or other employment-related discrimination "years or even decades after the alleged discrimination occurred," the administration said. Moreover, the House bill "far exceeds the stated purpose of undoing the court's decision" by "extending the expanded statute of limitations to any 'other practice' that remotely affects an individual's wages, benefits, or other compensation in the future."

Eric Dreiband, a former EEOC general counsel in the Bush administration, told this year's hearing on the Senate bill that the measure would subject state and local governments, unions, employers and others to potentially unlimited penalties and could expose pension funds to "potentially staggering liability."[58]

Still, women's advocates remain sanguine about the measure's prospects. "My hope is that the bill will move expeditiously [this] spring" in the Senate and that "the president will reconsider and recognize how important this fix to the law is," says Samuels of the National Women's Law Center.

The other two main bills on gender pay equity could have rougher sledding.

Sen. Clinton's Paycheck Fairness Act is similar to a bill by the same name proposed during her husband's presidential administration. As of early March, the bill had garnered 22 co-sponsors in the Senate and 226 in the House.

Among other things, it would strengthen penalties on employers who violate the Equal Pay Act, make it harder for companies to use the law's defense for wage differences based on factors "other than sex," and bar employers from retaliating against workers who share wage information with each other. It also calls for the Labor Department to draw up guidelines aimed at helping employers voluntarily evaluate job categories and compare wages paid for different jobs with the aim of eliminating unfair wage differences between male- and female-dominated occupations.

The bill has drawn enthusiastic support from some women's advocates, but it also has opponents. Washington lawyer Brown said the goal of the provision on voluntary guidelines was "nothing more than the discredited 'comparable-worth' theory in new clothing."[59]

The Fair Pay Act, proposed by Sen. Harkin and Del. Eleanor Holmes Norton, D-D.C., a former EEOC chair, steps even closer to embracing the comparable-worth theory and thus, many observers believe, is likely to face stiff headwinds. The main ideas have circulated in Congress for years.

Is the Equal Rights Amendment to the Constitution still needed?

YES
Idella Moore
Executive Officer, 4ERA

Written for *CQ Researcher*, February 2008

We still need the Equal Rights Amendment (ERA) because sex discrimination is still a problem in our country. Like race or religious discrimination, gender discrimination is intended to render its victims economically, socially, legally and politically disadvantaged. But unlike racism and religious intolerance — whose practice against certain groups is localized within countries or regions — sex discrimination is universal. Why, then, in our court system are race and religious discrimination considered more serious offenses?

Today, American women — of all races and religions — are still fighting to achieve equal opportunity, pay, status and recognition in all realms of our society. At this moment, the largest class-action lawsuit in the history of this country is being argued on behalf of 1.6 million women who were discriminated against purely because of their gender. If the ERA had been ratified back in the 1970s, by now these types of lawsuits would be extinct.

We still need the ERA because ratification of the amendment will elevate "sex" to, in legal terms, a so-called suspect class. A suspect class has the advantage in discrimination cases. Gender, as yet, is not afforded that advantage. As we've seen with race, suspect class status increased the chance of favorable outcomes in discrimination cases. This, in turn, served as a deterrent. Consequently, in our society racism is now socially unacceptable. Sex discrimination, however, is not.

We still need the ERA because the continuing struggle for legal equality for women should be seen as a shameful and embarrassing condition of our society. Yet today lawmakers — sworn to represent all their constituents — proudly voice their objections to granting legal equality to women and without any fear of consequences to their political careers. How different our reactions would be if they were espousing racism.

The Equal Rights Amendment will perfect our Constitution by explicitly guaranteeing that the privileges, laws and responsibilities it contains apply equally to men and women. As it stands today the Constitution is sometimes interpreted that way, but women, as a universally and historically disadvantaged group, cannot rely on such interpretations. We have seen these "interpretations" vary and change, often due to the whims of the political climate. Therefore, without the ERA any gains women make will always be tenuous.

I see the Equal Rights Amendment, too, as a pledge to ourselves and posterity that we recognize that sexism exists and that we as a country are determined to continue perfecting our democracy by proudly and unequivocally guaranteeing that one's gender will no longer be a detriment to achieving the American dream.

NO
Phyllis Schlafly
President, Eagle Forum

Written for *CQ Researcher*, February 2008

The Equal Rights Amendment (ERA) was fiercely debated across America for 10 years (1972-1982) and was rejected. ERA has been reintroduced into the current Congress under a slightly different name, but it's the same old amendment with the same bad effects.

The principal reason ERA failed is that although it was marketed as a benefit to women, its advocates were never able to prove it would provide any benefit whatsoever to women. ERA would put "sex" (not women) in the Constitution and just make all our laws sex-neutral.

ERA advocates used their massive access to a friendly media to suggest that ERA would raise women's wages. But ERA would have no effect on wages because our employment laws are already sex-neutral. The equal-pay-for-equal-work law was passed in 1963, and the Equal Employment Opportunity Act — with all its enforcement mechanisms — was passed in 1972.

Supreme Court Justice Ruth Bader Ginsburg's book *Sex Bias in the U.S. Code* spells out the changes ERA would require, and it proves ERA would take away benefits from women. For example, the book states that the "equality principle" would eliminate the concept of "dependent women." This would deprive wives and widows of their Social Security dependent-wife benefits, on which millions of mothers and grandmothers depend.

Looking at the experience of states that have put ERA language into their constitutions, we see that ERA would most probably require taxpayer funding of abortions. The feminists aggressively litigate this issue. Their most prominent victory was in the New Mexico Supreme Court, which accepted the notion that since only women undergo abortions, the denial of taxpayer funding is sex discrimination.

ERA would also give the courts the power to legalize same-sex marriages. Courts in four states have ruled that the ERA's ban on gender discrimination requires marriage licenses to be given to same-sex couples. In Maryland and Washington, those decisions were overturned by a higher court by only a one-vote margin. The ERA would empower the judges to rule either way.

If all laws are made sex-neutral, the military draft-registration law would have to include women. We don't have a draft today, but we do have registration, and those who fail to register immediately lose their college grants and loans and will never be able to get a federal job.

As Harkin describes it, the bill requires employers to provide equal pay for jobs that are comparable in skill, effort, responsibility and working conditions, regardless of sex, race or national origin, and it bars companies from reducing other employees' wages to achieve pay equity.[60]

Again, advocates such as Samuels are hopeful Congress will pass both the Paycheck Fairness and Fair Pay Act and that the president won't veto them if they do make it to his desk. "The hope would be that the level of support for these bills both in Congress and among the public is so substantial, and they so clearly are a necessary step toward ensuring true equality of wages, that the president would understand the necessity for them and sign them," she says.

But business opposition is likely to be strong. Eastman at the U.S. Chamber of Commerce lists a variety of complaints about both bills, such as their provisions for punitive damages and their allowances for class-action suits against employers. "The case has not been made that these bills are justified," he says.

State Action

While women's advocates hold out hope for congressional action, they also are turning their attention to the states in hopes of pressing legislatures to stiffen laws on pay equity and make local economies friendlier to gender issues. As of April 2007, all but 11 states and the District of Columbia had laws on equal pay.[61]

Minnesota has had a system of comparable worth, or "pay equity," for public employees since the 1980s, and last year proposals were made to expand the system to private employers that do business with the state. The Minnesota program gave smaller raises to public workers in male-dominated jobs and bigger raises to those in female-dominated ones, according to a former staff member of the Minnesota Commission on the Economic Status of Women. The system shrank the pay gap from 72 percent to nearly equal pay.[62]

A report by the Institute for Women's Policy Research said in 2006 that while women's wages had risen in all states in inflation-adjusted terms since 1989, "in no state does the typical full-time woman worker earn as much as the typical man." It would take 50 years "at the present rate of progress" for women to achieve wage parity with men nationwide, it said.[63]

Some advocates are unwilling to wait that long. In Colorado, for example, a Pay Equity Commission appointed by Donald J. Mares, executive director of the state Department of Labor and Employment, worked since last June to formulate policy recommendations to curb gender and racial pay inequities in the private and public sectors. The 12-member commission includes policy analysts, business and labor union representatives, academics and advocates for women and minorities.[64]

Meric, the 9to5 director and a Colorado resident, said her group was instrumental in getting the state to appoint the commission. Although the panel has no authority to force employers to alter pay practices, Meric hopes the commission's work leads to change. One key recommendation, she says, is that employers do more to create flexible policies so that workers — especially women with caregiving responsibilities — aren't penalized for meeting both work and family responsibilities.

Mares told the Colorado Women's Legislative Breakfast in February that another recommendation calls for making the commission permanent, so it can continue to monitor gender pay equity in the state and help educate businesses on good practices.

In Colorado, he said, the average woman makes 79 cents for every dollar earned by the average man. "Every day you as a community walk in the door," he told the gathering of women, "your pay is being discounted. That's not good."[65]

Better negotiating skills could help narrow the gender wage gap, in the view of women's advocates. The Clinton/DeLauro bill calls for grants to help women and girls "strengthen their negotiation skills to allow the girls and women to obtain higher salaries and the best compensation packages possible for themselves."

It's a talent that many women don't exercise, says Linda Babcock, an economist at Carnegie Mellon University in Pittsburgh and co-author of the recent book *Women Don't Ask: Negotiation and the Gender Divide*. Babcock found in a study of Carnegie Mellon students graduating with master's degrees in public policy that only 12.5 percent of females tried to negotiate for better pay when they received a job offer, while 51.5 percent of males did. Afterward, the females earned 8.5 percent less than the males.

Babcock sees several reasons why women are not inclined to negotiate more, including that they have been socialized by American culture to be less assertive than men. And, she says, women who do try to bargain for better wages often are subjected to "backlash" by employers and peers.

Not that women are incapable of negotiating, Babcock stresses. While they may not always stand up for themselves in seeking higher wages, women outperform men when negotiating on behalf of somebody else, she has found.

"It's really striking," she says. "If we were missing some gene, we wouldn't really be able to turn it on on behalf of somebody else."

OUTLOOK

Pressure for Change

Some women's advocates are not especially sanguine about the possibility of big strides on the gender-wage front, at least in the near future.

"I don't think five years is long enough [for there] to be much change, particularly if we don't see much concerted effort among employers," says Lovell of the Institute for Women's Policy Research.

Big change would require a "push from the federal government" or "some dramatic effort on the part of socially conscious employers," she says. "That hasn't happened before, and I don't think it will in the next few years."

Still, observers believe that social and political shifts will produce new pressure for changes in the way employers deal with wage equity.

Meric says 9to5's "long-term agenda" is to have the theory of comparable worth enshrined in law as well as to have "guaranteed minimum labor standards" for all workers that include paid sick leave and expanded coverage under the Family and Medical Leave Act. In Colorado, she hopes the recommendations outlined by the Pay Equity Commission will serve as a model for other states and "move us closer" to that long-term goal. "Basic protections should apply to workers wherever they live in the United States."

"In the last five or 10 years we have seen progress stall in [achieving] gender equality," says Philip Cohen, a sociologist at the University of North Carolina at Chapel Hill who studies gender inequity. But in coming years, he says he is inclined to think that college-educated women will exert increasing pressure on federal and state lawmakers and employers to make policy changes that can narrow the wage gap.

"If you look back to feminism in the '60s," Cohen says, "a lot of women had college degrees but weren't able to take advantage of their skills in the marketplace, and that became the 'feminine mystique' " explored in Betty Friedan's groundbreaking 1963 book.

Today, "Women are outnumbering men in college graduation rates, and I think we are going to see more and more women looking around for better opportunities. If they don't see gender equality resulting, they're going to be very dissatisfied."

And that dissatisfaction, Cohen says, could well show up in the political arena.

Samuels of the National Women's Law Center hopes the debate in Congress and fallout from the Supreme Court's *Ledbetter* decision will spur further gains in wage equity for women.

"Unfortunately, over the course of the last several years things have pretty much stagnated," she says. "I do hope that the recent public attention paid to wage disparity will cause employers to take a look at their pay scales and try to do the right thing."

NOTES

1. Testimony of Lilly Ledbetter before the Committee on Education and Labor, U.S. House of Representatives, on the Amendment of Title VII, June 12, 2007.

2. Testimony of Lilly Ledbetter before Senate Committee on Health, Education, Labor and Pensions, Jan. 24, 2008.

3. *Ledbetter v. Goodyear Tire & Rubber Co. Inc.*, 550 U.S. __ (May 29, 2007).

4. Ledbetter testimony, *op. cit.*, June 12, 2007.

5. Diana Furchtgott-Roth, testimony on the Paycheck Fairness Act before House Committee on Education and Labor, April 24, 2007.

6. "Highlights of Women's Earnings in 2006," U.S. Department of Labor, September 2007, Table 1, p. 7. Data are for median usual weekly earnings of full-time wage and salary workers ages 16 and older. For the Census Bureau data, see www.census.gov/compendia/statab/tables/08s0628.pdf. The Census Bureau data represent median full-time, year-round earnings for male and female workers 15 years old and older as of March 2006.

7. Testimony before Senate Committee on Health, Education, Labor and Pensions, April 12, 2007.

8. "Women's Earnings: Work Patterns Partially Explain Difference between Men's and Women's

Earnings," U.S. General Accounting Office, October 2003.

9. U.S. Equal Employment Opportunity Commission, Office of Inspector General, "Semiannual Report to Congress," April 1, 2007-Sept. 30, 2007, Oct. 30, 2007, p. 7.

10. "Highlights of Women's Earnings," *op. cit.*, p. 1.

11. *Ibid.*, Table 1, p. 7.

12. Judy Goldberg Dey and Catherine Hill, "Behind the Pay Gap," American Association of University Women Educational Foundation, 2007.

13. Paycheck Fairness Act, HR 1338, S 766.

14. Fair Pay Act, S 1087 and HR 2019, sponsored in the House of Representatives by Del. Eleanor Holmes Norton, D-D.C.

15. Lilly Ledbetter Fair Pay Act, HR 2831 and Fair Pay Restoration Act, S 1843.

16. David Neumark, with the assistance of Roy J. Bank and Kyle D. Van Nort, "Sex Discrimination in Restaurant Hiring: An Audit Study," *The Quarterly Journal of Economics*, August 1996.

17. Furchtgott-Roth testimony, *op. cit.*

18. Dey and Hill, *op. cit.*

19. Testimony of Heather Boushey before House Committee on Education and Labor, April 24, 2007, p. 4.

20. John F. Kennedy, Remarks Upon Signing the Equal Pay Act, June 10, 1963, quoted in John T. Woolley and Gerhard Peters, *The American Presidency Project* [online], Santa Barbara, Calif., University of California (hosted), Gerhard Peters (database), www .presidency.ucsb.edu/ws/?pid=9267.

21. Testimony of Jocelyn Samuels before Senate Committee on Health, Education, Labor and Pensions, "Closing the Gap: Equal Pay for Women Workers," April 12, 2007, p. 6.

22. Carrie Lukas, "A Bargain At 77 Cents To a Dollar," *The Washington Post*, April 3, 2007, p. 23A.

23. Testimony of Barbara Berish Brown before Senate Committee on Health, Education, Labor and Pensions, April 12, 2007.

24. Quoted in Justine Andronici, "Court Gives OK To Unequal Pay," *Ms. Magazine*, summer 2007, accessed at www.msmagazine.com/summer2007/ledbetter.asp.

25. Quoted in Michael Doyle, "Justices Put Bias Lawsuits on Tight Schedule," *Kansas City Star*, May 30, 2007, p. 1A.

26. Testimony of Margot Dorfman before Senate Committee on Health, Education, Labor and Pensions on the "The Fair Pay Restoration Act: Ensuring Reasonable Rules in Pay Discrimination Cases," Jan. 24, 2008, pp. 2-3.

27. "Employment Discrimination: Post-Ledbetter Discrimination," "Justice Talking," National Public Radio, Oct. 22, 2007, accessed at www.justicetalking .org/transcripts/071022_EqualPay_transcript.pdf.

28. Richard B. Morris, ed., "The U.S. Department of Labor Bicentennial History of the American Work," U.S. Department of Labor, 1976, p. 67.

29. Claudia Goldin, *Understanding the Gender Gap: An Economic History of American Women* (1990), p. 50.

30. *Ibid.*, Figure 3.1, p. 62, and text pp. 63, 66.

31. *Ibid.*, p. 67.

32. Quoted in *ibid.*, p. 209.

33. *Ibid.*, Table 2.1, p. 17, citing U.S. Census data.

34. "Setting a New Course," *CQ Researcher*, May 10, 1985, citing Julie A. Matthaei, *An Economic History of Women in America* (1982), p. 222.

35. Goldin, *op. cit.*, p. 17.

36. Michael McGerr, *A Fierce Discontent: The Rise and Fall of the Progressive Movement in America, 1870-1920* (2003), pp. 295-296.

37. For background, see Richard Boeckel, "Sex Equality and Protective Laws," *Editorial Research Reports*, July 13, 1926; and Richard Boeckel, "The Woman's Vote in National Elections," *Editorial Research Reports*, May 31, 1927, both available at *CQ Researcher Plus Archive*, www.cqpress.com.

38. *Ibid.*

39. American Association of University Women, "A Brief History of the Wage Gap, Pay Inequity, and the Equal Pay Act," www.aauw.org/advocacy/laf/ lafnetwork/library/payequity_hist.cfm. For background, see K. R. Lee, "Women in War Work," *Editorial Research Reports*, Jan. 26, 1942, available at *CQ Researcher Plus Archive*, www.cqpress.com.

40. *Ibid.*

41. Goldin, *op. cit.*, Table 2.2, p. 18.

42. *Ibid.*, Table 3.1, p. 60.

43. John F. Kennedy, *op. cit.*

44. Juliene James, "The Equal Pay Act in the Courts: A De Facto White-Collar Exemption," *New York University Law Review*, Vol. 79, November 2004, p. 1875.

45. Clare Cushman, *Supreme Court Decisions and Women's Rights*, CQ Press (2000), p. 146. The case is *County of Washington v. Gunther* (1981). For background, see Sandra Stencel, "Equal Pay Fight," *Editorial Research Reports*, March 20, 1981, and R. Thompson, "Women's Economic Equity," *Editorial Research Reports*, May 10, 1985, both available at *CQ Researcher Plus Archive*, www.cqpress.com.

46. Deborah Churchman, "Comparable Worth: The Equal-Pay Issue of the '80s," *The Christian Science Monitor*, July 22, 1982, p. 15.

47. Cushman, *op. cit.*

48. James Warren, "Fight for Pay Equity Produces Results, But Not Parity," *Chicago Tribune*, Sept. 8, 1985, p. 13.

49. Judy Mann, "New Victory in Women's Pay," *The Washington Post*, Aug. 27, 1986, p. 3B.

50. Linda Greenhouse, "Judge Roberts, the Committee Is Interested in Your View On . . . ," *The New York Times*, Sept. 11, 2005, p. 1A.

51. "Women Dealt Setback on 'Comparable Worth,' " *Chicago Tribune*, June 18, 1985, p. 1.

52. "Highlights of Women's Earnings in 2006," *op. cit.*, Table 13, p. 28.

53. Martha Graybow, "Female U.S. corporate directors out-earn men: study," Reuters, Nov. 7, 2007. The study of more than 25,000 directors at more than 3,200 U.S. companies was done by the Corporate Library. It found that female directors earned median compensation of $120,000 compared with $104,375 for male board members.

54. Libby George, "House Democrats Prevail in Effort to Clarify Law on Wage Discrimination," *CQ Weekly*, Aug. 6, 2007, p. 2381.

55. Sen. Edward Kennedy, statement on S 1843, "Statements on Introduced Bills and Joint Resolutions," Senate, July 20, 2007, accessed at www.thomas.gov.

56. U.S. Chamber of Commerce, "Letter Opposing HR 2831, the Ledbetter Fair Pay Act," July 27, 2007, accessed at www.uschamber.com/issues/letters/2007/070727_ledbetter.htm.

57. "Statement of Administration Policy: HR 2831, Lilly Ledbetter Fair Pay Act of 2007," Executive Office of the President, Office of Management and Budget, July 27, 2007, accessed at www.whitehouse.gov/omb/legislative/sap/110-1/hr2831sap-r.pdf.

58. Statement of Eric S. Dreiband before Senate Committee on Health, Education, Labor and Pensions, Jan. 24, 2008, pp. 11-13.

59. Barbara Berish Brown testimony, *op. cit.*

60. Statement of Sen. Tom Harkin at the Health, Education, Labor and Pensions Committee Hearing on Equal Pay for Women Workers, April 12, 2007, accessed at www.harkin.senate.gov/pr/p.cfm?i=272330.

61. National Conference of State Legislatures, "State Laws on Equal Pay," April 2007.

62. H.J. Cummins, "Legislature will look at closing the gender gap," *Star Tribune*, April 23, 2007, p. 1D.

63. Heidi Hartmann, Olga Sorokina and Erica Williams, "The Best and Worst State Economies for Women," Institute for Women's Policy Research, December 2006.

64. "Pay Equity Commission holds first meeting," *Denver Business Journal*, June 26, 2007.

65. Remarks of Donald Mares, Colorado Women's Legislative Breakfast, Feb. 12, 2008, accessed at www.youtube.com/watch?v=UIO0mlHb6b8&feature=related.

BIBLIOGRAPHY

Books

Cushman, Clare, *Supreme Court Decisions and Women's Rights*, CQ Press, 2000.
In clear prose, the director of publications for the Supreme Court Historical Society covers the waterfront of Supreme Court cases and issues involving women's rights, including those related to pay equity and discrimination in the workplace.

Farrell, Warren, *Why Men Earn More,* **AMACOM, 2005.**

The only man elected three times to the board of directors of the National Organization for Women's New York chapter argues that the pay gap can no longer be ascribed to discrimination, and he seeks "to give women ways of earning more rather than suing more."

Goldin, Claudia, *Understanding the Gender Gap: An Economic History of American Women,* **Oxford University Press, 1990.**

A Harvard University economics professor traces the evolution of female workers and gender differences in occupations and earnings from the early days of the republic to the modern era.

Murphy, Evelyn, with E.J. Graff, *Getting Even: Why Women Don't Get Paid Like Men — and What to Do About It,* **Touchstone, 2005.**

The former Massachusetts lieutenant governor writes in this anecdote-filled book that the "gender wage gap is unfair" and "it's not going away on its own."

Articles

Hymowitz, Carol, "On Diversity, America Isn't Putting Its Money Where Its Mouth Is," *The Wall Street Journal,* **Feb. 25, 2008.**

Progress for women and minorities in business has stalled or moved backward at many of the nation's largest companies, and the inequality shapes perceptions about who can or should fill leadership roles.

Murphy, Cait, "Obama flunks Econ 101," *Fortune, CNNMoney.com,* **June 6, 2007, http://money.cnn .com/2007/06/04/magazines/fortune/muphy_ payact.fortune/index.htm.**

The presidential candidate is "flirting with a very bad idea" by co-sponsoring the Fair Pay Act, "a bill that would bureaucratize most of the labor market," Murphy argues.

Parloff, Roger, and Susan M. Kaufman, "The War Over Unconscious Bias," *Fortune,* **Oct. 15, 2007.**

Wal-Mart and other companies are facing accusations of gender pay bias and other forms of job discrimination, "but the biggest problem isn't their policies, it's their managers' unwitting preferences."

Reports and Studies

Dey, Judy Goldberg, and Catherine Hill, "Behind the Pay Gap," *American Association of University Women Educational Foundation,* **April 2007, www.aauw.org/ research/upload/behindPayGap.pdf.**

A study of college graduates concludes that one year out of college women working full time earn only 80 percent as much as their male colleagues and that a decade after graduation the proportion falls to 69 percent.

Foust-Cummings, Heather, Laura Sabattini and Nancy Carter, "Women in Technology: Maximizing Talent, Minimizing Barriers," *Catalyst,* **2008, www .catalyst.org/files/full/2008%20Women%20in%20 High%20Tech.pdf.**

Technology companies are making progress at creating more diverse work environments, but women in the high-technology field still face barriers to advancement, such as a lack of role models, mentors and access to networks.

Hartmann, Heidi, Olga Sorokina and Erica Williams, *et al.,* **"The Best and Worst State Economies for Women,"** *Institute for Women's Policy Research,* **IWPR No. R334, December 2006, www.iwpr.org/ pdf/R334_BWState Economies2006.pdf.**

The advocacy group concludes that women's wages have risen in all states since 1989 after adjusting for inflation, but that in "no state does the typical full-time woman worker earn as much as the typical man."

U.S. General Accounting Office, **"Women's Earnings: Work Patterns Partially Explain Difference between Men's and Women's Earnings," October 2003, www .gao.gov/new.items/d0435.pdf.**

This statistical study concludes that "work patterns are key" among the factors that account for earnings differences between men and women, but that some differences remain unexplained.

U.S. Department of Labor, U.S. Bureau of Labor Statistics, **"Highlights of Women's Earnings in 2006," Report 1002, September 2007, www.bls.gov/cps/ cpswom2006.pdf.**

Among this report's conclusions: The earnings gap between men and women narrowed for most major age groups between 1979 and 2006 and was largest among those ages 45 to 64, with women earning about 73 percent as much as men in that age range.

For More Information

Eagle Forum, PO Box 618, Alton, IL 62002; (618) 462-5415; www.eagleforum.org. Conservative social-policy organization opposed to ratification of the Equal Rights Amendment.

4ERA, 4355J Cobb Parkway, #233, Atlanta, GA 30339; (678) 793-6965; www.4era.org. Single-issue organization advocating ratification of the Equal Rights Amendment.

Institute for Women's Policy Research, 1707 L St., N.W., Suite 750, Washington, DC 20036; (202) 785-5100; www.iwpr.org. Research organization that focuses on gender pay as well as other issues affecting women, including poverty and education.

National Committee on Pay Equity, c/o AFT, 555 New Jersey Ave., N.W., Washington, DC 20001-2029; (703) 920-2010; www.pay-equity.org. Coalition of women's and civil rights organizations, labor unions, religious, professional, legal and educational associations and others focused on pay-equity issues.

National Women's Law Center, 11 Dupont Circle, N.W., Suite 800, Washington, DC 20036; (202) 588-5180; www.nwlc.org. Advocacy group that focuses on employment, health, education and economic-security issues affecting women and girls.

9to5, National Association of Working Women, 207 E. Buffalo St., #211, Milwaukee, WI 53202; (414) 274-0925; www.9to5.org. Grassroots organization focusing on economic-justice issues for women.

U.S. Chamber of Commerce, 1615 H St., N.W., Washington, DC 20062-2000;.(202) 659-6000; www.uschamber.com. Represents business interests before Congress, government agencies and the courts.

Women's Rights

Are Violence and Discrimination Against Women Declining?

Karen Foerstel

8

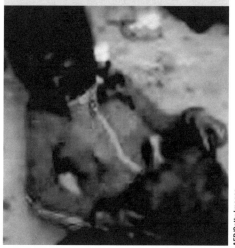

Iraqi teenager Du'a Khalil Aswad lies mortally wounded after her "honor killing" by a mob in the Kurdish region of Iraq. No one has been prosecuted for the April 2007 murder, even though a cell-phone video of the incident was posted on the Internet. Aswad's male relatives are believed to have arranged her ritualistic execution because she had dated a boy from outside her religious sect. The United Nations estimates that 5,000 women and girls are murdered in honor killings around the globe each year.

From *CQ Researcher*, May 2008.

She was 17 years old. The blurry video shows her lying in a dusty road, blood streaming down her face, as several men kick and throw rocks at her. At one point she struggles to sit up, but a man kicks her in the face forcing her back to the ground. Another slams a large, concrete block down onto her head. Scores of onlookers cheer as the blood streams from her battered head.[1]

The April 7, 2007, video was taken in the Kurdish area of northern Iraq on a mobile phone. It shows what appear to be several uniformed police officers standing on the edge of the crowd, watching while others film the violent assault on their phones.

The brutal, public murder of Du'a Khalil Aswad reportedly was organized as an "honor killing" by members of her family — and her uncles and a brother allegedly were among those in the mob who beat her to death. Her crime? She offended her community by falling in love with a man outside her religious sect.[2]

According to the United Nations, an estimated 5,000 women and girls are murdered in honor killings each year, but it was only when the video of Aswad's murder was posted on the Internet that the global media took notice.[3]

Such killings don't only happen in remote villages in developing countries. Police in the United Kingdom estimate that up to 17,000 women are subjected to some kind of "honor"-related violence each year, ranging from forced marriages and physical attacks to murder.[4]

But honor killings are only one type of what the international community calls "gender based violence" (GBV). "It is universal," says Taina Bien-Aimé, executive director of the New York-based

Only Four Countries Offer Total Equality for Women

Costa Rica, Cuba, Sweden and Norway receive the highest score (9 points) in an annual survey of women's economic, political and social rights. Out of the world's 193 countries, only 26 score 7 points or better, while 28 — predominantly Islamic or Pacific Island countries — score 3 or less. The United States rates 7 points: a perfect 3 on economic rights but only 2 each for political and social rights. To receive 3 points for political rights, women must hold at least 30 percent of the seats in the national legislature. Women hold only 16.6 percent of the seats in the U.S. Congress. The U.S. score of 2 on social rights reflects what the report's authors call "high societal discrimination against women's reproductive rights."

Status of Women's Rights Around the Globe

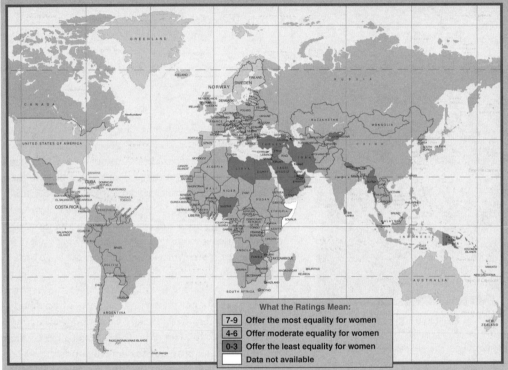

What the Ratings Mean:

7-9	Offer the most equality for women
4-6	Offer moderate equality for women
0-3	Offer the least equality for women
	Data not available

Source: Cingranelli-Richards Human Rights Dataset, http://ciri.binghamton.edu/, based on Amnesty International's annual reports and U.S. State Department annual Country Reports on Human Rights. The database is co-directed by David Louis Cingranelli, a political science professor at Binghamton University, SUNY, and David L. Richards, an assistant political science professor at the University of Memphis.

women's-rights group Equality Now. "There is not one country in the world where violence against women doesn't exist."

Thousands of women are murdered or attacked around the world each day, frequently with impunity. In Guatemala, where an estimated 3,000 women have been killed over the past seven years, most involving some kind of misogynistic violence, only 1 percent of the perpetrators were convicted.[5] In India, the United Nations estimates that five women are burned to death each day by husbands upset that they did not receive sufficient dowries from their brides.[6] In Asia, nearly 163 million females are "missing" from the population — the result of sex-selective abortions, infanticide or neglect.

And since the 1990s some African countries have seen dramatic upsurges in rapes of very young girls by men who believe having sex with a virgin will protect or cure them from HIV-AIDS. After a 70-year-old man allegedly raped a 3-year-old girl in northern Nigeria's commercial hub city of Kano, Deputy Police Chief Suleiman Abba told reporters in January, "Child rape is becoming rampant in Kano." In the last six months of 2007, he said, 54 cases of child rape had been reported. "In some cases the victims are gang-raped."[7]

Epidemics of sexual violence commonly break out in countries torn apart by war, when perpetrators appear to have no fear of prosecution. Today, in Africa, for instance, UNICEF says there is now a "license to rape" in eastern regions of the Democratic Republic of the Congo, where some human-rights experts estimate that up to a quarter of a million women have been raped and often sexually mutilated with knives, branches or machetes.[8] Several of the Congolese rapists remorselessly bragged to an American filmmaker recently about how many women they had gang-raped.[9]

"The sexual violence in Congo is the worst in the world," said John Holmes, the United Nations under secretary general for humanitarian affairs. "The sheer numbers, the wholesale brutality, the culture of impunity — it's appalling."[10]

In some cultures, the female victims themselves are punished. A report by the Human Rights Commission of Pakistan found that a woman is gang-raped every eight hours in that country. Yet, until recently, rape cases could not be prosecuted in Pakistan unless four Muslim men "all of a pious and trustworthy nature" were willing to testify that they witnessed the attack. Without their testimony the victim could be prosecuted for fornication and alleging a false crime, punishable by stoning, lashings or prison.[11] When the law was softened in 2006 to allow judges to decide whether to try rape cases in Islamic courts or criminal courts, where such witnesses are not required, thousands took to the streets to protest the change.[12]

Honor killings are up 400 percent in Pakistan over the last two years, and Pakistani women also live in fear of being blinded or disfigured by "acid attacks" — a common practice in Pakistan and a handful of other countries — in which attackers, usually spurned suitors, throw acid on a woman's face and body.

Women's Suffering Is Widespread

More than two decades after the U.N. Decade for Women and 29 years after the U.N. adopted the Convention on the Elimination of All Forms of Discrimination against Women (CEDAW), gender discrimination remains pervasive throughout the world, with widespread negative consequences for society.

According to recent studies on the status of women today:

- Violence against women is pervasive. It impoverishes women, their families, communities and nations by lowering economic productivity and draining resources. It also harms families across generations and reinforces other violence in societies.
- Domestic violence is the most common form of violence against women, with rates ranging from 8 percent in Albania to 49 percent in Ethiopia and Zambia. Domestic violence and rape account for 5 percent of the disease burden for women ages 15 to 44 in developing countries and 19 percent in developed countries.
- Femicide — the murder of women — often involves sexual violence. From 40 to 70 percent of women murdered in Australia, Canada, Israel, South Africa and the United States are killed by husbands or boyfriends. Hundreds of women were abducted, raped and murdered in and around Juárez, Mexico, over the past 15 years, but the crimes have never been solved.
- At least 160 million females, mostly in India and China, are "missing" from the population — the result of sex-selective abortions.
- Rape is being used as a genocidal tool. Hundreds of thousands of women have been raped and sexually mutilated in the ongoing conflict in Eastern Congo. An estimated 250,000 to 500,000 women were raped during the 1994 genocide in Rwanda; up to 50,000 women were raped during the Bosnian conflict in the 1990s. Victims are often left unable to have children and are deserted by their husbands and shunned by their families, plunging the women and their children into poverty.
- Some 130 million girls have been genitally mutilated, mostly in Africa and Yemen, but also in immigrant communities in the West.
- Child rape has been on the increase in the past decade in some African countries, where some men believe having sex with a virgin will protect or cure them from HIV-AIDS. A study at the Red Cross children's hospital in Cape Town, South Africa, found that 3-year-old girls were more likely to be raped than any other age group.
- Two million girls between the ages of 5 and 15 are forced into the commercial sex market each year, many of them trafficked across international borders.
- Sexual harassment is pervasive. From 40 to 50 percent of women in the European Union reported some form of sexual harassment at work; 50 percent of schoolgirls surveyed in Malawi reported sexual harassment at school.
- Women and girls constitute 70 percent of those living on less than a dollar a day and 64 percent of the world's illiterate.
- Women work two-thirds of the total hours worked by men and women but earn only 10 percent of the income.
- Half of the world's food is produced by women, but women own only 1 percent of the world's land.
- More than 1,300 women die each day during pregnancy and childbirth — 99 percent of them in developing countries.

Sources: "Ending violence against women: From words to action," United Nations, October, 2006; www.un.org/womenwatch/daw/public/VAW_Study/VAW studyE.pdf; www.womankind.org.uk; www.unfp.org; www.oxfam.org.uk; www.ipu.org; www.unicef.org; www.infant-trust.org.uk; "State of the World Population 2000;" http://npr.org; http://asiapacific.amnesty.org; http://news.bbc.co.uk

Negative Attitudes Toward Women Are Pervasive

Negative attitudes about women are widespread around the globe, among women as well as men. Rural women are more likely than city women to condone domestic abuse if they think it was provoked by a wife's behavior.

Location	Percentage of women in selected countries who agree that a man has good reason to beat his wife if:						Women who agree with:	
	Wife does not complete housework	Wife disobeys her husband	Wife refuses sex	Wife asks about other women	Husband suspects infidelity	Wife is unfaithful	One or more of the reasons mentioned	None of the reasons mentioned
Bangladesh city	13.8	23.3	9.0	6.6	10.6	51.5	53.3	46.7
Bangladesh province	25.1	38.7	23.3	14.9	24.6	77.6	79.3	20.7
Brazil city	0.8	1.4	0.3	0.3	2.0	8.8	9.4	90.6
Brazil province	4.5	10.9	4.7	2.9	14.1	29.1	33.7	66.3
Ethiopia province	65.8	77.7	45.6	32.2	43.8	79.5	91.1	8.9
Japan city	1.3	1.5	0.4	0.9	2.8	18.5	19.0	81.0
Namibia city	9.7	12.5	3.5	4.3	6.1	9.2	20.5	79.5
Peru city	4.9	7.5	1.7	2.3	13.5	29.7	33.7	66.3
Peru province	43.6	46.2	25.8	26.7	37.9	71.3	78.4	21.6
Samoa	12.1	19.6	7.4	10.1	26.0	69.8	73.3	26.7
Serbia and Montenegro city	0.6	0.97	0.6	0.3	0.9	5.7	6.2	93.8
Thailand city	2.0	0.8	2.8	1.8	5.6	42.9	44.7	55.3
Thailand province	11.9	25.3	7.3	4.4	12.5	64.5	69.5	30.5
Tanzania city	24.1	45.6	31.1	13.8	22.9	51.5	62.5	37.5
Tanzania province	29.1	49.7	41.7	19.8	27.2	55.5	68.2	31.8

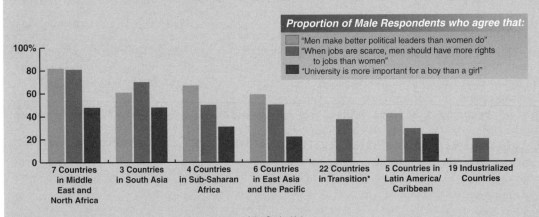

Proportion of Male Respondents who agree that:

- "Men make better political leaders than women do"
- "When jobs are scarce, men should have more rights to jobs than women"
- "University is more important for a boy than a girl"

7 Countries in Middle East and North Africa · 3 Countries in South Asia · 4 Countries in Sub-Saharan Africa · 6 Countries in East Asia and the Pacific · 22 Countries in Transition* · 5 Countries in Latin America/Caribbean · 19 Industrialized Countries

*Countries in transition are generally those that were once part of the Soviet Union.

Sources: World Health Organization, www.who.int/gender/violence/who_multicountry_study/Chapter3-Chapter4.pdf; "World Values Survey," www.worldvaluessruvey.org

But statistics on murder and violence are only a part of the disturbing figures on the status of women around the globe. Others include:

- Some 130 million women have undergone female genital mutilation, and another 2 million are at risk every year, primarily in Africa and Yemen.
- Women and girls make up 70 percent of the world's poor and two-thirds of its illiterate.
- Women work two-thirds of the total hours worked by men but earn only 10 percent of the income.
- Women produce more than half of the world's food but own less than 1 percent of the world's property.
- More than 500,000 women die during pregnancy and childbirth every year — 99 percent of them in developing countries.
- Two million girls between the ages of 5 and 15 are forced into the commercial sex market each year.[13]
- Globally, 10 million more girls than boys do not attend school.[14]

Despite these alarming numbers, women have made historic progress in some areas. The number of girls receiving an education has increased in the past decade. Today 57 percent of children not attending school are girls, compared to two-thirds in the 1990s.[15]

And women have made significant gains in the political arena. As of March, 2008, 14 women are serving as elected heads of state or government, and women now hold 17.8 percent of the world's parliamentary seats — more than ever before.[16] And just three months after the brutal killing of Aswad in Iraq, India swore in its first female president, Pratibha Patil, who vows to eliminate that country's practice of aborting female fetuses because girls are not as valued as boys in India. (See *"At Issue," p. 187.*)[17]

Last October, Argentina elected its first female president, Cristina Fernández de Kirchner,* the second woman in two years to be elected president in South America. Michelle Bachelet, a single mother, won the presidency in Chile in 2006.[18] During her inaugural speech Kirchner

admitted, "Perhaps it'll be harder for me, because I'm a woman. It will always be harder for us."[19]

Indeed, while more women than ever now lead national governments, they hold only 4.4 percent of the world's 342 presidential and prime ministerial positions. And in no country do they hold 50 percent or more of the national legislative seats.[20]

"Women make up half the world's population, but they are not represented" at that level, says Swanee Hunt, former U.S. ambassador to Austria and founding director of the Women and Public Policy Program at Harvard's Kennedy School of Government.

While this is "obviously a fairness issue," she says it also affects the kinds of public policies governments pursue. When women comprise higher percentages of officeholders, studies show "distinct differences in legislative outputs," Hunt explains. "There's less funding of bombs and bullets and more on human security — not just how to defend territory but also on hospitals and general well-being."

Today's historic numbers of women parliamentarians have resulted partly from gender quotas imposed in nearly 100 countries, which require a certain percentage of women candidates or officeholders.[21]

During the U.N.'s historic Fourth World Conference on Women — held in Beijing in 1995 — 189 governments adopted, among other things, a goal of 30 percent female representation in national legislatures around the world.[22] But today, only 20 countries have reached that goal, and quotas are often attacked as limiting voters' choices and giving women unfair advantages.[23]

Along with increasing female political participation, the 5,000 government representatives at the Beijing conference — one of the largest gatherings in U.N. history — called for improved health care for women, an end to violence against women, equal access to education for girls, promotion of economic independence and other steps to improve the condition of women around the world.[24]

"Let Beijing be the platform from which our global crusade will be carried forward," Gertrude Mongella, U.N. secretary general for the conference, said during closing ceremonies. "The world will hold us accountable for the implementation of the good intentions and decisions arrived at in Beijing."[25]

---

* Isabel Martínez Perón assumed the presidency of Argentina on the death of her husband, Juan Perón, in 1974 and served until she was deposed in a coup d'etat in 1976; but she was never elected.

Spain's visibly pregnant new Defense minister, Carme Chacón, reviews troops in Madrid on April 14, 2008. She is the first woman ever to head Spain's armed forces. Women hold nine out of 17 cabinet posts in Spain's socialist government, a reflection of women's entrance into the halls of power around the world.

But more than 10 years later, much of the Beijing Platform still has not been achieved. And many question whether women are any better off today than they were in 1995.

"The picture's mixed," says June Zeitlin, executive director of the Women's Environment & Development Organization (WEDO). "In terms of violence against women, there is far more recognition of what is going on today. There has been some progress with education and girls. But the impact of globalization has exacerbated differences between men and women. The poor have gotten poorer — and they are mostly women."

Liberalized international trade has been a two-edged sword in other ways as well. Corporations have been able to expand their global reach, opening new businesses and factories in developing countries and offering women unprecedented employment and economic opportunities. But the jobs often pay low wages and involve work in dangerous conditions because poor countries anxious to attract foreign investors often are willing to ignore safety and labor protections.[26] And increasingly porous international borders have led to growing numbers of women and girls being forced or sold into prostitution or sexual slavery abroad, often under the pretense that they will be given legitimate jobs overseas.[27]

Numerous international agreements in recent years have pledged to provide women with the same opportunities and protections as men, including the U.N.'s Millennium Development Goals (MDGs) and the Convention on the Elimination of All Forms of Discrimination Against Women (CEDAW). But the MDGs' deadlines for improving the conditions for women have either been missed already or are on track to fail in the coming years.[28] And more than 70 of the 185 countries that ratified CEDAW have filed "reservations," meaning they exempt themselves from certain parts.[29] In fact, there are more reservations against CEDAW than against any other international human-rights treaty in history.[30] The United States remains the only developed country in the world not to have ratified it.[31]

"There has certainly been progress in terms of the rhetoric. But there are still challenges in the disparities in education, disparities in income, disparities in health," says Carla Koppell, director of the Cambridge, Mass.-based Initiative for Inclusive Security, which advocates for greater numbers of women in peace negotiations.

"But women are not just victims," she continues. "They have a very unique and important role to play in solving the problems of the developing world. We need to charge policy makers to match the rhetoric and make it a reality. There is a really wonderful opportunity to use the momentum that does exist. I really think we can."

Amidst the successes and failures surrounding women's issues, here are some of the questions analysts are beginning to ask:

Has globalization been good for women?

Over the last 20 years, trade liberalization has led to a massive increase of goods being produced and exported from developing countries, creating millions of manufacturing jobs and bringing many women into the paid workforce for the first time.

"Women employed in export-oriented manufacturing typically earn more than they would have in traditional sectors," according to a World Bank report. "Further, cash income earned by women may improve their status and bargaining power in the family."[32] The report cited a study of 50 families in Mexico that found "a significant proportion of the women reported an improvement in their 'quality of life,' due mainly to their income from working outside their homes, including in (export-oriented) factory jobs."

But because women in developing nations are generally less educated than men and have little bargaining power, most of these jobs are temporary or part-time, offering no health-care benefits, overtime or sick leave.

Women comprise 85 percent of the factory jobs in the garment industry in Bangladesh and 90 percent in Cambodia. In the cut flower industry, women hold 65 percent of the jobs in Colombia and 87 percent in Zimbabwe. In the fruit industry, women constitute 69 percent of temporary and seasonal workers in South Africa and 52 percent in Chile.[33]

Frequently, women in these jobs have no formal contract with their employers, making them even more vulnerable to poor safety conditions and abuse. One study found that only 46 percent of women garment workers in Bangladesh had an official letter of employment.[34]

"Women are a workforce vital to the global economy, but the jobs women are in often aren't covered by labor protections," says Thalia Kidder, a policy adviser on gender and sustainable livelihoods with U.K.-based Oxfam, a confederation of 12 international aid organizations. Women lack protection because they mostly work as domestics, in home-based businesses and as part-time workers. "In the global economy, many companies look to hire the most powerless people because they cannot demand high wages. There are not a lot of trade treaties that address labor rights."

In addition to recommending that countries embrace free trade, Western institutions like the International Monetary Fund and the World Bank during the 1990s recommended that developing countries adopt so-called structural adjustment economic reforms in order to qualify for certain loans and financial support. Besides opening borders to free trade, the neo-liberal economic regime known as the Washington Consensus advocated privatizing state-owned businesses, balancing budgets and attracting foreign investment.

But according to some studies, those reforms ended up adversely affecting women. For instance, companies in Ecuador were encouraged to make jobs more "flexible" by replacing long-term contracts with temporary, seasonal and hourly positions — while restricting collective bargaining rights.[35] And countries streamlined and privatized government programs such as health care and education, services women depend on most.

Globalization also has led to a shift toward cash crops grown for export, which hurts women farmers, who produce 60 to 80 percent of the food for household consumption in developing countries.[36] Small women farmers are being pushed off their land so crops for exports can be grown, limiting their abilities to produce food for themselves and their families.

While economic globalization has yet to create the economic support needed to help women out of poverty, women's advocates say females have benefited from the broadening of communications between countries prompted by globalization. "It has certainly improved access to communications and helped human-rights campaigns," says Zeitlin of WEDO. "Less can be done in secret. If there is a woman who is condemned to be stoned to death somewhere, you can almost immediately mobilize a global campaign against it."

Homa Hoodfar, a professor of social anthropology at Concordia University in Montreal, Canada, and a founder of the group Women Living Under Muslim Laws, says women in some of the world's most remote towns and villages regularly e-mail her organization. "Globalization has made the world much smaller," she says. "Women are getting information on TV and the Internet. The fact that domestic violence has become a global issue [shows globalization] provides resources for those objecting locally."

But open borders also have enabled the trafficking of millions of women around the world. An estimated 800,000 people are trafficked across international borders each year — 80 percent of them women and girls — and most are forced into the commercial sex trade. Millions more are trafficked within their own countries.[37] Globalization has sparked a massive migration of women in search of better jobs and lives. About 90 million women — half of the world's migrants and more than ever in history — reside outside their home countries. These migrant women — often unable to speak the local language and without any family connections — are especially susceptible to traffickers who lure them with promises of jobs abroad.[38]

And those who do not get trapped in the sex trade often end up in low-paying or abusive jobs in foreign factories or as domestic maids working under slave-like conditions.

Female Peacekeepers Fill Vital Roles

Women bring a different approach to conflict resolution.

The first all-female United Nations peacekeeping force left Liberia in January after a year's mission in the West African country, which is rebuilding itself after 14 years of civil war. Comprised of more than 100 women from India, the force was immediately replaced by a second female team.

"If anyone questioned the ability of women to do tough jobs, then those doubters have been [proven] wrong," said U.N. Special Representative for Liberia Ellen Margrethe Løj, adding that the female peacekeepers inspired many Liberian women to join the national police force.[1]

Women make up half of the world's refugees and have systematically been targeted for rape and sexual abuse during times of war, from the 200,000 "comfort women" who were kept as sex slaves for Japanese soldiers during World War II[2] to the estimated quarter-million women reportedly raped and sexually assaulted during the current conflict in the Democratic Republic of the Congo.[3] But women account for only 5 percent of the world's security-sector jobs, and in many countries they are excluded altogether.[4]

In 2000, the U.N. Security Council unanimously adopted Resolution 1325 calling on governments — and the U.N. itself — to include women in peace building by adopting a variety of measures, including appointing more women as special envoys, involving women in peace negotiations, integrating gender-based policies in peacekeeping missions and increasing the number of women at all decision-making levels.[5]

But while Resolution 1325 was a critical step in bringing women into the peace process, women's groups say more women should be sent on field missions and more data collected on how conflict affects women around the world.[6]

"Women are often viewed as victims, but another way to view them is as the maintainers of society," says Carla Koppell, director of the Cambridge, Mass.-based Initiative for Inclusive Security, which promotes greater numbers of women in peacekeeping and conflict resolution. "There must be a conscious decision to include women. It's a detriment to promote peace without including women."

Women often comprise the majority of post-conflict survivor populations, especially when large numbers of men have either fled or been killed. In the wake of the 1994 Rwandan genocide, for example, women made up 70 percent of the remaining population.

And female peacekeepers and security forces can fill vital roles men often cannot, such as searching Islamic women wearing burkas or working with rape victims who may be reluctant to report the crimes to male soldiers.

But some experts say the real problem is not migration and globalization but the lack of labor protection. "Nothing is black and white," says Marianne Mollmann, advocacy director for the Women's Rights Division of Human Rights Watch. "Globalization has created different employment opportunities for women. Migration flows have made women vulnerable. But it's a knee-jerk reaction to say that women shouldn't migrate. You can't prevent migration. So where do we need to go?" She suggests including these workers in general labor-law protections that cover all workers.

Mollmann said countries can and should hammer out agreements providing labor and wage protections for domestic workers migrating across borders. With such protections, she said, women could benefit from the jobs and incomes promised by increased migration and globalization.

Should governments impose electoral quotas for women?

In 2003, as Rwanda struggled to rebuild itself after the genocide that killed at least 800,000 Hutus and Tutsis, the country adopted an historic new constitution that, among other things, required that women hold at least 30 percent of posts "in all decision-making organs."[39]

Today — ironically, just across Lake Kivu from the horrors occurring in Eastern Congo — Rwanda's lower house of parliament now leads the world in female representation, with 48.8 percent of the seats held by women.[40]

"Women bring different experiences and issues to the table," says Koppell. "I've seen it personally in the Darfur and Uganda peace negotiations. Their priorities were quite different. Men were concerned about power- and wealth-sharing. Those are valid, but you get an entirely different dimension from women. Women talked about security on the ground, security of families, security of communities."

In war-torn countries, women have been found to draw on their experiences as mothers to find nonviolent and flexible ways to solve conflict.[7] During peace negotiations in Northern Ireland, for example, male negotiators repeatedly walked out of sessions, leaving a small number of women at the table. The women, left to their own, found areas of common ground and were able to keep discussions moving forward.[8]

"The most important thing is introducing the definition of security from a woman's perspective," said Orzala Ashraf, founder of Kabul-based Humanitarian Assistance for the Women and Children of Afghanistan. "It is not a man in a uniform standing next to a tank armed with a gun. Women have a broader term — human security — the ability to go to school, receive health care, work and have access to

The first all-female United Nations peacekeeping force practices martial arts in New Delhi as it prepares to be deployed to Liberia in 2006.

AP Photo/Mustafa Quraishi

justice. Only by improving these areas can threats from insurgents, Taliban, drug lords and warlords be countered."[9]

[1] "Liberia: UN envoy welcomes new batch of female Indian police officers," U.N. News Centre, Feb. 8, 2008, www.un.org/apps/news/story.asp?NewsID=25557&Cr=liberia&Cr1=.

[2] "Japan: Comfort Women," European Speaking Tour press release, Amnesty International, Oct. 31, 2007.

[3] "Film Documents Rape of Women in Congo," "All Things Considered," National Public Radio, April 8, 2008, www.npr.org/templates/story/story.php?storyId=89476111.

[4] "Ninth Annual Colloquium and Policy Forum," Hunt Alternatives Fund, Jan. 22, 2008, www.huntalternatives.org/pages/7650_ninth_annual_colloquium_and_policy_forum.cfm. Also see Elizabeth Eldridge, "Women cite utility in peace efforts," _The Washington Times_, Jan. 25, 2008, p. A1.

[5] "Inclusive Security, Sustainable Peace: A Toolkit for Advocacy and Action," International Alert and Women Waging Peace, 2004, p. 15, www.huntalternatives.org/download/35_introduction.pdf.

[6] _Ibid._, p. 17.

[7] Jolynn Shoemaker and Camille Pampell Conaway, "Conflict Prevention and Transformation: Women's Vital Contributions," Inclusive Security: Women Waging Peace and the United Nations Foundation, Feb. 23, 2005, p. 7.

[8] The Initiative for Inclusive Security, www.huntalternatives.org/pages/460_the_vital_role_of_women_in_peace_building.cfm.

[9] Eldridge, _op. cit._

Before the civil war, Rwandan women never held more than 18 percent of parliament. But after the genocide, the country's population was 70 percent female. Women immediately stepped in to fill the vacuum, becoming the heads of households, community leaders and business owners. Their increased presence in leadership positions eventually led to the new constitutional quotas.[41]

"We see so many post-conflict countries going from military regimes to democracy that are starting from scratch with new constitutions," says Drude Dahlerup, a professor of political science at Sweden's Stockholm University who studies the use of gender quotas. "Today, starting from scratch means including women. It's seen as a sign of modernization and democratization."

Both Iraq and Afghanistan included electoral quotas for women in their new constitutions, and the number of women in political office in sub-Saharan Africa has increased faster than in any other region of the world, primarily through the use of quotas.[42]

But many point out that simply increasing the numbers of women in elected office will not necessarily expand women's rights. "It depends on which women and which positions they represent," says Wendy Harcourt, chair of Women in Development Europe (WIDE), a feminist network in Europe, and editor of _Development_, the journal of the Society for International Development, a global network of individuals and institutions working on development issues. "It's positive, but I don't see yet what it means [in terms of addressing] broader gender issues."

Few Women Head World Governments

Fourteen women currently serve as elected heads of state or government including five who serve as both. Mary McAleese, elected president of Ireland in 1997, is the world's longest-serving head of state. Helen Clark of New Zealand has served as prime minister since 1999, making her the longest-serving female head of government. The world's first elected female head of state was Sirimavo Bandaranaike of Sri Lanka, in 1960.

Current Female Elected Heads of State and Government

Heads of both state and government:

 Gloria Macapagal-Arroyo — President, the Philippines, since 2001; former secretary of Defense (2002) and secretary of Foreign Affairs (2003 and 2006-2007).

 Ellen Johnson-Sirleaf — President, Liberia, since 2006; held finance positions with the government and World Bank.

 Michelle Bachelet Jeria — President, Chile, since 2006; former minister of Health (2000-2002) and minister of Defense (2002-2004).

 Cristina E. Fernández — President, Argentina, since 2007; succeeded her husband, Nestor de Kirchner, as president; former president, Senate Committee on Constitutional Affairs.

 Rosa Zafferani — Captain Regent, San Marino, since April 2008; secretary of State of Public Education, University and Cultural Institutions (2004 to 2008); served as captain regent in 1999; San Marino elects two captains regent every six months, who serve as co-heads of both state and government.

Heads of Government:

 Helen Clark — Prime Minister, New Zealand, since 1999; held government posts in foreign affairs, defense, housing and labor.

 Luísa Días Diogo — Prime Minister, Mozambique, since 2004; held several finance posts in Mozambique and the World Bank.

 Angela Merkel — Chancellor, Germany, since 2005; parliamentary leader of Christian Democratic Union Party (2002-2005).

 Yuliya Tymoshenko — Prime Minister, Ukraine, since 2007; chief of government (2005) and designate prime minister (2006).

 Zinaida Grecianîi — Prime Minister, Moldova, since March 2008; vice prime minister (2005-2008).

Heads of State:

 Mary McAleese — President, Ireland, since 1997; former director of a television station and Northern Ireland Electricity.

Tarja Halonen — President, Finland, since 2000; former minister of foreign affairs (1995-2000).

Pratibha Patil — President, India, since 2007; former governor of Rajasthan state (2004-2007).

Borjana Kristo — President, Bosnia and Herzegovina, since 2007; minister of Justice of Bosniak-Croat Federation, an entity in Bosnia and Herzegovina (2003-2007).

Source: www.guide2womenleaders.com

While Afghanistan has mandated that women hold at least 27 percent of the government's lower house seats and at least 17 percent of the upper house, their increased representation appears to have done little to improve women's rights.[43] Earlier this year, a student journalist was condemned to die under Afghanistan's strict Islamic sharia law after he distributed articles from the Internet on women's rights.[44] And non-governmental groups in Afghanistan report that Afghan women and girls have begun killing themselves in record numbers, burning themselves alive in order to escape widespread domestic abuse or forced marriages.[45]

Having gender quotas alone doesn't necessarily ensure that women's rights will be broadened, says Hoodfar of Concordia University. It depends on the type of quota a government implements, she argues, pointing out that in Jordan, for example, the government has set aside parliamentary seats for the six women who garner the most votes of any other female candidates in their districts — even if they do not win more votes than male candidates.[46] Many small, conservative tribes that cannot garner enough votes for a male in a countrywide victory are now nominating their sisters and wives in the hope that the lower number of votes needed to elect a woman will get them one of the reserved seats. As a result, many of the women moving into the reserved seats are extremely conservative and actively oppose providing women greater rights and freedoms.

And another kind of quota has been used against women in her home country of Iran, Hoodfar points out. Currently, 64 percent of university students in Iran are women. But the government recently mandated that at least 40 percent of university enrollees be male, forcing many female students out of school, Hoodfar said.

"Before, women didn't want to use quotas for politics because of concern the government may try to use it against women," she says. "But women are beginning to look into it and talk about maybe developing a good system."

Quotas can be enacted by constitutional requirements, such as those enacted in Rwanda, by statute or voluntarily by political parties. Quotas also can vary in their requirements: They can mandate the number of women each party must nominate, how many women must appear on the ballot (and the order in which they appear, so women are not relegated to the bottom of the list), or the number of women who must hold government office. About 40 countries now use gender quotas in national parliamentary elections, while another 50 have major political parties that voluntarily use quotas to determine candidates.

Aside from questions about the effectiveness of quotas, others worry about the fairness of establishing quotas based on gender. "That's something feminists have traditionally opposed," says Harcourt.

"It's true, but it's also not fair the way it is now," says former Ambassador Hunt. "We are where we are today through all kinds of social structures that are not fair. Quotas are the lesser of two evils."

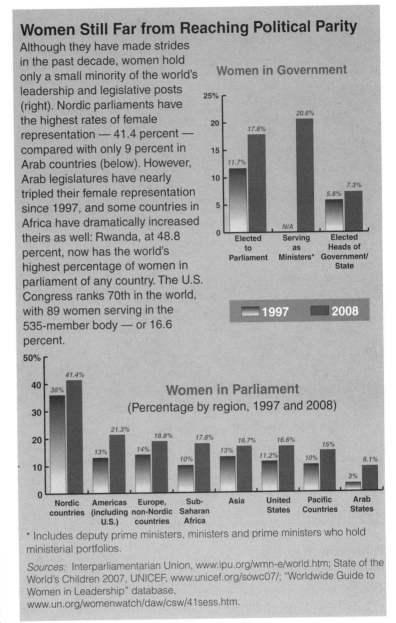

Women Still Far from Reaching Political Parity

Although they have made strides in the past decade, women hold only a small minority of the world's leadership and legislative posts (right). Nordic parliaments have the highest rates of female representation — 41.4 percent — compared with only 9 percent in Arab countries (below). However, Arab legislatures have nearly tripled their female representation since 1997, and some countries in Africa have dramatically increased theirs as well: Rwanda, at 48.8 percent, now has the world's highest percentage of women in parliament of any country. The U.S. Congress ranks 70th in the world, with 89 women serving in the 535-member body — or 16.6 percent.

Women in Government

Elected to Parliament: 11.7% (1997), 17.8% (2008)
Serving as Ministers*: N/A (1997), 20.6% (2008)
Elected Heads of Government/State: 5.8% (1997), 7.3% (2008)

■ 1997 ■ 2008

Women in Parliament
(Percentage by region, 1997 and 2008)

Nordic countries: 36% (1997), 41.4% (2008)
Americas (including U.S.): 13% (1997), 21.3% (2008)
Europe, non-Nordic countries: 14% (1997), 18.8% (2008)
Sub-Saharan Africa: 10% (1997), 17.8% (2008)
Asia: 13% (1997), 16.7% (2008)
United States: 11.2% (1997), 16.6% (2008)
Pacific Countries: 10% (1997), 15% (2008)
Arab States: 3% (1997), 9.1% (2008)

* Includes deputy prime ministers, ministers and prime ministers who hold ministerial portfolios.

Sources: Interparliamentarian Union, www.ipu.org/wmn-e/world.htm; State of the World's Children 2007, UNICEF, www.unicef.org/sowc07/; "Worldwide Guide to Women in Leadership" database, www.un.org/womenwatch/daw/csw/41sess.htm.

Stockholm University's Dahlerup says quotas are not "discrimination against men but compensation for discrimination against women." Yet quotas are not a panacea for women in politics, she contends. "It's a mistake to think this is a kind of tool that will solve all problems. It doesn't solve problems about financing campaigns, caring for families while being in politics or removing patriarchal attitudes. It would be nice if it wasn't necessary, and hopefully sometime in the future it won't be."

Until that time, however, quotas are a "necessary evil," she says.

AP Photo/Rajesh Kumar Singh

National Geographic/Getty Images/Melvyn Goldstein

Women's Work: From Hauling and Churning . . .

Women's work is often back-breaking and monotonous, such as hauling firewood in the western Indian state of Maharashtra (top) and churning yogurt into butter beside Lake Motsobunnyi in Tibet (bottom). Women labor two-thirds of the total hours worked around the globe each year but earn only 10 percent of the income.

Do international treaties improve women's rights?

In recent decades, a variety of international agreements have been signed by countries pledging to improve women's lives, from the 1979 Convention for the Elimination of All Forms of Discrimination Against Women to the Beijing Platform of 1995 to the Millennium Development Goals (MDGs) adopted in 2000. The agreements aimed to provide women with greater access to health, political representation, economic stability and social status. They also focused attention on some of the biggest obstacles facing women.

But despite the fanfare surrounding the launch of those agreements, many experts on women's issues say on-the-ground action has yet to match the rhetoric. "The report is mixed," says Haleh Afshar, a professor of politics and women's studies at the University of York in the United Kingdom and a nonpartisan, appointed member of the House of Lords, known as a crossbench peer. "The biggest problem with Beijing is all these things were stated, but none were funded. Unfortunately, I don't see any money. You don't get the pay, you don't get the job done."

The Beijing Platform for Action, among other things, called on governments to "adjust budgets to ensure equality of access to public sector expenditures" and even to "reduce, as appropriate, excessive military expenditure" in order to achieve the Platform goals.

But adequate funding has yet to be provided, say women's groups.[47] In a report entitled "Beijing Betrayed," the Women's Environment & Development Organization says female HIV cases outnumber male cases in many parts of the world, gender-related violence remains a pandemic and women still make up the majority of the world's poor — despite pledges in Beijing to reverse these trends.[48]

And funding is not the only obstacle. A 2004 U.N. survey revealed that while many countries have enacted laws in recent years to help protect women from violence and discrimination, long-standing social and cultural traditions block progress. "While constitutions provided for equality between women and men on the one hand, [several countries] recognized and gave precedent to customary law and practice in a number of areas . . . resulting in discrimination against women," the report said. "Several countries noted that statutory, customary and religious law coexist, especially in regard to family, personal status and inheritance and land rights. This perpetuated discrimination against women."[49]

While she worries about the lack of progress on the Beijing Platform, WEDO Executive Director Zeitlin says international agreements are nevertheless critical in raising global awareness on women's issues. "They have a major impact on setting norms and standards," she says. "In many countries, norms and standards are very important in setting goals for women to advocate for. We complain about lack of implementation, but if we didn't have the norms and standards we couldn't complain about a lack of implementation."

Like the Beijing Platform, the MDGs have been criticized for not achieving more. While the U.N. says promoting women's rights is essential to achieving the millenium goals — which aim to improve the lives of all the world's populations by 2015 — only two of the eight specifically address women's issues.[50]

One of the goals calls for countries to "Promote gender equality and empower women." But it sets only one measurable target: "Eliminate gender disparity in primary and secondary education, preferably by 2005, and in all levels of education" by 2015.[51] Some 62 countries failed to reach the 2005 deadline, and many are likely to miss the 2015 deadline as well.[52]

Another MDG calls for a 75 percent reduction in maternal mortality compared to 1990 levels. But according to the human-rights group ActionAid, this goal is the "most off track of all the MDGs." Rates are declining at less than 1 percent a year, and in some countries — such as Sierra Leone, Pakistan and Guatemala — maternal mortality has increased since 1990. If that trend continues, no region in the developing world is expected to reach the goal by 2015.[53]

Activist Peggy Antrobus of Development Alternatives with Women for a New Era (DAWN) — a network of feminists from the Southern Hemisphere, based currently in Calabar, Cross River State, Nigeria — has lambasted the MDGs, quipping that the acronym stands for the "Most Distracting Gimmick."[54] Many feminists argue that the goals are too broad to have any real impact and that the MDGs should have given more attention to women's issues.

But other women say international agreements — and the public debate surrounding them — are vital in promoting gender equality. "It's easy to get disheartened, but Beijing is still the blueprint of where we need to be," says Mollmann of Human Rights Watch. "They are part of a political process, the creation of an international culture. If systematically everyone says [discrimination against women] is a bad thing, states don't want to be hauled out as systematic violators."

In particular, Mollmann said, CEDAW has made real progress in overcoming discrimination against women. Unlike the Beijing Platform and the MDGs, CEDAW legally obliges countries to comply. Each of the 185 ratifying countries must submit regular reports to the U.N. outlining their progress under the convention. Several

AP Photo/Sergei Grits

AFP/Getty Images/Ali Burafi

. . . to Gathering and Herding

While many women have gotten factory jobs thanks to globalization of trade, women still comprise 70 percent of the planet's inhabitants living on less than a dollar a day. Women perform a variety of tasks around the world, ranging from gathering flax in Belarus (top) to shepherding goats in central Argentina (bottom).

countries — including Brazil, Uganda, South Africa and Australia — also have incorporated CEDAW provisions into their constitutions and legal systems.[55]

Still, dozens of ratifying countries have filed official "reservations" against the convention, including Bahrain, Egypt, Kuwait, Morocco and the United Arab Emirates, all of whom say they will comply only within the bounds of Islamic sharia law.[56] And the United States has refused to ratify CEDAW, with or without reservations, largely because of conservatives who say it would, among other things, promote abortion and require the government to pay for such things as child care and maternity leave.

Indian women harvest wheat near Bhopal. Women produce half of the food used domestically worldwide and 60 to 80 percent of the household food grown in developing countries.

BACKGROUND

'Structural Defects'

Numerous prehistoric relics suggest that at one time matriarchal societies existed on Earth in which women were in the upper echelons of power. Because early societies did not understand the connection between sexual relations and conception, they believed women were solely responsible for reproduction — which led to the worship of female goddesses.[57]

In more modern times, however, women have generally faced prejudice and discrimination at the hands of a patriarchal society. In about the eighth century B.C. creation stories emerged describing the fall of man due to the weakness of women. The Greeks recounted the story of Pandora who, through her opening of a sealed jar, unleashed death and pain on all of mankind. Meanwhile, similar tales in Judea eventually were recounted in Genesis, with Eve as the culprit.[58]

In ancient Greece, women were treated as children and denied basic rights. They could not leave their houses unchaperoned, were prohibited from being educated or buying or selling land. A father could sell his unmarried daughter into slavery if she lost her virginity before marriage. If a woman was raped, she was outcast and forbidden from participating in public ceremonies or wearing jewelry.[59]

The status of women in early Rome was not much better, although over time women began to assert their voices and slowly gained greater freedoms. Eventually, they were able to own property and divorce their husbands. But early Christian leaders later denounced the legal and social freedom enjoyed by Roman women as a sign of moral decay. In the view of the early church, women were dependent on and subordinate to men.

In the 13th century, the Catholic priest and theologian St. Thomas Aquinas helped set the tone for the subjugation of women in Western society. He said women were created solely to be "man's helpmate" and advocated that men should make use of "a necessary object, woman, who is needed to preserve the species or to provide food and drink."[60]

From the 14th to 17th centuries, misogyny and oppression of women took a step further. As European societies struggled against the Black Plague, the 100 Years War and turmoil between Catholics and Reformers, religious leaders began to blame tragedies, illnesses and other problems on witches. As witch hysteria spread across Europe — instituted by both the religious and non-religious — an estimated 30,000 to 60,000 people were executed for allegedly practicing witchcraft. About 80 percent were females, some as young as 8 years old.[61]

"All wickedness is but little to the wickedness of a woman," Catholic inquisitors wrote in the 1480s. "What else is woman but a foe to friendship, an unescapable punishment, a necessary evil, a natural temptation, a desirable calamity. . . . Women are . . . instruments of Satan, . . . a structural defect rooted in the original creation."[62]

Push for Protections

The Age of Enlightenment and the Industrial Revolution in the 18th and 19th centuries opened up job opportunities for women, released them from domestic confines and provided them with new social freedoms.

In 1792 Mary Wollstonecraft published *A Vindication of the Rights of Women*, which has been hailed as "the feminist declaration of independence." Although the book had been heavily influenced by the French Revolution's notions of equality and universal brotherhood, French revolutionary leaders, ironically, were not sympathetic to feminist causes.[63] In 1789 they had refused to accept a Declaration of the Rights of Women when it was presented at the National Assembly. And Jean Jacques Rousseau, one of the philosophical founders of the revolution, had written in 1762:

"The whole education of women ought to be relative to men. To please them, to be useful to them, to make themselves loved and honored by them, to educate them when young, to care for them when grown, to counsel them, to make life sweet and agreeable to them — these are the duties of women at all times, and what should be taught them from their infancy."[64]

As more and more women began taking jobs outside the home during the 19th century, governments began to pass laws to "protect" them in the workforce and expand their legal rights. The British Mines Act of 1842, for instance, prohibited women from working underground.[65] In 1867, John Stuart Mill, a supporter of women's rights and author of the book *Subjection of Women*, introduced language in the British House of Commons calling for women to be granted the right to vote. It failed.[66]

But by that time governments around the globe had begun enacting laws giving women rights they had been denied for centuries. As a result of the Married Women's Property Act of 1870 and a series of other measures, wives in Britain were finally allowed to own property. In 1893, New Zealand became the first nation to grant full suffrage rights to women, followed over the next two decades by Finland, Norway, Denmark and Iceland. The United States granted women suffrage in 1920.[67]

One of the first international labor conventions, formulated at Berne, Switzerland, in 1906, applied exclusively to women — prohibiting night work for women in industrial occupations. Twelve nations signed on to it. During the second Berne conference in 1913, language was proposed limiting the number of hours women and children could work in industrial jobs, but the outbreak of World War I prevented it from being enacted.[68] In 1924 the U.S. Supreme Court upheld a night-work law for women.[69]

In 1946, public attention to women's issues received a major boost when the United Nations created the Commission on the Status of Women to address urgent problems facing women around the world.[70] During the 1950s, the U.N. adopted several conventions aimed at improving women's lives, including the Convention on the Political Rights of Women, adopted in 1952 to ensure women the right to vote, which has been ratified by 120 countries, and the Convention on the Nationality of Married Women, approved in 1957 to ensure that marriage to an alien does not automatically affect the nationality of the woman.[71] That convention has been ratified by only 73 countries; the United States is not among them.[72]

In 1951 The International Labor Organization (ILO), an agency of the United Nations, adopted the Convention on Equal Remuneration for Men and Women Workers for Work of Equal Value, to promote equal pay for equal work. It has since been ratified by 164 countries, but again, not by the United States.[73] Seven years later, the ILO adopted the Convention on Discrimination in Employment and Occupation to ensure equal opportunity and treatment in employment. It is currently ratified by 166 countries, but not the United States.[74] U.S. opponents to the conventions claim there is no real pay gap between men and women performing the same jobs and that the conventions would impose "comparable worth" requirements, forcing companies to pay equal wages to men and women even if the jobs they performed were different.[75]

In 1965, the Commission on the Status of Women began drafting international standards articulating equal rights for men and women. Two years later, the panel completed the Declaration on the Elimination of Discrimination Against Women, which was adopted by the General Assembly but carried no enforcement power.

The commission later began to discuss language that would hold countries responsible for enforcing the declaration. At the U.N.'s first World Conference on Women in Mexico City in 1975, women from around the world called for creation of such a treaty, and the commission soon began drafting the text.[76]

Women's 'Bill of Rights'

Finally in 1979, after many years of often rancorous debate, the Convention on the Elimination of All Forms of Discrimination Against Women (CEDAW) was adopted by the General Assembly — 130 to none, with 10 abstentions. After the vote, however, several countries said their "yes" votes did not commit the support of their governments. Brazil's U.N. representative told the assembly, "The signatures and ratifications necessary to make this effective will not come easily."[77]

Despite the prediction, it took less than two years for CEDAW to receive the required number of ratifications to enter it into force — faster than any human-rights convention had ever done before.[78]

CHRONOLOGY

1700s-1800s *Age of Enlightenment and Industrial Revolution lead to greater freedoms for women.*

1792 Mary Wollstonecraft publishes *A Vindication of the Rights of Women,* later hailed as "the feminist declaration of independence."

1893 New Zealand becomes first nation to grant women full suffrage.

1920 Tennessee is the 36th state to ratify the 19th Amendment, giving American women the right to vote.

1940s-1980s *International conventions endorse equal rights for women. Global conferences highlight need to improve women's rights.*

1946 U.N. creates Commission on the Status of Women.

1951 U.N. International Labor Organization adopts convention promoting equal pay for equal work, which has been ratified by 164 countries; the United States is not among them.

1952 U.N. adopts convention calling for full women's suffrage.

1960 Sri Lanka elects the world's first female prime minister.

1974 Maria Estela Martínez de Perón of Argentina becomes the world's first woman president, replacing her ailing husband.

1975 U.N. holds first World Conference on Women, in Mexico City, followed by similar conferences every five years. U.N. launches the Decade for Women.

1979 U.N. adopts Convention on the Elimination of All Forms of Discrimination against Women (CEDAW), dubbed the "international bill of rights for women."

1981 CEDAW is ratified — faster than any other human-rights convention.

1990s *Women's rights win historic legal recognition.*

1993 U.N. World Conference on Human Rights in Vienna, Austria, calls for ending all violence, sexual harassment and trafficking of women.

1995 Fourth World Conference on Women in Beijing draws 30,000 people, making it the largest in U.N. history. Beijing Platform outlining steps to grant women equal rights is signed by 189 governments.

1996 International Criminal Tribunal convicts eight Bosnian Serb police and military officers for rape during the Bosnian conflict — the first time sexual assault is prosecuted as a war crime.

1998 International Criminal Tribunal for Rwanda recognizes rape and other forms of sexual violence as genocide.

2000s *Women make political gains, but sexual violence against women increases.*

2000 U.N. calls on governments to include women in peace negotiations.

2006 Ellen Johnson Sirleaf of Liberia, Michelle Bachelet of Chile and Portia Simpson Miller of Jamaica become their countries' first elected female heads of state. . . . Women in Kuwait are allowed to run for parliament, winning two seats.

2007 A woman in Saudi Arabia who was sentenced to 200 lashes after being gang-raped by seven men is pardoned by King Abdullah. Her rapists received sentences ranging from 10 months to five years in prison, and 80 to 1,000 lashes. . . . After failing to recognize any gender-based crimes in its first case involving the Democratic Republic of the Congo, the International Criminal Court hands down charges of "sexual slavery" in its second case involving war crimes in Congo. More than 250,000 women are estimated to have been raped and sexually abused during the country's war.

2008 Turkey lifts 80-year-old ban on women's headscarves in public universities, signaling a drift toward religious fundamentalism. . . . Former housing minister Carme Chacón — 37 and pregnant — is named defense minister of Spain, bringing to nine the number of female cabinet ministers in the Socialist government. . . . Sen. Hillary Rodham Clinton becomes the first U.S. woman to be in a tight race for a major party's presidential nomination.

Often described as an international bill of rights for women, CEDAW defines discrimination against women as "any distinction, exclusion or restriction made on the basis of sex which has the effect or purpose of impairing or nullifying the recognition, enjoyment or exercise by women, irrespective of their marital status, on a basis of equality of men and women, of human rights and fundamental freedoms in the political, economic, social, cultural, civil or any other field."

Ratifying countries are legally bound to end discrimination against women by incorporating sexual equality into their legal systems, abolishing discriminatory laws against women, taking steps to end trafficking of women and ensuring women equal access to political and public life. Countries must also submit reports at least every four years outlining the steps they have taken to comply with the convention.[79]

CEDAW also grants women reproductive choice — one of the main reasons the United States has not ratified it. The convention requires signatories to guarantee women's rights "to decide freely and responsibly on the number and spacing of their children and to have access to the information, education and means to enable them to exercise these rights."[80]

While CEDAW is seen as a significant tool to stop violence against women, it actually does not directly mention violence. To rectify this, the CEDAW committee charged with monitoring countries' compliance in 1992 specified gender-based violence as a form of discrimination prohibited under the convention.[81]

In 1993 the U.N. took further steps to combat violence against women during the World Conference on Human Rights in Vienna, Austria. The conference called on countries to stop all forms of violence, sexual harassment, exploitation and trafficking of women. It also declared that "violations of the human rights of women in situations of armed conflicts are violations of the fundamental principles of international human rights and humanitarian law."[82]

Shortly afterwards, as fighting broke out in the former Yugoslavia and Rwanda, new legal precedents were set to protect women against violence — and particularly rape — during war. In 1996, the International Criminal Tribunal in the Hague, Netherlands, indicted eight Bosnian Serb police officers in connection with the mass rape of Muslim women during the Bosnian war, marking the first time sexual assault had ever been prosecuted as a war crime.[83]

Two years later, the U.N.'s International Criminal Tribunal for Rwanda convicted a former Rwandan mayor for genocide, crimes against humanity, rape and sexual violence — the first time rape and sexual violence were recognized as acts of genocide.[84]

"Rape is a serious war crime like any other," said Regan Ralph, then executive director of Human Rights Watch's Women's Rights Division, shortly after the conviction. "That's always been true on paper, but now international courts are finally acting on it."[85]

Today, the International Criminal Court has filed charges against several Sudanese officials for rape and other crimes committed in the Darfur region.[86] But others are demanding that the court also prosecute those responsible for the rapes in the Eastern Congo, where women are being targeted as a means of destroying communities in the war-torn country.[87]

Beijing and Beyond

The U.N. World Conference on Women in Mexico City in 1975 produced a 44-page plan of action calling for a decade of special measures to give women equal status and opportunities in law, education, employment, politics and society.[88] The conference also kicked off the U.N.'s Decade for Women and led to creation of the U.N. Development Fund for Women (UNIFEM).[89]

Five years later, the U.N. held its second World Conference on Women in Copenhagen and then celebrated the end of the Decade for Women with the third World Conference in Nairobi in 1985. More than 10,000 representatives from government agencies and NGOs attended the Nairobi event, believed to be the largest gathering on women's issues at the time.[90]

Upon reviewing the progress made on women's issues during the previous 10 years, the U.N. representatives in Nairobi concluded that advances had been extremely limited due to failing economies in developing countries, particularly those in Africa struggling against drought, famine and crippling debt. The conference developed a set of steps needed to improve the status of women during the final 15 years of the 20th century.[91]

Ten years later, women gathered in Beijing in 1995 for the Fourth World Conference, vowing to turn the rhetoric of the earlier women's conferences into action. Delegates from 189 governments and 2,600

Women Suffer Most in Natural Disasters

Climate change will make matters worse.

In natural disasters, women suffer death, disease and hunger at higher rates then men. During the devastating 2004 tsunami in Asia, 70 to 80 percent of the dead were women.[1] During cyclone-triggered flooding in Bangladesh that killed 140,000 people in 1991, nearly five times more women between the ages of 20 and 44 died than men.[2]

Gender discrimination, cultural biases and lack of awareness of women's needs are part of the problem. For instance, during the 1991 cyclone, Bangladeshi women and their children died in higher numbers because they waited at home for their husbands to return and make evacuation decisions.[3] In addition, flood warnings were conveyed by men to men in public spaces but were rarely communicated to women and children at home.[4]

And during the tsunami, many Indonesian women died because they stayed behind to look for children and other family members. Women clinging to children in floodwaters also tired more quickly and drowned, since most women in the region were never taught to swim or climb trees.[5] In Sri Lanka, many women died because the tsunami hit early on a Sunday morning when they were inside preparing breakfast for their families. Men were generally outside where they had earlier warning of the oncoming floods so they were better able to escape.[6]

Experts now predict global climate change — which is expected to increase the number of natural disasters around the world — will put women in far greater danger than men because natural disasters generally have a disproportionate impact on the world's poor. Since women comprise 70 percent of those living on less than $1 a day, they will be hardest hit by climate changes, according to the Intergovernmental Panel on Climate Change.[7]

"Climate change is not gender-neutral," said Gro Harlem Brundtland, former prime minister of Norway and now special envoy to the U.N. secretary-general on climate change. "[Women are] more dependent for their livelihood on natural resources that are threatened by climate change.... With changes in climate, traditional food sources become more unpredictable and scarce. This exposes women to loss of harvests, often their sole sources of food and income."[8]

Women produce 60 to 80 percent of the food for household consumption in developing countries.[9] As drought, flooding and desertification increase, experts say women and their families will be pushed further into poverty and famine.

Women also suffer more hardship in the aftermath of natural disasters, and their needs are often ignored during relief efforts.

In many Third World countries, for instance, women have no property rights, so when a husband dies during a natural disaster his family frequently confiscates the land from his widow, leaving her homeless and destitute.[10] And because men usually dominate emergency relief and response agencies, women's specific needs, such as contraceptives and sanitary napkins, are often overlooked. After floods in Bangladesh in 1998, adolescent girls reported high rates of rashes and urinary tract infections because they had

NGOs attended. More than 30,000 women and men gathered at a parallel forum organized by NGOs, also in Beijing.[92]

The so-called Beijing Platform that emerged from the conference addressed 12 critical areas facing women, from poverty to inequality in education to inadequate health care to violence. It brought unprecedented attention to women's issues and is still considered by many as the blueprint for true gender equality.

The Beijing Conference also came at the center of a decade that produced historic political gains for women around the world — gains that have continued, albeit at a slow pace, into the new century. The 1990s saw more women entering top political positions than ever before. A record 10 countries elected or appointed women as presidents between 1990 and 2000, including Haiti, Nicaragua, Switzerland and Latvia. Another 17 countries chose women prime ministers.[93]

In 2006 Ellen Johnson Sirleaf of Liberia became Africa's first elected woman president.[94] That same year, Chile elected its first female president, Michelle Bachelet, and Jamaica elected Portia Simpson Miller

no clean water, could not wash their menstrual rags properly in private and had no place to hang them to dry.[11]

"In terms of reconstruction, people are not talking about women's needs versus men's needs," says June Zeitlin, executive director of the Women's Environment and Development Organization, a New York City-based international organization that works for women's equality in global policy. "There is a lack of attention to health care after disasters, issues about bearing children, contraception, rape and vulnerability, menstrual needs — things a male programmer is not thinking about. There is broad recognition that disasters have a disproportionate impact on women. But it stops there. They see women as victims, but they don't see women as agents of change."

Women must be brought into discussions on climate change and emergency relief, say Zeitlin and others. Interestingly, she points out, while women are disproportionately affected by environmental changes, they do more than men to protect the environment. Studies show women emit less climate-changing carbon dioxide than men because they recycle more, use resources more efficiently and drive less than men.[12]

"Women's involvement in climate-change decision-making is a human right," said Gerd Johnson-Latham, deputy director of the Swedish Ministry for Foreign Affairs. "If we get more women in decision-making positions, we

The smell of death hangs over Banda Aceh, Indonesia, which was virtually destroyed by a tsunami on Dec. 28, 2004. From 70 to 80 percent of the victims were women.

will have different priorities, and less risk of climate change."[13]

[1] "Tsunami death toll," CNN, Feb. 22, 2005. Also see "Report of High-level Roundtable: How a Changing Climate Impacts Women," Council of Women World Leaders, Women's Environment and Development Organization and Heinrich Boll Foundation, Sept. 21, 2007, p. 21, www.wedo.org/files/Roundtable%20Final%20Report%206%20Nov.pdf.

[2] *Ibid.*

[3] "Cyclone Jelawat bears down on Japan's Okinawa island," CNN. com, Aug. 7, 2000, http://archives.cnn.com/2000/ASIANOW/east/08/07/asia.weather/index.html.

[4] "Gender and Health in Disasters," World Health Organization, July 2002, www.who.int/gender/other_health/en/genderdisasters.pdf.

[5] "The tsunami's impact on women," Oxfam briefing note, March 5, 2005, p. 2, www.oxfam.org/en/files/bn050326_tsunami_women/download.

[6] "Report of High-level Roundtable," *op. cit.*, p. 5.

[7] "Gender Equality" fact sheet, Oxfam, www.oxfam.org.uk/resources/issues/gender/introduction.html. Also see *ibid.*

[8] *Ibid.*, p. 4.

[9] "Five years down the road from Beijing: Assessing progress," *News and Highlights*, Food and Agriculture Organization, June 2, 2000, www.fao.org/News/2000/000602-e.htm.

[10] "Gender and Health in Disasters," *op. cit.*

[11] *Ibid.*

[12] "Women and the Environment," U.N. Environment Program, 2004, p. 17, www.unep.org/Documents.Multilingual/Default.asp?DocumentID=468&ArticleID=4488&l=en. Also see "Report of High-level Roundtable," *op. cit.*, p. 7.

[13] *Ibid.*

as its first female prime minister.[95] Also that year, women ran for election in Kuwait for the first time. In Bahrain, a woman was elected to the lower house of parliament for the first time.[96] And in 2007, Fernández de Kirchner became the first woman to be elected president of Argentina.

Earlier, a World Bank report had found that government corruption declines as more women are elected into office. The report also cited numerous studies that found women are more likely to exhibit "helping" behavior, vote based on social issues, score higher on "integrity tests," take stronger stances on ethical behavior and behave more generously when faced with economic decisions.[97]

"Increasing the presence of women in government may be valued for its own sake, for reasons of gender equality," the report concluded. "However, our results suggest that there may be extremely important spinoffs stemming from increasing female representation: If women are less likely than men to behave opportunistically, then bringing more women into government may have significant benefits for society in general."[98]

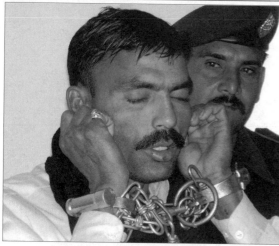

Honor Killings on the Rise

Women in Multan, Pakistan, demonstrate against "honor killings" in 2003 (top). Although Pakistan outlawed such killings years ago, its Human Rights Commission says 1,205 women were killed in the name of family honor in 2007 — a fourfold jump in two years. Nazir Ahmed Sheikh, a Punjabi laborer (bottom), unrepentantly told police in December 2005 how he slit the throats of his four daughters one night as they slept in order to salvage the family's honor. The eldest had married a man of her choice, and Ahmed feared the younger daughters would follow her example.

CURRENT SITUATION

Rise of Fundamentalism

Despite landmark political gains by women since the late 1990s, violence and repression of women continue to be daily occurrences — often linked to the global growth of religious fundamentalism.

In 2007, a 21-year-old woman in Saudi Arabia was sentenced to 200 lashes and ordered jailed for six months after being raped 14 times by a gang of seven men. The Saudi court sentenced the woman — who was 19 at the time of the attack — because she was alone in a car with her former boyfriend when the attack occurred. Under Saudi Arabia's strict Islamic law, it is a crime for a woman to meet in private with a man who is not her husband or relative.[99]

After public outcry from around the world, King Abdullah pardoned the woman in December. A government spokesperson, however, said the king fully supported the verdict but issued the pardon in the "interests of the people."[100]

Another Saudi woman still faces beheading after she was condemned to death for "witchcraft." Among her accusers is a man who claimed she rendered him impotent with her sorcery. Despite international protest, the king has yet to say if he will pardon her.[101]

In Iraq, the rise of religious fundamentalism since the U.S. invasion has led to a jump in the number of women being killed or beaten in so-called honor crimes. Honor killings typically occur when a woman is suspected of unsanctioned sexual behavior — which can range from flirting to "allowing" herself to be raped. Her relatives believe they must murder her to end the family's shame. In the Kurdish region of Iraq, the stoning death of 17-year-old Aswad is not an anomaly. A U.N. mission in October 2007 found that 255 women had been killed in Iraqi Kurdistan in the first six months of 2007 alone — most thought to have been murdered by their communities or families for allegedly committing adultery or entering into a relationship not sanctioned by their families.[102]

The rise of fundamentalism is also sparking a growing debate on the issue of women wearing head scarves, both in Iraq and across the Muslim world. Last August Turkey elected a conservative Muslim president whose wife wears a head scarf, signaling the emergence of a new ruling elite that is more willing to publicly display religious beliefs.[103] Then in February, Turkey's parliament voted to ease an

80-year ban on women wearing head scarves in universities, although a ban on head scarves in other public buildings remains in effect.

"This decision will bring further pressure on women," Nesrin Baytok, a member of parliament, said during debate over the ban. "It will ultimately bring us Hezbollah terror, al Qaeda terror and fundamentalism."[104]

But others said lifting the ban was actually a victory for women. Fatma Benli, a Turkish women's-rights activist and lawyer, said the ban on head scarves in public buildings has forced her to send law partners to argue her cases because she is prohibited from entering court wearing her head scarf. It also discourages religiously conservative women from becoming doctors, lawyers or teachers, she says.[105]

Many women activists are quick to say that it is unfair to condemn Islam for the growing abuse against women. "The problem women have with religion is not the religion but the ways men have interpreted it," says Afshar of the University of York. "What is highly negative is sharia law, which is made by men. Because it's human-made, women can unmake it. The battle now is fighting against unjust laws such as stoning."

She says abuses such as forced marriages and honor killings — usually linked in the Western media to Islamic law — actually go directly against the teachings of the *Koran*. And while the United Nations estimates that some 5,000 women and girls are victims of honor killings each year, millions more are abused and killed in violence unrelated to Islam. Between 10 and 50 percent of all women around the world have been physically abused by an intimate partner in their lifetime, studies show.[106]

"What about the rate of spousal or partner killings in the U.K. or the U.S. that are not called 'honor killings'?" asks Concordia University's Hoodfar. "Then it's only occasional 'crazy people' [committing violence]. But when it's present in Pakistan, Iran or Senegal, these are uncivilized people doing 'honor killings.' "

And Islamic fundamentalism is not the only brand of fundamentalism on the rise. Christian fundamentalism is also growing rapidly. A 2006 Pew Forum on Religion and Public Life poll found that nearly one-third of all Americans feel the Bible should be the basis of law across the United States.[107] Many women's-rights activists say Christian fundamentalism threatens women's rights, particularly with regard to reproductive issues. They also condemn the Vatican's opposition to the use of condoms, pointing

Getty Images/Paula Bronstein

Pakistani acid attack survivors Saira Liaqat, right, and Sabra Sultana are among hundreds, and perhaps thousands, of women who are blinded and disfigured after being attacked with acid each year in Pakistan, Bangladesh, India, Cambodia, Malaysia, Uganda and other areas of Africa. Liaqat was attacked at age 18 during an argument over an arranged marriage. Sabra was 15 when she was burned after being married off to an older man who became unsatisfied with the relationship. Only a small percentage of the attacks — often perpetrated by spurned suitors while the women are asleep in their own beds — are prosecuted.

out that it prevents women from protecting themselves against HIV.

"If you look at all your religions, none will say it's a good thing to beat up or kill someone. They are all based on human dignity," says Mollmann of Human Rights Watch. "[Bad things] are carried out in the name of religion, but the actual belief system is not killing and maiming women."

In response to the growing number of honor-based killings, attacks and forced marriages in the U.K., Britain's Association of Chief Police Officers has created an honor-based violence unit, and the U.K.'s Home Office is drafting an action plan to improve the response of police and other agencies to such violence. Legislation going into effect later this year will also give U.K. courts greater guidance on dealing with forced marriages.[108]

Evolving Gender Policies

This past February, the U.N. Convention on the Elimination of All Forms of Discrimination Against Women issued a report criticizing Saudi Arabia for its repression of women. Among other things, the report attacked Saudi Arabia's ban on women drivers and its

Female farmworkers in Nova Lima, Brazil, protest against the impact of big corporations on the poor in March 2006, reflecting the increasing political activism of women around the globe.

system of male guardianship that denies women equal inheritance, child custody and divorce rights.[109] The criticism came during the panel's regular review of countries that have ratified CEDAW. Each government must submit reports every four years outlining steps taken to comply with the convention.

The United States is one of only eight countries — among them Iran, Sudan and Somalia — that have refused to ratify CEDAW.[110] Last year, 108 members of the U.S. House of Representatives signed on to a resolution calling for the Senate to ratify CEDAW, but it still has not voted on the measure.[111] During a U.N. vote last November on a resolution encouraging governments to meet their obligations under CEDAW, the United States was the lone nay vote against 173 yea votes.[112]

American opponents of CEDAW — largely pro-life Christians and Republicans — say it would enshrine the right to abortion in *Roe v. Wade* and be prohibitively expensive, potentially requiring the U.S. government to provide paid maternity leave and other child-care services to all women.[113] They also oppose requirements that the government modify "social and cultural patterns" to eliminate sexual prejudice and to delete any traces of gender stereotypes in textbooks — such as references to women's lives being primarily in the domestic sector.[114] Many Republicans in Congress also have argued that CEDAW would give too much control over U.S. laws to the United Nations and that it could even require the legalization of prostitution and the abolition of Mother's Day.[115]

The last time the Senate took action on CEDAW was in 2002, when the Senate Foreign Relations Committee, chaired by Democratic Sen. Joseph Biden of Delaware, voted to send the convention to the Senate floor for ratification. The full Senate, however, never took action. A Biden spokesperson says the senator "remains committed" to the treaty and is "looking for an opportune time" to bring it forward again. But Senate ratification requires 67 votes, and there do not appear to be that many votes for approval.

CEDAW proponents say the failure to ratify not only hurts women but also harms the U.S. image abroad. On this issue, "the United States is in the company of Sudan and the Vatican," says Bien-Aimé of Equality Now.

Meanwhile, several countries are enacting laws to comply with CEDAW and improve the status of women. In December, Turkmenistan passed its first national law guaranteeing women equal rights, even though its constitution had addressed women's equality.[116] A royal decree in Saudi Arabia in January ordered an end to a long-time ban on women checking into hotels or renting apartments without male guardians. Hotels can now book rooms to women who show identification, but the hotels must register the women's details with the police.[117] The Saudi government has also said it will lift the ban on women driving by the end of the year.[118]

And in an effort to improve relations with women in Afghanistan, the Canadian military, which has troops stationed in the region, has begun studying the role women play in Afghan society, how they are affected by military operations and how they can assist peacekeeping efforts. "Behind all of these men are women who can help eradicate the problems of the population," said Capt. Michel Larocque, who is working with the study. "Illiteracy, poverty, these things can be improved through women."[119]

In February, during the 52nd session of the Commission on the Status of Women, the United Nations kicked off a new seven-year campaign aimed at ending violence against women. The campaign will work with international agencies, governments and individuals to increase funding for anti-violence campaigns and pressure policy makers around the world to enact legislation to eliminate violence against women.[120]

But women's groups want increased U.N. spending on women's programs and the creation of a single unified

Should sex-selective abortions be outlawed?

YES

Nicholas Eberstadt
*Henry Wendt Chair in Political Economy,
American Enterprise Institute Member,
President's Council on Bioethics*

Written for *CQ Global Researcher*, April 2008

The practice of sex-selective abortion to permit parents to destroy unwanted female fetuses has become so widespread in the modern world that it is disfiguring the profile of entire countries — transforming (and indeed deforming) the whole human species.

This abomination is now rampant in China, where the latest census reports six boys for every five girls. But it is also prevalent in the Far East, South Korea, Hong Kong, Taiwan and Vietnam, all of which report biologically impossible "sex ratios at birth" (well above the 103-106 baby boys for every 100 girls ordinarily observed in human populations). In the Caucasus, gruesome imbalances exist now in Armenia, Georgia and Azerbaijan; and in India, the state of Punjab tallies 126 little boys for every 100 girls. Even in the United States, the boy-girl sex ratio at birth for Asian-Americans is now several unnatural percentage points above the national average. So sex-selective abortion is taking place under America's nose.

How can we rid the world of this barbaric form of sexism? Simply outlawing sex-selective abortions will be little more than a symbolic gesture, as South Korea's experience has shown: Its sex ratio at birth continued a steady climb for a full decade after just such a national law was passed. As long as abortion is basically available on demand, any legislation to abolish sex-selective abortion will have no impact.

What about more general restrictions on abortion, then? Poll data consistently demonstrate that most Americans do not favor the post-*Roe* regimen of unconditional abortion. But a return to the pre-*Roe* status quo, where each state made its own abortion laws, would probably have very little effect on sex-selective abortion in our country. After all, the ethnic communities most tempted by it are concentrated in states where abortion rights would likely be strongest, such as California and New York.

In the final analysis, the extirpation of this scourge will require nothing less than a struggle for the conscience of nations. Here again, South Korea may be illustrative: Its gender imbalances began to decline when the public was shocked into facing this stain on their society by a spontaneous, homegrown civil rights movement.

To eradicate sex-selective abortion, we must convince the world that destroying female fetuses is horribly wrong. We need something akin to the abolitionist movement: a moral campaign waged globally, with victories declared one conscience at a time.

NO

Marianne Mollmann
*Advocacy Director,
Women's Rights Division,
Human Rights Watch*

Written for *CQ Global Researcher*, April 2008

Medical technology today allows parents to test early in pregnancy for fetal abnormalities, hereditary illnesses and even the sex of the fetus, raising horrifying questions about eugenics and population control. In some countries, a growing number of women apparently are terminating pregnancies when they learn the fetus is female. The resulting sex imbalance in countries like China and India is not only disturbing but also leads to further injustices, such as the abduction of girls for forced marriages.

One response has been to criminalize sex-selective abortions. While it is tempting to hope that this could safeguard the gender balance of future generations, criminalization of abortion for whatever reason has led in the past only to underground and unsafe practices. Thus, the criminalization of sex-selective abortion would put the full burden of righting a fundamental wrong — the devaluing of women's lives — on women.

Many women who choose to abort a female fetus face violence and exclusion if they don't produce a boy. Some see the financial burden of raising a girl as detrimental to the survival of the rest of their family. These considerations will not be lessened by banning sex-selective abortion. Unless one addresses the motivation for the practice, it will continue — underground.

So what is the motivation for aborting female fetuses? At the most basic level, it is a financial decision. In no country in the world does women's earning power equal men's. In marginalized communities in developing countries, this is directly linked to survival: Boys may provide more income than girls.

Severe gaps between women's and men's earning power are generally accompanied by severe forms of gender-based discrimination and rigid gender roles. For example, in China, boys are expected to stay in their parental home as they grow up, adding their manpower (and that of a later wife) to the family home. Girls, on the other hand, are expected to join the husbands' parental home. Thus, raising a girl is a net loss, especially if you are only allowed one child.

The solution is to remove the motivation behind sex-selective abortion by advancing women's rights and their economic and social equality. Choosing the blunt instrument of criminal law over promoting the value of women's lives and rights will only serve to place further burdens on marginalized and often vulnerable women.

agency addressing women's issues, led by an under-secretary general.[121] Currently, four different U.N. agencies address women's issues: the United Nations Development Fund for Women, the International Research and Training Institute for the Advancement of Women (INSTRAW), the Secretary-General's Special Advisor on Gender Issues (OSAGI) and the Division for the Advancement of Women. In 2006, the four agencies received only $65 million — a fraction of the more than $2 billion budget that the U.N.'s children's fund (UNICEF) received that year.[122]

"The four entities that focus on women's rights at the U.N. are greatly under-resourced," says Zeitlin of the Women's Environment & Development Organization. "If the rhetoric everyone is using is true — that investing in women is investing in development — it's a matter of putting your money where your mouth is."

Political Prospects

While the number of women leading world governments is still miniscule compared to their male counterparts, women are achieving political gains that just a few years ago would have been unthinkable.

While for the first time in U.S. history a woman is in a tight race for a major party's nomination as its candidate for president, South America — with two sitting female heads of state — leads the world in woman-led governments. In Brazil, Dilma Rousseff, the female chief of staff to President Luiz Inacio Lula da Silva, is the top contender to take over the presidency when da Silva's term ends in 2010.[123] In Paraguay, Blanca Ovelar was this year's presidential nominee for the country's ruling conservative Colorado Party, but she was defeated on April 20.[124]

And in Europe, Carme Chacón was named defense minister of Spain this past April. She was not only the first woman ever to head the country's armed forces but also was pregnant at the time of her appointment. In all, nine of Spain's 17 cabinet ministers are women.

In March, Pakistan's National Assembly overwhelmingly elected its first female speaker, Fahmida Mirza.[125] And in India, where Patil has become the first woman president, the two major political parties this year pledged to set aside one-third of their parliamentary nominations for women. But many fear the parties will either not keep their pledges or will run women only in contests they are unlikely to win.[126]

There was also disappointment in Iran, where nearly 600 of the 7,000 candidates running for parliament in March were women.[127] Only three won seats in the 290-member house, and they were conservatives who are not expected to promote women's rights. Several of the tallies are being contested. Twelve other women won enough votes to face run-off elections on April 25; five won.[128]

But in some countries, women running for office face more than just tough campaigns. They are specifically targeted for violence. In Kenya, the greatest campaign expense for female candidates is the round-the-clock security required to protect them against rape, according to Phoebe Asiyo, who served in the Kenyan parliament for more than two decades.[129] During the three months before Kenya's elections last December, an emergency helpdesk established by the Education Centre for Women in Democracy, a nongovernmental organization (NGO) in Nairobi, received 258 reports of attacks against female candidates.[130]

The helpdesk reported the attacks to police, worked with the press to ensure the cases were documented and helped victims obtain medical and emotional support. Attacks included rape, stabbings, threats and physical assaults.[131]

"Women are being attacked because they are women and because it is seen as though they are not fit to bear flags of the popular parties," according to the center's Web site. "Women are also viewed as guilty for invading 'the male territory' and without a license to do so!"[132]

"All women candidates feel threatened," said Nazlin Umar, the sole female presidential candidate last year. "When a case of violence against a woman is reported, we women on the ground think we are next. I think if the government assigned all women candidates with guns…we will at least have an item to protect ourselves when we face danger."[133]

Impunity for Violence

Some African feminists blame women themselves, as well as men, for not doing enough to end traditional attitudes that perpetuate violence against women.

"Women are also to blame for the violence because they are the gatekeepers of patriarchy, because whether educated or not they have different standards for their sons and husbands [than for] their daughters," said Njoki Wainaina, founder of the African Women Development

Communication Network (FEMNET). "How do you start telling a boy whose mother trained him only disrespect for girls to honor women in adulthood?"[134]

Indeed, violence against women is widely accepted in many regions of the world and often goes unpunished. A study by the World Health Organization found that 80 percent of women surveyed in rural Egypt believe that a man is justified in beating a woman if she refuses to have sex with him. In Ghana, more women than men — 50 percent compared to 43 percent — felt that a man was justified in beating his wife if she used contraception without his consent.[135] (*See survey results, p. 168.*)

Such attitudes have led to many crimes against women going unpunished, and not just violence committed during wartime. In Guatemala, no one knows why an estimated 3,000 women have been killed over the past seven years — many of them beheaded, sexually mutilated or raped — but theories range from domestic violence to gang activity.[136] Meanwhile, the government in 2006 overturned a law allowing rapists to escape charges if they offered to marry their victims. But Guatemalan law still does not prescribe prison sentences for domestic abuse and prohibits abusers from being charged with assault unless the bruises are still visible after 10 days.[137]

In the Mexican cities of Chihuahua and Juárez, more than 400 women have been murdered over the past 14 years, with many of the bodies mutilated and dumped in the desert. But the crimes are still unsolved, and many human-rights groups, including Amnesty International, blame indifference by Mexican authorities. Now the country's 14-year statute of limitations on murder is forcing prosecutors to close many of the unsolved cases.[138]

Feminists around the world have been working to end dismissive cultural attitudes about domestic violence and other forms of violence against women, such as forced marriage, dowry-related violence, marital rape, sexual harassment and forced abortion, sterilization and prostitution. But it's often an uphill battle.

After a Kenyan police officer beat his wife so badly she was paralyzed and brain damaged — and eventually died — media coverage of the murder spurred a nationwide debate on domestic violence. But it took five years of protests, demonstrations and lobbying by both women's advocates and outraged men to get a family protection bill enacted criminalizing domestic violence. And the bill passed only after legislators removed a provision outlawing

marital rape. Similar laws have languished for decades in other African legislatures.[139]

But in Rwanda, where nearly 49 percent of the elected representatives in the lower house are female, gender desks have been established at local police stations, staffed mostly by women trained to help victims of sexual and other violence. In 2006, as a result of improved reporting, investigation and response to rape cases, police referred 1,777 cases for prosecution and convicted 803 men. "What we need now is to expand this approach to more countries," said UNIFEM's director for Central Africa Josephine Odera.[140]

Besides criticizing governments for failing to prosecute gender-based violence, many women's groups also criticize the International Criminal Court (ICC) for not doing enough to bring abusers to justice.

"We have yet to see the investigative approach needed to ensure the prosecution of gender-based crimes," said Brigid Inder, executive director of Women's Initiatives for Gender Justice, a Hague-based group that promotes and monitors women's rights in the international court.[141] Inder's group released a study last November showing that of the 500 victims seeking to participate in ICC proceedings, only 38 percent were women. When the court handed down its first indictments for war crimes in the Democratic Republic of the Congo last year, no charges involving gender-based crimes were brought despite estimates that more than 250,000 women have been raped and sexually abused in the country. After an outcry from women's groups around the world, the ICC included "sexual slavery" among the charges handed down in its second case involving war crimes in Congo.[142]

The Gender Justice report also criticized the court for failing to reach out to female victims. It said the ICC has held only one consultation with women in the last four years (focusing on the Darfur conflict in Sudan) and has failed to develop any strategies to reach out to women victims in Congo.[143]

OUTLOOK

Economic Integration

Women's organizations do not expect — or want — another international conference on the scale of Beijing. Instead, they say, the resources needed to launch such a

Seaweed farmer Asia Mohammed Makungu in Zanzibar, Tanzania, grows the sea plants for export to European companies that produce food and cosmetics. Globalized trade has helped women entrepreneurs in many developing countries improve their lives, but critics say it also has created many low-wage, dangerous jobs for women in poor countries that ignore safety and labor protections in order to attract foreign investors.

conference would be better used to improve U.N. oversight of women's issues and to implement the promises made at Beijing.

They also fear that the growth of religious fundamentalism and neo-liberal economic policies around the globe have created a political atmosphere that could actually set back women's progress.

"If a Beijing conference happened now, we would not get the type of language or the scope we got 10 years ago," says Bien-Aimé of Equity Now. "There is a conservative movement, a growth in fundamentalists governments — and not just in Muslim countries. We would be very concerned about opening up debate on the principles that have already been established."

Dahlerup of Stockholm University agrees. "It was easier in the 1990s. Many people are afraid of having big conferences now, because there may be a backlash because fundamentalism is so strong," she says. "Neo-liberal trends are also moving the discourse about women toward economics — women have to benefit for the sake of the economic good. That could be very good, but it's a more narrow discourse when every issue needs to be adapted into the economic discourse of a cost-benefit analysis."

For women to continue making gains, most groups say, gender can no longer be treated separately from

broader economic, environmental, health or other political issues. While efforts to improve the status of women have historically been addressed in gender-specific legislation or international treaties, women's groups now say women's well-being must now be considered an integral part of all policies.

Women's groups are working to ensure that gender is incorporated into two major international conferences coming up this fall. In September, the Third High-Level Forum on Aid Effectiveness will be hosted in Accra, Ghana, bringing together governments, financial institutions, civil society organizations and others to assess whether assistance provided to poor nations is being put to good use. World leaders will also gather in November in Doha, Qatar, for the International Conference on Financing for Development to discuss how trade, debt relief and financial aid can promote global development.

"Women's groups are pushing for gender to be on the agenda for both conferences," says Zeitlin of WEDO. "It's important because . . . world leaders need to realize that it really does make a difference to invest in women. When it comes to women's rights it's all micro, but the big decisions are made on the macro level."

Despite decades of economic-development strategies promoted by Western nations and global financial institutions such as the World Bank, women in many regions are getting poorer. In Malawi, for example, the percentage of women living in poverty increased by 5 percent between 1995 and 2003.[144] Women and girls make up 70 percent of the world's poorest people, and their wages rise more slowly than men's. They also have fewer property rights around the world.[145] With the growing global food shortage, women — who are the primary family caregivers and produce the majority of crops for home consumption in developing countries — will be especially hard hit.

To help women escape poverty, gain legal rights and improve their social status, developed nations must rethink their broader strategies of engagement with developing countries. And, conversely, female activists say, any efforts aimed at eradicating poverty around the world must specifically address women's issues.

In Africa, for instance, activists have successfully demanded that women's economic and security concerns be addressed as part of the continent-wide development plan known as the New Partnership for Africa's Development (NEPAD). As a result, countries participating in NEPAD's peer review process must now show they are taking measures to promote and protect women's rights. But, according to Augustin Wambo, an agricultural specialist at the NEPAD secretariat, lawmakers now need to back up their pledges with "resources from national budgets" and the "necessary policies and means to support women."[146]

"We have made a lot of progress and will continue making progress," says Zeitlin. "But women's progress doesn't happen in isolation to what's happening in the rest of the world. The environment, the global economy, war, peace — they will all have a major impact on women. Women all over world will not stop making demands and fighting for their rights."

NOTES

1. http://ballyblog.wordpress.com/2007/05/04/warning-uncensored-video-iraqis-stone-girl-to-death-over-loving-wrong-boy/.

2. Abdulhamid Zebari, "Video of Iraqi girl's stoning shown on Internet," Agence France Presse, May 5, 2007.

3. *State of the World Population 2000*, United Nations Population Fund, Sept. 20, 2000, Chapter 3, "Ending Violence against Women and Girls," www.unfpa.org/swp/2000/english/ch03.html.

4. Brian Brady, "A Question of Honour," *The Independent on Sunday*, Feb. 10, 2008, p. 8, www.independent.co.uk/news/uk/home-news/a-question-of-honour-police-say-17000-women-are-victims-every-year-780522.html.

5. Correspondance with Karen Musalo, Clinical Professor of Law and Director of the Center for Gender & Refugee Studies at the University of California Hastings School of Law, April 11, 2008.

6. "Broken Bodies, Broken Dreams: Violence Against Women Exposed," United Nations, July 2006, http://brokendreams.wordpress.com/2006/12/17/dowry-crimes-and-bride-price-abuse/.

7. Various sources: www.womankind.org.uk, www.unfpa.org/gender/docs/studies/summaries/reg_exe_summary.pdf, www.oxfam.org.uk. Also see "Child rape in Kano on the increase," IRIN Humanitarian News and Analysis, United Nations, www.irinnews.org/report.aspx?ReportId=76087.

8. "UNICEF slams 'licence to rape' in African crisis," Agence France-Press, Feb. 12, 2008.

9. "Film Documents Rape of Women in Congo," "All Things Considered," National Public Radio, April 8, 2008, www.npr.org/templates/story/story.php?storyId=89476111.

10. Jeffrey Gettleman, "Rape Epidemic Raises Trauma Of Congo War," *The New York Times*, Oct. 7, 2007, p. A1.

11. Dan McDougall, "Fareeda's fate: rape, prison and 25 lashes," *The Observer*, Sept. 17, 2006, www.guardian.co.uk/world/2006/sep/17/pakistan.theobserver.

12. Zarar Khan, "Thousands rally in Pakistan to demand government withdraw rape law changes," The Associated Press, Dec. 10, 2006.

13. *State of the World Population 2000, op. cit.*

14. Laura Turquet, Patrick Watt, Tom Sharman, "Hit or Miss?" ActionAid, March 7, 2008, p. 10.

15. *Ibid.*, p. 12.

16. "Women in Politics: 2008" map, International Parliamentary Union and United Nations Division for the Advancement of Women, February 2008, www.ipu.org/pdf/publications/wmnmap08_en.pdf.

17. Gavin Rabinowitz, "India's first female president sworn in, promises to empower women," The Associated Press, July 25, 2007. Note: India's first female prime minister was Indira Ghandi in 1966.

18. Monte Reel, "South America Ushers In The Era of La Presidenta; Women Could Soon Lead a Majority of Continent's Population," *The Washington Post*, Oct. 31, 2007, p. A12. For background, see Roland Flamini, "The New Latin America," *CQ Global Researcher*, March 2008, pp. 57-84.

19. Marcela Valente, "Cristina Fernandes Dons Presidential Sash," Inter Press Service, Dec. 10, 2007.

20. "Women in Politics: 2008" map, *op. cit.*

21. *Ibid.*; Global Database of Quotas for Women, International Institute for Democracy and Electoral Assistance and Stockholm University, www.quotaproject.org/country.cfm?SortOrder=Country.

22. "Beijing Betrayed," Women's Environment and Development Organization, March 2005, p. 10, www.wedo.org/files/gmr_pdfs/gmr2005.pdf.

23. "Women in Politics: 2008" map, *op. cit.*

24. Gertrude Mongella, address by the Secretary-General of the 4th World Conference on Women, Sept. 4, 1995, www.un.org/esa/gopher-data/conf/fwcw/conf/una/950904201423.txt. Also see Steven Mufson, "Women's Forum Sets Accord; Dispute on Sexual Freedom Resolved," *The Washington Post*, Sept. 15, 1995, p. A1.

25. "Closing statement," Gertrude Mongella, U.N. Division for the Advancement of Women, Fourth World Conference on Women, www.un.org/esa/gopher-data/conf/fwcw/conf/una/ closing.txt.

26. "Trading Away Our Rights," Oxfam International, 2004, p. 9, www.oxfam.org.uk/resources/policy/trade/downloads/trading_rights.pdf.

27. "Trafficking in Persons Report," U.S. Department of State, June 2007, p. 7, www.state.gov/g/tip/rls/tiprpt/2007/.

28. Turquet, *et al.*, *op. cit.*, p. 4.

29. United Nations Division for the Advancement of Women, www.un.org/womenwatch/daw/cedaw/.

30. Geraldine Terry, *Women's Rights* (2007), p. 30.

31. United Nations Division for the Advancement of Women, www.un.org/womenwatch/daw/cedaw/.

32. "The impact of international trade on gender equality," The World Bank PREM notes, May 2004, http://siteresources.worldbank.org/INTGENDER/Resources/premnote86.pdf.

33. Thalia Kidder and Kate Raworth, " 'Good Jobs' and hidden costs: women workers documenting the price of precarious employment," *Gender and Development*, July 2004, p. 13.

34. "Trading Away Our Rights," *op. cit.*

35. Martha Chen, *et al.*, "Progress of the World's Women 2005: Women, Work and Poverty," UNIFEM, p. 17, www.unifem.org/attachments/products/PoWW2005_eng.pdf.

36. Eric Neumayer and Indra de Soys, "Globalization, Women's Economic Rights and Forced Labor," London School of Economics and Norwegian University of Science and Technology, February 2007, p. 8, http://papers.ssrn.com/sol3/papers.cfm?abstract_id=813831. Also see "Five years down

the road from Beijing — assessing progress," *News and Highlights*, Food and Agriculture Organization, June 2, 2000, www.fao.org/News/2000/000602-e.htm.

37. "Trafficking in Persons Report," *op. cit.*, p. 13.

38. "World Survey on the Role of Women in Development," United Nations, 2006, p. 1, www.un.org/womenwatch/daw/public/WorldSurvey2004-Women&Migration.pdf.

39. Julie Ballington and Azza Karam, eds., "Women in Parliament: Beyond the Numbers," International Institute for Democracy and Electoral Assistance, 2005, p. 155, www.idea.int/publications/wip2/upload/WiP_inlay.pdf.

40. "Women in Politics: 2008," *op. cit.*

41. Ballington and Karam, *op. cit.*, p. 158.

42. *Ibid.*, p. 161.

43. Global Database of Quotas for Women, *op. cit.*

44. Jerome Starkey, "Afghan government official says that student will not be executed," *The Independent*, Feb. 6, 2008, www.independent.co.uk/news/world/asia/afghan-government-official-says-that-student-will-not-be-executed-778686.html?r=RSS.

45. "Afghan women seek death by fire," BBC, Nov. 15, 2006, http://news.bbc.co.uk/1/hi/world/south_asia/6149144.stm.

46. Global Database for Quotas for Women, *op. cit.*

47. "Beijing Declaration," Fourth World Conference on Women, www.un.org/womenwatch/daw/beijing/beijingdeclaration.html.

48. "Beijing Betrayed," *op. cit.*, pp. 28, 15, 18.

49. "Review of the implementation of the Beijing Platform for Action and the outcome documents of the special session of the General Assembly entitled 'Women 2000: gender equality, development and peace for the twenty-first century,' " United Nations, Dec. 6, 2004, p. 74.

50. "Gender Equality and the Millennium Development Goals," fact sheet, www.mdgender.net/upload/tools/MDGender_leaflet.pdf.

51. *Ibid.*

52. Turquet, *et al.*, *op. cit.*, p. 16.

53. *Ibid.*, pp. 22-24.

54. Terry, *op. cit.*, p. 6.

55. "Inclusive Security, Sustainable Peace: A Toolkit for Advocacy and Action," International Alert and Women Waging Peace, 2004, p. 12, www.huntalternatives .org/download/35_introduction.pdf.

56. "Declarations, Reservations and Objections to CEDAW," www.un.org/womenwatch/daw/cedaw/ reservations-country.htm.

57. Merlin Stone, *When God Was a Woman* (1976), pp. 18, 11.

58. Jack Holland, *Misogyny* (2006), p. 12.

59. *Ibid.*, pp. 21-23.

60. Holland, *op. cit.*, p. 112.

61. "Dispelling the myths about so-called witches" press release, Johns Hopkins University, Oct. 7, 2002, www.jhu.edu/news_info/news/home02/ oct02/witch.html.

62. The quote is from the *Malleus maleficarum* (*The Hammer of Witches*), and was cited in "Case Study: The European Witch Hunts, c. 1450-1750," *Gendercide Watch*, www.gendercide.org/case_witch-hunts.html.

63. Holland, *op. cit.*, p. 179.

64. Cathy J. Cohen, Kathleen B. Jones and Joan C. Tronto, *Women Transforming Politics: An Alternative Reader* (1997), p. 530.

65. *Ibid.*

66. Holland, *op. cit*, p. 201.

67. "Men and Women in Politics: Democracy Still in the Making," IPU Study No. 28, 1997, http:// archive.idea.int/women/parl/ch6_table8.htm.

68. "Sex, Equality and Protective Laws," *CQ Researcher*, July 13, 1926.

69. The case was *Radice v. People of State of New York*, 264 U. S. 292. For background, see F. Brewer, "Equal Rights Amendment," *Editorial Research Reports*, April 4, 1946, available at *CQ Researcher Plus Archive*, www.cqpress.com.

70. "Short History of the CEDAW Convention," U.N. Division for the Advancement of Women, www .un.org/womenwatch/daw/cedaw/history.htm.

71. U.N. Women's Watch, www.un.org/womenwatch/ asp/user/list.asp-ParentID=11047.htm.

72. United Nations, http://untreaty.un.org/ENGLISH/ bible/englishinternetbible/partI/chapterXVI/ treaty2.asp.

73. International Labor Organization, www.ilo. org/ public/english/support/lib/resource/subject/ gender.htm.

74. *Ibid.*

75. For background, see "Gender Pay Gap," *CQ Researcher*, March 14, 2008, pp. 241-264.

76. "Short History of the CEDAW Convention" *op. cit.*

77. "International News," The Associated Press, Dec. 19, 1979.

78. "Short History of the CEDAW Convention" *op. cit.*

79. "Text of the Convention," U.N. Division for the Advancement of Women, www.un.org/women-watch/daw/cedaw/cedaw.htm.

80. Convention on the Elimination of All Forms of Discrimination against Women, Article 16, www .un.org/womenwatch/daw/ cedaw/text/econvention .htm.

81. General Recommendation made by the Committee on the Elimination of Discrimination against Women No. 19, 11th session, 1992, www.un.org/women watch/daw/cedaw/recommendations/recomm .htm#recom19.

82. See www.unhchr.ch/huridocda/huridoca.nsf/ (Symbol)/A.CONF.157.23.En.

83. Marlise Simons, "For First Time, Court Defines Rape as War Crime," *The New York Times*, June 28, 1996, www.nytimes.com/specials/bosnia/ context/0628warcrimes-tribunal.html.

84. Ann Simmons, "U.N. Tribunal Convicts Rwandan Ex-Mayor of Genocide in Slaughter," *Los Angeles Times*, Sept. 3, 1998, p. 20.

85. "Human Rights Watch Applauds Rwanda Rape Verdict," press release, Human Rights Watch, Sept. 2, 1998, http://hrw.org/english/ docs/1998/09/02/ rwanda1311.htm.

86. Frederic Bichon, "ICC vows to bring Darfur war criminals to justice," Agence France-Presse, Feb. 24, 2008.

87. Rebecca Feeley and Colin Thomas-Jensen, "Getting Serious about Ending Conflict and Sexual Violence in Congo," Enough Project, www.enoughproject.org/reports/congoserious.

88. "Women; Deceived Again?" *The Economist*, July 5, 1975.

89. "International Women's Day — March 8: Points of Interest and Links with UNIFEM," UNIFEM New Zealand Web site, www.unifem.org.nz/IWDPointsofinterest.htm.

90. Joseph Gambardello, "Reporter's Notebook: Women's Conference in Kenya," United Press International, July 13, 1985.

91. "Report of the World Conference to Review and Appraise the Achievements of the United Nations Decade for Women: Equality Development and Peace," United Nations, 1986, paragraph 8, www.un.org/womenwatch/confer/nfls/Nairobi1985report.txt.

92. U.N. Division for the Advancement of Women, www.un.org/womenwatch/daw/followup/background.htm.

93. "Women in Politics," Inter-Parliamentary Union, 2005, pp. 16-17, www.ipu.org/PDF/publications/wmn45-05_en.pdf.

94. "Liberian becomes Africa's first female president," Associated Press, Jan. 16, 2006, www.msnbc.msn.com/id/10865705/.

95. "Women in the Americas: Paths to Political Power," *op. cit.*, p. 2.

96. "The Millennium Development Goals Report 2007," United Nations, 2007, p. 12, www.un.org/millenniumgoals/pdf/mdg2007.pdf.

97. David Dollar, Raymond Fisman, Roberta Gatti, "Are Women Really the 'Fairer' Sex? Corruption and Women in Government," The World Bank, October 1999, p. 1, http://siteresources.worldbank.org/INTGENDER/Resources/wp4.pdf.

98. *Ibid.*

99. Vicky Baker, "Rape victim sentenced to 200 lashes and six months in jail; Saudi woman punished for being alone with a man," *The Guardian*, Nov. 17, 2007, www.guardian.co.uk/world/2007/nov/17/saudiarabia.international.

100. Katherine Zoepf, "Saudi King Pardons Rape Victim Sentenced to Be Lashed, Saudi Paper Reports," *The New York Times*, Dec. 18, 2007, www.nytimes.com/2007/12/18/world/middleeast/18saudi.html.

101. Sonia Verma, "King Abdullah urged to spare Saudi 'witchcraft' woman's life," *The Times* (Of London), Feb. 16, 2008.

102. Mark Lattimer, "Freedom lost," *The Guardian*, Dec. 13, 2007, p. 6.

103. For background, see Brian Beary, "Future of Turkey," *CQ Global Researcher*, December, 2007, pp. 295-322.

104. Tracy Clark-Flory, "Does freedom to veil hurt women?" *Salon.com*, Feb. 11, 2008.

105. Sabrina Tavernise, "Under a Scarf, a Turkish Lawyer Fighting to Wear It," *The New York Times*, Feb. 9, 2008, www.nytimes.com/2008/02/09/world/europe/09benli.html?pagewanted=1&sq=women&st=nyt&scp=96.

106. Terry, *op. cit.*, p. 122.

107. "Many Americans Uneasy with Mix of Religion and Politics," The Pew Forum on Religion and Public Life, Aug. 24, 2006, http://pewforum.org/docs/index.php?DocID=153.

108. Brady, *op. cit.*

109. "Concluding Observations of the Committee on the Elimination of Discrimination against Women: Saudi Arabia," Committee on the Elimination of Discrimination against Women, 40th Session, Jan. 14-Feb. 1, 2008, p. 3, www2.ohchr.org/english/bodies/cedaw/docs/co/CEDAW.C.SAU.CO.2.pdf.

110. Kambiz Fattahi, "Women's bill 'unites' Iran and US," BBC, July 31, 2007, http://news.bbc.co.uk/2/hi/middle_east/6922749.stm.

111. H. Res. 101, Rep. Lynn Woolsey, http://thomas.loc.gov/cgi-bin/bdquery/z?d110:h.res.00101.

112. "General Assembly Adopts Landmark Text Calling for Moratorium on Death Penalty," States News Service, Dec. 18, 2007, www.un.org/News/Press/docs//2007/ga10678.doc.htm.

113. Mary H. Cooper, "Women and Human Rights," *CQ Researcher*, April 30, 1999, p. 356.

114. Christina Hoff Sommers, "The Case against Ratifying the United Nations Convention on the Elimination of All Forms of Discrimination against Women," testimony before the Senate Foreign Relations Committee, June 13, 2002, www.aei.org/publications/filter.all,pubID.15557/pub_detail.asp.

115. "CEDAW: Pro-United Nations, Not Pro-Woman" press release, U.S. Senate Republican Policy Committee, Sept. 16, 2002, http://rpc.senate.gov/_files/FOREIGNje091602.pdf.

116. "Turkmenistan adopts gender equality law," BBC Worldwide Monitoring, Dec. 19, 2007.

117. Faiza Saleh Ambah, "Saudi Women See a Brighter Road on Rights," *The Washington Post*, Jan. 31, 2008, p. A15, www.washingtonpost.com/wp-dyn/content/article/2008/01/30/AR2008013003805.html.

118. Damien McElroy, "Saudi Arabia to lift ban on women drivers," *The Telegraph*, Jan. 1, 2008.

119. Stephanie Levitz, "Lifting the veils of Afghan women," *The Hamilton Spectator* (Ontario, Canada), Feb. 28, 2008, p. A11.

120. "U.N. Secretary-General Ban Ki-moon Launches Campaign to End Violence against Women," U.N. press release, Feb. 25, 2008, http://endviolence.un.org/press.shtml.

121. "Gender Equality Architecture and U.N. Reforms," the Center for Women's Global Leadership and the Women's Environment and Development Organization, July 17, 2006, www.wedo.org/files/Gender%20Equality%20Architecture%20and%20UN%20Reform0606.pdf.

122. Bojana Stoparic, "New-Improved Women's Agency Vies for U.N. Priority," Women's eNews, March 6, 2008, www.womensenews.org/article.cfm?aid=3517.

123. Reel, *op. cit.*

124. Eliana Raszewski and Bill Faries, "Lugo, Ex Bishop, Wins Paraguay Presidential Election," Bloomberg, April 20, 2008.

125. Zahid Hussain, "Pakistan gets its first woman Speaker," *The Times* (of London), March 20, p. 52.

126. Bhaskar Roy, "Finally, women set to get 33% quota," *Times of India*, Jan. 29, 2008.

127. Massoumeh Torfeh, "Iranian women crucial in Majlis election," BBC, Jan. 30, 2008, http://news.bbc.co.uk/1/hi/world/middle_east/7215272.stm.

128. "Iran women win few seats in parliament," Agence-France Presse, March 18, 2008.

129. Swanee Hunt, "Let Women Rule," *Foreign Affairs*, May-June 2007, p. 109.

130. Kwamboka Oyaro, "A Call to Arm Women Candidates With More Than Speeches," Inter Press Service, Dec. 21, 2007, http://ipsnews.net/news.asp?idnews=40569.

131. Education Centre for Women in Democracy, www.ecwd.org.

132. *Ibid.*

133. Oyaro, *op. cit.*

134. *Ibid.*

135. Mary Kimani, "Taking on violence against women in Africa," *AfricaRenewal*, U.N. Dept. of Public Information, July 2007, p. 4, www.un.org/ecosocdev/geninfo/afrec/vol21no2/212-violence-aganist-women.html.

136. Correspondence with Karen Musalo, Clinical Professor of Law and Director of the Center for Gender & Refugee Studies, University of California Hastings School of Law, April 11, 2008.

137. "Mexico and Guatemala: Stop the Killings of Women," Amnesty International USA Issue Brief, January 2007, www.amnestyusa.org/document.php?lang=e&id=engusa20070130001.

138. Manuel Roig-Franzia, "Waning Hopes in Juarez," *The Washington Post*, May 14, 2007, p. A10.

139. Kimani, *op. cit.*

140. *Ibid.*

141. "Justice slow for female war victims," *The Toronto Star*, March 3, 2008, www.thestar.com/News/GlobalVoices/article/308784p.

142. Speech by Brigid Inder on the Launch of the "Gender Report Card on the International Criminal Court," Dec. 12, 2007, www.iccwomen.org/news/docs/Launch_GRC_2007.pdf

143. "Gender Report Card on the International Criminal Court," Women's Initiatives for Gender Justice,

November 2007, p. 32, www.iccwomen.org/publi-cations/resources/docs/GENDER_04-01-2008_FINAL_TO_PRINT.pdf.

144. Turquet, *et al.*, *op. cit.*, p. 8.

145. Oxfam Gender Equality Fact Sheet, www.oxfam.org.uk/resources/issues/gender/introduction.html.

146. Itai Madamombe, "Women push onto Africa's agenda," *AfricaRenewal*, U.N. Dept. of Public Information, July 2007, pp. 8-9.

BIBLIOGRAPHY

Books

Holland, Jack, *Misogyny: The World's Oldest Prejudice*, Constable & Robinson, 2006.
The late Irish journalist provides vivid details and anecdotes about women's oppression throughout history.

Stone, Merlin, *When God Was a Woman*, Harcourt Brace Jovanovich, 1976.
The book contends that before the rise of Judeo-Christian patriarchies women headed the first societies and religions.

Terry, Geraldine, *Women's Rights*, Pluto Press, 2007.
A feminist who has worked for Oxfam and other non-governmental organizations outlines major issues facing women today — from violence to globalization to AIDS.

***Women and the Environment*, UNEP, 2004.**
The United Nations Environment Programme shows the integral link between women in the developing world and the changing environment.

Articles

Brady, Brian, "A Question of Honour," *The Independent on Sunday*, Feb. 10, 2008, p. 8.
"Honor killings" and related violence against women are on the rise in the United Kingdom.

Kidder, Thalia, and Kate Raworth, " 'Good Jobs' and hidden costs: women workers documenting the price of precarious employment," *Gender and Development*, Vol. 12, No. 2, p. 12, July 2004.
Two trade and gender experts describe the precarious working conditions and job security experienced by food and garment workers.

Reports and Studies

"Beijing Betrayed," *Women's Environment and Development Organization*, March 2005, www.wedo.org/files/ gmr_pdfs/gmr2005.pdf.
A women's-rights organization reviews the progress and shortcomings of governments in implementing the commitments made during the Fifth World Congress on Women in Beijing in 1995.

"The Millennium Development Goals Report 2007," *United Nations*, 2007, www.un.org/millenniumgoals/pdf/mdg2007.pdf.
International organizations demonstrate the progress governments have made — or not — in reaching the Millennium Development Goals.

"Trafficking in Persons Report," *U.S. Department of State*, June 2007, www.state.gov/documents/organization/82902.pdf.
This seventh annual report discusses the growing problems of human trafficking around the world.

"The tsunami's impact on women," *Oxfam briefing note*, March 5, 2005, www.oxfam.org/en/files/bn050326_tsunami_women/download.
Looking at how the 2004 tsunami affected women in Indonesia, India and Sri Lanka, Oxfam International suggests how governments can better address women's issues during future natural disasters.

"Women in Politics," *Inter-Parliamentary Union*, 2005, www.ipu.org/PDF/publications/wmn45-05_en.pdf.
The report provides detailed databases of the history of female political representation in governments around the world.

Ballington, Julie, and Azza Karam, "Women in Parliament: Beyond the Numbers," *International Institute for Democracy and Electoral Assistance*, 2005, www.idea.int/publications/wip2/upload/WiP_inlay.pdf.
The handbook provides female politicians and candidates information and case studies on how women have overcome obstacles to elected office.

Chen, Martha, Joann Vanek, Francie Lund, James Heintz, Renana Jhabvala and Christine Bonner, "Women, Work and Poverty," *UNIFEM*, 2005, www.unifem.org/attachments/products/PoWW2005_eng.pdf.
The report argues that greater work protection and security is needed to promote women's rights and reduce global poverty.

Larserud, Stina, and Rita Taphorn, "Designing for Equality," *International Institute for Democracy and Electoral Assistance*, 2007, www.idea.int/publications/designing_for_equality/upload/Idea_Design_low.pdf.
The report describes the impact that gender quota systems have on women's representation in elected office.

Raworth, Kate, and Claire Harvey, "Trading Away Our Rights," *Oxfam International*, 2004, www.oxfam.org.uk/resources/policy/trade/downloads/trading_rights.pdf.
Through exhaustive statistics, case studies and interviews, the report paints a grim picture of how trade globalization is affecting women.

Turquet, Laura, Patrick Watt and Tom Sharman, "Hit or Miss?" *ActionAid*, March 7, 2008.
The report reviews how governments are doing in achieving the U.N.'s Millennium Development Goals.

For More Information

Equality Now, P.O. Box 20646, Columbus Circle Station, New York, NY 10023; www.equalitynow.org. An international organization working to protect women against violence and promote women's human rights.

Global Database of Quotas for Women; www.quotaproject.org. A joint project of the International Institute for Democracy and Electoral Assistance and Stockholm University providing country-by-country data on electoral quotas for women.

Human Rights Watch, 350 Fifth Ave., 34th floor, New York, NY 10118-3299; (212) 290-4700; www.hrw.org. Investigates and exposes human-rights abuses around the world.

Hunt Alternatives Fund, 625 Mount Auburn St., Cambridge, MA 02138; (617) 995-1900; www.huntalternatives.org. A private foundation that provides grants and technical assistance to promote positive social change; its Initiative for Inclusive Security promotes women in peacekeeping.

Inter-Parliamentary Union, 5, Chemin du Pommier, Case Postale 330, CH-1218 Le Grand-Saconnex, Geneva, Switzerland; +(4122) 919 41 50; www.ipu.org. An organization of parliaments of sovereign states that maintains an extensive database on women serving in parliaments.

Oxfam International, 1100 15th St., N.W., Suite 600, Washington, DC 20005; (202) 496-1170; www.oxfam.org. Confederation of 13 independent nongovernmental organizations working to fight poverty and related social injustice.

U.N. Development Fund for Women (UNIFEM), 304 East 45th St., 15th Floor, New York, NY 10017; (212) 906-6400; www.unifem.org. Provides financial aid and technical support for empowering women and promoting gender equality.

U.N. Division for the Advancement of Women (DAW), 2 UN Plaza, DC2-12th Floor, New York, NY 10017; www.un.org/womenwatch/daw. Formulates policy on gender equality, implements international agreements on women's issues and promotes gender mainstreaming in government activities.

Women's Environment & Development Organization (WEDO), 355 Lexington Ave., 3rd Floor, New York, NY 10017; (212) 973-0325; www.wedo.org. An international organization that works to promote women's equality in global policy.

9

Future of Marriage

Is Traditional Matrimony Going Out of Style?

David Masci

Businesswoman Dimitra Hengen of Alexandria, Va., has no desire to remarry after divorcing her husband of 18 years. She is among a growing number of Americans who no longer see marriage as necessary to their happiness.

From *CQ Researcher*,
May 7, 2004.

For Washington-area lawyers Melissa Jurgens and Jim Reed, their four years of marriage has meant greater happiness and stability than they have ever known.

"After I married Jim, I had someone I could talk to all the time and who could support me in ways that my friends, as wonderful as they are, just can't," Jurgens says. "He's more than a friend: He's committed to me, like I am to him, and that makes all the difference. We're going to be living together for the next 50 years, and so he needs to ensure that I'm OK and happy."

Marriage has been equally transforming for Reed. "I was 35 at the time and had already been a lawyer for 10 years," he says. "But I didn't really feel established until I married my wife. Marrying Melissa committed me to certain things — like my career path and staying here in Washington . . . [in part] because we want to have children."

But Dimitra Hengen, a successful businesswoman in Alexandria, Va., wants no part of the institution. "A lot of people do much better living on their own rather than in a marriage," says Hengen, who divorced her husband in 2001 after 18 years of marriage. "I'm not saying that we don't need companionship, but you need to assess this need against the things you have to give up when you marry, like your independence — and I don't want to give those things up."

Instead, she sees herself having long-term relationships that may be permanent, but will never lead to matrimony. "I have a wonderful boyfriend right now, and he wants to marry me, but I don't even want to live with him," she says. "I want to be able to come and go as I please, to travel, to see friends and family, all without the

Courtesy Dimitra Hengen

199

More Americans Remaining Unmarried

The percentage of American men and women who remain unmarried jumped dramatically in the past 30 years. In the 25-29 age group, the percentage of unmarried men rose from just 19 percent in 1970 to 54 percent in 2002; among women it nearly quadrupled, to 40 percent. Marriage experts say increased wealth and new freedoms are largely behind the trend.

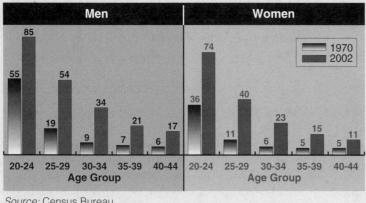

Percentage of Americans Who Never Married

Source: Census Bureau

Nonetheless, many marriage scholars argue that despite current trends, marriage will never really go out of style.

"We're a pair-bonding species, and we have a deep need at the species level to love and be loved by another and a need to pass on a part of ourselves to the next generation," says David Blankenhorn, founder and president of the Institute for American Values, a marriage advocacy group. "Marriage is the institution that encompasses these two great needs."

The case for marriage is further bolstered by research showing that married men and women are healthier, wealthier and happier than their single or divorced counterparts, Blankenhorn says. Children, too, are more likely to do better in school and less likely to have disciplinary trouble if they live in homes with married parents, he says.[2]

compromises that come with marriage. I don't need marriage."

Hengen is among a growing number of Americans who see marriage as more of an option than a necessity, according to Laura Kipnis, a professor of media studies at Northwestern University and author of *Against Love: A Polemic.* "As the economic necessity of it has become less pressing, people have discovered that they no longer need marriage," she says. "It restricts our choices and is too confining, which is why fewer people are marrying."

Indeed, in the last 50 years, the percentage of American households headed by married couples has fallen from nearly 80 percent to an all-time low of 50.7 percent, according to the Census Bureau. Meanwhile, the percentage of marriages that end in divorce jumped from roughly 25 percent to 45 percent.

As a result of the changing marriage and divorce statistics, married couples with children now comprise only 25 percent of all American households. The number is expected to fall to 20 percent by 2010. By that time, the bureau predicts, single adults will make up 30 percent of all households.[1]

Marriage advocates note there is already some evidence to support the institution's resiliency, pointing out that divorce rates, for instance, have leveled off and even declined slightly in recent years.

And compared with Northern Europeans, Americans are still marriage fanatics. In Denmark, for instance, 60 percent of all children are born out of wedlock, compared to 34 percent in the United States. [3] (*See sidebar, p. 206.*)

But many experts worry that the United States eventually will become more like Europe, with cohabitation and single parenthood replacing marriage as the dominant social institution. They point out that between 1996 and 2002, the number of cohabiting couples rose from 2.8 million to nearly 4.3 million, a trend that is expected to continue in the coming years.

Marriage advocates say that increasing rates of cohabitation can and should be stopped with education and other measures. They support a recent $1.5 billion Bush administration proposal to promote marriage among the poor, who are more likely to have children outside wedlock than middle-class Americans.

Fewer Couples Are Marrying . . .

The percentage of U.S. households headed by married couples has declined steadily from more than three-quarters of all homes in 1950 to barely more than half today.

Percentage of Married-Couple Households

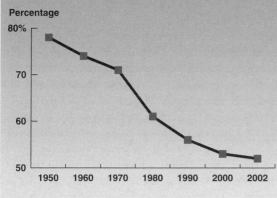

. . . And More Couples Are Cohabiting

More than 4 million unmarried American couples were living together in 2002, a 50 percent increase over the number just six years earlier.

Households Headed by Opposite-Sex Unmarried Couples

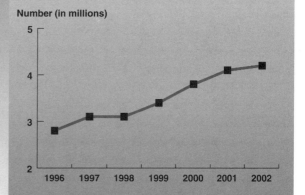

Source: U.S. Census Bureau, Current Population Survey, March 2002

But critics of the initiative, which is part of a planned reauthorization of the nation's welfare law, argue that marriage-promotion schemes are unlikely to work. And even if they did prove successful, they say, the money could better be spent on education or job training, which help women become more self-sufficient and less in need of a husband.

The Bush administration also has promoted marriage by rewriting the tax code to eliminate the so-called marriage penalty — which required some married couples to pay higher taxes together than if they had remained single. The penalty was eliminated when Congress passed the first Bush tax cuts in 2001, but the provisions expire at the end of the year. On April 28, the House passed legislation making the elimination of the penalty permanent. However, prospects in the Senate are uncertain.[4]

The drive to promote marriage, especially through taxes or benefits, has angered many singles. Unmarried workers, for instance, complain that pension and health benefits favor those with spouses and children. If the percentage of single employees in America were to surpass the percentage of married employees, singles could begin demanding equal treatment with regard to employee benefits and other government policies that currently favor married workers. Already, 40 percent of the nation's largest 500 companies have re-examined their "marriage-centric" benefit policies. For instance, Bank of America has redefined "family" to include non-traditional household members — such as domestic partners or adult children living at home.[5]

At the same time, many widows and widowers feel that they can't afford to remarry because they will lose health, pension and other benefits tied to their deceased spouses.

Ironically, concerns over the state of heterosexual marriage come as more gay couples are forming families by adopting children, and national debate flares over

Single mothers would be encouraged to marry under President Bush's Healthy Marriage Initiative, which would provide $1.5 billion to establish marriage-education programs, advertise pro-marriage messages and teach marriage skills.

whether same-sex partners should be allowed to wed. While most polls show that roughly two-thirds of Americans oppose same-sex unions, several state courts in recent years have expanded marriage rights for homosexuals — ranging from civil unions now allowed in Vermont to matrimony approved by the Massachusetts Supreme Court in November. In addition, mayors and other public officials in several cities — notably San Francisco and Portland, Ore. — have issued thousands of marriage licenses to gay couples.[6]

Social conservatives have responded by proposing an amendment to the U.S. Constitution defining marriage as the union of a man and a woman — effectively banning same-sex marriages. Supporters of the amendment, including President Bush, argue a constitutional amendment is needed to prevent liberal judges and officials from watering down the millennia-old definition of marriage until it loses all meaning and significance.

But the amendment's critics, including gay-rights groups and even some conservatives, note that the Constitution is usually changed to expand rights rather than take them away.

As the experts debate the future of one of the most important social institutions in human history, here are some of the questions they are asking:

Is the future bleak for traditional marriage?

Nearly half of all American marriages end in divorce, according to the U.S. Census Bureau. Meanwhile, the nation's marriage rate has been steadily dropping, from an annual rate of 9.9 marriages per thousand population in 1987 to 8.4 per thousand in 2001.[7] Today, just over 50 percent of all households are headed by married couples, compared with three-quarters of the same group 40 years ago.

At the same time, the number of households with unmarried couples has risen dramatically. In 1977, only 1 million Americans were cohabiting; today, it's 5 million. The percentage of children raised in single-parent households also has jumped, tripling in the last 40 years — from 9.4 percent in 1960 to 28.5 percent in 2002.

Some scholars say that while traditional marriage will not disappear, it will never again be the country's preeminent social arrangement. "Ozzie and Harriet are moving out of town, and they're not coming back," says Stephanie Coontz, a professor of history at The Evergreen State College in Olympia, Wash., and author of the upcoming *A History of Marriage.* "Americans now have too many choices — due to new technologies and economic and social opportunities — and it would take a level of repression unacceptable to nearly everyone to force us to begin marrying and stay married at the same levels we once did."

Coontz argues that marriage thrived during more socially and economically restricted times. "For thousands of years, marriage has been humanity's most important economic and social institution," she says. "It gave women economic security and helped men financially, through dowry payments and socially by connecting them to another family."

However, Coontz continues, the recent expansion of individual wealth and freedom — especially among women — makes the economic argument for marriage much less compelling. "We no longer need a spouse for economic security or to [financially] take care of us when we get old," she says. "We can do these things for ourselves now."

Northwestern's Kipnis agrees. "Look, marriage has essentially been an economic institution with some romantic aspects tacked onto it," she says. "Once you take away the economic need, marriage becomes different, and for many people it becomes confining."

Indeed, only 38 percent of Americans in first marriages say that they are happy, Kipnis says, citing a 1999 survey by the National Marriage Project at Rutgers University.[8] "What does that say about everyone else?" she asks.

Some of the unhappiness may be stoked by a consumer culture that emphasizes choice and happiness, raising unrealistic expectations, she says. "Our emotional and physical needs have expanded a lot, and people now expect that the person they are with is going to meet those needs," Kipnis says. "That makes it much harder to find someone else they feel they can marry."

It also increases the likelihood of divorce, says Diane Sollee, founder and director of the Coalition for Marriage, Family and Couples Education, in Washington. "There are a lot of unnecessary divorces because when married people feel unhappy, they assume that they married the wrong person. So they find someone new, and when that person doesn't make them happy, they move on to the next one."

But supporters of marriage believe that the institution is beginning to make a comeback, in part because of the very changes that the pessimists cite. "All of this mobility and freedom also means that we're living in an increasingly impersonal mass society," says William Doherty, director of the Marriage and Family Therapy Program at the University of Minnesota in St. Paul. "Marriage will continue to be important. We will continue to need someone who is permanently and unquestionably in our corner."

Others argue that marriage will survive and thrive because it is still the best way to organize society on a personal level. "People are beginning to see how much we need marriage, because it is the only effective way to raise children," says Tom Minnery, vice president of public policy at Focus on the Family, a Christian-oriented advocacy group in Colorado Springs. "And this isn't something that just religious people are saying; it's accepted by the scientific community too." (*For debate over research on the impact of marriage, see "Current Situation," p. 213.*)

There are already signs of a reverse in the trend away from marriage, Minnery says. He points out, for instance,

that the annual divorce rate has dropped, from five per 1,000 people in 1982 to four in 2002.[9]

In addition, the upward trend toward working motherhood has halted, Minnery says, noting that the percentage of working mothers (72 percent) has held steady since 1997, after rising dramatically during the previous 20 years.[10]

Optimists also contend that young people today take marriage more seriously than people did 20 or 30 years ago. "I actually think the institution of marriage is going to become more stable in the future," Doherty says. "People are becoming more sober and serious about marriage and doing more to prepare themselves for it, like taking marriage-education classes before they wed."

He attributes the trend, in part, to painful memories. "The children of divorce are keen to not make the same mistakes as their parents," he says. "Even those whose parents stayed together feel this way, because they were able to witness the divorce revolution in other families."

"When you look at young people, they're more conservative and religious than their parents," Minnery agrees. "Just like the Baby Boomers rebelled against what their parents did, the children of Baby Boomers are rebelling against their parents and their lifestyle choices."

Finally, marriage optimists say the institution will endure because it fulfills basic human needs. "There's still a great hunger for stable, loving, intimate relationships," Doherty says, "and marriage is still the best way to have them."

Would President Bush's initiative to promote marriage improve poor people's lives?

In 2001, President Bush proposed a new initiative to promote marriage for lower-income Americans as part of the reauthorization of the federal government's welfare program. Recently, the president expanded his proposal.

Bush's new Healthy Marriage Initiative would provide $1.5 billion over five years in grants to states, local governments and private charities for a variety of activities, including establishing marriage-education programs in schools and community centers and advertising pro-marriage messages. Funds could also be spent to teach marriage skills to people preparing to tie the knot or to mentor troubled married couples.[11]

The proposal was included in the massive Welfare Reform Reauthorization bill passed by the House last

Students' Views of Unmarried Parenthood

The idea of having a child outside of marriage has become more acceptable to high school seniors over the past 20 years — especially among girls.

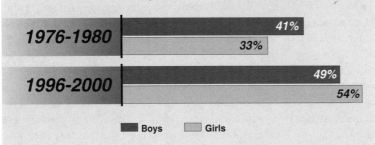

Percentage of high school seniors who said having a child without being married is experimenting with a "worthwhile lifestyle or not affecting anyone else"

	Boys	Girls
1976-1980	41%	33%
1996-2000	49%	54%

Source: "Monitoring the Future," Survey Research Center, University of Michigan

year. But the measure stalled in the Senate in March after legislators could not agree on several issues unrelated to the marriage proposal.[12] No new date has been set for the Senate to revisit the bill.

The University of Minnesota's Doherty says Bush's initiative is "well worth the effort" because research shows that marriage can dramatically improve the physical and emotional health of adults and especially children. "We know that the best way to raise children is in a healthy marriage, and yet our welfare policy hasn't reflected that," he says.

Blankenhorn agrees. "We know that in most cases children do better in a home where their biological parents are married," he says. "So we have a real interest in seeing that people with children marry and making those marriages work."

But opponents of the initiative contend it will neither be effective nor do much social good.

"You can't really push people into a decision this big, absent fraud or coercion," Evergreen State's Coontz says. "This will change a few minds, but not many."

Moreover, says Barbara Risman, a sociology professor at North Carolina State University in Raleigh, changing minds in one direction should not be the focus of federal efforts. Instead, she says, the government should better equip people to make their own decisions.

"The focus on marriage, as opposed to self-sufficiency, is a negative issue for women," she says. "The government shouldn't promote marriage. It should promote the ability of anyone to live with the kind of families they want to have."

There are better ways to spend welfare funds, critics like Risman and Coontz add. "No matter what [supporters] say, this is going to divert funds, because at the end of the day you only have so much welfare money to spend," Coontz says. "And we have such pressing needs in child care, health and in so many other areas."

In fact, opponents say, marriage promotion could do more harm than good, in part because the pool of good husbands is likely to be much smaller among the group the administration would target — poor women. "Poor people have a lot of barriers in their lives, like economic instability, substance abuse and a history of being on the giving or receiving end of domestic violence," says Lisalyn Jacobs, vice president for government relations at the National Organization for Women's (NOW) Legal Defense and Education Fund, a women's-rights group. "It may not be such a great idea to get or keep some of these people together, because they're not very stable."

Marriage also might not be such a good idea if the potential marriage partner is abusive, she adds. "Sixty percent of all women on welfare have been victims of domestic violence at some time in their lives," Jacobs says, "so encouraging people to stay together might put many of them at greater risk of injury."

But Sollee disputes that argument. "Women generally get beat up when they're single or in a cohabiting relationship," she says, pointing to the National Crime Victimization Survey, which shows that two-thirds of the violence against women is not committed by husbands but by casual dating partners.[13] "Marriage stabilizes relationships and makes domestic violence much less likely."

Indeed, the idea that those on welfare are different with regard to marriage is "insulting to the poor," she says. "Poor people pair up for the same reason everyone else does: Because they're human, and they want to form love relationships. Given that, we need to give them the skills needed to make the right choices."

Should the Constitution be amended to define marriage as the union of a man and a woman?

After months of pressure from religious and conservative groups alarmed at what they saw as a rising tide of same-sex marriages, President Bush on Feb. 24 publicly endorsed amending the U.S. Constitution to define marriage as the union of a man and a woman. Bush said the decision had been forced upon him by "activist judges and local officials" in Massachusetts, California and elsewhere, who had made "an aggressive attempt" to redefine marriage.

"On a matter of such importance, the voice of the people must be heard," the president said at a White House press conference announcing his support for the amendment. "Activist courts have left the people with one recourse. If we are to prevent the meaning of marriage from being changed forever, our nation must enact a constitutional amendment to protect marriage in America." (*See "At Issue," p. 214.*)

The federal government had addressed the issue in 1996, when Congress passed, and President Bill Clinton signed, the Defense of Marriage Act, which defined marriage as being between a man and a woman for purposes of federal law. It explicitly prevents any jurisdiction from being forced to accept another's definition of marriage. In other words, if Massachusetts legalizes same-sex unions and marries two men, New York is not required to acknowledge the marriage if the couple subsequently moves there.

But because the Constitution's Full Faith and Credit Clause requires each state to recognize the lawful actions of other states, Bush and gay marriage opponents say federal courts might overrule the law and require other states to recognize gay marriages performed outside their jurisdiction.

The amendment, proposed by Colorado Republicans Sen. Wayne Allard and Rep. Marilyn M. Musgrave, would still allow states to pass civil-union or domestic-partnership laws that could grant same-sex partners and others the same rights as married couples. But marriage would be strictly limited to heterosexual couples.

Gay-rights activists, civil-liberties organizations and even some conservatives and libertarians oppose the amendment, albeit for different reasons.

"It's a perversion to use our founding document to discriminate against a group of people when it has traditionally been used to expand liberties and rights," says Kevin M. Cathcart, executive director of the Lambda Legal Defense and Education Fund, a gay-rights group in New York. "It's ironic that on the 50th anniversary of the *Brown v. Board of Education* decision, which struck down the doctrine of 'separate but equal,' we're on the verge of writing it back into the Constitution."[14]

Other opponents of gay marriage say a Constitutional amendment is heavy-handed and unnecessary. Former Rep. Bob Barr, who as a conservative Republican from Georgia helped write the 1996 Defense of Marriage Act, calls the amendment an unwarranted intrusion into an area traditionally left to the states.

"Changing the Constitution is just unnecessary — even after the Massachusetts decision, the San Francisco circus and the Oregon licenses," Barr told the House Judiciary Subcommittee on the Constitution on March 30. "We have a perfectly good law on the books that defends marriage on the federal level and protects states from having to dilute their definitions of marriage by recognizing other states' same-sex marriage licenses."[15]

Opponents also call the administration's claim that the amendment process was "forced" on them by activist judges and mayors a cynical election-year ploy. "I don't think this is really about gay marriage at all, but is a distraction meant to focus attention away from the [Iraq] war and the deficit and all of the other problems this administration is dealing with," Cathcart says.

Moreover, by trying to amend the Constitution, they say, conservatives are trying to cut off the emerging national debate on same-sex marriage. "You know, we're really just beginning this debate all over the country, and already they want to amend the Constitution," Cathcart says. "They accuse liberals of trying to use the courts and local officials to circumvent debate, but that's actually what they're doing."

"Amending the Constitution is something that you traditionally do when you've run out of remedies," agrees NOW's Jacobs. "It seems to me that we've only just begun to try to work this one out."

But supporters point out that to become law any constitutional amendment must first win the support of two-thirds of Congress and three-quarters of the nation's state legislatures. "This would be a wonderful way to

Will U.S. Follow Europe's Cohabitation Trend?

In their 21 years together, Stig Skovlind and Malene Breining Nielsen of Denmark have dated, lived together and raised three children — but they never got married. "We trust each other. We don't need a document," Malene said.[1]

More and more Europeans — particularly from the Nordic countries — are cohabiting. Nineteen percent fewer Europeans got married in 2002 than in 1980, compared to a 5.7 percent drop in the United States during the same period. Meanwhile, almost 20 percent of young Europeans — and 40 percent of Swedes — are cohabiting, compared to 7.7 percent in the United States.[2]

Northern European demographics could be headed toward a point at which "marriage and cohabitation have become indistinguishable," says Kathleen Kiernan, a professor of social policy and demography at the London School of Economics. And many cohabiting Nordic couples are having children. More than half of Swedish mothers ages 25 to 29 give birth to their first child out of wedlock, and more than a quarter of Norwegian mothers.

About 80 percent of those surveyed in Sweden, Finland and Denmark consider cohabiting couples with children a "family." But in mostly Catholic Southern European countries like Italy, Spain and Portugal, attitudes about cohabitation are more conservative; only 44 percent of Italians, for instance, view unwed couples with children as a family.

Still, many European courts now accommodate the emerging class of cohabiting partners and parents. In Sweden, Finland and Denmark, "family law has come to be applied to married and cohabiting couples in the same way," writes Kiernan. And in 1998 the Netherlands began recognizing both homosexual and heterosexual partnerships as if they were "functionally equivalent to marriage," she notes.

In France, so-called PACS (pacte civil de solidarite) offer unwed heterosexual and homosexual couples some of the same rights accorded to marriage; more than 130,000 couples have signed PACS. Even in Italy, the government is considering granting some legal rights to unmarried couples.

Some experts worry the United States may be headed in the same direction as Europe. "Our marriage rate continues to drop, our divorce rate is high and our cohabitation rate continues to climb," says David Popenoe, professor of sociology and co-director of the National Marriage Project at Rutgers University.

And recent U.S. demographic surveys support his concerns: The number of American couples cohabiting rose by 72 percent in the 1990s, with nearly half of them raising children.[3]

debate the issue, given all of the hurdles that have to be jumped before it became part of the Constitution," says Minnery of Focus on the Family. "There would be a debate in Congress and then in every state legislature in the country. That seems pretty thorough to me."

And while proponents admit the amendment process has traditionally been about expanding rights, they argue that gay marriage presents a unique challenge to American society that calls for a unique solution.

"Our founding documents, like the Declaration of Independence, tell us that our rights have come from our creator or, to put it another way, they are part of natural law," says Ed Vitagliano, pastor of Harvester Church in Pontotoc, Miss., and a spokesman for the conservative American Family Association. "When you talk about redefining marriage, you're really talking about an overthrow of this natural order or natural law, because

marriage is something that predates government. So this is a big deal, a once-in-a-lifetime debate about whether to overturn the natural order upon which our rights are based. That requires a big response."

BACKGROUND

Origins of Marriage

Marriage has meant very different things in different places at different times. "Marriage has been continually evolving through the centuries, and it's still doing so," the University of Minnesota's Doherty says.

For instance, the idea of choosing a mate or "marrying for love" became commonplace only in the 18th and 19th centuries and only in some cultures. Marriages are still arranged by families throughout much of Asia, Africa and

The rising numbers worry experts because cohabiting couples tend to break up more than married couples, Popenoe says, and there is no safety net in the United States for kids who slip through the financial cracks when parents separate.

"We can't agree that . . . welfare provisions are proper, and we don't [want] to give up our hard-earned taxes in times of need," says Popenoe.

But in Northern Europe, expansive welfare measures provide a safety net for children when relationships break down, Kiernan says. Unwed European mothers — cohabiting or not — have the same rights as married mothers and, although the law is less clear-cut for men, unwed fathers generally have a financial duty to their children once paternity is established.

Critics of cohabitation note that children of single parents have a higher incidence of psychiatric problems, Popenoe says. "Kids are much better off when raised by two married biological parents than . . . by a single parent or a broken cohabiting couple and are then thrust on the welfare state," he says.

Protesters in Paris oppose the Civil Solidarity Pacts (PACS) being considered by the French National Assembly, which would give traditional rights to homosexual and unwed couples.

But Stephanie Coontz, national co-chair of the Council on Contemporary Families, notes that while "transitions can be hard on kids," the effects of a parental breakup on kids can be exaggerated.

Americans will have to get used to a broader definition for family, she concludes. "There's been a worldwide transformation of marriage — it will never again have a monopoly on organized child care or on the caring for dependents," she says.

— Benton Ives-Halperin

[1] Jennie James, *et al.*, "All In The Family . . . or Not," *Time*, Sept. 17, 2001, p. 54.

[2] Marriage statistics are from "Demography: EU Population Up by 0.3% in 2002," *European Report*, Sept. 3, 2003, and National Center for Health Statistics; cohabiting statistics — which are for those ages 25 to 34 — come from the U.S. Census Bureau and Kathleen Kiernan, "Unmarried Cohabitation and Parenthood: Here to Stay?" Conference on Public Policy and the Future of the Family, Oct. 25, 2002. Unless otherwise noted, other data are from Kiernan.

[3] Laurent Belsie, "More Couples Living Together, Roiling Debate on Family," *The Christian Science Monitor*, March 13, 2003, p. 1.

the Middle East today. Moreover polygamy — long rejected by Western cultures — is common in many Muslim countries and among various ethnic groups.

Still, there have been some constants, especially in the West. For instance, until recently, most people saw marriage as a necessary right of passage into adulthood, rather than a choice. Moreover, definitions of marriage were — and to some degree still are — largely dictated by the Judeo-Christian ethic, which sees the institution as a permanent, unbreakable union between a man and a woman.

While polygamy was allowed in early Jewish life, ancient Hebrew laws on marriage eventually came to stress monogamy. Laws strictly forbade adultery (the prohibition is one of the Ten Commandments) and incest. Restrictions against divorce also were enforced, making it almost impossible for Jewish couples to legally separate.

Christian thinkers built on this tradition. St. Mark, in his New Testament gospel, echoed the Old Testament when he said "from the beginning of the creation God made them male and female. For this cause shall a man leave his father and mother, and cleave to his wife; and the twain shall be one flesh: so then they are no more twain, but one flesh. What therefore God hath joined together, let no man put asunder."[16]

But the Christian emphasis on monogamy and fidelity was more than a reaffirmation of ancient Jewish traditions or the teachings of the new church's founders; it also was a reaction to what Christians viewed as the weak marriage laws of Rome, which allowed couples to separate and gave women an unusual amount of personal freedom.

While many of today's matrimonial traditions — such as the wearing of bridal veils and the exchange of wedding rings — date back to ancient Rome, the Christian church

CHRONOLOGY

17th-19th Centuries *Less-restrictive ideas and laws concerning marriage develop among colonies and later states in the New World.*

1620 Puritans arrive in America and establish more liberal marriage and divorce laws than those in England.

1770s The struggle for U.S. independence spurs debate on the rights of women and the obligations of marriage.

1800s Western territories pass liberal divorce laws in the transition to statehood to attract settlers.

1867 All but three states have abolished the most restrictive divorce laws.

1870 Only 3 percent of all U.S. marriages end in divorce.

1900-1960 *War and social changes bring women more freedom.*

1900 U.S. divorce rate stands at 8 percent.

August 1920 American women get the right to vote after Tennessee becomes the 36th state to ratify the 19th Amendment.

1925 The divorce rate is 25 percent.

1941 U.S. entry into World War II brings millions of American women into the work force.

1947 California Supreme Court rules that the state's miscegenation law violates the state Constitution, making California the first state to abolish limits on interracial marriage.

1960-Present *The civil and women's rights movements and the sexual revolution dramatically change the institution of marriage.*

1960 Nation's divorce rate is 26 percent.

1964 Civil Rights Act prohibits discrimination based on gender.

1967 U.S. Supreme Court overturns state miscegenation laws.

1969 Gov. Ronald Reagan, R-Calif., signs the first no-fault divorce law.

1972 The launch of *Ms.* magazine heralds the arrival of the woman's movement. Jessie Barnard's book *The Future of Marriage* argues that marriage is often detrimental to women.

1974 The number of American children whose parents divorce in a year reaches 1 million.

1980 Divorce rate hits 50 percent.

1989 Psychologist Judith Wallerstein argues in her book *Second Chances* that the impact of divorce on kids is worse than previously thought.

1992 Vice President Dan Quayle criticizes decision by TV sitcom character "Murphy Brown" not to wed her child's father.

1996 Hawaiian Supreme Court rules the state cannot ban gay marriages.

1998 Hawaiians amend Constitution to permit ban on gay marriage.

1999 Vermont Supreme Court rules that gay couples are entitled to the same benefits as married people.

2000 *The Case for Marriage* attempts to counter earlier arguments that many people don't need marriage.

Nov. 18, 2003 Massachusetts Supreme Court rules the state's law prohibiting gay marriage violates the state Constitution.

April 29, 2004 Massachusetts legislature adopts constitutional amendment banning gay marriage but allowing civil unions.

May 17, 2004 Massachusetts must begin issuing marriage licenses to same-sex couples, according to a state Supreme Court order; governor is seeking an emergency stay of the deadline until action is completed on new constitutional amendment.

2006 Massachusetts gay-marriage amendment would take effect, if approved again by the legislature and by a statewide voter referendum.

rejected Rome's lax marriage laws and instead transformed marriage into a divinely ordained sacrament. Separation or divorce were strictly forbidden, although widows could remarry after a spouse's death. Jesus himself condemned divorce, calling men who leave their wives for others — even if legally sanctioned — adulterers.[17]

The only option for irreconcilable couples was to petition the church for an annulment, which did not dissolve the marriage but declared that it had been invalid from the start and hence had never actually taken place. Annulments were usually employed in cases of bigamy or when a husband and wife were closely related. Otherwise, annulments were difficult to obtain. Petitions, even from kings — like England's Henry VIII — were routinely denied.

Besides fidelity, the church emphasized the dominant role of the husband, continuing the tradition of the Jews and of most other ancient cultures at the time. Ironically, Christian teachings held that men and women were equal in God's eyes but, nonetheless, women were to be "in submission" to their husbands.[18]

Sweeping Changes

Sweeping changes in the state of marriage did not begin to occur until the 16th-century Protestant Reformation, which rejected much of the institutionalization of religion and stressed individual choice. Many Protestants, including the movement's founder, Martin Luther, cast off the notion that marriage was a holy sacrament to be regulated entirely by the church. Instead, Luther wrote, marriage was "a secular and outward thing having to do with wife and children, house and home and with other matters that belong to the realm of the government, all of which have been completely subjected to reason." According to Luther, the laws of marriage and divorce "should be left to the lawyers and made by secular government."[19]

Luther's new attitudes set the stage for ensuing changes. "The Reformation set out a bunch of new ingredients on the table . . . but it took the Enlightenment and the spread of wage labor to bring these ingredients together," says Coontz, who is writing a book on the history of marriage.

The 17th- and 18th-century Enlightenment "brought to dominance the notion that people have the right to organize their lives as they see fit," says Coontz, leading to the belief that "marriage should be a love match."

Indeed, during subsequent centuries, more and more people, especially among the educated classes, eschewed family obligation and chose their own mates.

If the Enlightenment gave people the intellectual justification for choosing their spouses, wage labor — brought on by the Industrial Revolution — gave them the means. Having a job with a steady wage disconnected people from rural life and its familial and other controls, giving them both geographic and social mobility.

"If you didn't want to marry the person your parents had chosen for you, you could leave and find someone else," Coontz says.

New World Flexibility

The first European settlers in the New World took with them many of their old laws and customs, including rules on matrimony. But as the late historian Daniel Boorstin has pointed out, views on marriage in England's American colonies and eventually the United States were always more flexible than those in Europe, in large part to fit the needs of a less socially rigid and more mobile society.

"The rights of married women and their powers to carry on business and to secure divorce were much enlarged," he wrote about matrimony laws in early North America. "The law protected women in ways unprecedented in the English common law."[20]

In colonial Massachusetts and Connecticut, Puritan-influenced law even allowed for divorce if a spouse could prove that the other had neglected a fundamental duty of the marriage, such as providing food and shelter. However, by modern standards, divorce was difficult to obtain in colonial America, so it was rare. In many jurisdictions, divorce could only be granted by legislative action (the passage of a private bill), which meant that usually only the rich had the resources to legally separate.

Between the American Revolution and the Civil War, most states greatly liberalized their divorce laws as part a trend to expand individual freedoms, such as voting and other rights and eventually to abolish slavery. For instance, by 1867, 34 of the 37 states had abolished legislative divorce, giving the courts authority to grant divorces. Still, the divorce rate remained relatively low — about 3 percent in 1870.

During the great westward migration just before and after the Civil War, many of the new states carved out of the Western territories, such as Nevada, passed liberal divorce laws in an effort to attract more settlers.

Can You Click to Find Your Soul Mate?

"**A**ren't you just a little curious?" purrs an ad on Match.com's Web site. "With 8 million profiles to choose from, imagine the possibilities."

Match.com and other large Internet dating services do offer singles many choices, but most online services are not concerned with whether a potential customer is seeking a friend, a date or a spouse.

But a relatively new online service, eHarmony.com, actively plays cupid, limiting its clientele to those looking to get married. Its Web site is full of pictures of happily married or engaged couples who met through the service. Potential subscribers are urged to join "when you're ready to find the love of your life."

Americans have been seeking love online for more than a decade, but in recent years Internet dating has become much more widespread and socially acceptable.

"The traditional institutionalized means for getting people together are not working as well as they did previously," says Norville Glenn, a sociology professor at the University of Texas. "There's a need for something new, and the Internet is filling it."

Last year an estimated 21 million Americans spent $313 million on Internet dating, a figure likely to more than double by 2008, according to the Internet market research firm Jupiter Communications.[1]

Most of the biggest services, like Match.com and Yahoo Personals, are largely search engines with millions of profiles, each usually containing a photo and personal information ranging from height and weight and likes and dislikes to "latest book read." Users scroll through the results, e-mailing anyone who catches their fancy.

But eHarmony works differently. Founded in August 2000 by clinical psychologist Neil Clark Warren, it does not allow users to choose whom they're going to contact. Instead, it matches people based on an exhaustive personality survey completed by each subscriber.

"I saw more than 7,000 patients over the years, and so many of these troubled people were in bad marriages," he says of his 35 years as a therapist. "It struck me that the most important need we have is to get marriage right."

Warren conducted more than 500 "divorce autopsies," usually interviewing both spouses and sometimes even the children of the divorced parents. He found that most people in failed marriages chose their spouse for the wrong reasons, such as physical appearance, sense of humor or financial status, while neglecting more important concerns. "The true things people need for a happy marriage are on the inside, like character and intellect, rather than the shape of their nose," he says.

Moreover, he found, those who succeed at marriage are usually paired with someone who shares most of their basic values and beliefs. "We're told that opposites attract, but that's not so," he says. "When people have a lot in common, they have much less to negotiate, fewer things to compromise on."

For instance, different work ethics or attitudes about how to raise children might not be a problem on a first or second date, but they can breed resentment when a couple is living together.

eHarmony tries to match clients by requiring all new subscribers to fill out an extensive 436-item questionnaire based on what Warren says are "29 dimensions for compatibility," such as spirituality, education, sexual desire and kindness.

After computers match candidates with similar traits, early communication is limited to e-mail, but eventually, matched singles can move on to calling and then meeting for dates.

Around the same time, some Mormon settlers in the Western territory began practicing polygamy. The practice, which only involved men marrying multiple wives, began after the religion's leader, Brigham Young, declared it acceptable in 1852.

While most Mormon marriages involved only two people, a substantial minority of Mormon men had two or three wives. Still, the practice was relatively short-lived due to pressure from the rest of the country. As a condition for Utah joining the Union, the Mormon Church banned new polygamous marriages in 1890.*

Meanwhile, many states, new and old, were also passing so-called miscegenation laws, prohibiting marriage between members of different races. Support for miscegenation goes back to the earliest English settlements and, while largely prompted by a desire to prevent whites

---

* Polygamy is still practiced illegally in parts of Utah.

Although eHarmony is less than four years old, it has attracted 4 million users and is adding roughly 10,000 new customers per day, Warren says. It has also spawned imitators, such as TrueBeginnings. Even Match.com, the nation's largest online service, is offering a short personality test to its registered users.

Warren claims his company has already connected at least 2,500 couples who have gotten married, and who, for the most part, are "doing very well so far."

Paul Consbruck, 42, of Jacksonville Fla., says he is now happily married to someone he met through eHarmony. "A lot of dating takes place on a very superficial level," he says. "eHarmony tells you to step back and look at what's really important to you before you get attracted to someone physically."[2]

About 15 percent of all eHarmony applicants are turned away for a variety of reasons including emotional difficulties, substance abuse or concerns about truthfulness.

"A lot of people are not ready for marriage, and so we encourage them to get better and then reapply," Warren says. "It's painful, but the alternative is to match them when

Neil Clark Warren, founder of eHarmony.com, says most marriages fail because people choose mates for the wrong reasons.

they're not ready and bring other, healthier people down with them. That's not fair."

Because eHarmony takes such an active interest in who and even how its subscribers meet, it has been likened to an old-fashioned matchmaker, a comparison Warren does not reject. In today's increasingly urban and mobile society, he says, a service like eHarmony can provide "the kind of wisdom that people might have found with their family or community in the past," when most people lived in small towns.

Indeed, Warren believes that it is "extremely hard" for a single person to find a suitable partner without a service like eHarmony. "You need to tap into a large pool of people so that you can increase your odds of finding your soul mate," he says. "That's why I think that in 10 or 15 years, virtually every person will find their husband or wife on the Internet."

---

[1] Figures cited in Adrienne Mand, "Dr. Love is In," ABCNEWS.com, March 26, 2004.

[2] Quoted in Anna Kuchment, "The Internet: Battle of the Sexes," *Newsweek International*, Dec. 2, 2003.

---

and African-Americans from marrying, was not confined to the South or to black-white couplings. Indeed, the first miscegenation law in North America was passed by the Maryland Assembly in 1664. Later, Western states prohibited Asians and whites from marrying.

Miscegenation laws were still common until the second half of the 20th century. In 1948, the California Supreme Court became the first state court to strike down a law banning racial mixing, arguing that it violated the Constitution's 14th Amendment guaranteeing equal protection under the law.[21] Nearly 20 years later, in *Loving v. Virginia*, the Supreme Court followed suit, repealing miscegenation laws nationwide.

Shifting Views

The last 50 years have witnessed dramatic, indeed unprecedented, changes in perceptions of marriage in the United States. During the 1960s and '70s, the general purposes and

Library of Congress

Female shipyard workers in California were among the millions of women brought into the work force by World War II, helping to plant the seeds for social changes in the 1960s that experts say have contributed to the decline in marriage nationwide.

characteristics of marriage changed, reflecting new attitudes about freedom and self-realization, especially for women.

"In the middle of the 20th century, we began to shift away from an institutional view of marriage — that it is based on economic viability, child rearing and a sense that this is something that all adults should do," Doherty says. "And we moved into what is called psychological marriage — the notion that you marry by choice and that you do so primarily for romantic reasons."

This change was prompted by several important social and cultural changes, beginning with the entry of married women into the workplace during World War II. In 1930, only 12 percent of all married women worked outside the home.[22] By 1944, with the labor demands of rising war production and 12 million young men taken out of the work force for military service, that figure had nearly tripled, to 35 percent.[23]

Work brought many women, regardless of their marital status, more than financial independence. "In the absence of men, women found doors suddenly open to them in higher education and in the professions; the Army and Navy admitted women for the first time," writes Nancy F. Cott, a professor of history at Yale University and author of *Public Vows: A History of Marriage and the Nation.*[24]

After the war, as men returned home to resume their lives and jobs, many women — both married and unmarried — left work to focus on starting and raising families. While many women wanted to leave their war jobs, some were pressured or even forced to do so in order to make room for returning soldiers.

But during the 1960s, social changes erupted that significantly altered the marital landscape. The civil rights and anti-war movements, the rise of youth culture and the sexual revolution wrought profound changes in society's mores. Old ideas, such as female premarital chastity or the notion that a woman went to college only to find a suitable husband, began to collapse.

By the early 1970s, the women's-liberation movement was demanding equality in the workplace and elsewhere and advising American women that they had other choices in life besides getting married and having children. Indeed, many feminists declared housework demeaning and encouraged women to "find themselves."

By the 1980s, a majority of married women with children at home were working, driven not only by new freedoms and a desire for a career but also by financial need, as the cost of housing, education and health care rose faster than wages. Today, about 72 percent of married women with children are in the labor force, a figure that has held steady since the mid-1990s.

A dramatic rise in the rate of divorce accompanied these societal changes. From 1955 to 1975, the divorce rate more than doubled, from 23 percent of all marriages to 48 percent, where it has remained, give or take a few points, ever since.[25]

Many say that the advent of no-fault divorce laws across the country in the late 1960s and early '70s helped spur the divorce trend. No-fault laws made it significantly easier to untie the knot. "These new laws did a terrible disservice to couples, but they especially hurt the weakest people in our society — children," says Richard Land, a spokesman for the Southern Baptist Convention. "No-fault divorce sent people the message that it was easy and alright to breakup their families."

But others disagree. "When you look at almost all the reputable research, you find that no-fault divorce was not the cause of the divorce rate rising," says Coontz of Evergreen State. "It was the rising demand for divorce that caused the rising divorce rate. No-fault only speeded up what was going to happen anyway."

CURRENT SITUATION

Gay-Marriage Push

Although the debate over same-sex marriage has only recently risen to national attention, the issue is by no means new. Gay-rights advocates have been working for decades to secure matrimonial and related rights for same-sex couples. And they have had some successes, as well as setbacks.

In 1996, Hawaii's Supreme Court ruled that the state government's ban on same-sex marriages violated the state Constitution. Then in 1998, Hawaiian voters amended their state constitution giving the legislature authority to ban same-sex marriage. In 1999, Vermont's highest court ruled that gay couples were entitled to the same benefits as married people, although it stopped short of giving same-sex couples the right to marry.

But developments in the last six months have clearly pushed gay marriage to the top of the national domestic agenda, making it one of the most hotly debated topics of the year and an important issue in the 2004 presidential election.

New interest in same-sex unions was sparked by a Nov. 18 Massachusetts Supreme Court decision holding that denying gay couples the right to marry violated the state constitution's equal protection clause. The court gave the state until May 17 to start allowing same-sex couples to wed.

For gay advocates and some civil-rights proponents, the decision was long overdue. "A court finally had the courage to say that this really is an issue about human equality and human dignity, and it's time that the government treats these people fairly," said Mary Bonauto, the lawyer for Gay & Lesbian Advocates & Defenders who argued the case.

But the decision also prompted a substantial backlash and not just in Massachusetts, where both Republicans

Married life improves the physical and psychological health of both adults and children, according to many social researchers. But skeptics contend that bad marriages aren't counted in the statistics.

and Democrats proposed amending the state Constitution to reverse the court's ruling. President Bush also weighed in, arguing that "activist judges" had no right to rewrite the rules of marriage.

After several months of negotiation, Massachusetts Gov. Mitt Romney, a Republican, and legislative leaders proposed amending the state Constitution defining marriage as a union of a man and a woman, but allowing for gay civil unions. It was approved by the legislature on April 29, but it must be passed again next year and then approved by voters before it would take effect — in 2006 at the earliest.

Meanwhile, Romney has asked the state legislature to pass emergency legislation allowing him to seek a stay of the Massachusetts Supreme Court's May 17 marriage order. The governor contends the state should not be required to issue marriage licenses to same-sex couples while the legislature and citizens are debating the issue.

"This is a decision that is so important it should be made by the people," Romney said on April 15, the day the legislation was filed.[26]

But the Massachusetts court ruling has emboldened gay-marriage supporters elsewhere. On Feb. 12, newly elected San Francisco Mayor Gavin Newsom authorized the city government to begin issuing marriage licenses to gay and lesbian couples. The response surprised even supporters: In the first five days 2,500 couples came from

Will same-sex marriage hurt traditional marriage?

YES
Maggie Gallagher
President, Institute for
Marriage and Public Policy

Written for *The CQ Researcher*, April 2004

Same-sex marriage divides people into two camps: Those who say that gay marriage will affect only gays and those who believe that court-ordered same-sex marriage will dramatically alter the legal, shared public understanding of marriage.

If the Massachusetts court had decided the "right to marry" includes the right to polygamy, would that affect only those who want a polygamous marriage? Of course not. The entire marriage culture would shift if polygamy were to become a "normal" marriage variant. Monogamy would no longer be a core part of our definition of marriage.

I'm not saying that same-sex marriage will lead to polygamy. I'm pointing out that legally changing the definition of marriage affects everyone and would radically transform what marriage is. It would hurt the traditional form of marriage by:

- Sending a terrible message to the next generation: The law will say that two men or two women raising children are just the same as a mom and dad; thus, social institutions would be bending to adult sexual desires, regardless of who gets hurt.
- Creating an abyss between "civil" and "religious" marriage. Civil marriage would be divorced from religious traditions that gave rise to marriage and which continue to sustain marriage as a social institution. Government should be more modest about redefining marriage to make it unrecognizable to most religious traditions in this country.
- Neutering our shared language about parenting. You won't be able to say "children need mothers and fathers, and marriage has something important to do with getting this for children" because it will no longer be true, and because the government will be committed to the idea that two mothers or two fathers are just as good as a mom and a dad in raising children.
- Marginalizing or silencing traditional advocates of marriage. Marriage is a public act. Faith-based organizations that fail to endorse and accept same-sex marriage may find themselves driven from the public square: their broadcasting licenses, tax-exempt statuses and school accreditation at risk.

If gay marriage is a civil right, then people who believe that children need moms and dads will be treated like bigots. How will we raise young men to become reliable husbands and fathers in a society that officially promotes the idea that fathers don't matter?

NO
Kevin Cathcart
Executive Director, Lambda Legal
Defense and Education Fund

Written for *The CQ Researcher*, April 2004

The current, unprecedented dialogue about marriage for same-sex couples isn't an abstract discussion about politics or religion — it's about real people's lives and the human cost of denying basic equality to an entire group of Americans.

Denying marriage to same-sex couples blocks hundreds of thousands of families nationwide from the critical rights and protections that others take for granted. Same-sex couples are left vulnerable and scrambling to cobble together a patchwork of legal documents that still don't provide them with the security and protections they want and need.

To understand why marriage is so important for so many same-sex couples, look no further than Lydia Ramos. Lydia's partner of 14 years died in a car accident, triggering a legal and emotional nightmare. The coroner refused to turn the body over to Lydia, and the daughter they raised together was taken away by her partner's relatives after the funeral. Mother and daughter were kept apart for months — at a time when they most needed each other.

Mother and daughter were finally reunited after Lambda Legal fought a long and gut-wrenching legal battle on their behalf. But if Lydia and her late partner had been able to marry, their daughter would never have been put through such a nightmare.

If they had been able to marry, it would not have changed the marriages of their heterosexual neighbors and co-workers. Heterosexual marriages are not on such shaky ground that they will fall apart simply because loving, committed same-sex couples are given equal access to the rights and protections provided by marriage.

The nation is about to see that in Massachusetts, where lesbian and gay couples will soon begin getting married. That state's highest court — with six of seven justices appointed by Republican governors — ruled that only marriage can fix the inequalities in how the state treats same-sex couples.

Within the next year, our lawsuit on behalf of seven New Jersey couples is expected to reach that state's high court. Our cases in New York, Washington state and California will ask the same fundamental questions addressed by the Massachusetts court.

The courts and political leaders are beginning to recognize that anything less than marriage treats same-sex couples differently, and that separate is never equal. Loving couples are being kept from our nation's promise of fairness, and we'll fight for them for as long as it takes to win equality.

around the country to marry, often waiting for hours in long lines for their chance to get hitched.

In 2000, California voters had approved a ballot measure — Proposition 22 — which defined marriage as a union between a man and a woman, leading many, including the state's popular new governor, Arnold Schwarzenegger, to criticize Newsom for ignoring the law. "If the people change their minds and they want to overrule [Proposition 22], that's fine with me," Schwarzenegger said on March 2. "But right now, that's the law, and I think that every mayor and everyone should abide by the law."[27]

But Newsom justified his decision as an attempt to abide by the U.S. Constitution's requirement to treat all people equally. "I've got an obligation that I took seriously to defend the Constitution," he said. "There is simply no provision that allows me to discriminate."[28]

However, as a result of a court order, San Francisco stopped granting licenses on March 12. The state Supreme Court is currently deciding whether the 4,000 same-sex marriage licenses eventually issued by the city are valid, and a decision is expected in the next month or so.[29]

Even if the marriage licenses are ultimately invalidated, supporters of same-sex unions see developments in San Francisco as the event that most fully energized the gay-marriage movement. "Sometimes when you're in a civil rights struggle, you reach a tipping point," Lambda Legal's Cathcart says. "We reached it in San Francisco when you saw thousands of couples lining up to pay for their license and legally get married. Mayor Newsom lit the spark, and what happened afterwards inspired others around the country to act."

Indeed, mayors in a handful of other cities — including New Paltz, N.Y., and Asbury Park, N.J. — and the commissioners of Multnomah County, Ore., (which includes Portland) followed Newsom's lead.

But conservatives and marriage traditionalists view events in San Francisco differently. "The mayor of San Francisco has done more than anyone else to solidify support behind a federal amendment," the Southern Baptist Convention's Land says. "When a public official defies not only public opinion but [also] the law he's sworn to uphold, average people get outraged."

As evidence, Land points to a March CBS poll showing that 59 percent of Americans favor a constitutional amendment allowing marriage only between a man and woman. In December, only 35 percent favored such an amendment.[30]

Two newly married women ignore demonstrators protesting same-sex marriages at San Francisco City Hall on Feb. 20. San Francisco Mayor Gavin Newsom authorized the city government to begin issuing marriage licenses to gay and lesbian couples on Feb. 12, helping put the controversial issue at the top of the national agenda.

AFP Photo/Hector Mata

But gay-marriage advocates remain optimistic, in part because polls show that resistance to same-sex unions increases with age. "I'm very confident that we're going to win on this issue over the long term, because most young people just don't think this is a big deal," Cathcart says. "The young are already largely with us, and the coming generations will be even more supportive."

Impact of Marriage

In the last decade, a consensus has begun to emerge that marriage can provide tangible benefits, both for adults and their children. Indeed, most sociologists now agree that a good marriage generally improves the lives of all involved.

"Even the skeptics have come around on this," the University of Minnesota's Doherty says. "Marriage, and by that we mean a good marriage, is good for people."

Linda J. Waite, a sociology professor at the University of Chicago and co-author of *The Case for Marriage*, agrees, claiming the amount of evidence available on the benefits of matrimony is "fairly overwhelming."

Waite and Doherty point to a host of studies by sociologists and others, which found that men and women in happy marriages enjoy better mental and physical health and more financial security than similarly situated unmarried counterparts.

For instance, a 1990 study published in the *Journal of Marriage and Family* found that singles had higher mortality rates than their married counterparts. The mortality rate for single men was a whopping 250 percent higher, while unmarried women had a 50 percent higher rate. The study found the unwed were particularly at greater risk for those diseases that hinged on behavior, like lung cancer and cirrhosis of the liver.[31]

"It helps to have someone around who is a stakeholder in your health," Doherty says. "People do better when there is someone who can push you or nag you to keep to your diet, exercise regularly, take your medicine and just take proper care of yourself."

Likewise, married people are happier — almost twice as happy as singles and roughly three times happier than those who are widowed or divorced, according to studies.[32]

Children growing up in homes where both biological parents are present also do better than those living with a single parent. For instance, they are more likely to graduate from high school and have fewer discipline problems if raised by a married biological mother and father.

But some experts question at least parts of the consensus, arguing that much of the research paints too rosy a picture of the institution. "We need to remember that when we're talking about the benefits of marriage, we're really only talking about good marriages," Evergreen State's Coontz says. "When researchers tell you that marriage in general is good even after they average together good and bad marriages, it's very skewed," she continues. "Good marriages tend to last, while many bad marriages aren't being counted because they end in divorce. So good marriages tend to get overcounted, and the bad ones are undercounted."

Skeptics also contend that while marriage may provide real benefits for men, it is not necessarily as beneficial for women. "Men are used to being taken care of by women, first their mothers and then their wives," North Carolina State's Risman says. "So, sure, marriage is a good deal for men because without a wife to replace their mother they usually don't take good care of themselves."

But women marry for very different reasons, she says. "In a world where women still earn 75 cents or less for every dollar earned by men, they still have to marry for economic and other practical reasons."

According to Risman, a woman might be mentally or physically healthier in marriage because her husband's income provides health insurance or better food and shelter. "So, yes, maybe she's better off, but only because society's inequalities force her to marry," she says. "We need to construct a society where women don't need to marry in order to have these things."

But researchers who tout the benefits of marriage dispute both arguments. Coontz's contention of research bias is incorrect, they say, because the studies account for both divorce and bad marriages. "The best research looks at transitions into and out of marriage and includes all marriages experienced by respondents, including those prior bad marriages that ended in divorce," Waite says.

Others question the contention that marriage favors men more than women. "There are some areas where men benefit more and some where women are better off," says Maggie Gallagher, president of the Institute for Marriage and Public Policy. "But in all cases, in every area, both benefit."

OUTLOOK

Privatizing Marriage

The 20th century witnessed unparalleled changes in married life, at least in rich countries. In the United States, Europe and elsewhere, a trend away from obligations to parents, community and other forms of collective responsibility gave way to more of an emphasis on individual choice. People had more freedom to decide whom to marry and whether to stay in that marriage.

Many scholars predict even more dramatic changes for married life in the coming decades. "People will view their marriages more and more as a private relationship, rather than as a public institution as they did in the past," Blankenhorn of the Institute for American Values says. "You already see this happening, as, for instance, many couples no longer speak the traditional vows but write their own."

Blankenhorn contends that the "privatization of marriage," as he calls it, will make marriages less stable, because the built-in guideposts and expectations that once accompanied married life are disappearing. "Without a set of public expectations — like permanence or total commitment to each other — people are more on their own, and that's more risky."

> "Men will look more and more at a woman's earning potential when they decide on a mate and be less interested in women as homemakers."
>
> — William Doherty
> Director, Marriage and Family Therapy Program
> University of Minnesota

The University of Minnesota's Doherty agrees that marriage will continue shifting from a public to a private institution. But he sees a positive side to such a change: Future marriages, to some degree, will be on a more equal footing. "Men will look more and more at a woman's earning potential when they decide on a mate and be less interested in women as homemakers," Doherty says. "They're going to want someone who can bring in as much as they do or just a little less. This will bring a new level of gender equality to marriage."

Still, Doherty adds, husbands will continue to want to make at least as much as their wives. "Men are hard-wired by evolution to be 'providers,' and this is still backed up by our culture," he says. "So a man who makes less than his wife will still feel 'inadequate.' "

However, some scholars see few, if any, additional changes occurring in married life — at least in the United States. "We've had so much happen in the last 100 years that I think we're entering a period of stability," North Carolina State's Risman says. "We're going to spend the next 50 years absorbing the changes of the last 50."

Risman says several trends bolster her belief. "The divorce rate has stabilized over the last few decades, which to me is a sign that the rate of change has slowed significantly." Moreover, things don't seem to be changing all that much for women at home, she points out.

"Housework remains women's work, and there is no sign that that is changing." The fact that the number of women staying at home with their children and putting off a career is holding steady is a sign that "women still don't think they can have it all, because their husbands still haven't been told that they can't have it all."

Evergreen State's Coontz agrees that marriage won't change too much in the immediate future, but neither will it remain static. "We're going to spend the next few decades sorting through the enormous changes we've seen in marriage."

In particular, she says, couples will have to work through the consequences of gender equality in marriage. "Men can no longer count on being the boss anymore," she says. "So, we're going to see more efforts to develop new habits, new emotional expectations, new time schedules and new negotiating skills as we sort out the details of this new reality."

NOTES

1. Figures cited at www.census.gov/population/www/socdemo/ms-la.html.

2. For background, see David Masci, "Children and Divorce," *The CQ Researcher*, Jan. 19, 2001, pp. 25-40.

3. Figures cited in *National Vital Statistics Reports*, Centers for Disease Control and Prevention, Vol. 52, No. 10, Dec. 17, 2003, p. 1.

4. Amy Fagen, "Permanent Tax Cut OK'd," *The Washington Times*, April 29, 2004, p. A1.

5. Michelle Conlin, "Unmarried America," *Business Week*, Oct. 20, 2003, p. 106.

6. For background, see Kenneth Jost, "Gay Marriage," *The CQ Researcher*, Sept. 5, 2003, 721-748.

7. Figures cited in *National Vital Statistics Reports*, Centers for Disease Control and Prevention, Vol. 50, No. 14, Sept. 11, 2002, p. 1.

8. Barbara Defoe Whitehead and David Popenoe, "The State of Our Unions: The Social Health of Marriage in America," National Marriage Project, Rutgers University, 1999.

9. Figures cited in *National Vital Statistics Reports*, *op. cit.*, and "U.S. Per Capita Divorce Rates Every Year: 1940-1990," Centers for Disease Control and Prevention, www.cdc.gov/nchs/fastats/pdf/43-9s-t1.pdf.

10. Figures cited in Claudia Wallis, "The Case for Staying at Home," *Time*, March 22, 2004, p. 51. See also, Sarah Glazer, "Mothers' Movement," *The CQ Researcher*, April 4, 2003, pp. 297-320.

11. Amy Fagen, "Senate Mulls Pro-Marriage Funds," *The Washington Times*, April 1, 2004, p. A5.

12. Bill Swindell, "Welfare Reauthorization Becomes Another Casualty in Congress' Partisan Crossfire," *CQ Weekly*, April 3, 2004, p. 805.

13. Figures cited in Ronet Bachman and Linda E. Saltzman, "Violence Against Women: Estimates from the Redesigned Survey," *National Crime Victimization Survey Special Report*, August 1996, p. 4.

14. For background, see Kenneth Jost, "School Desegregation," *The CQ Researcher*, April 23, 2004, pp. 345-372.

15. Barr's testimony available at www.house.gov/ judiciary/barr033004.pdf.

16. *The Gospel According to St. Mark*, 10:6-9.

17. *The Gospel According to St. Mark*, 10:11-12.

18. *1 Corinthians.*, 14:34-35.

19. Quoted in Daniel J. Boorstin, *The Americans: The Colonial Experience* (1958), p. 67.

20. *Ibid.*, p. 187.

21. Nancy Cott, *Public Vows: A History of Marriage and the Nation* (2000), p. 184.

22. *Ibid.*, p. 167.

23. *Ibid.*, p. 187.

24. Quoted in *ibid.*, p. 185.

25. Figures available from the Bureau of the Census at www.census.gov/prod/2004pubs/03statab/vitstat. pdf.

26. Cheryl Wetzstein, "Romney Moves to Get Vote on Same-Sex 'Marriage,'" *The Washington Times*, April 16, 2004, p. A3.

27. Quoted in Dean E. Murphy, "Scwharzenegger Backs Off His Stance Against Gay Marriage," *The New York Times*, March 2, 2004, p. A11.

28. Quoted in Dean E. Murphy, "San Francisco Mayor Exults in Move on Gay Marriage," Feb. 18, 2004, p. A18.

29. Maura Dolan, "State High Court Seeks Briefs on Validity of Gay Marriage," *Los Angeles Times*, April 15, 2004, p. B6.

30. CBS News Poll, March 15, 2004, at www.cbsnews .com/stories/2004/03/15/opinion/polls/ main606453.shtml.

31. Catherine E. Ross, John Mirowsky and Karen Goodsteen, "The Impact of the Family on Health: Decade in Review," *Journal of Marriage and the Family 52* (1990), p. 1061.

32. Cited in Linda J. Waite and Maggie Gallagher, *The Case for Marriage: Why Married People are Happier, Healthier and Better Off Financially* (2000), p. 67.

BIBLIOGRAPHY

Books

Cott, Nancy F., *Public Vows: A History of Marriage and the Nation, Harvard University Press*, 2000.
A professor of history and American studies at Yale University examines marriage in the United States from European settlement in the New World through social revolutions following World War II.

Waite, Linda J., and Maggie Gallagher, *The Case for Marriage: Why Married People Are Happier, Healthier and Better Off Financially, Doubleday*, 2000.
A sociology professor of at the University of Chicago (Waite) and the director of the Marriage Program at the Institute for American Values (Gallagher) review recent research showing the benefits of marriage.

Articles

"The Case for Gay Marriage," *The Economist*, Feb. 28, 2004, p. 9.
The venerable English news weekly argues that same-sex couples should be allowed to marry, asking: "Why should one set of loving, consenting adults be denied a right that other such adults have and which, if exercised, will do no damage to anyone else?"

Belsie, Laurent, "More Couples Living Together, Roiling Debate on Family," *The Christian Science Monitor*, March 13, 2003, p. A1.
The author explores the social impact of the growth of cohabiting couples in the United States.

Conlin, Michelle, "UnMarried America," *Business Week*, Oct. 20, 2003, pp. 106-116.
Conlin's cover story reports that the dramatic decline in traditional families has significant implications for American business and society.

Crary, David, "Will the Institution of Marriage Continue? It's Debatable," *Los Angeles Times*, Feb. 10, 2002, p. A20.
The article explores the marriage movement, which seeks to promote the social benefits of matrimony.

Gallagher, Maggie, "Massachusetts vs. Marriage," *The Weekly Standard*, Dec. 1, 2003.

A pro-marriage writer argues that expanding the definition of marriage to include same-sex couples will make the institution largely meaningless.

Hubler, Shawn, "Nothing But 'I Do' Will Do Now for Many Gays," *Los Angeles Times*, March 21, 2004, p. A1.

Hubler explores the evolution of attitudes about gay marriage within the homosexual community.

Jost, Kenneth, "Gay Marriage," *The CQ Researcher*, Sept. 5, 2003, pp. 721-748.

The author provides a broad overview of the debate over gay marriage.

Kmiec, Douglas R., "Marriage is Based on Procreation, a Fact No Claim of Gay 'Equality' Can Avoid," *Los Angeles Times*, March 14, 2004, p. M1.

A professor of constitutional law at Pepperdine University argues that marriage is not a matter of rights but a question of public policy.

Lyall, Sarah, "In Europe, Lovers Now Propose: Marry Me a Little," *The New York Times*, Feb. 15, 2004, p. A3.

Lyall looks at legal arrangements short of marriage that are gaining popularity in Europe.

Munro, Neil, "Supporting Marriage, But for What Goal?" *National Journal*, Jan. 3, 2004.

An overview of the debate over the benefits of marriage.

Reich, Robert B., "Marriage Aid That Misses the Point," *The Washington Post*, Jan. 22, 2004, p. A25.

The former secretary of Labor argues that the money President Bush would like to spend on marriage promotion would be better used on job training and education.

Wallis, Claudia, "The Case for Staying At Home," *Time*, March 22, 2004, p. 50.

Wallis examines the trend among professional women who put their careers on hold to care for their children.

Reports and Studies

Coontz, Stephanie, and Nancy Folbre, *Marriage, Poverty and Public Policy, Council on Contemporary Families*, April 2002.

A professor of history and family studies at The Evergreen State College (Coontz) and an economics professor at the University of Massachusetts (Folbre) argue that money devoted to promoting marriage among welfare recipients would be better spent on education and other social services.

Why Marriage Matters: Twenty-One Conclusions from the Social Sciences, Center for the American Experiment, Coalition for Marriage, Family and Couples Education and Institute for American Values, 2002.

Three marriage-advocacy groups catalog many of the arguments traditionally given in favor of marriage.

For More Information

Coalition for Marriage, Family and Couples Education, 5310 Belt Rd., N.W., Washington, DC 20015-1961; (202) 362-3332; www.smartmarriages.com. Nonpartisan group that promotes marriage education.

Council on Contemporary Families, 208 E. 51st St., Suite 315, New York, NY 10022; www.contemporaryfamilies.org. Left-leaning think tank that researches marriage and other family issues.

Focus on the Family, 8685 Explorer Dr., Colorado Springs, CO, 80995; (719) 531-3400; www.family

.org. Christian advocacy group that opposes gay marriage.

Institute for American Values, 1841 Broadway, Suite 211, New York, NY 10023; (212) 246-3942; www.americanvalues.org. Promotes the renewal of marriage and family life.

Lambda Legal Defense and Education Fund, 120 Wall St., Suite 1500, New York, NY 10005-3904; (212) 809-8585; www.lambdalegal.org. National gay-rights organization that represents plaintiffs in a number of gay-marriage cases.

10

Student Aid

*Will Many Low-Income
Students Be Left Out?*

Marcia Clemmitt

AP Photo/Elise Amendola

Harvard graduates celebrate at commencement on June 7, 2007. In December the university cut education costs by up to 50 percent for families that earn $120,000 to $180,000; tuition is waived for students from families earning under $60,000. Meanwhile, as merit-based scholarships replace need-based aid across the country, many low-income and minority students are finding it hard to finance their college dreams.

From *CQ Researcher*,
January 25, 2008.

Each month, Lucia DiPoi, a 24-year-old graduate of Tufts University in Boston, pays $900 toward her college debt — $65,000 in private loans, $19,000 in federal loans.

The first in her family to attend college, DiPoi blames some of her plight on her lack of financial knowledge. When grants and federal loans didn't cover enough of her fees, she took out private loans with interest rates of more than 13 percent. "How bad could it be?" she figured.[1]

Now she knows. DiPoi's loan burden is well above the average $20,000 in debt that American college graduates face, but it's not unusual. Many students from low-income families — and graduate students in particular — also face large burdens.

As the cost of higher education rises, grants for needy students have lagged behind, and more students are dependent on loans to finance their education. At the same time, worries about college costs have been reaching higher up the socioeconomic scale. In response, states and private colleges have launched new merit-based scholarships that shift some aid from the neediest students to middle- and even upper-income families.[2]

In-state tuition and fees (excluding room and board) for public, four-year schools average $6,185 for the 2007-2008 school year, up 6.6 percent from 2006-2007; out-of-state tuition averages $16,640. At private four-year schools, the average 2007-2008 tuition and fees is $23,712, up 6.3 percent from 2006-2007.[3] The cost of college has nearly doubled over the past 20 years, in inflation-adjusted dollars, and college tuition and fees have risen faster than inflation, personal income, consumer prices or even the cost of prescription drugs and health insurance.[4]

Student Debt Highest in New Mexico

College graduates in New Mexico in 2006 had an average of $28,770 in student debt, the highest of any state and almost 50 percent higher than the national average. Hawaii had the lowest average debt: $11,758 per student.

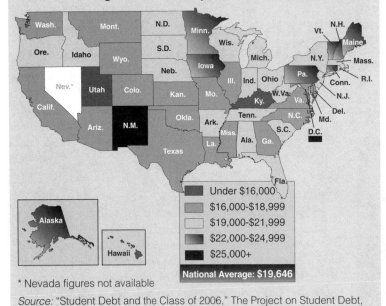

Average Student Debt by State, Class of 2006

Legend:
- Under $16,000
- $16,000-$18,999
- $19,000-$21,999
- $22,000-$24,999
- $25,000+

National Average: $19,646

* Nevada figures not available

Source: "Student Debt and the Class of 2006," The Project on Student Debt, September 2007

The increasing difficulty of paying for college and the importance of lending are bringing new attention to financial-aid issues. Congress recently passed legislation boosting the value of grants for low-income students and trimming subsidies for private education lenders. New York State Attorney General Andrew M. Cuomo reached legal settlements last year with more than two-dozen colleges where financial-aid officials had accepted kickbacks in exchange for steering students toward particular private lenders; similar investigations are ongoing in other states.

In December, pricey Harvard University — which had already waived tuition payments for students whose families earn less than $60,000 a year, announced a boon for middle-earners, cutting costs by as much as 50 percent for families that earn between $120,000 to $180,000.[5]

If the main job of financial aid is to make college education more accessible to lower-income and minority students, current financial-aid and cost trends aren't helping, some analysts say.

Three decades ago, there was a 30-percentage-point gap in college attendance between low-income students and other students, says Donald E. Heller, director of the Center for the Study of Higher Education at Pennsylvania State University. "And since that time they haven't gained any ground," he adds.

"Only 7 percent of high-school sophomores from the lowest quartile of socioeconomic status eventually earn a bachelor's degree, compared with 60 percent of those from the highest quartile," according to Associate Professor of Public Policy Susan M. Dynarski and doctoral candidate Judith Scott-Clayton, both at Harvard. "Only 12 percent of Hispanics and 16 percent of African-Americans eventually earn a B.A., compared with 33 percent of non-Hispanic whites." Moreover, the gaps persist "even among well-prepared students," so difficulties paying for college are at least partly to blame, the researchers say.[6]

Federal Pell Grants for low-income students — a big part of the money problem — have lagged behind rising costs for decades. "The purchasing power of the Pell Grant is less than half what it was in the 1970s," Heller says.

Need-based aid offered by states has been dropping in many places over the past decade, says Ross Rubenstein, associate professor of public administration at Syracuse University. While need-based grants are still the largest share overall, a growing proportion of state grants are merit-based scholarships, a "huge middle-class entitlement" that shifts the focus of aid programs away from expanding access for the neediest, he says.

Merit aid has virtues, though. Reserving some scholarship aid for top students might improve student achievement in high school, Rubenstein explains, noting it has been shown to have modest positive effects on student achievement in Georgia, for example, which launched the first state merit program, HOPE, in 1993. "I wouldn't want to see a merit-based system replace a need-based one, but it has its place," he says.

A certain level of merit aid might not be a problem, Heller says; he cites Indiana's program, which has been successful in promising aid to all students who graduate from high school with a C average. "But I would caution against" including additional criteria "such as SAT scores or requiring higher grade averages," he adds, because of racial gaps in SAT scores and the general discouraging effect such criteria have on low-income students who are hesitant about their college chances anyway.

Merit aid may create heavier loan burdens for low-income students. The University of Maryland recently discovered that low-income students were graduating with more debt than middle- and high-income students and concluded that its grant program — which had 60 percent merit-based awards — should be revamped to include more need-based grants.[7]

The biggest trend in college financing is the heavy reliance on loans, which make up about 70 percent of higher-education financing today. The loans include both federally guaranteed and subsidized loan programs and, increasingly, completely private, non-federally subsidized loans with much higher interest rates. (*See graphs, p. 225.*)

"Loans are ubiquitous as more and more students are borrowing," says Laura W. Perna, an associate professor at the University of Pennsylvania Graduate School of Education.

The federal government makes some loans directly and subsidizes private lenders to handle others. The private lenders came under fire last year, as reports surfaced of outsized bonuses paid even at many nonprofit lenders and sweetheart deals that some lenders gave college financial-aid officials who steered students toward their services.

"Ninety percent of students who receive loans choose their lender based on their schools' recommendation," according to the Center for American Progress, a liberal think tank. With education debt high and

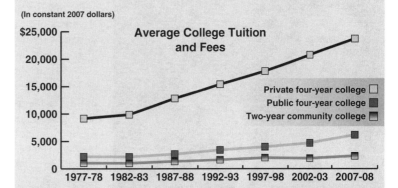

Costs Rose for State and Private Colleges

The average tuition and fees for private four-year colleges more than doubled in the past 30 years, to nearly $24,000 in the 2007-08 academic year.* During the same period, costs also doubled at public two-year schools and nearly tripled at public four-year institutions.

(In constant 2007 dollars)

Average College Tuition and Fees

Private four-year college ☐
Public four-year college ■
Two-year community college ▦

1977-78 1982-83 1987-88 1992-93 1997-98 2002-03 2007-08

* Tuition and fees constitute about two-thirds of the total budget for students at private four-year colleges but are just over one-third of the total budget for in-state students in public four-year colleges and less than 20 percent for public two-year college students.

Source: "Trends in College Pricing 2007," The College Board, 2007

rising, "students should be able to count on their schools for impartial and helpful advice as they navigate a complicated and stressful process." Instead, "kickbacks" and "conflict of interest" put students at risk of doing business only with lenders who have offered the most blandishments to their colleges, the center said.[8]

But many college officials argue that students are getting excellent service from lenders and could suffer if they switched to direct government loans or if the number of private lenders was cut. "Private lenders tend to be more efficient, have better technology and are able to provide better services that aren't available from the government," said Seamus Harreys, dean of student financial services at Boston's Northeastern University.[9]

Disputes over private vs. public lending aside, however, mushrooming education debt is a trend that troubles many because of the financial burdens loans place on graduates.

Students don't understand the implications of the loans they apply for, according to the Federation of State PIRGs (public interest research groups). "The loan industry recommends that graduates . . . dedicate no more than 8 percent of their income to student-loan

New York State Attorney General Andrew M. Cuomo, right, announces on May 31, 2007, an agreement between Columbia University and the National Association of Student Financial Aid Administrators to curtail improper student-aid practices. Investigations by Cuomo last year forced 10 private lenders to stop kickbacks to aid officials at two-dozen colleges in exchange for steering students to their firms. Similar investigations are ongoing in other states, including Missouri, Iowa and Pennsylvania.

repayment," according to PIRG analysts Tracey King and Ivan Frishberg, but students themselves "expected to contribute an average of 10.7 percent of future income." Furthermore, "students with larger debts more significantly overestimated the percentage of their income they could afford" for repayment.[10]

As students, college officials and political leaders wonder how future students will pay for higher education, here are some of the questions being asked:

Has the right balance been struck between merit-based and need-based aid?

Over the past decade and a half, public and private student-aid programs have increased the proportion of scholarships based on academic and athletic merit, many going to middle- and upper-class students. As a result, financially needy students have seen their share of the student-aid pie shrink.

Critics of the shift argue that achieving equity in education demands mostly need-based aid, since low-income and minority students still lag far behind in college attendance. But supporters of increased merit-based aid say merit scholarships spur students to work harder in high school and that, with college costs soaring, middle-class students deserve financial help, too.

Low-income students still get the most financial aid, but the share of aid claimed by students from wealthier families has increased in recent years. The Indianapolis-based Lumina Foundation for Education found that between 1995 and 2000 grants to students from families earning $40,000 or less increased by 22 percent, compared with a 45 percent boost for families making $100,000 or more.[11]

Many education analysts say merit-based grants, unlike need-based grants, don't increase low-income and minority access to higher education.

An adequate pool of need-based college aid actually improves high-school graduation rates, says Ed St. John, a professor of higher education at the University of Michigan. "Living in a state with more need-based aid increases the chances a low-income student will graduate," he says. "If a sophomore sees that there's aid, there's a bigger chance he'll finish."

"Poor kids don't think they can go to college, and their schools don't have the counselors" needed to explain how they can, St. John explains. A widely publicized need-based aid program can counter that perception. "Knowing you can afford to go allows people to do things they wouldn't do otherwise," he adds.

Merit-based aid does little to expand access for low-income students, says Rubenstein of Syracuse. "By targeting people with a B average, you're mostly targeting students who would go to college anyway," he says. State merit-based aid mainly helps the state keep its better students at in-state colleges, he says.

Often financed by lotteries, many state merit programs draw their funding from the mostly low-income people who play the games, while the aid flows mostly toward middle- and upper-income students, says Rubenstein. "There's no question that it's highly regressive."

Under a recent Massachusetts proposal, for example, more than 50 percent of students in the state's wealthiest school districts would qualify for scholarships compared to less than 10 percent in the poorest districts, according to Penn State's Heller.[12]

Similarly, private colleges' merit aid also boosts mainly the middle and upper classes. "If the private colleges don't refocus more dollars on students with high-level needs, they are going to become places that are totally closed to low-income students," said Sandy Baum, an economics professor at Skidmore College in Saratoga Springs, N.Y., and senior policy analyst for the New York-based College Board.[13]

Many low-income students are shut out from merit-based grants mainly because they attend very-low-performing schools, says Sara Goldrick-Rab, an assistant professor of education policy at the University of Wisconsin, Madison. "Essentially it's holding kids responsible for the poor K-12 program," she says.

Indeed, a recent analysis of a Michigan test that determines who gets state merit aid found that "schools with high numbers of minority students didn't even have the curriculum for the kids to take the whole test that would qualify them for the merit aid," says the University of Michigan's St. John.

Merit-based aid advocates contend that it exists for good reasons — to help deserving students get an education without plunging their families into debt. The high cost of college increasingly forces middle-class families to seek financial aid.

"The reality is even parents who work hard to save are coming up short" of college costs, said Mary Beth Moran, a financial adviser in Bloomingdale, Ill.[14]

"Middle-class families are being squeezed out," said Sharon Williams, college adviser at Elgin Academy in

Loans Are Now Biggest Source of Aid

In the past three decades federal education grants have been overtaken by federally guaranteed low-interest loans. Today, the ratio of loans to grants is seven to two, compared to just three to two in 1975-76 (top). Meanwhile, high-interest, wholly private loans have emerged in the past decade, largely in response to heavy marketing and the long, confusing federal loan applications. They represented 18 percent of all student loans in 2004-05, or double the percentage four years earlier (bottom).

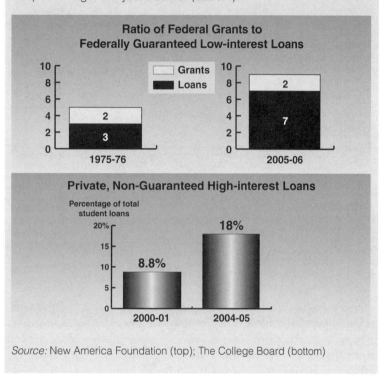

Ratio of Federal Grants to Federally Guaranteed Low-interest Loans

Private, Non-Guaranteed High-interest Loans

Source: New America Foundation (top); The College Board (bottom)

Elgin, Ill. "Need-based scholarships will not be there for them."[15]

Because of a growing perceived need among the middle class, but also because the public often doesn't like government programs that offer assistance without requiring something in return, merit-based aid is more politically popular, says William Doyle, an assistant professor of higher education at Vanderbilt University. That attitude shows up in public-opinion polls, he says. "People want reciprocity; they don't like entitlement-based grants."

"Mainly in response to middle-class demands, there was a huge surge in the 1990s and early 2000s of merit-based aid," says Rubenstein at Syracuse. The trend "has

States Providing More Grants

State grants to students have risen markedly over the past decade. Need-based grants totaled nearly $5 billion during the 2005-06 school year, more than a 50 percent increase over 1995-96. Merit-based grants more than tripled, totaling about $1.9 billion in 2005-06.

Need-based and Merit-based Education Grants

(in $ billions, in constant 2006 dollars)

Need-based grants   Merit-based grants

Source: "37th Annual Survey Report on State-Sponsored Student Financial Aid: 2005-2006 Academic Year," National Association of State Student Grant and Aid Programs, 2006

tapered off a bit, but I don't expect it to end. It's an extremely popular entitlement and hard to take away."

Some argue that merit-based aid ensures universities attract higher-quality students. "The incremental or marginal students that we have gained through [need-based] federal programs likely have extremely poor records with respect to college completion and probably shouldn't have been in college in the first place," said Richard Vedder, a former professor of economics at Ohio University and director of the Washington, D.C.-based Center on College Affordability and Productivity.[16]

While merit-based aid is skewed toward the middle class, it has helped students from poor families, Rubenstein points out. Georgia's merit-based HOPE Scholarship, for example, still increases aid to low-income students. With HOPE money in the system, all students receive substantially more aid than they did previously, he says.[17]

Merit-based programs have set criteria and specified award amounts, compared with most need-based

scholarships. Because their requirements are clear and simple, merit programs may encourage even low-income students to persist in preparing for college, explains the University of Pennsylvania's Perna. By comparison, the lengthy, complex applications of need-based programs "just leave a big question mark," as students remain in doubt for many months about whether they'll qualify for enough aid to allow them to afford their chosen school, she adds.

Georgia's HOPE Scholarships "provide an incentive for kids to work harder" in high school because they know for a fact they can claim the money if they attain the required grade average, says Rubenstein. High-school grades increased and not just because of grade inflation after the scholarship was launched, he says. In Georgia, many lower-income "kids really thought they couldn't go before the HOPE Scholarship established a simple rule: 'If I get that B, I can go,' " he says.

Merit-based scholarships can be designed to be more equitable, says St. John. For example, Texas has a plan in which state colleges must accept the top 10 percent of the graduating class from any state high school, he says. Besides giving students from low-income and high-minority schools an equal shot, the program indirectly encourages the schools to improve, he says.

Has the right balance been struck between grants and loans?

Since the late 1970s, federal student aid has been shifting from grants toward loans, most provided by private lenders that the government guarantees against loss should students default. Furthermore, over the last decade limits on the amount of guaranteed-loan funding each student can claim have lagged behind college costs, and more lenders have sought out student customers. As a result, more students also have turned to fully private loans — offered at regular consumer interest rates, unsubsidized by the government — as part of their college-funding package.

Forty years ago, the belief that expanding college access resulted in substantial public benefit led to the establishment of most federal student-aid grant programs, says James C. Hearn, a professor of higher education at the University of Georgia. Educators and officials reasoned that students shouldn't have to bear all their college costs because education serves many other goals, such as creating smarter voters and a more civilized and cultured citizenry, he says.

Since then, however, the reigning political philosophy has shifted toward economic conservatism. Today, the financial-aid landscape is shaped by the idea that education delivers primarily personal, not public, benefits, and that idea favors loans, Hearn says.

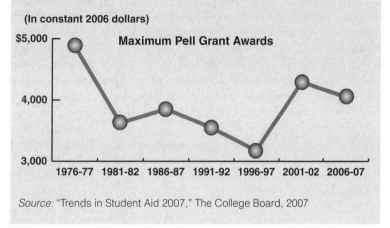

Size of Maximum Pell Grant Has Declined

The maximum Pell Grant award has declined nearly 20 percent in the past 30 years. The top grant today covers about 32 percent of average tuition, fees and room and board at public four-year colleges and 13 percent at private four-year schools. In 1986-87, maximum Pell Grants covered 52 percent of the costs at public four-year schools and 21 percent at private four-year institutions. Pell Grants are for low-income students and do not have to be repaid.

(In constant 2006 dollars)

Maximum Pell Grant Awards

Source: "Trends in Student Aid 2007," The College Board, 2007

Nevertheless, many analysts say that expanding college access to those who couldn't afford it on their own should remain the key goal of financial aid and that grants, not loans, are the best tool.

"The fact is that loans of any kind don't improve college access for low-income students," says Heller of Penn State. For one thing, lower-income families tend to be more loan-averse and often lack the experience and information to successfully navigate a loan-based system, he explains.

"Limits on borrowing also are among the factors that make the loan-heavy aid system not work for low-income kids," whose total college-funding needs are greater and often outstrip the slow-rising limit on federally guaranteed loans, says the University of Michigan's St. John.

Today's graduates carry an average of about $20,000 in education debt, says Amaury Nora, a professor of education leadership and cultural studies at the University of Houston. (*See map, p. 222.*) Add to that credit-card debt for books and meals, and "a lot of times students say, 'I can't keep this up,'" he adds. Debt often becomes a compelling additional reason students leave college, initially intending to return. But in a 10-to-15-year study of such "stop outs," only a handful of students who left actually returned, contrary to conventional wisdom in education-policy circles, Nora says. "Once they're out, they're pretty much gone."

Nonetheless, loans really expand the reach of federal aid in dramatic ways simply by adding more money to the student-aid system than taxpayers would ever pony up in the form of grants, Hearn says. Because there are finite resources for grants, other funding must be found, he says.

Loans also "provide a better opportunity for choice of college for upper- and middle-income students," says Heller.

The current loan-grant balance "is still a pretty good system for middle-class kids," says St. John.

Graduates who complain that education debt is an unfair burden ignore the benefits they've reaped, argues Radley Balko, a senior editor at the libertarian *Reason* magazine, citing two medical students profiled on CBS News as burdened with a "mountain" of debt. The two were shown drinking Starbucks coffee and Vitaminwater, pricey habits, in Balko's opinion, for a couple so worried about expenses. He estimated their lifetime earnings would be $8.2 million, or "a tidy $7.7 million profit on the investment they made in their education. If only everyone had it so rough."[18]

Many analysts say a thoughtful balancing of loans and grants, plus greater flexibility on loan repayment, is the fairest way to distribute education funds. To make paybacks fairer, Congress passed the College Cost Reduction and Access Act in 2007, establishing new payback schedules for direct federal student loans that are "income-contingent," or based on graduates' earnings.

Basing the system mainly on income-contingent loans, instead of grants or non-income-contingent loans, could ultimately simplify applying for aid and perhaps do away with the complicated needs-analysis financial-aid application that scares off some students today, says Donald Hossler, a professor of educational leadership and policy studies at Indiana University, Bloomington. Need would be taken into account after graduation, instead of before enrollment, with high-earning graduates required to pay back their loans fully and low earners entitled to reduced payment.

An income-contingent payment scheme also might enable more students to select lower-paying, service-oriented careers like teaching, which some avoid because of worries about paying off their debts, Hossler says.

If money is tight, federal aid heavy on grants and income-contingent loans might be reserved for students in fields deemed to produce more public benefit — such as inner-city teaching — while students in fields that yield more individual benefit, like business, could rely more on traditional loans, says the University of Georgia's Hearn. "You can try to target your subsidies to public good" so taxpayers are on the hook only for educations that provide needed public services at a modest cost, he adds.

However, "some research shows we're not very good at playing the market" with targeted subsidies, Hearn says. In the 1960s and '70s, for example, the government used financial aid to lure people into nursing and engineering. But by the time the subsidies kicked in, market forces had erased the personnel shortages.

Do private lenders deliver good value for taxpayers and students?

Government-backed student loans come in two varieties: direct loans and loans offered by private lenders the government pays to participate in the program and guarantees against loss should students default. Debate has long raged over whether the government should remedy

perceived shortcomings in the federally backed loan program by beefing up its own direct-loan programs and making it tougher and less lucrative for private lenders to participate.

Supporters of retaining a large number of private lenders to students say the lenders aren't getting rich at the program's expense.

"No abnormal profits are being made in student loans," said Kevin Bruns, executive director of the industry group America's Student Loan Providers. Trimming federal subsidies to lenders could hurt students if banks respond by eliminating bonuses like discounted interest rates for payments that are made electronically or on time or waivers of some borrowers' fees, he said.[19]

Replacing loans made by private lenders with more direct government loans "would result in massive waste as well as burdensome red tape for students and parents," said John Berlau, a fellow in economic policy at the Competitive Enterprise Institute, a free-enterprise-oriented think tank. "The reason schools, with both programs to choose from, have stayed with the banks is that the banks offer the private sector's level of service," he said. "Some 600 schools have stopped participating in direct lending, and surveys of schools cite poor service as the primary reason."[20]

Some economic conservatives argue that all student aid should come in the form of private-sector loans or other private financial arrangements.

"The intellectual justification for expanded federal student-loan programs is extremely weak" because they've attracted only marginal students and encouraged colleges to be bigger spenders, said Vedder, at the Center on College Affordability and Productivity. "It is not clear that higher education has major positive spillover effects that justify government subsidies in the first place, and the private loan market that can handle anything from automobile loans to billion-dollar government bond sales can handle providing financial assistance to students."[21]

Late last year, lawmakers approved requiring lenders to bid for a limited number of spots in the loans-to-parents program, but private lenders generally oppose the move. Such an auction could backfire because unreliable lenders who give bad service might lowball the bidding and end up with all the business, said Bruns. "An auction creates that problem where everything is based on price," he said.[22]

But advocates of a strengthened government role argue that allowing large numbers of private lenders to offer federally backed loans costs a lot and is of questionable benefit.

The large number of lenders competing for student-loan business is evidence that private lenders have long received "excess subsidies," according to Michael Dannenberg, director of education policy at the liberal New America Foundation "There are reasons Sallie Mae's [Student Loan Marketing Association's] stock has increased by 2,000 percent in the last decade, and those reasons are a government guarantee against risk and very large government subsidies."[23]

Furthermore, the fact that the government guarantees private lenders 99 percent of a loan's value if a student defaults gives lenders "little reason to put resources into collecting payments from delinquent borrowers," he said.[24]

The idea that having more lenders in the program improves customer service through competition doesn't make sense, says Jason Delisle, research director for education policy at the New America Foundation. The loans are an identical commodity offered under government rules, so lenders have little room to customize loans or services, he says. In addition, lenders generally hold the loans on their own books for only a few months before selling them.

Because the subsidies are not market-based but negotiated by lenders and members of Congress, the private loan program has been plagued with "influence peddling" and "a dangerous amount of political influence," says Delisle, who sat in on discussions with lenders as a budget analyst for Sen. Judd Gregg, R-N.H.

"Without a market mechanism to set student-loan provider subsidy rates, banks will continue to inundate Congress and its staff with papers, meetings and phone calls pleading that a cut in that arbitrary subsidy rate would be 'catastrophic' to the lending business and that a lender-subsidy reduction means loans will no longer be made available to students," said Delisle.[25]

Education lenders complain to Congress that higher subsidies are necessary for their business to be profitable but often say the opposite to business associates, Delisle charges. "When Congress began considering this most recent round of subsidy cuts, Sallie Mae representatives told me and other congressional staffers that for the company to continue making federal student loans at the proposed lower subsidy rates, Sallie Mae would have to make the loans through its charity organization," he says.

"In other words, the proposed and now enacted subsidy cut" — in September 2007 Congress cut the subsidies by about half — "makes the federal student-loan business unprofitable, according to Sallie Mae," Delisle says. "But try to square that position" with recent Sallie Mae comments to a group that wanted to buy out the company. The potential buyers were worried that lower government subsidies would harm profitability, but "reports have those close to Sallie Mae" saying that the subsidy cuts' "effects on the company's earnings will be 'de minimis,' " Delisle says.

The buyout deal ultimately collapsed, however, and the company has recently reiterated to the U.S. Securities and Exchange Commission that Congress' 2007 subsidy cuts will substantially reduce profits from new loans and could potentially make the loans unprofitable.[26]

BACKGROUND

Gradual Beginnings

Financial aid to needy students dates back to 13th-century Europe.[27] But scholarship aid always has had mixed goals, which have complicated its development over the years. Some view student aid as mainly a boost for gifted but low-income individuals; others see it as a broad, public initiative for expanding access to education to the poor and to minorities. Many private and state universities have viewed it primarily as a means of boosting their own reputations by attracting higher-caliber students.

The first private colleges in America opened in the 17th century. Most then gave many grants to expand their enrollment and reputations. At Harvard in the 1700s endowed scholarships paid about half of school expenses for between a quarter and a third of students. New York University gave substantial grants to about half the students from the time it opened in 1831 through the mid-19th century; the grants enabled many to attend tuition-free.

But as private universities gathered students and solidified their reputations — and college education remained confined to the elite few — the days of generous grants waned.

Meanwhile, in the early days of state-funded colleges, financial aid as such was scarce, says Hearn of the University of Georgia. Instead, the states provided a virtually free education to all comers as a public good, he says.

C H R O N O L O G Y

1960s–1970s *As college prices rise, Congress creates loans and grants.*

1965 Federal Family Education Loan Program — then called the Guaranteed Student Loan Program — is launched.

1972 Congress creates Sallie Mae — the Student Loan Marketing Association — and establishes federal grants for low-income students, later named for Sen. Claiborne Pell, D-R.I.

1976 Congress allows states to issue tax-exempt bonds for education lending.

1977 College loans total $1.8 billion.

1978 Congress expands eligibility for Pell Grants to some middle-income students and makes federally guaranteed loans available to all income levels.

1979 Congress removes the cap on its subsidies to private education lenders, ensuring them good returns.

1980s *College costs rise, and pressure builds for aid to the middle class. With private lenders assured of high subsidies, loan volume explodes.*

1980 Parent Loans for Undergraduate Students — PLUS loans — allow parents at all income levels to borrow.

1986 Federal payments reimbursing lenders for education-loan defaults total more than $1 billion. . . . Private lenders barred from offering inducements to borrow, such as free appliances.

1989 College loans total $12 billion.

1990s *Private lenders begin marketing directly to students and parents. Merit-based scholarships proliferate in states.*

1990 A large guarantee organization, the Higher Education Assistance Foundation, collapses from loan defaults at for-profit colleges.

1992 Congress and the Clinton administration establish the federal Direct Lending Program to compete with private education lenders.

1993 Georgia uses lottery receipts to fund new merit-based HOPE scholarships for families with incomes below $66,000.

1994 College attendance peaks among welfare recipients.

1995 Georgia removes income restriction on HOPE Scholarships.

1996 Student loans total $30 billion. . . . Welfare-reform law directs low-income women toward jobs and away from college.

1997 Congress increases middle-class aid with tax credits, tax-free college savings accounts and tax deductions for student-loan interest.

2000s *More private lenders offer student loans at market rates without government guarantees, as rising tuitions see wealthier families seeking loans.*

2000 Private, non-government-backed loans total $4 billion. . . . Georgia allows HOPE Scholarship recipients to accept Pell Grants.

2002 Need-blind merit grants account for 24 percent of state education grants, up from 10 percent in 1992.

2005 Congress bars students from discharging most fully private education loans in bankruptcy.

2006 Non-federally guaranteed private loans for college education total over $16 billion. . . . Harvard waives tuition payments for families earning less than $60,000.

2007 New York Attorney General Andrew Cuomo investigates colleges that accept favors to steer students toward specific lenders. . . . Harvard expands tuition breaks to families earning up to $180,000. . . . Congress cuts subsidies to private lenders, ups Pell Grant funding and institutes income-dependent repayment for direct government loans. . . . Private investors back out of deal to buy student-lending giant Sallie Mae, which faces rising defaults after extending loans to risky borrowers.

2008 University Financial Services in Clearwater, Fla., settles charges it deceptively used Ohio University's logo and mascot to market loans.

While the picture varies from school to school, by the 1929-30 academic year, grant aid amounted to only 2.5 percent of U.S. college costs, including both tuition and living costs. Even after the Depression, when college charges fell and grant aid rose, grants amounted only to 3.6 percent of college costs by 1939-40.

Expanding Access

With the end of World War II, a new boom in financial aid began, along with new ideas about who would go to college. By the mid-20th century, large numbers of students became the first in their families to attend college.

In 1944, as the war was approaching its end, Congress enacted the Servicemen's Readjustment Act — the so-called G.I. Bill of Rights. As returning veterans signed up for classes, student aid quickly grew as a proportion of college costs, although the spurt was temporary. By 1949-50, student grants amounted to 55 percent of costs. That share fell to 14 percent by the 1959-60 school year, after G.I. benefits tailed off.

At the same time, colleges, especially state universities, were raising their fees — albeit gradually — after many decades of subsidizing all students by charging very low tuition.

"In the 1950s and 1960s, people began saying, 'We're subsidizing people who would go to college anyway. So why not charge something and aid students who can't afford it?'" says Hearn. Accordingly, at public colleges a new era of higher tuitions offset by financial aid for lower-income students replaced the old system under which states heavily subsidized lower tuition for all students in public universities.

With tuitions on the rise, the federal government as well as individual colleges launched new aid programs in the late 1950s. The new programs were influenced by the growing interest in seeing more students attend college, especially after the Soviet Union highlighted its national scientific prowess with the launch of the Sputnik satellite in 1957.

In 1954, a group of 95 private colleges and universities, mostly in the Northeast, formed the College Scholarship Service. Based on the philosophy that college aid should be largely based on need, the group set about developing standards and tools the schools could use to collect and assess a family's financial information and determine how much aid students required.[28]

In 1958, the National Defense Education Act financed low-interest college loans, with debt cancellation for students who became teachers after graduation. In 1965, Congress launched several aid programs, including a talent search to identify low-income students with academic ability and College Work-Study to subsidize schools' employment of needy students.[29]

By 1966-67, the poorest quarter of college students were getting 94 percent of their college costs paid for, 44 percent of it through grants, says Rupert Wilkinson, a former professor of history and American studies at England's University of Sussex. The aid dropped off steeply for the second-poorest quarter of students, however; only 38 percent of that group's needs were covered, 15 percent of it by grants, he says.

In 1972, Congress created Basic Grants, later renamed for Sen. Claiborne Pell, D-R.I. Initially proposed by the Nixon administration, the grants were designed to offer low-income students a basic subsidy large enough to ensure that they saw college as a possibility, but not so large that students wouldn't need to tap other sources.

Borrowers and Lenders

In addition to the various aid programs launched in 1965, Congress created the Federal Family Education Loan Program (FFELP) — then called the Guaranteed Student Loan Program — to offer federally backed loans through private lenders whom the government insured against default.

In the 1960s and early '70s student-lending programs didn't garner much attention, since college costs were still relatively low. Today, however, about 70 percent of federal student aid is in the form of government-guaranteed loans, while only about 20 percent is in the form of grants. Another 5 percent of aid is in the form of tax benefits, and the rest comes through various channels such as support for work-study programs.[30]

The growing importance of loans to students has become controversial over the last two decades. A "funding gap" between tuition and grant aid began in the early 1980s, "and that's where you begin to see students relying more on loans," says Karen Miksch, an assistant professor of higher education and law at the University of Minnesota. Since then, the funding gap has grown exponentially, she says.

Furthermore, with tuition rising and more students aspiring to college, middle-class families, which didn't

Low-Income Students Unaware of Aid

Clearer information is needed

College dreams for U.S. students — rich and poor — are at the highest levels ever. But when it comes to attendance and graduation, low-income students lag as far behind the middle class as they did 30 years ago.

More than 90 percent of students in all demographic groups now hope to attend college — the same expectation level for high-school graduation just a few years ago, says James C. Hearn, a professor of higher education at the University of Georgia. But while minority and low-income kids have quickly caught up to middle-class expectations, "big attainment gaps remain between what minority and low-income students aspire to do and what they actually do," he says.

For some of the students who hope to go to college but don't, the availability of aid is not the real problem, says Hearn. "Given the aid that's out there, there are still fewer low-income and minority students attending college than you'd expect," he says.

Some of the barriers between low-income students and college, such as bad schools and difficult family situations, are well-known and intractable. But many researchers are pointing to a hitherto unnoticed problem that's easier to fix: Low-income students and their families are less likely to know that aid is available for them, possibly causing many to give up on their college dreams.

"It really relates to how people grow up and whether they think of themselves as being able to go or not," says Ed St. John, a professor of higher education at the University of Michigan.

While middle- and upper-class students and their families believe college is in the future and prepare for it, many lower-income students doubt they can make it. "And if you can't imagine being able to pay for college, why would you prepare for it?" asks Karen Miksch, an assistant professor of higher education and law at the University of Minnesota.

Research shows that "middle- and upper-class kids get information from a whole variety of sources," says Sara Goldrick-Rab, an assistant professor of education policy at the University of Wisconsin, Madison, including the Internet, high-school counselors and college-educated family friends. But lower-income students, most of whom attend schools lacking guidance counselors, rely mainly on friends and family with little college experience, she says.

qualify for public need-based grants, were clamoring for aid.

In 1979-80, Pell Grants covered 99 percent of costs. Today a Pell Grant covers only 36 percent of tuition and on-campus room and board at the average four-year public institution, By the 1990s the gap between a Pell Grant and tuition at a public school was $4,000; today, it's more than $6,000, Miksch says.

"You're looking at two-thirds of the cost that now has to come from someplace else," Miksch adds.

For most, that "someplace else" is the private sector — specifically federally guaranteed private loans offered to both students and parents through several programs, such as federal Parent Loans for Undergraduate Students (PLUS), created in 1980.

Questions about how private lenders work in this arena stem mainly from the way the government pays them to participate, says the New America Foundation's Delisle.

When the program started, no banks were willing to lend to college students, he says, so Congress guaranteed lenders against default and also provided generous subsidies to entice lenders into the game. Because the government isn't set up to do the work of a bank, it uses private lenders' infrastructure to make and manage the loans.

Driven by concerns that not all students would get access to loans, Congress allowed as many lenders as possible to participate in the program and set subsidies high enough to attract many lenders.

Beginning in the 1980s, however, concerns grew in Congress about whether private lenders manipulated the system to push students into inadvisable loans. In 1986, Congress barred banks and other lenders from offering students "inducements" to borrow — such as toasters or other appliances.[31]

Despite tighter rules, however, by the early 1990s Congress and President Bill Clinton remained convinced

Informing students when they're in middle school that college aid will be available to them is crucial, because when students know they can go to college, they're more likely to stay in school and take courses that will prepare them for it, says Donald E. Heller, director of the Center for the Study of Higher Education at Pennsylvania State University.

He points to Indiana's successful Twenty-first Century Scholars Program targeting low-income eighth-graders. They are told that they can attend state colleges tuition-free or receive aid to attend a private college in Indiana if they graduate high school with a 2.0 average, use no illegal drugs or alcohol, commit no crimes and enroll in college within two years of graduation.[1]

Among 2,202 students enrolled in the program, 1,752 — nearly 80 percent — enrolled in a college in the state within a year of graduation, according to the Indianapolis-based Lumina Foundation for Education.[2]

Indiana also has a program to increase parental involvement, and it has worked, says Heller. It "creates a culture of college-going" and "gives the kids an incentive to prepare themselves," he says.

Congress could enact a similar early-commitment program, pledging federal Pell Grants to eligible middle-schoolers, says Heller. "By a conservative estimate, over 75 percent of students who get a free or reduced-price school lunch ultimately will be eligible for Pell," he says. "If they

knew about that and kept working in school because of it, they'd get a grant that's more than the average tuition at a community college," he says.

The current system of applying for aid is complicated for a good reason: to target aid to the neediest students. But the complexity itself puts low-income families at a disadvantage because they are more likely to be daunted by the form and less likely to find good help to prepare it, according to a study by Susan M. Dynarski and Judith E. Scott-Clayton of Harvard University's Kennedy School of Government. Dropping the federal-aid application form from 72 questions to a more manageable 14 would result in virtually no change in eligibility for Pell Grants, they found in a recent analysis.[3]

Furthermore, programs such as Georgia's HOPE Scholarship and Social Security student benefits provide plenty of evidence that simplified aid programs can increase college enrollments significantly, they found.

---

[1] For background, see "Meeting the Access Challenge: Indiana's Twenty-first Century Scholars Program," Lumina Foundation for Education, August 2002, www.luminafoundation.org.

[2] *Ibid.*

[3] Susan M. Dynarski and Judith E. Scott-Clayton, "College Grants on a Postcard," The Hamilton Project, The Brookings Institution, February 2007. Dynarski is an associate professor of public policy at Harvard; Scott-Clayton is a doctoral student.

---

that lenders were unfairly getting rich off subsidies they received for offering federally guaranteed loans. In 1992, Congress created the Federal Direct Lending Program, which makes student loans without going through private middlemen, to compete with the private lenders in FFELP. In 1993, Congress called for direct government lending to make up 60 percent of federal student loans by 1998-99.

These moves didn't eliminate the private sector's role in the student-aid business. In fact, critics see education lenders' hot competition for borrowers as evidence that the federal subsidies remain too high.

Furthermore, nonprofit student lenders have been set up in most states, says Delisle, and "members of Congress — including Republicans — say we have to keep the subsidy high because we have to keep the nonprofits in business," although there's little evidence that those organizations serve the program any better than the hordes of private lenders.

Lenders Fight Back

When Republicans took control of Congress in 1995, they expressed dismay at the burgeoning government-run Federal Direct Lending Program. By then, 1,300 colleges — about a third of the nation's total — had switched to direct lending, taking that market share away from private lenders. Congressional leaders pressured Clinton to slow the program's growth.

They told him he would get no other legislation passed "if he continued to push direct lending," said Robert M. Shireman, a former top aide to Sen. Paul Simon, D-Ill., who sponsored the 1993 direct-lending expansion.[32]

Clinton agreed to back off, and private lenders stepped up their efforts to regain business. For example, private lenders such as Sallie Mae offered student borrowers discounted fees and interest rates that the federal government wasn't legally permitted to match.

Non-Traditional Students Face Pitfalls

Many feel disconnected from campus life

As more Americans aspire to higher education, the number of non-traditional students — active military, older workers, immigrants — has skyrocketed, causing problems for the system and the students.

The earliest large-scale federal student-aid program was the G.I. Bill of 1944, which helped World War II vets attend college by providing money and new education programs to suit these non-traditional, often older students.

Today, the Pentagon still touts education funding as a benefit of military service, but today's system may not be working as planned. Of the veterans who entered four-year colleges in 1995, only 3 percent had graduated by 2001, compared with about 30 percent of students overall. Several reasons account for this low rate. Universities don't have to refund tuition for soldiers who are pulled out in mid-semester for overseas deployment, and the schools are allowed to terminate veterans' student status if they don't immediately re-enroll when they return from deployment.[1]

Military assurances that soldiers can attend college online from their bases also may be unrealistic for many. "I don't know how they expect us to take classes in Iraq," said Alejandro Rocha, 23, a Marine from Los Angeles. "I manned Humvees and rolled around in Humvees. . . . When we were back in the U.S., we were just training and training." Consistent study wasn't possible, Rocha said.[2]

Growing numbers of today's students are older and often work to support families, but financial-aid rules and practices sometimes make it difficult for them.

"A good 40 percent of Latino students are working off-campus," for example, says Amaury Nora, a professor of education leadership and cultural studies at the University of Houston. Working off-campus — which is often a side effect of inadequate aid — distances students from college life, with serious negative effects on their ability to persist in school, Nora says. "Students who work off-campus are 36 percent more likely to drop out."

Stress and disconnection are part of the reason, he adds. Emotional stress related to juggling off-campus jobs and classes and feeling strapped for money keeps students from fully participating in class and the other activities important to college life, he says. The less connected students feel, the more likely they are to drop out.

For most middle- and upper-income students, rising tuitions will only change which schools they attend, says William Doyle, an assistant professor of higher education at Vanderbilt University in Nashville, Tenn. "But for lower-income students, the price will cause them not to go to school, to work more or to go to school part time," all non-traditional

To attract universities, private lenders launched incentives such as the "school-as-lender" programs. In these deals, universities agree to stop offering federal direct loans to their professional- and graduate-school students and lend the money themselves, backed by a private lender. Then the university sells the loans back to the private lender for a profit.

In 2004, for example, the University of Nebraska made such a deal with the National Education Loan Network (Nelnet) and dropped direct federal loans. "The government could not match Nelnet's offer, which would provide the university with dollars it could apply to more-generous financial-aid packages for its students," said Stephen Burd, a fellow at the New America Foundation.[33]

Realizing that banks would pay for the right to be named a preferred lender, some universities actively sought out favorable deals from private lenders in return for abandoning the federal direct loan program. In 2003, for example, Michigan State University, the second-largest participant in the government-run program, asked private lenders to compete for its business.

The university openly asked lenders, "What will you do for us if we leave direct lending?" said Barmak Nassirian, associate executive director of the American Association of Collegiate Registrars and Admissions Officers. Other colleges soon followed suit, soliciting benefits from lenders such as staff support for financial-aid offices in return for dropping out of federal direct loans. "A giant sucking sound was inaugurated with this deal," Nassirian said.[34]

paths that lower students' chance of graduating, he says.

"Once you see yourself as somebody who's working and also going to school, your chances of graduating decrease," says Doyle. Part-time study "pushes down your credit hours, and that robs you of momentum," partly because it makes graduation seem very far off indeed. Taking six credits per semester means it takes between eight and 10 years to finish college, a daunting prospect for most, he adds.

On-campus work-study jobs are an exception to the rule, says Karen Miksch, an assistant professor of higher education and law at the University of Minnesota. Work-study students tend to say, "My job, it was great. The people were so helpful," she says. Bosses on work-study jobs tend to be another voice helping students navigate the shoals of college, she adds.

Financial-aid rules sometimes trip up working students in unexpected ways, says Sara Goldrick-Rab, an assistant professor of education policy at the University of Wisconsin, Madison. Today, students' earnings count in the income-qualification calculation for most aid, so that low-income

Military veteran Marc Edgerly, a sophomore at George Mason University, in Fairfax, Va., says he will have $50,000 in student loans when he graduates despite federal education assistance for vets.

AP Photo/Jacquelyn Martin

students could literally work their way out of qualifying for aid, even though school would still be unaffordable, she says. "You can begin to work and find yourself without aid very quickly, and if you switch to part-time school, then your aid is cut."

The financial-aid system can work against non-traditional students in other ways, particularly if they are low-income students, Goldrick-Rab said. "We do not reward their choices. We penalize students who withdraw from school temporarily by stopping their financial aid and making it hard to restart. . . . We often fail to award full credit for courses our students take at other institutions, forcing them to repeat courses two or three times." College policies "are designed with traditional students, engaged in traditional attendance patterns, in mind."[3]

---

[1] Aaron Glantz, "Military Recruitment Lie: Pentagon's Education Pitch Is a Scam," *The Nation*, Nov. 29, 2007.

[2] Quoted in *ibid.*

[3] Sara Goldrick-Rab, "Connecting College Access With Success," *Wisconsin School Boards Magazine*, September 2005, p. 26.

Lenders also offered discounts, university staff support and other benefits as inducements for colleges to name them "preferred lenders." Students aren't required to pick a lender on a college's preferred list, but with students and parents often overwhelmed by the complexity of college applications and financing, the vast majority do.

"Lenders learned that it's much easier and more effective to market their goods and services to a couple of people at an institution than to thousands of customers," said Craig Munier, director of financial aid at the University of Nebraska, Lincoln, and chair of the National Direct Student Loan Coalition.[35]

With private lenders competing hard, direct government loan growth stalled, and private loans once again became the bulk of federal student lending. In 2005,

$287 billion in guaranteed private loans were outstanding, compared with $95 billion in direct loans.[36]

Meanwhile, as college costs continued to rise, private lenders also began offering non-government-guaranteed, wholly private education loans.

In 2005-06, lenders issued $17.3 billion in such loans, whose terms reflect those in regular consumer lending, including much higher interest rates than federally guaranteed loans.[37] In the past few years, the number of these loans has skyrocketed, making up 18 percent of all education loans by 2005, compared with 8.8 percent four years earlier.[38] Also in 2005, lenders won a provision in a new federal bankruptcy law that makes it extremely difficult to discharge the loans through bankruptcy, leaving many students with huge outstanding debt that will dog them for decades.

Only 12 percent of Hispanics and 16 percent of African-Americans eventually earn a B.A. degree, compared with 33 percent of non-Hispanic whites, according to researchers, who say difficulty in paying for college is partly to blame. Above, Hispanic students at Hayes High School in Birmingham, Ala. Federal Pell Grants for low-income students have lagged behind rising costs for decades.

The non-guaranteed private loans aren't subject to federal rules like the 1986 ban on offering "inducements" to borrow, and lenders have taken advantage of this to expand their markets. For example, Sallie Mae offered New York's Pace University a $4-million private-loan fund if it would agree to make Sallie Mae its exclusive lender.[39]

Analysts say two things are driving the growth of the private loan business: limits on the size of federally guaranteed loans that fall far short of covering annual college costs and the complexity of qualifying for federally backed loans.

"Borrowers have the private people coming to them" rather than having to seek them out, says Goldrick-Rab at the University of Wisconsin, Madison. Private lenders do direct mailings, and their application forms are faster and easier to complete than the Free Application for Federal Student Aid (FAFSA) forms, she says.

But she sees a big downside. "Financial-aid officers are seeing people getting private loans before they've maxed out" their eligibility for subsidized federal loans, and they could end up owing much more than they need to, she says.

Borrowers are "drawn to the '30 seconds and you'll be approved' type of approach we're seeing more and more

of now," said Robert Shireman, director of the nonprofit advocacy group Project on Student Debt.[40]

Also in the last decade-and-a-half, states have increasingly created mainly merit-based grant programs. The proportion of state aid that is merit-based rather than need-based had grown from 13 percent in 1994-95 to 27 percent by 2004-05.[41]

State merit-based aid wins political favor because it gives more support to the middle class.

The merit-based programs' main purpose may be state economic development, however. Such grants are pitched as economic development to attract good students to the state's schools and attract new residents, says Rubenstein of Syracuse. "We don't know if it's working," he says. "But now we have states competing against each other" to offer merit aid, which may reduce the grants' effectiveness in attracting upwardly mobile families to states.

Both state and private universities are currently in a merit arms race that may be cutting down what they're willing to spend on need-based aid, says Hossler of Indiana University. College rankings such as those published by *U.S. News & World Report* give big incentives for colleges to spend heavily on financial aid that could bring in students with higher SAT scores and a higher class rank, he says. Incoming freshmen with higher scores and grades mean "you automatically go up in the rankings," while with need-based scholarships or increased spending on faculty you don't, he says.

CURRENT SITUATION

In Congress

A full slate of financial-aid issues faces the new Democratic Congress and the Bush administration. Late in 2007, lawmakers modestly raised funding for Pell Grants, reduced interest rates on direct government loans, cut subsidies for private lenders and eased some students' loan burdens by making payments dependent on their incomes.[42]

The new law increases student financial aid by more than $20 billion over five years, paid for by cutting subsidies to lenders offering government-backed student loans.

The maximum annual Pell Grant for low-income students is slated to increase from $4,310 to $5,400 over

Should Congress do more to keep private lenders in the student-loan business?

YES
John Berlau
Director, Center for Entrepreneurship,
Competitive Enterprise Institute

From the CEI Web site

Current federal student-loan programs are not perfect. They are a mishmash of subsidies and regulations that cause distortions, not the least of which is to raise the sticker price of tuition. That doesn't mean that it's not possible for the student-aid system to get worse. And that's what a bill from Massachusetts Sen. Ted Kennedy would likely do.

Under the plan, the government would basically bribe schools with extra federal aid to participate only in the Direct Lending program — rather than have subsidized dealings with private banks — a plan that would result in massive waste as well as burdensome red tape.

When signed into law by President Bill Clinton in 1993, "direct lending" was sold as a way to actually make money for the government by cutting out the "middleman." Since the 1960s, the government has subsidized banks and other firms that lend to students to make student loans more affordable. Advocates argued that the government would spend less money and could even profit if it made the loans and collected the interest itself.

As with many claims for government programs, direct lending hasn't yielded the benefits it promised. Not only is the program not making a profit, but over the decade it has been in existence the costs of direct lending have been coming in higher than the government's initial estimates, and these cost differences have been increasing.

At the same time, subsidized loans from banks have cost the government less than their estimated costs. The White House budget for fiscal year 2006 reported that direct lending has cost the government $7 billion more than initially predicted over the last decade, while subsidized loans have cost the government $5 billion less than they were estimated to cost. In the words of a report from the respected spending watchdog Citizens Against Government Waste, "the Direct Loan program flunks out."

Direct Lending has also failed in its promise to greatly reduce default rates. Department of Education statistics show a projected default rate above 15 percent for direct loans in 2005. The rate for subsidized loans was 13 percent.

In addition, some 600 schools have stopped participating in direct lending, and surveys of schools cite poor service as the primary reason.

If we were to make federal student loans totally government run, we would lose the innovation that private firms can show in servicing their customers, even in a situation that's not the free-market ideal.

NO
Sen. Edward M. Kennedy, D-Mass.
Chairman, Senate Health, Education,
Labor and Pensions Committee

Written for *CQ Researcher*, January 2008

Millions of students face staggering tuition bills, and recent graduates juggle an average of about $20,000 in student debt. Congress should do more — not to keep private lenders in the loan business but to help students afford college and deal with debt. We should reduce unnecessary subsidies to private lenders and use the savings to increase aid to the neediest students.

Last year, Congress passed a bipartisan law raising federal aid to college students by $20 billion — the largest increase since the G.I. Bill. That's significant progress, but far from enough. Each year, 400,000 qualified students are still unable to attend a four-year college because of cost.

We should redouble efforts to ensure that the loan program is run as efficiently as possible. I'm eager to see the results of a pilot program that requires private lenders to bid for the right to offer federally subsidized parent loans. This mechanism will ensure that lenders compete for subsidies. The Congressional Budget Office estimates it will save $2 billion. If it's successful, we should expand it to all federal student loans made by private lenders.

We should examine the role of loan-guaranty agencies, which too often focus on aggressively pursuing borrowers who have defaulted, rather than preventing borrowers from defaulting. Colleges should be encouraged to switch from the privately funded loan program to the Direct Loan program, which time and again has been shown to be cheaper to taxpayers, and is untainted by the recent student-loan scandals.

Predictably, lenders claim the new law cut subsidies too deeply and is causing some to leave the program. Recently, the industry issued a flawed analysis suggesting that the privately funded loan program is less expensive than the Direct Loan program. Lenders have made these arguments before. The reality is that today more than 3,000 lenders participate in the federal program. Lenders make millions of dollars in profits from higher-interest private loans, which have grown tenfold in a decade and now account for almost a quarter of education loans.

The new law has begun to restore balance to the grossly unfair system by directing funds to students, not to banks. This year, we'll continue reform by finalizing new ethics rules. We'll ensure that banks treat students who take out private loans fairly and give them good terms and service. Most important, we'll keep the focus on students, so more can afford college and have a genuine chance at the American dream.

Joy Mueller displays her sense of humor during graduation at New York University. Mueller, who earned a master's degree in digital design, says the computer and the piggy bank signify the money she is going to have to earn with her computer skills to pay off her student loan. The average American owes $20,000 after graduation. Many students from low-income families — and graduate students in particular — also face large debt burdens.

five years. The reaction to the increase from college insiders was mixed.

"The increase in the Pell Grants is unprecedented in terms of its size," said Richard Doherty, president of the Association of Independent Colleges and Universities in Massachusetts.[43]

But Shelley Steinback, former general counsel of the American Council on Education, said that the increase would not have a really significant impact, given the likelihood of inflationary costs in everything that goes into providing higher education.[44]

Furthermore, whether the maximum authorized Pell Grant amounts are available each year will depend on whether funds are available in the federal budget. Already a budget squeeze has limited the funds for the first year of the expansion. For the 2008-09 school year, the new Pell maximum was set at $4,800, but in December Congress approved an omnibus spending bill that trims it to $4,731.[45]

For student borrowers, the new law temporarily cuts interest rates on both direct and federally guaranteed subsidized undergraduate loans — so-called Stafford loans — from 6.8 percent to 3.4 percent over four years. After the four years are up, it reverts to a higher rate, because Congress could not find a way to pay for a longer-lasting cut. And that draws a skeptical response from many congressional Republicans.

"Reducing student-loan interest rates is a good sound bite," but the per-student savings will be small; it "may be one latte, it may be two lattes. It's kind of hard to tell with today's market for coffee," said Wyoming Sen. Michael Enzi, top-ranking Republican on the Senate Health, Education, Labor and Pensions Committee.[46]

The new law also establishes the income-contingent loan repayment advocated by groups such as the Project on Student Debt for direct federal loans to students — not parents — and offers loan forgiveness for some graduates in public-service-oriented jobs in schools, charities and some government service.

Under income-based repayment, set to begin in July 2009, student-loan repayments in the direct lending program will be capped at a percentage of income, and most remaining balances will be canceled after 25 years. No payments will be required for people earning less than 150 percent of the federal poverty level — about $31,000 for a family of four — and payments will increase on a sliding scale above that level, with the highest earners' payments capped at 15 percent of family income. The plan will be available to all students, including those already holding the loans.

Public-service workers in both government and nongovernment organizations can have their loan debt from the government direct-lending program canceled after 10 years if they've made all the payments and been consistently employed in public service.

Meanwhile, loan experts say that even loan programs with lower student burdens probably won't improve college access much for lower-income students. And that still remains the goal of many higher-education analysts and advocates.

Willingness to borrow varies across groups, with low-income and minority families and students worrying more about the risks of loans, says the University of Pennsylvania's Perna.

In addition to concerns about filling out complicated loan applications and being able to repay in the future, many low-income and minority students know neighbors and family members who didn't complete college and are saddled with hefty loans they must repay on a high-school graduate's wages.

"The biggest problem is this uncertainty about the return on investment," and "part of this concern is definitely justified," Perna says. "Low-income and minority students are less likely to finish," while those who graduate often leave their old neighborhoods, skewing perceptions among remaining families, who only see dropouts around them.

Loan Scandals

Law enforcement authorities as well as lawmakers continue to take a keen interest in education aid following reports of illegal activities by lenders.[47]

Beginning in late 2006 and accelerating last year, reports surfaced of financial-aid administrators accepting payoffs and luxury gifts — like hefty consulting fees and Caribbean trips — from education lenders, presumably in exchange for listing the lenders as "preferred" in college financial-aid material. The flurry of stories stemmed from a massive investigation conducted by New York Attorney General Cuomo into what he called "an unholy alliance" between college officials and lenders involving kickbacks and other inducements for colleges to steer students toward certain lenders.

Indeed, financial-aid directors at Columbia University and the University of Southern California reportedly sat on the governing board of a New Jersey company, Student Loan Xpress, while reaping profits from selling company stock.[48]

Ironically, some of the revelations came from student journalists at the University of Texas, Austin. They found evidence that their school's financial-aid office kept tabs on how many gifts, "lunches, breakfasts and extracurricular functions" loan companies offered the university staff and factored the gifts into its decisions about which lenders to recommend. The office's director

was placed on leave in April 2007 when he was found to own at least 1,500 shares in the parent company of Student Loan Xpress, one of the lenders the university recommended.[49]

In April 2007 Cuomo announced the first settlements related to his investigation. More than two-dozen colleges and 10 lenders have signed pledges to abide by a new code of conduct that increases disclosure requirements on college-lending practices and bans gifts and compensation from lenders to college officials.[50] Colleges and lenders also have paid monetary settlements of more than $15 million as restitution to borrowers or as seed money for loan-education funds.

Investigations are ongoing in some other states.

Alarmed that Iowa students have a high debt burden — estimated at an average $22,926 for 2006 graduates, sixth-highest in the country — officials in 2007 investigated the nonprofit Iowa Student Loan Liquidity Corp., which the state created in 1979 to help make guaranteed loans available.[51] By 2007, the corporation was the state's dominant lender, with $3.3 billion in outstanding loans. Last fall, with the state attorney general investigating the agency's business practices, some Iowa lawmakers threatened to bar it from raising money for loans by issuing tax-free bonds.[52]

A similar nonprofit, the Pennsylvania Higher Education Assistance Agency, is also under scrutiny by state auditors after giving its senior employees $7 million in bonuses since 2004; the agency's chief executive resigned in October.

The Missouri Higher Education Agency also has paid millions of dollars in perks to senior executives. "They did not act like a state agency at all" but had "the mindset that they were a for-profit business," said state auditor Susan Montee.[53]

OUTLOOK

Generation Debt

As college costs soar, helping students pay for college will remain high on the agenda in the states and in Washington.

"We're at the intersection of two really important trends bumping against each other" that may change approaches to education finance, says Vanderbilt's Doyle. "One is the increasing importance of postsecondary education to attain a middle-class lifestyle." The other is a "bulge in the demographic pipeline," with the children of baby boomers — the baby-boom "echo" — set to create by 2009 the biggest generation ever with college aspirations.

Public pressure from all income groups for relief on college costs will only grow, and that likely means pressure on colleges to slow cost increases, says Doyle. "One thing we know is that states won't be doing a big tax increase" any time soon, he says. "So the question will be asked, 'Do the costs have to go up this fast?'"

"A couple of years of double-digit tuition increases and kids not getting the classes they need" will bring political pressure on the government and colleges to act, Doyle adds.

"Can institutions keep raising tuition without resistance? Probably not," says the University of Georgia's Hearn. "We may be hitting the point where public opposition" becomes a big factor in tuition decisions. "There may be some strong, strong resistance from students and families."

Already, the burden of education loans is heavy for many graduates in all income brackets, says Doyle. "I'm always surprised at the level of debt" even for "students from higher-income families," he says.

Most political trends point to a continuing evolution toward a system of loans and merit-based aid that helps the middle class, says Penn State's Heller. Political, not financial, imperatives are driving that shift, he emphasizes. "The merit-based aid that has cropped up in states in recent years could just as easily go to need-based aid, but political considerations work in the opposite direction."

The prospect of heavier reliance on loans and merit-based grants to help families with ever-rising college costs raises troubling questions about the future of college access for low-income, minority and immigrant students. The two fastest-growing parts of college aid — loans and merit-based grants — do little for those populations, as costs continue to rise.

Even colleges that currently make strong efforts to boost need-based grants and other assistance for low-income students probably won't do so indefinitely.

"Paying for 100 percent of need is expensive," said Yvonne Hubbard, director of student financial services at the University of Virginia. The school's need-based assistance program, AccessUVa, pays for everything for qualified students and costs the university about $20 million a year.[54]

The University of Illinois at Urbana-Champaign also has a strong program of need-based grants to improve access. Still, "What happens four years or 10 years down the road?" asked Daniel Mann, the school's director of financial aid, who doubts that future administrations will support the current ambitious aid program.[55]

The nation's burgeoning population of Hispanic immigrants also may mean that college attendance and graduation rates would decline, since Hispanics have "relatively low rates" of college attendance, says Penn's Perna.

"I'm a little skeptical of simple correspondences between college-graduation rates and money, but if nothing changes, we are going to see fewer college graduates" in years to come, says Indiana's Hossler.

NOTES

1. Quoted in Diana Jean Schemo, "Private Loans Deepen a Crisis in Student Debt," *The New York Times*, June 10, 2007.

2. For background, see Tom Price, "Rising College Costs," *CQ Researcher*, Dec. 5, 2003, pp. 1013-1044.

3. "Trends in College Pricing, 2007," The College Board, 2007, www.collegeboard.com/prod_downloads/about/news_info/trends/trends_pricing_07.pdf.

4. Jane V. Wellman, "Costs, Prices and Affordability," Commission on the Future of Higher Education, www.ed.gov/about/bdscomm/list/hiedfuture/reports/wellman.pdf.

5. Matthew Keenan and Brian Kladko, "Harvard Targets Middle Class With Student Cost Cuts," Bloomberg.com, Dec. 10, 2007.

6. Susan M. Dynarski and Judith E. Scott-Clayton, "College Grants on a Postcard: A Proposal for Simple and Predictable Federal Student Aid," The Hamilton Project, The Brookings Institution, February 2007.

7. Steven Pearlstein, "Cost-Conscious Colleges," *The Washington Post*, Nov. 16, 2007, p. D1.

8. Kate Sabatini and Pedro de la Torre III, "Federal Aid Fails Needy Students," Center for American Progress, May 16, 2007, www.americanprogress.org.

9. "What Financial Aid Officers Say," America's Student Loan Providers, www.aslp.us.

10. Tracey King and Ivan Frishberg, "Big Loans, Bigger Problems: A Report on the Sticker Shock of Student Loans," The State PIRGs, March 2001.

11. Quoted in Jay Mathews, "As Merit-Aid Race Escalates, Wealthy Often Win," *The Washington Post*, April 19, 2005, p. A8.

12. Donald Heller and Patricia Marin, "State Merit Scholarship Programs and Racial Inequality," The Civil Rights Project, 2004.

13. Quoted in Mathews, *op. cit.*

14. Quoted in Tara Malone, "Rising Tuition Hits Middle Class Hardest," [Chicago] *Daily Herald*, Nov. 21, 2004, www.collegeparents.org.

15. Quoted in *ibid.*

16. Richard Vedder, "The Real Costs of Federal Aid to Higher Education," *Heritage Lectures*, Jan. 12, 2007.

17. Ross Rubenstein, "Helping Outstanding Pupils Educationally," Education Finance and Accountability Project, 2003.

18. Radley Balko, "Government May Be Cause, Not Solution, to Gen Y Economic Woes," Cato Institute, July 12, 2006, www.cato.org.

19. Quoted in Larry Abramson, "Student Loan Industry Struggles Amid Controversy," "Morning Edition," National Public Radio, June 26, 2007.

20. John Berlau, "Ted Kennedy Says Eliminate Private Sector From Student Loans," Competitive Enterprise Institute Web site, April 4, 2007, www.cei.org/utils/printer.cfm?AID=5854.

21. Vedder, *op. cit.*

22. Quoted in Abramson, *op. cit.*

23. Michael Dannenberg, "A College Access Contract," New America Foundation Web site, www.newamerica.net.

24. Dannenberg, *op. cit.*

25. Jason Delisle, "The Business of Sallie Mae — Political Risk for Investors and Taxpayers," Higheredwatch.org, New America Foundation, Oct. 2, 2007, www.newamerica.net.

26. "Sallie Mae Decides To Be More Selective In Pursuing Loan Origination Activity," RTT News Global Financial Newswires, Jan. 8, 2008, www.rttnews.com/sp/breaking news.asp?date=01/04/2008&item=103&vid=0.

27. For background, see Rupert Wilkinson, *Buying Students: Financial Aid in America* (2005), and Tom Price, "Rising College Costs," *CQ Researcher*, Dec. 5, 2003, pp. 1013-1044.

28. "History of Financial Aid," Center for Higher Education Support Services (Chess Inc.), www.chessconsulting.org/financialaid/history.htm.

29. Lawrence E. Gladieux, "Federal Student Aid Policy: A History and an Assessment," in *Financing Postsecondary Education: The Federal Role* (1995), U.S. Department of Education, www.ed.gov/offices/OPE/PPI/Fin-PostSecEd/gladieux.html.

30. "Higher Education," New America Foundation, www.newamerica.net/programs/education_policy/federal_education_budget_project/higher_ed.

31. For background, see Kelly Field, "The Selling of Student Loans," *The Chronicle of Higher Education*, June 1, 2007.

32. Quoted in *ibid.*

33. Stephen Burd, "Direct Lending in Distress," *The Chronicle of Higher Education*, July 8, 2005.

34. Quoted in Field, *op. cit.*

35. Quoted in *ibid.*

36. Deborah Lucas and Damien Moore, "Guaranteed vs. Direct Lending: The Case of Student Loans," paper prepared for National Bureau of Economic Research conference, January 2007, www.newamerica.net/files/Guaranteed%20vs.%20Direct%20Lending.pdf.

37. "Private Loan Policy Agenda," Project on Student Debt, http://projectonstudentdebt.org/initiative_view.php?initiative_idx=7.

38. Aleksandra Todorova, "The Best Rates on Private Loans," *Smart Money*, June 20, 2006, www.smartmoney.com/college/finaid/index.cfm?story=privateloans.

39. Field, *op. cit.*

40. Quoted in Sandra Block, "Private Student Loans Pose Greater Risk," *USA Today*, Oct. 25, 2006.

41. "37th Annual Survey Report on State-Sponsored Student Financial Aid," National Association of State Student Aid and Grant Programs, July 2007, www.nassgap.org.

42. For background, see "Summary of The College Cost Reduction And Access Act (H.R. 2669)," News from NASFAA, National Association of Student Financial Aid Administrators, www.nasfaa.org/Publications/2007/G2669summary091007.html.

43. Alex Wirzbicki, "$20.2 Billion Boost in Student Aid Approved," *The Boston Globe*, Sept. 8, 2007, www.boston.com.111.

44. Quoted in *ibid.*

45. Jason Delisle, "Pell Grants Cut," Higheredwatch.com, New America Foundation, Dec. 18, 2007.

46. Quoted in Libby George, "Broad Student Aid Overhaul Clears," *CQ Weekly*, Sept. 10, 2007, p. 2620.

47. For background, see "Special Report: Student Loan Scandal," Education Policy Program, New America Foundation, www.newamerica.net.

48. John Hechinger, "Probe Into College-Lender Ties Widens," *The Wall Street Journal*, April 5, 2007.

49. Josh Keller, "University of Texas Financial-Aid Office Took Gifts From Lenders, Student Journalists Report," *The Chronicle of Higher Education*, May 11, 2007.

50. Meyer Eisenberg and Ann H. Franke, "Financial Scandals and Student Loans," *The Chronicle of Higher Education*, June 29, 2007.

51. "Student Debt and the Class of 2006," The Project on Student Debt, September 2007.

52. Jonathan D. Glater, "College Loans by States Face Fresh Scrutiny," *The New York Times*, Dec. 9, 2007.

53. *Ibid.*

54. Quoted in Karin Fischer, "Student-Aid Officials Say Efforts to Expand Access Need Widespread Backing," *The Chronicle of Higher Education*, Sept. 22, 2006.

55. Quoted in *ibid.*

BIBLIOGRAPHY

Books

Getz, Malcolm, *Investing in College: A Guide for the Perplexed*, Harvard University Press, 2007.
An associate professor of economics at Vanderbilt University outlines the questions parents and students should ask about colleges and their financial-aid programs.

Vedder, Richard, *Going Broke by Degree: Why College Costs So Much*, AEI Press, American Enterprise Institute, 2004.
A former Ohio University economics professor argues that for-profit universities can provide badly needed price competition for traditional colleges, where tuitions are skyrocketing because the schools are inefficient and spend too much subsidizing non-instructional programs like sports.

Wilkinson, Rupert, *Buying Students: Financial Aid in America*, Vanderbilt University Press, 2005.
A former professor of American studies and history at Britain's University of Sussex details the social and economic history of student financial aid.

Articles

Field, Kelly, "The Selling of Student Loans," *The Chronicle of Higher Education*, June 1, 2007.
Beginning with the creation of the direct federal-loan program in the early 1990s, which set up a government competitor to private student-loan firms, lenders competed to be colleges' "preferred" loan sources, offering discounts, gifts and other favors to woo financial-aid officers.

Fischer, Karin, "Student-Aid Officials Say Efforts to Expand Access Need Widespread Backing," *The Chronicle of Higher Education*, Sept. 22, 2006.
Officials at selective universities say need-based grants and personal support are needed to expand enrollment of low-income students in top schools.

Schemo, Diana Jean, "Private Loans Deepen a Crisis in Student Debt," *The New York Times*, June 10, 2007.
Non-government-guaranteed loans are becoming a bigger part of the college financing picture as costs climb, and, unlike guaranteed loans, interest rates can be as high as 20 percent.

Reports and Studies

"Course Corrections: Experts Offer Solutions to the College Cost Crisis," Lumina Foundation for Education, October 2005, www.collegecosts.info/pdfs/solution_papers/Collegecosts_Oct2005.pdf.
Analysts assembled by a nonprofit group suggest technological and organizational changes to control rising costs.

Recession, Retrenchment, and Recovery: State Higher Education Funding and Student Financial Aid, **Center for the Study of Education Policy, Illinois State University, October 2006, www.coe.ilstu.edu/eafdept/centerforedpolicy/downloads/3R%20report10272006/3R_Final_Oct06_Updated%5B1%5D.pdf.**
Analysts explore the consequences for higher-education funding and student aid in an era where state governments face recurring severe budget shortfalls.

"Student Debt and the Class of 2006," The Project on Student Debt, September 2007, http://projectonstudentdebt.org/files/pub/State_by_State_report_FINAL.pdf.
An advocacy group finds District of Columbia students have the highest debt — an average of $27,757 — and Oklahoma grads the lowest: $17,680 on average.

Cook, Bryan J., and Jacqueline E. King, "2007 Status Report on the Pell Grant Program," American Council on Education, June 2007, www.acenet.edu/AM/Template.cfm?Section=Home&TEMPLATE=/CM/ContentDisplay.cfm&CONTENTID=23271.
Analysts for a university membership alliance trace trends in the federal need-based grant program, finding that Pell grantees' median incomes are around $18,000, compared to $55,000 for other undergraduates.

Dynarski, Susan M., and Judith E. Scott-Clayton, "College Grants on a Postcard: A Proposal for Simple and Predictable Student Aid," The Hamilton Project, The Brookings Institution, February 2007, www.brookings.edu/~/media/Files/rc/papers/2007/02education_dynarski/200702dynarski%20scott%20clayton.pdf.
Public-policy analysts at Harvard argue that a drastically simplified federal aid-application process would significantly increase college enrollment among low-income and minority students.

Haycock, Kati, "Promise Abandoned: How Policy Choices and Institutional Practices Restrict College Opportunities," The Education Trust, August 2006, www2.edtrust.org/NR/rdonlyres/B6772F1A-116D-4827-A326-F8CFAD33975A/0/PromiseAbandonedHigherEd.pdf.
A nonprofit group says higher-education and financial-aid policies aren't helping low-income and minority students catch up to their higher-income white peers.

Wolfram, Gary, "Making College More Expensive: The Unintended Consequences of Federal Tuition Aid," Policy Analysis No. 531, Cato Institute, January 2005, www.cato.org/pubs/pas/pa531.pdf.
An analysis prepared for a libertarian think tank argues that college costs would decrease if private aid, rather than government aid, helped students fund their studies.

For More Information

American Association of Collegiate Registrars and Admissions Officers, One Dupont Circle, N.W., Suite 520, Washington, DC 20036; (202) 293-9161; www.aacrao.org. Provides information and professional education on college-admissions issues and policies.

America's Student Loan Providers, www.studentloanfacts.org. Advocacy group of banks and other organizations that make federally guaranteed college loans.

Education Sector, 1201 Connecticut Ave., N.W., Suite 850, Washington, DC 20036; (202) 552-2840; www.educationsector.org. A think tank providing research and analysis on education issues, including financial aid.

Lumina Foundation for Education, P.O. Box 1806, Indianapolis, IN 46206-1806; (317) 951-5300; www.luminafoundation.org. Supports higher-education research and projects to improve college access.

National Association for College Admission Counseling, 1050 N Highland St., Suite 400, Arlington, VA 22201; (703) 836-2222; www.nacacnet.org. Provides information and news updates related to college admissions.

National Association of State Student Grant and Aid Programs, www.nassgap.org. Provides information on state financial-aid programs.

National Association of Student Financial Aid Administrators, 1101 Connecticut Ave., N.W., Suite 1100, Washington, DC 20036; (202) 785-0453; www.nasfaa.org. Sets voluntary standards and provides professional education on student financial aid.

New America Foundation, 1899 L St., N.W., Suite 400, Washington, DC 20036; (202) 986-2700; www.newamerica.net. Liberal think tank provides research and analysis of higher-education issues, with a focus on financial aid.

Pell Institute for the Study of Opportunity in Higher Education, 1025 Vermont Ave., N.W., Suite 1020, Washington, DC 20005; (202) 638-2887; www.pellinstitute.org. Think tank that conducts research on college access and financial aid for low-income students.

Project on Student Debt, 2054 University Ave., Suite 500, Berkeley, CA 94704; (510) 559-9509; http://projectonstudentdebt.org. Provides advocacy and information on the growing debt load carried by U.S. students.

U.S. PIRG, 218 D St., S.E., Washington, DC 20003; (202) 546-9707; www.uspirg.org/higher-education. Consumer group advocating for more need-based financial aid and simplified aid-application processes.

11

Religious Fundamentalism

Does It Lead to Intolerance and Violence?

Brian Beary

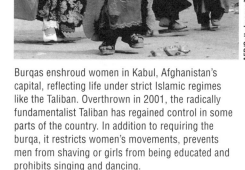

Burqas enshroud women in Kabul, Afghanistan's capital, reflecting life under strict Islamic regimes like the Taliban. Overthrown in 2001, the radically fundamentalist Taliban has regained control in some parts of the country. In addition to requiring the burqa, it restricts women's movements, prevents men from shaving or girls from being educated and prohibits singing and dancing.

From *CQ Researcher*, February 2009.

L ife is far from idyllic in Swat, a lush valley once known as "the Switzerland of Pakistan." Far from Islamabad, the capital, a local leader of the Taliban — the extremist Islamic group that controls parts of the country — uses radio broadcasts to coerce residents into adhering to the Taliban's strict edicts.

"Un-Islamic" activities that are now forbidden — on pain of a lashing or public execution — range from singing and dancing to watching television or sending girls to school. "They control everything through the radio," said one frightened Swat resident who would not give his name. "Everyone waits for the broadcast." And in case any listeners in the once-secular region are considering ignoring Shah Duran's harsh dictates, periodic public assassinations — 70 police officers beheaded in 2008 alone — provide a bone-chilling deterrent.[1]

While the vast majority of the world's religious fundamentalists do not espouse violence as a means of imposing their beliefs, religious fundamentalism — in both its benign and more violent forms — is growing throughout much of the world. Scholars attribute the rise to various factors, including a backlash against perceived Western consumerism and permissiveness. And fundamentalism — the belief in a literal interpretation of holy texts and the rejection of modernism — is rising not only in Muslim societies but also among Christians, Hindus and Jews in certain countries. (*See graph, p. 246.*)

Religious Fundamentalism Spans the Globe

Fundamentalists from a variety of world religions are playing an increasingly important role in political and social life in countries on nearly every continent. Generally defined as the belief in a literal interpretation of holy texts and a rejection of modernism, fundamentalism is strongest in the Middle East and in the overwhelmingly Christian United States.

Where Fundamentalism Influences Social and Political Life

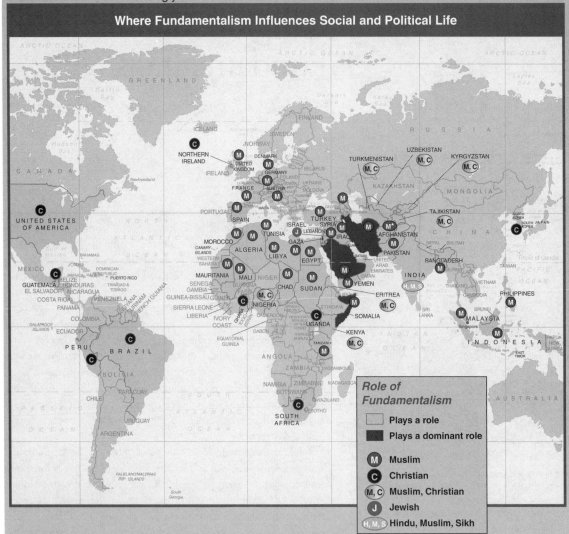

Role of Fundamentalism

☐ Plays a role
■ Plays a dominant role

Ⓜ Muslim
Ⓒ Christian
ⓂⒸ Muslim, Christian
Ⓙ Jewish
Ⓗ,Ⓜ,Ⓢ Hindu, Muslim, Sikh

* The ultra-conservative Taliban ruled from 1996-2001 and are fighting to regain control.

Sources: U.S. National Counter Terrorism Center, Worldwide Incidents Tracking System, http://wits.nctc.gov; David Cingranelli and David Richards, Cingranelli-Richards (CIRI) Human Rights Dataset, CIRI Human Rights Project, 2007, www.humanrightsdata.org; The Association of Religious Data Archives at Pennsylvania State University, www.thearda.com; Office of the Coordinator for Counterterrorism, Country Reports on Terrorism, United States Department of State, April 2008, www.state.gov/documents/organization/105904.pdf; Peter Katel, "Global Jihad," CQ Researcher, Oct. 14, 2005

Islamic fundamentalism is on the rise in Pakistan, Afghanistan, the Palestinian territories and European nations with large, often discontented Muslim immigrant populations — notably the United Kingdom, Germany, Denmark, Spain and France, according to Maajid Nawaz, director of the London-based Quilliam Foundation think tank.

In the United States — the birthplace of Christian fundamentalism and the world's most populous predominantly Christian nation — 90 percent of Americans say they believe in God, and a third believe in a literal interpretation of the Bible.[2] Perhaps the most extreme wing of U.S. Christian fundamentalism are the Christian nationalists, who believe the scriptures "must govern every aspect of public and private life," including government, science, history, culture and relationships, according to author Michelle Goldberg, who has studied the splinter group.[3] She says Christian nationalists are "a significant and highly mobilized minority" of U.S. evangelicals that is gaining influence.[4] TV evangelist Pat Robertson is a leading Christian nationalist and "helped put dominionism — the idea that Christians have a God-given right to rule — at the center of the movement to bring evangelicals into politics," she says.[5]

Although the number of the world's Christians who are fundamentalists is not known, about 20 percent of the 2 billion Christians are conservative evangelicals, according to the World Evangelical Alliance (WEA).[6] Evangelicals reject the "fundamentalist" label, and most do not advocate creating a Christian theocracy, but they are the socially conservative wing of the Christian community, championing "family values" and opposing abortion and gay marriage. In recent decades they have exercised considerable political power on social issues in the United States.

Many Religions Have Fundamentalist Groups

Religious fundamentalism comes in many forms around the globe, and many different groups have emerged to push their own type of fundamentalism — a handful through violence. The term "Islamist" is often used to describe fundamentalist Muslims who believe in a literal interpretation of the Koran and want to implement a strict form of Islam in all aspects of life. Some also want to have Islamic law, or sharia, imposed on their societies.

Christian Fundamentalists

- Lord's Resistance Army (LRA), a rebel group in Uganda that wants to establish a Christian nation — **violent**
- Various strands within the evangelical movement worldwide, including the U.S.-based Christian nationalists, who insist the United States was founded as a Christian nation and believe that all aspects of life (including family, religion, education, government, media, entertainment and business) should be taken over by fundamentalist Christians — **rarely violent**
- Society of St. Pius X, followers of Catholic Archbishop Marcel Lefebvre, who reject the Vatican II modernizing reforms — **nonviolent**

Islamic Fundamentalists

- Jihadists, like al Qaeda and its allies across the Muslim world — **violent***
- Locally focused Islamist groups Hezbollah (Lebanon) and Hamas (Gaza) — **violent**
- Revolutionary Islamists, like Hizb-ut-Tahrir (HT), a pan-Islamic Sunni political movement that wants all Muslim countries combined into a unitary Islamic state or caliphate, ruled by Islamic law; has been involved in some coup attempts in Muslim countries and is banned in some states — **sometimes violent**
- Political Islamists, dedicated to the "social and political revivification of Islam" through nonviolent, democratic means. Some factions of the Muslim Brotherhood — the world's largest and oldest international Islamist movement — espouse using peaceful political and educational means to convert Muslim countries into sharia-ruled states, re-establishing the Muslim caliphate. Other factions of the group have endorsed violence from time to time.
- Post-Islamists, such as the AKP, the ruling party in Turkey, which has Islamist roots but has moderated its fundamentalist impulses — **nonviolent**

Judaism

- Haredi, ultra-orthodox Jews — **mostly nonviolent**
- Gush Emunim, aim to reoccupy the biblical Jewish land including Palestinian territories — **sometimes violent**
- Chabad missionaries, who support Jewish communities across the globe — **nonviolent**

Indian subcontinent

- Sikh separatists — **sometimes violent**
- Hindu extremists, anti-Christian/Muslim — **sometimes violent**

*For an extensive list of global jihadist groups, see "Inside the Global Jihadist Network," pp. 860-861, in Peter Katel, "Global Jihad," CQ Researcher, Oct. 14, 2005, pp. 857-880.

Sources: Encyclopedia of Fundamentalism; "Foreign Terrorist Organizations," U.S. Department of State

Christians Are a Third of the World's Population

About 20 percent of the world's 2 billion Christians are evangelicals or Pentecostals — many of whom are fundamentalists. But statistics on the number of other fundamentalists are not available. Christians and Muslims together make up more than half the world's population.

World Population by Religious Affiliation

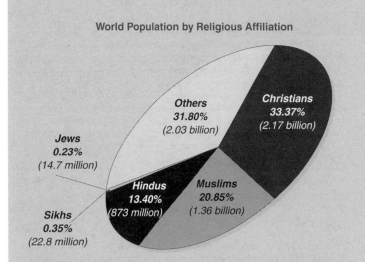

- **Christians** 33.37% (2.17 billion)
- **Muslims** 20.85% (1.36 billion)
- **Hindus** 13.40% (873 million)
- **Sikhs** 0.35% (22.8 million)
- **Jews** 0.23% (14.7 million)
- **Others** 31.80% (2.03 billion)

Major Concentrations of Religious Denominations
(in millions)

Christians		Hindus		Jews	
United States	247	India	817	United States	5.3
Brazil	170	Nepal	19	Israel/Palestine	5.3
Russia	115	Bangladesh	15	France	0.6
China	101	Indonesia	7	Argentina	0.5
Mexico	100	Sri Lanka	2.5	Canada	0.4

Muslims		Sikhs	
Indonesia	178	India	21
India	155	United Kingdom	0.4
Pakistan	152	Canada	0.3
Bangladesh	136	United States	0.3
Turkey	71	Thailand	0.05

Sources: World Christian Database, Center for the Study of Global Christianity, Gordon-Conwell Theological Seminary, www.worldchristiandatabase.org/wcd/home.asp; John L. Allen Jr., "McCain's choice a nod not only to women, but post-denominationalists," National Catholic Reporter, Aug. 30, 2008, http://ncrcafe.org/node/2073

Anglicans and Baptists — very active in evangelizing," says James Nkansah, a Ghanaian-born Baptist minister who teaches at the Nairobi Evangelical Graduate School of Theology in Kenya. "Even the Catholics are doing it, although they do not call themselves evangelists." A similar trend is occurring in Latin America, especially in Brazil, Guatemala and Peru among the Pentecostals, who stress the importance of the Holy Spirit, faith healing and "speak in tongues" during services.

Both evangelicals and Catholics in Latin America have adopted the basic tenets of U.S.-style evangelicalism, according to Valdir Steuernagel, a Brazilian evangelical Lutheran pastor who is vice president at World Vision International, a Christian humanitarian agency. Like U.S. evangelicals, South American evangelicals passionately oppose gay marriage and abortion, but they do not use the term "fundamentalist," says Steuernagel, because the word "does not help us to reach out to the grassroots."

South Korea also has a thriving evangelical community. A visiting U.S. journalist describes a recent service for about 1,000 people at a popular Korean evangelical church: "It was part rock concert and part revival meeting," with the lead guitarist, "sometimes jumping up and down on the altar platform" like Mick Jagger, recalls Michael Mosettig.[8] Elsewhere in Asia — the world's most religiously diverse continent — Christian missionaries in China have grown their flocks from fewer than 2 million Christians in 1979 to more than 16 million Protestants alone in 2008.[9] It is unknown how many of those are fundamentalists.

Christian evangelicalism is booming in Africa — especially in Anglophone countries like Kenya, Uganda, Nigeria, Ghana and South Africa.[7] "We are all — Pentecostals,

Among the world's 15 million Jews, about 750,000 are ultra-Orthodox "Haredi" Jews who live in strict accordance with Jewish law. Half of them live in Israel, most of the rest in the United States, while there are small pockets in France, Belgium, the United Kingdom, Canada and Australia. About 80,000 live in the Palestinian territories on Israel's West Bank because they believe it is God's will.[10] The flourishing fundamentalist Chabad movement — whose adherents would prefer to live in a Jewish theocracy governed by religious laws — sends missionaries to support isolated Jewish communities in 80 countries.

"We accept the Israeli state, but we would have liked the Torah to be its constitution," says Belgian-based Rabbi Avi Tawil, an Argentine Chabad missionary. "But we are not Zionists, because we do not encourage every Jew to go to Israel. Our philosophy is, 'Don't run away from your own place — make it better.' "

In India, Hindu fundamentalists insist their vast country should be for Hindus only. In late 2008, a sudden upsurge in fundamentalist Hindu attacks against Christian minorities in the state of Orissa in eastern India ended with 60 Christians killed and 50,000 driven from their homes.[11] (*See p. 269.*)

Besides their rejection of Western culture, the faithful embrace fundamentalism out of fear of globalization and consumerism and anger about U.S. action — or inaction — in the Middle East, experts say. Some also believe a strict, religiously oriented government will provide better services than the corrupt, unstable, secular regimes governing their countries. Religious fundamentalism also thrives in societies formerly run by repressive governments. Both Christian and Muslim fundamentalism are spreading in Central Asian republics — particularly Uzbekistan, Kyrgyzstan and Tajikistan — that were once part of the repressive, anti-religious Soviet Union. (*See sidebar, p. 260.*)

Many fundamentalists — such as the Quakers, Amish and Jehovah's Witnesses — oppose violence for any reason. And fundamentalists who call themselves "political Islamists" pursue their goal of the "social and political revivification of Islam" through nonviolent, democratic means, according to Loren Lybarger, an assistant professor of classics and world religions at Ohio University and author of a recent book on Islamism in the Palestinian territories.[12]

In recent years radical Islamic extremists have perpetrated most violence committed by fundamentalists. From January 2004 to July 2008, for instance, Muslim militants killed 20,182 people, while Christian, Jewish and Hindu extremists together killed only 925, according to a U.S. government database.[13] Most of the Muslim attacks were between Sunni and Shia Muslims fighting for political control of Iraq. (*See chart, p. 251.*)[14]

Asmaa Abdol-Hamiz, a Muslim Danish politician and social worker, questions the State Department's statistics. "When Muslims are violent, you always see them identified as Muslims," she points out. "When Christians are violent, you look at the social and psychological reasons."

In addition, according to Radwan Masmoudi, president of the Center for the study of Islam and Democracy, such statistics do not address the "more than one million innocent people" killed in the U.S.-led wars in Iraq and Afghanistan, which, in his view, were instigated due to pressure from Christian fundamentalists in the United States. (*See "At Issue," p. 267.*)

Nevertheless, some radical Islamists see violence as the only way to replace secular governments with theocracies. The world's only Muslim theocracy is in Iran. While conservative Shia clerics exert ultimate control, Iranians do have some political voice, electing a parliament and president. In neighboring Saudi Arabia, the ruling royal family is not clerical but supports the ultra-conservative Sunni Wahhabi sect as the state-sponsored religion. Meanwhile, in the Palestinian territories, "there has been a striking migration from more nationalist groups to more self-consciously religious-nationalist groups," wrote Lybarger.[15]

Experts say Muslim militants recently have set their sights on troubled countries like Somalia and nuclear-armed Pakistan as fertile ground for establishing other Islamic states. Some extremist groups, such as Hizb-ut-Tahrir, want to establish a single Islamic theocracy — or caliphate — across the Muslim world, stretching from Indonesia to Morocco.

Still other Muslim fundamentalists living in secular countries such as Britain want their governments to allow Muslims to settle legal disputes in Islamic courts. Islamic law, called sharia, already has been introduced in some areas in Africa, such as northern Nigeria's predominantly Muslim Kano region.[16]

AP Photo/Sunday Alamba

AP Photo/Gurinder Osan

In the Wake of Fundamentalist Violence

Two days of fighting between Christians and fundamentalist Muslims in December destroyed numerous buildings in Jos, Nigeria, (top) and killed more than 300 people. In India's Orissa state, a Christian woman (bottom) searches through the remains of her house, destroyed during attacks by fundamentalist Hindus last October. Sixty Christians were killed and 50,000 driven from their homes.

Muslim extremists are not the only fundamentalists wanting to establish theocracies in their countries. The Jewish Israeli group Kach, for instance, seeks to restore the biblical state of Israel, according to the U.S. State Department's list of foreign terrorist organizations. Hindu fundamentalists want to make India — a secular country with a majority Hindu population that also has many Muslims and Christians — more "Hindu" by promoting traditional Hindu beliefs and customs.

While militant Christian fundamentalist groups are relatively rare, the Lord's Resistance Army (LRA) has led a 20-year campaign to establish a theocracy based on the Ten Commandments in Uganda. The group has abducted hundreds of children and forced them to commit atrocities as soldiers. The group has been blamed for killing hundreds of Ugandans and displacing 2 million people.[17]

In the United States, most Christian fundamentalists are nonviolent, although some have been responsible for sporadic incidents, primarily bombings of abortion clinics. "The irony," says John Green, a senior fellow at the Washington-based Pew Forum on Religion and Public Policy, "is that America is a very violent country where the 'regular' crime rates are actually higher than they are in countries where global jihad is being waged."

Support for violence by Islamic extremists has been declining in the Muslim world in the wake of al Qaeda's bloody anti-Western campaigns, which have killed more Muslims than non-Muslims. U.S. intelligence agencies concluded in November 2008 that al Qaeda "may decay sooner" than previously assumed because of "undeliverable strategic objectives, inability to attract broad-based support and self-destructive actions."[18]

But fundamentalist violence, especially Islamist-inspired, remains a serious threat to world peace. In Iraq, fighting between Sunni and Shia Muslims has killed tens of thousands since 2003 and forced more than 4 million Iraqis to flee their homes. And 20 of the 42 groups on the State Department's list of terrorist organizations are Islamic fundamentalist groups.[19] No Christian or Hindu fundamentalists are included on the terrorist list.

However, Somali-born writer Ayaan Hirsi Ali — herself a target of threats from Islamic fundamentalists — says that while "Christian and Jewish fundamentalists are just as crazy as the Islamists . . . the Islamists are more violent because 99 percent of Muslims think Mohammad is perfect. Christians do not see Jesus in as absolute a way."

As religious fundamentalism continues to thrive around the world, here are some of the key questions experts are grappling with:

Is religious fundamentalism on the rise?

Religious fundamentalism has been on the rise worldwide for 30 years and "remains strong," says Pew's Green.

Fundamentalism is growing throughout the Muslim and Hindu worlds but not in the United States, where its growth has slowed down in recent years, says Martin Marty, a religious history professor at the University of Chicago, who authored a multivolume series on fundamentalism.[20] Christian fundamentalism is strong in Africa and Latin America and is even being exported to industrialized countries. Brazilian Pastor Steuernagel says "evangelical missionaries are going from Brazil, Colombia and Argentina to Northern Hemisphere countries like Spain, Portugal and the United Kingdom. They are going to Asia and Africa too, but there they must combine their missionary activities with aid work."

Islamic fundamentalism, meanwhile, has been growing for decades in the Middle East and Africa. For example, in Egypt the Muslim Brotherhood — which seeks to make all aspects of life in Muslim countries more Islamic, such as by applying sharia law — won 20 percent of the seats in 2005 parliamentary elections — 10 times more than it got in the early 1980s.[21] In Somalia, the Islamist al-Shabaab militia threatens the fragile government.

More moderate Muslims who want to "reform" Islam into a more tolerant, modern religion face an uphill battle, says Iranian-born Shireen Hunter, author of a recent book on reformist voices within Islam. Reformers' Achilles' heel is the fact that "they are often secular and do not understand the Islamic texts as well as the fundamentalists so they cannot compete on the same level," she says.

In Europe, secularism is growing in countries like France and the Netherlands as Christian worship rates plummet, but Turkey has been ruled since 2002 by the Justice and Development Party, which is rooted in political Islam. Though it has vowed to uphold the country's secular constitution, critics say the party harbors a secret fundamentalist agenda, citing as evidence the

Radical Muslims Caused Most Terror Attacks

More than 6,000 religiously motivated terrorist attacks in recent years were perpetrated by radical Muslims — far more than any other group. The attacks by Christians were mostly carried out by the Lord's Resistance Army (LRA) in Uganda.

Religious Attacks, Jan. 1, 2004-June 30, 2008

	Killed	Injured	Incidents
Christian	917	371	101
Muslim*	20,182	43,852	6,180
Jewish	5	28	5
Hindu**	3	7	6
Total	**21,107**	**44,258**	**6,292**

* More than 90 percent of the reported attacks on civilians by Sunni and Shia terrorists were by Sunnis. Does not include the Muslim attacks in Mumbai, India, in December 2008, allegedly carried out by Muslim extremists from Pakistan.

** Uncounted are the Hindu extremist attacks on Christian minorities in late 2008 in India, which left more than 60 Christians dead.

Note: Perpetrators do not always claim responsibility, so attributing blame is sometimes impossible. Also, it is often unclear whether the attackers' motivation is purely political or is, in part, the result of criminality.

Sources: National Counter Terrorism Center's Worldwide Incidents Tracking System, http://wits.nctc.gov; Human Security Research Center, School for International Studies, Simon Fraser University, Vancouver, www.hsrgroup.org.

government's recent relaxation of restrictions on women wearing headscarves at universities.[22]

In Israel, the ultra-Orthodox Jewish population is growing thanks to an extremely high birthrate. Haredi Jews average 7.6 children per woman compared to an average Israeli woman's 2.5 children.[23] And ultra-Orthodox political parties have gained 15 seats in the 120-member Knesset (parliament) since the 1980s, when they had only five.[24] Secularists in the United States saw Christian fundamentalists grow increasingly powerful during the presidency of George W. Bush (2001-2009). Government policies limited access to birth control and abortions, and conservative religious elements in the military began to engage in coercive proselytizing. "From about 2005, I noticed a lot of religious activity: Bible study weeks, a multitude of religious services linked to public holidays that I felt were excessive," says U.S. Army Reserve intelligence officer Laure Williams. In February 2008, she

Moderate Islamist cleric Sheik Sharif Ahmed became Somalia's new president on Jan. 31, raising hope that the country's long war between religious extremists and moderates would soon end. But the hard-line Islamist al-Shabaab militia later took over the central Somali town of Baidoa and began imposing its harsh brand of Islamic law.

recalls, she was sent by her superiors to a religious conference called "Strong Bonds," where fundamentalist books advocating sexual abstinence, including one called *Thrill of the Chaste*, were distributed. Williams complained to her superiors but did not get a satisfactory response, she says.

In the battle for believers among Christian denominations, "Conservative evangelicals are doing better than denominations like Methodists and Lutherans, whose liberal ideology is poisonous and causing them to implode," says Tennessee-based Southern Baptist preacher Richard Land. "When you make the Ten Commandments the 'Ten Suggestions,' you've got a problem."

However, the tide may be turning, at least in some quarters, in part because the next generation appears to be less religious than its elders. Some see the November

2008 election of President Barack Obama — who got a lot of his support from young voters in states with large evangelical populations where the leaders had endorsed Obama's opponent — as evidence that the reign of the Christian right is over in the United States.

"The sun may be setting on the political influence of fundamentalist churches," wrote *Salon.com* journalist Mike Madden.[25] In fact, the fastest-growing demographic group in the United States is those who claim no religious affiliation; they make up 16 percent of Americans today, compared to 8 percent in the 1980s.[26]

And in Iran, while the Islamic theocracy is still in charge, "the younger generation is far less religious than the older," says Ahmad Dallal, a professor of Arab and Islamic studies at Georgetown University in Washington, D.C.

Moreover, support for fundamentalist violence — specifically by al Qaeda's global terrorist network — has been declining since 2004.[27] For example, 40 percent of Pakistanis supported suicide bombings in 2004 compared to 5 percent in 2007.[28] Nigeria is an exception: 58 percent of Nigerians in 2007 said they still had confidence in al Qaeda leader Osama bin Laden, who ordered the Sept. 11, 2001, terrorist attacks on the United States. Notably, al Qaeda has not carried out any terrorist attacks in Nigeria. Support for al Qaeda has plummeted in virtually all countries affected by its attacks.[29]

And while the Muslim terrorist group Jemaah Islamiyah remains active in Indonesia — the world's most populous Muslim-majority country — claims of rampant fundamentalism there are overstated, according to a report by the Australian Strategic Policy Institute. The study found that 85 percent of Indonesians oppose the idea of their country becoming an Islamic republic.[30]

Although there has been a "conspicuous cultural flowering of Islam in Indonesia," the report continued, other religions are booming, too. In September 2008, for example, authorities overrode Muslim objections and approved an application for a Christian megachurch that seats more than 4,500 people.[31]

Is religious fundamentalism a reaction to Western permissiveness?

Religious experts disagree about what attracts people to religious fundamentalism, but many say it is a response to rapid modernization and the spread of Western multiculturalism and permissiveness.

"Fundamentalism is a modern reaction against modernity," says Jerusalem-based journalist Gershom Goremberg. "They react against the idea that the truth is not certain. It's like a new bottle of wine with a label saying 'ancient bottle of wine.' "

Peter Berger, director of the Institute on Culture, Religion and World Affairs at Boston University, says fundamentalism is "an attempt to restore the taken-for-grantedness that has been lost as a result of modernization. We are constantly surrounded by people with other views, other norms, other lifestyles. . . . Some people live with this quite well, but others find it oppressive, and they want to be liberated from the liberation."[32]

Sayyid Qutb, founder of Egypt's Muslim Brotherhood, was repulsed by the sexual permissiveness and consumerism he found in the United States during a visit in 1948.[33] He railed against "this behavior, like animals, which you call 'Free mixing of the sexes'; at this vulgarity which you call 'emancipation of women'; at these unfair and cumbersome laws of marriage and divorce, which are contrary to the demands of practical life. . . . These were the realities of Western life which we encountered."[34]

A similar sentiment was felt by Mujahida, a Palestinian Islamic jihadist who told author Lybarger she worried that her people were losing their soul after the 1993 peace agreement with Israel. "There were bars, nightclubs, loud restaurants serving alcohol, satellite TV beaming American sitcoms, steamy Latin American soap operas [and] casinos in Jericho" to generate tax and employment.[35]

And opposition to abortion and gay rights remain the primary rallying call for U.S. evangelicals. In fact, the late American fundamentalist Baptist preacher Jerry Falwell blamed the 9/11 Islamic terrorist attacks in the United States on pagans, abortionists, feminists and homosexuals who promote an "alternative lifestyle" and want to "secularize America."[36]

In her account of the rise of Christian nationalism, journalist Goldberg said the things Islamic fundamentalists hate most about the West — "its sexual openness, its art, the possibilities for escaping the bonds of family and religion, for inventing one's own life — are what Christian nationalists hate as well."[37]

Pew's Green agrees fundamentalists are irritated by permissive Western culture. "There has always been sin in the world," he says, "but now it seems glorified."

But others say the U.S.-led invasion of Iraq in March 2003 triggered the global surge in violent Islamic militancy. The average annual global death toll between March 2003 to September 2006 from Muslim terrorist attacks jumped 237 percent from the toll between September 2001 to March 2003, according to a study published by Simon Fraser University in Canada.[38]

Moreover, when bin Laden declared war on the United States in a 1998 fatwa, he never mentioned Western culture. Instead, he objected to U.S. military bases in Saudi Arabia, the site of some of Islam's holiest shrines. "The Arabian Peninsula has never — since God made it flat, created its desert and encircled it with seas — been stormed by any forces like the crusader armies now spreading in it like locusts, consuming its riches and destroying its plantations." Bin Laden also railed against Israel — "the Jew's petty state" — and "its occupation of Jerusalem and murder of Muslims there."[39]

Some believe former President George W. Bush's habit of couching the "war on terror" in religious terms helped radical Islamic groups recruit jihadists. *An-Nuur* — a Tanzanian weekly Islamic magazine — noted: "Let us remember President Bush is a saved Christian. He is one of those who believe Islam should be destroyed."[40]

Nawaz, a former member of the revolutionary Islamist Hizb ut-Tahrir political movement, says fundamentalists' motivation varies depending on where they come from. "Some political Islamists are relatively liberal," says the English-born Nawaz. "It's the Saudis that are religiously conservative. The problem is their vision is being exported elsewhere."

Indeed, since oil prices first skyrocketed in the 1970s, the Saudi regime has used its growing oil wealth to build conservative Islamic schools (madrassas) and mosques around the world. As *New York Times* reporter Barbara Crossette noted, "from the austere Faisal mosque in Islamabad, Pakistan — a gift of the Saudis — to the stark Istiqlal mosque of Jakarta, Indonesia, silhouettes of domes and minarets reminiscent of Arab architecture are replacing Asia's once-eclectic mosques, which came in all shapes and sizes."[41]

Pew Forum surveys have found no single, predominant factor motivating people to turn to Islamic fundamentalism. Thirty five percent of Indonesians blame immorality for the growth in Islamic extremism; 40 percent of Lebanese blame U.S. policies and

What Is a Fundamentalist?

Few claim the tarnished label

With the word fundamentalism today conjuring up images of cold-blooded suicide bombers as well as anti-abortion zealots, it is hardly surprising that many religious people don't want to be tarred with the fundamentalist brush.

Yet there was a time when traditionalist-minded Christianity wore it as a badge of honor. Baptist clergyman Curtis Lee Laws coined the term in 1910 in his weekly newspaper *Watchman-Examiner*, when he said fundamentalists were those "who still cling to the great fundamentals and who mean to do battle royal for the faith."[1] Several years earlier, Christian theologians had published a series of pamphlets called "The Fundamentals," which defended traditional belief in the Bible's literal truth against modern ideas such as Charles Darwin's theory of evolution.

Essentially a branch within the larger evangelical movement, the fundamentalists felt that the Christian faith would be strengthened if its fundamental tenets were clearly spelled out. Today, while one in three U.S. Christians considers himself an evangelical, "a small and declining percentage would describe themselves as fundamentalist," says Southern Baptist minister Richard Land of Nashville, Tenn. "While most evangelicals support fundamentalist principles, it is unfair to compare them to the Islamists who take up arms and kill people," he says.

Although some may see the label "fundamentalist" as synonymous with radical Islamic extremists, Ahmad Dallal, a professor of Arab and Islamic Studies at Georgetown University in Washington, D.C., notes that the Arabic word for fundamental — *usul* — was never used in this context historically. "There is some logic to applying the word 'fundamental' in an Islamic context, however," he says, because "both the Muslim and Christian fundamentalists emphasize a literal interpretation of the holy texts."

Traditionalist Catholics do not call themselves fundamentalists either. But Professor Martin Marty, a religious history professor at the University of Chicago and author of a multivolume series on fundamentalism, says Catholic followers of French Archbishop Marcel Lefebvre are fundamentalists because they refuse to accept reforms introduced by the Second Vatican Council in 1965. But "theocons" — a group of conservative U.S. Catholic intellectuals — are

influence; 39 percent of Moroccans blame poverty and 34 percent of Turks blame a lack of education.[42]

Then there are those who just want to regain their lost power, notes Iranian-born author Hunter. "In Iran, Turkey, Tunisia and Egypt, there was a forced secularization of society," she says. "Religious people lost power — sometimes their jobs, too. They had to develop a new discourse to restore their standing."

Religious fundamentalists in Nigeria are largely motivated by anger at the government for frittering away the country's vast oil supplies through corruption and mismanagement. "When a government fails its people, they turn elsewhere to safeguard themselves and their futures, and in Nigeria . . . they have turned to religion," asserted American religion writer Eliza Griswold.[43]

Many Christian and Muslim leaders preach the "Gospel of prosperity," which encourages Nigerians to better themselves economically. But Kenyan-based Baptist preacher Nkansah says that "while the Gospel brings good health and prosperity," the message can be taken too far. "There are some people in the Christian movement who are too materialistic."

Nkansah argues that evangelism is growing in Africa because "as human beings we all have needs. When people hear Christ came onto this planet to save them, they tend to respond."

But a journalist in Tajikistan says poverty drives Central Asians to radical groups like the Hizb ut-Tahrir (HT). "In the poor regions, especially the Ferghana Valley on the Kyrgyz-Tajik-Uzbek border, HT is very active," says the journalist, who asks to remain unnamed for fear of reprisals. "Unemployment pushes people to find consolation in something else, and they find it in religion."

Should religious fundamentalists have a greater voice in government?

Religious fundamentalists who have taken the reins of government — in Iran (since 1979), Afghanistan

not fundamentalists, he says, because they accept the so-called Vatican II changes. Theocon George Weigel, a fellow at the Ethics and Public Policy Center in Washington, eschews the word "fundamentalist" because he says it is "a term used by secular people with prejudices, which doesn't illuminate very much."

Neither are religious Jews keen on the term. Rabbi Avi Tawil, director of the Brussels office of the Chabad Jewish missionary movement, says "fundamentalism is about forcing people. We don't do that. We strictly respect Jewish law, which says if someone would like to convert then you have to help them."

Jerusalem-based writer Gershom Gorenberg notes that unlike Christians and Muslims, fundamentalist Jews do not typically advocate reading holy texts literally because their tradition has always been to have multiple interpretations. The

AFP/Getty Images/MBC

Al Qaeda leader Osama bin Laden hails the economic losses suffered by the United States after the Sept. 11, 2001, terrorist attacks. "God ordered us to terrorize the infidels, and we terrorized the infidels," bin Laden's spokesman Suleiman Abu Ghaith said in the same video, which was broadcast soon after the attacks that killed nearly 3,000 people.

term is even harder to apply to Hinduism because — unlike Christianity, Judaism and Islam — whose "fundaments" are their holy texts, Hinduism's origins are shrouded in ancient history, and its core elements are difficult to define.[2]

Yet fundamentalists are united in their aversion to modernism.

As Seyyed Hossein Nasr, an Islamic studies professor at George Washington University, noted: "When I was a young boy in Iran, 50 or 60 years ago . . . the word fundamentalism hadn't been invented. Modernism was just coming into the country."[3]

---

[1] Brenda E. Brasher, *Encyclopedia of Fundamentalism* (2001), p. 50.
[2] *Ibid.*, p. 222.
[3] His comments were made at a Pew Forum discussion, "Between Relativism and Fundamentalism: Is There a Middle Ground?" March 4, 2008, in Washington, D.C., http://pewforum.org/events/?EventID=172.

(1996-2001) and the Gaza Strip (since 2007) — have either supported terrorism or have instituted repressive regimes. Grave human rights abuses have been documented, dissenters tortured, homosexuals hanged, adulterers stoned, music banned and education denied for girls.

Ayaan Hirsi Ali — a Somali-born feminist writer, a former Dutch politician and a fellow at the conservative American Enterprise Institute who has denounced her family's Muslim faith — says fundamentalists should be able to compete for the chance to govern. "But we must tell them a system based on Islamic theology is bad," she says. "The problem is that Muslims cannot criticize their moral guide. Mohammad is more than a pope, he is a king. As a classical liberal, I say not even God is beyond criticism."

However, Danish politician Abdol-Hamid, whose parents are Palestinian, argues that because most countries won't talk to Hamas, the ruling party in the Gaza Strip, because of its terrorist activities, "we failed the Palestinians

by never giving Hamas a chance." In Denmark, she continues, "We have Christian extremists, and I have to accept them." For instance, she explains, the far-right Danish Peoples Party (DPP) wants to ban the wearing of Muslim headscarves in the Danish parliament, and DPP member of parliament Soren Krarup, a Lutheran priest, says the hijab and the Nazi swastika are both symbols of totalitarianism. Abdol-Hamid hopes to become the first hijab-wearing woman elected to the parliament.

After interviewing Hamas' founding father, Sheikh Ahmed Yassin, Lebanese-born journalist Zaki Chebab wrote that Yassin "was confident that . . . Israel would disappear off the map within three decades," a belief he said came from the Koran.[44]

A Christian fundamentalist came to power in Northern Ireland without dire consequences after the Rev. Ian Paisley — the longtime leader of Ulster's Protestants, who established his own church stressing biblical literalism and once called the pope the "antichrist" — ultimately

Many Voice Concern About Islamic Extremism

A majority of respondents in nine out of 10 Western countries were "very" or "somewhat" concerned about Islamic extremism in a 2005 poll. Islam was playing a greater role in politics in five out of six Muslim nations, according to the respondents, and most blamed U.S. policies and poverty for the rise in Islamic extremism.

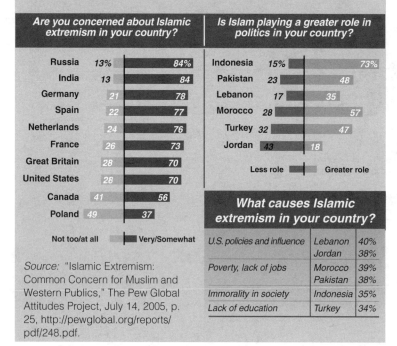

Are you concerned about Islamic extremism in your country?

Country	Not too/at all	Very/Somewhat
Russia	13%	84%
India	13	84
Germany	21	78
Spain	22	77
Netherlands	24	76
France	26	73
Great Britain	28	70
United States	28	70
Canada	41	56
Poland	49	37

Is Islam playing a greater role in politics in your country?

Country	Less role	Greater role
Indonesia	15%	73%
Pakistan	23	48
Lebanon	17	35
Morocco	28	57
Turkey	32	47
Jordan	43	18

What causes Islamic extremism in your country?

U.S. policies and influence	Lebanon	40%
	Jordan	38%
Poverty, lack of jobs	Morocco	39%
	Pakistan	38%
Immorality in society	Indonesia	35%
Lack of education	Turkey	34%

Source: "Islamic Extremism: Common Concern for Muslim and Western Publics," The Pew Global Attitudes Project, July 14, 2005, p. 25, http://pewglobal.org/reports/pdf/248.pdf.

"It is a delicate game," says fundamentalism expert Marty. "If you have a republican system with a secular constitution, then, yes [fundamentalists must be allowed to have a voice], because they have to respect that constitution. But it's very much a case of 'handle with care.' "

Conservative Catholic theologian George Weigel, a senior fellow at the Ethics and Public Policy Center in Washington, says religious people are entitled to be involved in politics, but "they should translate their religiously informed moral convictions into concepts and words that those who don't share their theological commitments can engage and debate. This is called 'democratic courtesy.' It's also political common sense."

Indeed, religious Muslims not only have the right but also the duty to participate in government, according to Rachid Ghannouchi, a Tunisian-born Islamic thinker. Denouncing countries like Tunisia and Algeria that repress Islamic fundamentalism, Ghannouchi said, "the real problem lies in convincing the ruling regimes... of the right of Islamists — just like other political groups — to form political parties, engage in political activities and compete for power or share in power through democratic means."[46]

But ex-Islamist Nawaz warns: "We should not be encouraging Islamists, because every terrorist group has grown out of a nonviolent Islamist group."

Israeli journalist Gorenberg also notes that radical Jewish fundamentalists have repeatedly resorted to violence, citing the case of Baruch Goldstein, a U.S.-born Israeli doctor and supporter of the Kach party who killed 29 Muslims at the tomb of Abraham in Hebron in 1994.

Washington-based Turkish journalist Tulin Daloglu is anxious about her country's future under the ruling Justice and Development Party. "Women are starting to cover their hair in order to get jobs in government," she claims. "The case is not at all proven that Islam and

reconciled his lifelong differences with Northern Irish Catholic leaders and has served amicably with them in government after they offered him political power.[45]

Kenyan-based evangelical Nkansah says "politics is part of life." If a religious person is called into politics in Kenya, he explains, "they should go because that is their vocation." He supports Kenya's model, in which many clergy members, including bishops, enter politics, even though the constitution bans parties based on religion. But evangelical pastor Steuernagel says that in Brazil, religious leaders are increasingly going into politics. "I do not think it is healthy," he says, "but it is happening."

In Central Asia, Islamic parties are only allowed in Tajikistan. But while the Islamic Revival Party has become a significant force there, the party "is neither dangerous nor radical," according to the Tajik journalist, and "does not dream about having a state like Iran."

democracy can live in harmony. Turkey is a swing state" in that regard.

Meanwhile, in some Asian and African countries where the rule of law is weak — Pakistan and Somalia for example — many are clamoring for Islamic law. Often the existing government is so dysfunctional that the quick, decisive administration of Islamic law, or sharia, is attractive. In Pakistan, says British journalist Jason Burke, "the choice between slow, corrupt and expensive state legal systems and the religious alternative — rough and ready though it may be — is not hard." Even educated, relatively wealthy women are demanding sharia, he said.[47]

For example, the Taliban has been able to seize control in Pakistan's Swat region because of "an ineffectual and unresponsive civilian government, coupled with military and security forces that, in the view of furious residents, have willingly allowed the militants to spread terror deep into Pakistan."[48]

BACKGROUND

'Great Awakening'

Christian fundamentalist movements trace their origins to the emergence of Protestantism in 16th-century Europe, when the German monk Martin Luther (1483-1546) urged people to return to the basics of studying the Bible.[49] In 1620 a group of fundamentalist Protestants known as the Pilgrims fleeing persecution in England settled in North America and, along with the Puritans who arrived shortly afterwards, greatly influenced the course of Christianity in New England.

In the 1700s, as science began to threaten religion's preeminence, North Americans launched a Protestant revival known as the "Great Awakening," from which the evangelical movement was born. Revivals held throughout the American colonies between 1739 and 1743, offered evangelical, emotionally charged sermons — often in open-air services before large groups — that stressed the need to forge a personal relationship with Jesus Christ. Leaders in the movement included preachers George Whitfield, Gilbert Tennent and Jonathan Edwards.[50]

A similar revival movement — the Sunday school movement — began in the late 18th century, becoming a

Evangelicals from Uganda's Born Again Church are spiritually moved last August while listening to a sermon by Pastor Robert Kayanja, one of Uganda's most prominent evangelical preachers. While Uganda has long been heavily Christian, many churchgoers have switched from mainstream to Pentecostal sects in recent years.

primary vehicle for evangelism.[51] The term "fundamentalist" originated in the United States when the first of a 12-volume collection of essays called *The Fundamentals* was published in 1910, outlining the core tenets of Christianity.[52] In 1925 fundamentalists were the driving force in the trial of Tennessee schoolteacher John Scopes, who was convicted of breaking a Tennessee law that forbade the teaching of evolution instead of the Bible's version of how the world was created. Even though the fundamentalists won the case, they were lampooned in the popular press, and their credibility and esteem suffered. They withdrew from the limelight and formed their own subculture of churches, Bible colleges, camps and seminaries.

By 1950, the charismatic American Baptist preacher Billy Graham had begun to broaden the fundamentalists' base, and they became masters at harnessing the mass media, especially radio and television. The 1973 U.S. Supreme Court's *Roe v. Wade* ruling legalizing abortion further galvanized evangelicals, leading Baptist preacher Falwell in 1979 to establish the Moral Majority — a conservative political advocacy group.

After his unsuccessful run for president of the United States in 1988, television evangelist and Christian nationalist Robertson formed the Christian Coalition to fight for "family-friendly" policies — specifically policies against homosexuality and abortion. By the mid-1990s

CHRONOLOGY

A.D. 70-1700s *The three great, monotheistic, text-based religions — Christianity, Islam and Judaism — spread worldwide.*

70 Romans destroy the second Jewish temple in Jerusalem, causing Jews to scatter across the globe.

319 Christianity becomes the official religion of the Roman Empire; pagan sacrifices are outlawed.

632 Mohammad dies in Medina, Arabia. . . . Islam begins to spread to the Middle East, Africa, India, Indonesia and Southern Europe.

1730s-40s Evangelical movement is born in the United States in a religious revival known as the "Great Awakening."

1800s-1920s *Fundamentalist impulses are triggered in reaction to scientific developments, modernization and — in the case of Islam — Western colonization.*

1859 British biologist Charles Darwin presents theory of evolution in *On the Origin of Species*, casting doubt on the Bible's account of creation.

1906 African-American evangelist William J. Seymour launches the Azusa Street revival in Los Angeles, sparking the worldwide Pentecostal movement.

1910 American Christian oil magnates Lyman and Milton Stewart commission The Fundamentals, promoting fundamentalist Protestant beliefs that the Bible contains no errors.

1921 Jailed Hindu nationalist Vinayak Damodar Savarkar writes *Hindutva: Who is a Hindu?* — laying the foundation for movements promoting Hindu identity, including the radical Bajrang Dal.

1928 Hasan al-Banna, a schoolteacher in Cairo, Egypt, establishes the Muslim Brotherhood, which calls for all Muslims to make their societies more Islamic.

1940s-1970s *Fundamentalism becomes a significant force in politics and society.*

1948 Israel declares independence, causing millions of Jews — both secular and religious — to return to their spiritual homeland.

1967 Fundamentalist Jews settle in Palestinian territories occupied after the Six-day War, triggering an explosion in Islamic fundamentalism among disgruntled Arabs.

1973 U.S. Supreme Court's *Roe v. Wade* ruling legalizes abortion, galvanizing Christian fundamentalists into political activism.

1979 Islamists overthrow the Shah of Iran and install the world's first Islamic theocracy in modern times.

1980s-2000s *Fundamentalists increasingly endorse violence to further their goals — especially in the Muslim world.*

1984 Indian government storms a Sikh temple, which Sikh militants had occupied, leading two of Prime Minister Indira Gandhi's Sikh bodyguards to murder her.

1994 American Jewish fundamentalist Baruch Goldstein kills 29 Muslims praying at a mosque in the Palestinian city of Hebron.

Sept. 11, 2001 Al Qaeda Islamists kill nearly 3,000 people by flying hijacked planes into the World Trade Center and Pentagon; a third hijacked plane crashes in Pennsylvania.

2002 Sectarian fighting between Hindus and Muslims in Gujarat, India, kills more than 800 people — mostly Muslims.

2006 Palestinians elect Hamas, a radical Islamic party, to lead the government.

2008 Sixty Christians die after outbreak of fundamentalist Hindu violence against Christians in India. . . . Pakistan-based Islamists launch coordinated attacks in Mumbai, India, killing 164 people. . . . Troops from Congo, Uganda and South Sudan launch ongoing joint offensive to crush Uganda's fundamentalist Lord's Resistance Army. . . . Israel launches major attack on Gaza in effort to weaken Hamas, resulting in 1,300 Palestinian deaths.

the coalition became the most prominent voice in the Christian movement, largely by publishing voter guides on how local politicians voted on specific social issues important to Christian fundamentalists. Many credit the coalition with helping the Republican Party, which had embraced their platform on social issues, to take majority control of the U.S. Congress in the 1994 midterm elections.[53]

Some U.S. fundamentalists segregated themselves from mainstream society — which they saw as immoral — and educated their children at home.[54] A strand of race-based fundamentalism also emerged, called the Christian Identity movement, which claimed the Bible was the history of the white race and that Jews were the biological descendants of Satan. A Christian Reconstructionist movement, led by preacher Mark Rushdoony, emerged as well, advocating local theocracies that would impose biblical law.[55] The reconstructionists oppose government schools and demand that civil disputes be settled in church courts and that taxes be limited to 10 percent of income (based on the tithe). Through its books, the movement has had a significant influence on other Christian political organizations.[56]

Meanwhile, a fundamentalist Catholic movement emerged in Europe after French Archbishop Marcel Lefebvre refused to accept changes introduced by the Vatican in the 1960s, notably saying Mass in languages other than Latin.[57] Other conservative Catholic movements include Opus Dei, founded by Spanish priest Josemaria Escriva in 1928. Today it is based in Rome, has 75,000 members in 50 countries and appeals to well-educated lay Catholics.[58] In the United States, a group of Catholic intellectuals — including Michael Novak, Weigel and Richard John Neuhaus — became known as the "theocons" and allied themselves with Protestant evangelicals in opposing abortion and gay rights.[59]

Bush's presidency was a high point for U.S. evangelicals. Bush announced during the 2000 campaign that he was a "born again" Christian whose favorite philosopher was Jesus Christ — "because he changed my heart." He also told a Texas evangelist that he felt God had chosen him to run for president, and he was accused of "creeping Christianization" of the federal government by establishing an Office for Faith-Based Initiatives, which critics claimed was just a vehicle for channeling tax dollars to conservative Christian groups.[60]

Bush liberally used religious rhetoric — declaring, for example, after the 9/11 attacks that his mission was "to rid the world of evil."[61] He named Missouri Sen. John Ashcroft, a fellow evangelist, as attorney general and filled his administration with Christian conservatives, such as Monica Goodling, a young Justice Department official who vetted candidates for executive appointments by checking their views on moral issues like abortion.[62]

Christian missionaries have been evangelizing — spreading their faith — since the 16th century, but fundamentalist strands have grown increasingly prominent in recent decades. Pentecostalism — which began in 1901 when a Kansas Bible studies student, Agnes Ozman, began "speaking in tongues" — is the dominant form of Protestantism in Latin America.[63] In Guatemala, evangelicalism began to overtake the Roman Catholic Church in the 1980s after Catholicism was seen as too European and elitist.[64] Although Pentecostals usually distinguish themselves from run-of-the-mill fundamentalists, both are part of the evangelical family.

In Africa, Christian fundamentalism developed its strongest base in sub-Saharan regions — particularly Nigeria, triggering rising tensions and sporadic violence between the country's Christian and Muslim populations. U.S. Christian fundamentalists have helped to spread an extreme brand of Christianity to Africa, according to Cedric Mayson, director of the African National Congress' Commission for Religious Affairs, in South Africa. "We are extremely concerned about the support given by the U.S. to the proliferation of right-wing Christian fundamentalist groups in Africa," Mayson wrote, as "they are the major threat to peace and stability in Africa."[65]

Uganda became home to the militant Christian fundamentalist Lord's Resistance Army. Its leader Joseph Kony — known as the "altar boy who grew up to be a guerrilla leader" — has transformed an internal Ugandan power struggle into an international conflict by roaming across Sudan and the Democratic Republic of Congo, kidnapping children en route for use as soldiers after slaughtering their parents.[66]

Patrick Makasi, the LRA's former director of operations, called Kony "a religious man" who "all the time…is talking about God. Every time he keeps calling many people to teach them about the legends and about God. That is how he leads people."[67]

Officials in the 'Stans' Uneasy About Islamization

Education is a key battleground

"The crowd in the airport parking lot was jubilant despite the cold, with squealing children, busy concession stands and a tangle of idling cars giving the impression of an eager audience before a rock concert," wrote journalist Sabrina Tavernise of a scene in Dushanbe, the capital of Tajikistan.[1]

"But it was religion, not rock 'n roll, that had drawn so many people," she wrote. The families were there to meet relatives returning from the Hajj — the pilgrimage to Mecca that Muslims strive to undertake at least once in their lifetime. Last year, 5,200 Tajiks participated — 10 times more than in 2000.

Since gaining independence from the anti-religious Soviet Union, Tajikistan has been re-embracing its Islamic roots, and a Westerner in the country — who asked to remain unnamed — worries the nation of 7.2 million people may adopt an extreme form of Islam. "Every day you can see on our streets more women wearing the veil and more men with beards," he says.

But while many women in Central Asia today do cover themselves from head to toe, it is "extremely rare" for them to cover their faces as well, which was not unusual in pre-Soviet Tajikistan and Uzbekistan, says Martha Brill Olcott, a senior associate at the Carnegie Endowment for International Peace in Washington, who has traveled there frequently since 1975.

The region is undergoing a wide mix of outside influences, not all of them Islamic, Olcott notes. For example, some women have begun wearing the hijab (a headscarf pinned tightly around the face so as to cover the hair) worn by modern Islamic women in the West, while others, notably in Uzbekistan, imitate secular Western fashions such as short skirts and visible belly piercing.

The Westerner in Tajikistan fears that the government's efforts to block the growing Islamization may be having the opposite effect. Government policies "are too severe," he says. "They give long prison sentences to young men and shut down unregistered mosques. This just strengthens people's resolution to resist an unfair system."

Further, he suggests, "If they developed local economies more, people would not think about radical Islam." Without economic development, "Tajikistan could become another Afghanistan or Iran."

Tajikistan, one of the poorer countries in the region, is in the midst of reverse urbanization due to economic decline, with

The refurbished Juma Mosque in Tashkent, Uzbekistan, reflects Islam's resurgence in Central Asia, where 18 years after the breakup of the former Soviet Union neighboring Iran and Saudi Arabia are exerting their influence on the vast region.

Getty Images/Uriel Sinai

77 percent of the population now living in rural areas compared to 63 percent in the mid-1980s.[2] A million Tajiks work in Russia.

In neighboring Uzbekistan, the picture is similar. Olcott likens the California-sized nation of 27 million people to an "ineffective police state. There are restrictions, but people can get around them and — more important — they are not afraid to get around them." She says the government's response is erratic: "If you do not draw attention to yourself, you can be an Islamist. But if you preach and open schools or wear very Islamic dress, you can get into trouble."

Christian missionaries are also active in Central Asia. Russian-dubbed broadcasts from U.S. televangelist Pat Robertson are aired throughout the region. According to the Tajikistan-based Westerner, after the 1991 fall of the Soviet Union "Jehovah's Witnesses, Baptists and Adventists came from Russia, Western Europe, South Korea and the United States. The locals were friendly to them because they provided humanitarian aid to poor people." However, authorities in the region have recently clamped down — especially on the Jehovah's Witnesses, he says.[3]

In Kazakhstan authorities have cracked down on Protestants and repressed the Hindu-based Hare Krishnas, while in Kyrgyzstan a new law makes it harder to register religious organizations.[4]

In Kyrgyzstan, the authorities are in a quandary about whether to allow a new political movement, the Union of Muslims, to be set up because bringing Islam into politics violates the constitution. Yet union co-founder Tursunbay Bakir Uulu argues that a moderately Islamic party would help stabilize the country. "Currently Hizb-ut-Tahrir is conquering the Issyk-Kul region," he warned. "Religious sects are stepping up their activities. We want moderate Islam, which has nothing to do with anti-religious teaching and which respects values of other world religions, to fill this niche."[5]

The Islamization began in the 1980s, when Soviet President Mikhail Gorbachev eased restrictions on religious worship that had been enforced by the communists for decades. After the Soviet Union's collapse, the relaxation accelerated as the Central Asian republics became independent nations. Muslim missionaries flocked to the region, and conservative Islamic schools, universities and mosques quickly sprang up, many financed by foundations in oil-rich Arab states like Saudi Arabia, where the ultra-fundamentalist Wahhabi Muslim sect is the state-sponsored religion.[6]

Many Central Asians see embracing conservative Islam as a way to define themselves and reject their Russian-dominated communist past. Curiously, their increasing exposure to secular culture through Russia-based migrant Tajik workers appears to be having a Westernizing influence on the society even as Islam is growing: "Five years ago, I could not wear shorts on the street," said the Westerner in Tajikistan. "Now in summer you can see a lot of Tajik men and even girls wearing shorts in the cities, although not in the villages."

The rise of Islam is strongest in Uzbekistan, Tajikistan and Kyrgyzstan, while Turkmenistan and Kazakhstan have stronger secular traditions. Uzbek authorities initially encouraged Islamization, believing it would help strengthen national identity. But by the late 1990s, they were afraid of losing control to radical elements and began repressing militant groups like the Islamic Movement of Uzbekistan and Hizb-ut-Tahrir.[7] A jailbreak by Islamists in Andijan, the Uzbek capital, in May 2005 triggered violent clashes between government forces and anti-corruption protesters — whom the government claimed were Islamic extremists — resulting in 187 deaths.[8]

Meanwhile, the Saudis are sending Islamic textbooks that promote their own conservative brand of Islam to schools in the region.[9] Saudi-Uzbek ties stretch back to the 1920s, when some Uzbeks fled to Saudi Arabia, according to Olcott.

But Saudi-inspired fundamentalism "is not a major factor" in Turkmenistan yet, says Victoria Clement, an assistant professor of Islamic world history at Western Carolina University, who has lived in Turkmenistan. "There are maybe a few individuals, but the government has not allowed madrasas [Islamic religious schools] since 2003." Even so, she notes, "when I went to the mosques, I saw clerics instructing the kids in the Koran, which technically they should not have been doing [under Turkmen law], but I do not think it was harmful."

Nevertheless, the Turkmen education system is growing more Islamic, Clement says, as new schools follow the model devised by Turkish preacher Fethullah Gulen. "They do not have classes in religion, but they teach a conservative moral code — no drinking, smoking, staying out late at night. I think it is a great alternative to the Islamic madrasas," she says.

Olcott says while the quality of education in the Gulen schools may be good, it is "still very Islamic." Gulen himself now lives in the United States, having left Turkey after being accused of undermining secularism.

The Westerner in Tajikistan notes, however, that in their efforts to stem the growth of radical Islam authorities have a bit of a blind spot when it comes to education. "In most Tajik villages, the children's only teacher is the person who can read the Koran in Arabic, and that is dangerous. The government makes demands about how students look — ties and suits for example — but does not care about what they have in their minds."

Islam Booming in the "Stans"

Several of the nations in Central Asia dubbed "the Stans" are rediscovering their Islamic roots, including Tajikistan and Uzbekistan. The Islamization began in the 1980s, when then Soviet President Mikhail Gorbachev eased restrictions on religious worship.

[1] Sabrina Tavernise, "Independent, Tajiks Revel in Their Faith," *The New York Times*, Jan. 3, 2009, www.nytimes.com/2009/01/04/world/asia/04tajik.html?emc=tnt&tntemail0=y.

[2] *Ibid.*

[3] Felix Corley, "Tajikistan: Jehovah's Witnesses Banned," Forum 18 News Service (Oslo, Norway), Oct. 18, 2007, www.forum18.org/Archive.php?article_id=1036; Felix Corley, "Turkmenistan: Fines, beatings, threats of rape and psychiatric incarceration," Forum 18 News Service (Oslo, Norway), Nov. 25, 2008, www.forum18.org/Archive.php?article_id=1221.

[4] Mushfig Bayram, "Kazakhstan: Police Struggle against Extremism, Separatism and Terrorism — and restaurant meals," Forum 18 News Service, Nov. 21, 2008, www.forum18.org/Archive.php?article_id=1220; and Mushfig Bayram, "Kyrgyzstan: Restrictive Religion Law passes Parliament Unanimously," Forum 18 News Service (Oslo, Norway), Nov. 6, 2008, www.forum18.org/Archive.php?article_id=1215.

[5] "Kyrgyz Experts Say Newly Set Up Union of Muslims Aims for Power," *Delo No* (Kyrgyzstan), BBC Monitoring International Reports, Dec. 9, 2008.

[6] See Martha Brill Olcott and Diora Ziyaeva, "Islam in Uzbekistan: Religious Education and State Ideology," Carnegie Endowment for International Peace, July 2008, www.carnegieendowment.org/publications/index.cfm?fa=view &id=21980&prog=zru.

[7] *Ibid.*, p. 2.

[8] For background, see Kenneth Jost, "Russia and the Former Soviet Republics," *CQ Researcher*, June 17, 2005, pp. 541-564.

[9] *Ibid.*, p. 19.

Getty Images/Chip Somodevilla

Anti-abortion demonstrators carry a statue of the Virgin Mary during the March for Life in Washington, D.C., on Jan. 22, 2009. The rally marked the 35th anniversary of the Supreme Court's landmark Roe v. Wade decision legalizing abortion in the United States. Fundamentalist Christians continue to exert significant influence on U.S. policies governing abortion, birth control and gay rights.

Islamic Fundamentalism

Originating in the 7th century with the Prophet Mohammad, Islam considers the Koran sacred both in content and form — meaning it should be read in the original language, Arabic. Muslims also follow the Hadith, Mohammad's more specific instructions on how to live, which were written down after he died. Though Islamic scholars have interpreted both texts for centuries, fundamentalists use the original texts.

The concept of a militant Islamic struggle was developed by scholar Taqi ad-Din Ahmad Ibn Taymiyyah (1263-1328), who called for "holy war" against the conquering, non-Muslim Mongols.[68] The Saudi-born Islamic scholar Muhammed Ibn Abd-al-Wahhab (1703-1792) criticized the Ottoman Empire for corrupting the purity of Islam. The descendants of one of Wahhab's followers, Muhammed Ibn Saud, rule Saudi Arabia today.[69]

Responding to the dominating influence of Western powers that were colonizing the Islamic world at the time, Egyptian schoolteacher Hasan Al-Banna set up the Muslim Brotherhood in 1928 to re-Islamize Egypt. The organization later expanded to other Arab countries and to Sudan.[70] "They copied what the Christian missionaries were doing in Africa by doing social work," notes Islamic studies Professor Dallal. "But they had no

vision for 'the state,' and they paid a price for this because the state ultimately suppressed them."

In the 1950s the extremist group Hizb-ut-Tahrir, which advocates a single Islamic state encompassing all predominantly Muslim countries, emerged and spread across the Islamic world. In the mid-1950s, while imprisoned in Egypt by the secular government, the U.S.-educated Egyptian scholar and social reformer Qutb (1906-1966) wrote *Milestones*, his diatribe against the permissiveness of the West, which persuaded many Muslims they needed to get more involved in politics in order to get their governments to make their societies more Islamic. In Pakistan, the politician Sayyid Abul A'la Mawdudi (1903-1979) urged Islamists to restore Islamic law by forming political parties and getting elected to political office, according to Dallal.

The 1973 oil crisis helped to spread conservative Islam by further enriching Saudi Arabia, which set up schools, universities and charities around the world advocating ultraconservative wahhabi Islam. And the 1979 Iranian Revolution — in which the pro-Western Shah Mohammad Reza Pahlavi was deposed in a conservative Shia Muslim revolt led by Ayatollah Ruhollah Khomeini — installed the first Islamic theocracy in the modern era.

In 1991 Islamists were voted into power in Algeria, but the military refused to let them govern, triggering a bloody civil war that the secularists eventually won. In Afghanistan, the ultraconservative Pakistan-sponsored Taliban seized power in 1996 and imposed their strict version of Islamic law — outlawing music, forbidding girls from going to school or leaving their homes without a male relative, forcing women to completely cover their bodies — even their eyes — in public, requiring men to grow beards and destroying all books except the Koran.[71] After the al Qaeda terrorist attacks of 9/11, the United States ousted the Taliban, which had been sheltering bin Laden.

Al Qaeda, a Sunni Muslim group that originated in Saudi Arabia, had been based in Afghanistan since the 1980s, when it helped eject Soviet occupiers, with U.S. aid. But in the 1990s bin Laden redirected his energies against the United States after American troops were stationed in his native Saudi Arabia, home to several sacred Muslim shrines.

After the U.S.-led invasion of Iraq in 2003, al Qaeda urged its followers to switch their attentions to Iraq, which became a magnet for Islamist jihadists. In 2007 al Qaeda attacks in Iraq escalated to such a level of violence — including attacking Shia mosques and repressing local Sunnis — that other Islamic groups like the Muslim Brotherhood repudiated them.[72]

In Europe, meanwhile, beginning in the 1980s the growing Muslim immigrant population began to attach greater importance to its religious identity, and some turned to violence. Algerian extremists set off bombs in Paris subways and trains in 1995-1996; Moroccan-born Islamic terrorists killed 191 people in train bombings in Madrid in 2004; and British-based al Qaeda operatives of mainly Pakistani origin killed 52 people in suicide train and bus bombings in London in 2005.[73] And an al Qaeda cell based in Hamburg, Germany, plotted the 9/11 attacks on the World Trade Center towers and the Pentagon.

The estimated 5 million Muslims in the United States — who are a mix of immigrants and African-Americans — are more moderate than their Western European counterparts.[74] Poverty is likely to have played a role in making European Muslims more radical: Whereas the average income of American Muslims is close to the national average, Muslims' average income lags well behind the national average in Spain, France, Britain and Germany.[75]

Meanwhile, the creation of Israel in 1948 — fiercely opposed by all of its Arab neighbors — and its successive expansions in the Gaza Strip and West Bank have helped to spur Islamic fundamentalism in the region. To Israel's north, the Shia-Muslim Hezbollah group emerged in the 1980s in Lebanon with the goal of destroying Israel and making Lebanon an Islamic state. The Sunni-Muslim group Hamas — an offshoot of the Muslim Brotherhood — won elections in the Palestinian territories in 2006. Hamas, which was launched during the Palestinian uprising against Israel of 1987, has forged strong links with Islamic fundamentalists in Iran and Saudi Arabia.[76]

Fundamentalist Jews

Predating both Islam and Christianity, Judaism takes the Torah and Talmud as its two holy texts and believes that the Prophet Moses received the Ten Commandments — inscribed on stone tablets — from God on Mount Sinai.[77] Fundamentalist Jews believe they are God's chosen people and that God gave them modern-day Israel as their homeland. A defining moment in this narrative is

Members of the ultra-Orthodox Chabad-Lubavitch Jewish fundamentalist movement attend the funeral in Israel of two members of the missionary sect killed last fall during Islamist militant attacks in Mumbai, India.

the destruction of the second Jewish Temple in Jerusalem in 70 A.D., which triggered the scattering of Jews throughout the world for nearly 2,000 years.

Jews began returning to their spiritual homeland in significant numbers in the early 1900s with the advent of Zionism — a predominantly secular political movement to establish a Jewish homeland, founded by the Austro-Hungarian journalist Theodor Herzl in the late 19th century in response to rising anti-Semitism in Europe. The migration was accelerated after Nazi Germany began persecuting the Jewish people in the 1930s in a racially motivated campaign that resulted in the Holocaust and the murder of 6 million Jews and millions of others.[78] Today, a third the world's 15 million Jews live in Israel; most of the rest live in the United States, with substantial Jewish communities in France, Argentina and Canada.

Fundamentalist Jews regret that Israel was established as a secular democracy rather than a theocracy. While most Israelis support the secular model, there is a growing minority of ultra-Orthodox (Haredi) Jews for whom the Torah and Talmud form the core of their identity. They try to observe 613 commandments and wear distinctive garb: long black caftans, side curls and hats for men and long-sleeve dresses, hats, wigs and scarves for women.[79] The Haredim dream of building a new Jewish temple in Jerusalem where the old ones stood, which also happens to be the site of the Dome on the Rock — one of Islam's most revered shrines. The fundamentalist

Islamic Fundamentalism Limits Women's Rights

But Muslim women disagree on the religion's impact

As a high official in Saudi Arabia, Ahmed Zaki Yamani crafted many of the kingdom's laws, basing them on Wahhabism, the strict form of Islam that is Saudi Arabia's state religion. Under those laws, Muslim judges "have affirmed women's competence in all civil matters," he has written, but "many of them have reservations regarding her political competence." In fact, he added, one of Islam's holiest texts, the Hadith, "considered deficiency a corollary of femaleness."[1]

Since the 1970s, the Saudis have used their vast oil wealth to spread their ultra-conservative form of Islam throughout the Middle East, North Africa and South and Central Asia, including its controversial view of women as unequal to men. Under Saudi Wahhabism, women cannot vote, drive cars or mix freely with men. They also must have a male guardian make many critical decisions on their behalf, which Human Rights Watch called "the most significant impediment to the realization of women's rights in the kingdom."[2]

The advocacy group added that "the religious establishment has consistently paralyzed any efforts to advance women's rights by applying only the most restrictive provisions of Islamic law, while disregarding more progressive interpretations."[3]

In her autobiography, *Infidel*, Somali-born writer and former Dutch politician Ayaan Hirsi Ali writes about how shocked she was as a young girl when her family moved from Somalia's less conservative Islamic society to Saudi Arabia, where females' lives were much more restricted. "Any girl who goes out unaccompanied is up for grabs," she says.

Raised a Muslim but today an outspoken critic of Islam, Hirsi Ali says Saudi Arabia has had a "horrific" influence on the Muslim world — especially on women. In Africa, she says, religious strictures against women going out in public can have dire consequences, because many women must work outside the home for economic reasons.

While Wahhabism is perhaps the most extreme form of Islam, Hirsi Ali doubts any form of Islam is compatible with women's rights. "Islamic feminism is a contradiction in terms," she says. "Islam means 'submission.' This is double for women: She must appeal to God before anyone else. Yet this same God tells your man he can beat you."

In 2004, Dutch filmmaker Theo Van Gogh was murdered by a Muslim man angered by a film he made portraying violence against women in Islamic societies. Hirsi Ali, then a member of the Dutch parliament, had written the script for the movie, and the assassin left a note on Van Gogh's body threatening her.

She believes the entire philosophical underpinnings of Islam are flawed. For example, she says, she had been taught that Muslim women must wear the veil so they will not corrupt men, yet, "when I came to Europe I could not understand how women were not covered, and yet the men were not jumping on them. Then I saw all it took was to educate boys to exercise self-control. They don't do that in Saudi Arabia, Iran and Pakistan."

But forcing women to cover themselves is not the only way conservative Muslim societies infringe on women's rights. Until recently in Pakistan, rape cases could not be prosecuted unless four pious Muslim men were willing to testify that they had witnessed the attack. Without their testimony the victim could be prosecuted for fornication and alleging a false crime, punishable by stoning, lashing or prison.[4]

Ali's views are not shared by Asmaa Abdol-Hamid, a young, Danish Muslim politician of Palestinian parentage who lived in the United Arab Emirates before moving to Denmark at age 6. Covering oneself, she says, "makes women more equal because there is less focus on her body.... When you watch an ad on television, it is always women in bikinis selling the car."

A social worker, local council member representing a left-wing party and former television-show host, Abdol-Hamid is a controversial figure in Denmark. She wears a hijab and refuses to shake hands with men. "I prefer to put my hand on my heart," she explains. "That's just my way of greeting them. It's not that shaking hands is un-Islamic."

She has her own view of Islam's emphasis on female submission. "If women want to obey their husbands, it's up to them." However, "I could not live the Arab lifestyle, where the men beat the women. That's not Islam — it's Arab." In a global study of women's rights, Arab states accounted for 10 of the 19 countries with the lowest ranking for women's equality.[5]

Many fundamentalist Muslims say the freedoms advocated by secular women's-rights advocates disrupt the complementary nature of male and female roles that have been the basis of social unity since the rise of Islam. A Palestinian Islamic jihadist, known only as Mujahida, said women should "return to their natural and [Koran-based] functions as child-bearers, home-keepers and educators of the next generation." She rejects women's-rights advocates who urge women to take their abusive husbands to secular courts.

Muslim "family mediators," she said, were best placed to resolve such disputes.[6]

According to the Washington-based Pew Research Center, more than a third of Jordanians and Egyptians oppose allowing women to choose whether or not to veil, although the percentage is falling.[7] Also on the decline: the number of those who support restrictions prohibiting men and women from working in the same workplace.[8] In Saudi Arabia, such restrictions limit women's employment, because employers must provide separate offices for women.[9]

However, Pew found considerable support in Muslim nations for restricting a woman's right to choose her husband. For example, 55 percent of Pakistanis felt the family, not the woman, should decide.[10]

In Nigeria, Islamic fundamentalism has hurt women's rights, according to Nigerian activist Husseini Abdu. "Although it is difficult separating the Hausa [Nigerian tribe] and Islam patriarchal structure, the reintroduction or politicization of sharia [Islamic law] in northern Nigeria has contributed in reinforcing traditional, religious and cultural prejudices against women," Abdu says.[11] This includes, among other things, the absence of women in the judiciary, discrimination in the standards of evidence in court cases (especially involving adultery) and restrictions in the freedom of association.[12]

Christian countries are not immune from criticism for limiting women's rights. Human Rights Watch found that in Argentina the Catholic Church has had a hand in establishing government policies that restrict women's access to modern contraception, sex education and abortion.[13] And fundamentalist Christian groups have played a significant role in restricting sex education and the availability of birth control and abortion services in the United States.

But while Islamic countries are often criticized for their treatment of women, the world's two most populous Muslim nations, Pakistan and Indonesia, have both elected female leaders in the past — the late Benazir Bhutto in Pakistan and Megawati Sukarnoputri in Indonesia. The world's largest Christian country, the United States, has never had a female president.

Ayaan Hirsi Ali (right), a Somali-born former member of the Dutch parliament, has been threatened with death for her outspoken criticism of Islam's treatment of women in Islam. But Danish Muslim politician and social worker Asmaa Abdol-Hamid (left) attributes repressive gender-based policies in Muslim countries to local culture, not the Koran.

In Iran, an Islamic theocracy since 1979, a debate is raging over whether to allow women to inherit real estate, notes Shireen Hunter, an Iranian-born author and visiting scholar at Georgetown University in Washington. "Reformers are also trying to have the age of [marriage] consent raised from 9 to 16 years. This will take time," she says, because "trying to blend Islam and modernity is hard. It is easier to just say, 'Let's go back to fundamentalism.' "

Yet Abdol-Hamid argues that "fundamentalism does not have to be a bad thing. In Islam, going back to the Koran and Hadith would be good."

Does Hirsi Ali see anything positive about a woman's life in Islamic societies? "I have never seen Muslim women doubt their femininity or sensuality," she says. "Western women question this more. They are less secure. They are always thinking, 'Am I really equal?' "

[1] Ahmed Zaki Yamani, "The Political Competence of Women in Islamic Law," pp. 170-177, in John J. Donohue and John L. Esposito, *Islam in Transition: Muslim Perspectives* (2007).

[2] "Perpetual Minors — Human Rights Abuses Stemming from Male Guardianship and Sex Segregation in Saudi Arabia," Human Rights Watch, April 19, 2008, p. 2, www.hrw.org/en/node/62251/section/1.

[3] *Ibid.*

[4] Karen Foerstel, "Women's Rights," *CQ Global Researcher*, May 2008, p. 118.

[5] *Ibid.*

[6] Loren D. Lybarger, *Identity and Religion in Palestine: The Struggle between Islamism and Secularism in the Occupied Territories* (2007), p. 105.

[7] In Jordan, 37 percent of respondents opposed women being allowed to choose whether to veil, compared to 33 percent in Egypt.

[8] The Pew Global Attitudes Project, "World Publics Welcome Global Trade — But Not Immigration," Pew Research Center, Oct. 4, 2007, p. 51, http://pewglobal.org/reports/pdf/258.pdf.

[9] "Perpetual Minors — Human Rights Abuses Stemming from Male Guardianship and Sex Segregation in Saudi Arabia," *op. cit.*, p. 3.

[10] Pew, *op. cit.*, p. 50.

[11] Carina Tertsakian, "Political Shari'a? Human Rights and Islamic Law in Northern Nigeria," Human Rights Watch, Sept. 21, 2004, p. 63, www.hrw.org/en/reports/2004/09/21/political-shari.

[12] *Ibid.*

[13] See Marianne Mollmann, "Decisions Denied: Women's Access to Contraceptives and Abortion in Argentina," Human Rights Watch, June 14, 2005, www.hrw.org/en/node/11694/section/1.

Haredim are represented by several different political parties in Israel — each with a distinct ideology.

A newer strain of Jewish fundamentalism, the Gush Eminum movement, grew out of the 1967 Israeli-Arab War, in which Israel captured large swathes of Syrian, Egyptian and Jordanian territory. Founded by Rabbi Zvi Yehuda Kook, it believes Israel's victory in that war was a sign that God wanted Jews to settle the captured territories. Israeli authorities initially opposed such actions but did a U-turn in 1977, setting up settlements to create a buffer to protect Israel from hostile Arab neighbors. There now are some 500,000 settlers, and they have become a security headache for the Israeli government, which protects them from attacks from Palestinians who believe they have stolen their land.[80]

Meanwhile the Chabad movement — founded in the 18th century in Lubavitch, Russia, by Rabbi Schoeur Zalman — operates outside of Israel.[81] "They are very religious communities that have become missionaries, even though Jews are not supposed to convert non-Jews, and conversion is very difficult and mostly refused," says Anne Eckstein, a Belgian Jewish journalist. "They are especially active in ex-Soviet countries where the Holocaust and Soviet power wiped out the Jewish community or reduced it to a bare minimum."

Fundamentalism in India

Unlike Christianity, Islam and Judaism, which are monotheistic, Hinduism has thousands of deities representing an absolute power. In addition, it is based not on a single text but the belief that the universe is impersonal and dominated by cosmic energy.[82] Hindu fundamentalism emerged in the early 20th century, partly in reaction to proselytizing by Muslim and Christian missionaries. Some Hindus came to believe that their country needed to be made more Hindu, and that only Hindus could be loyal Indians.

Indian politician Vinayak Damodar Savarkar wrote the book *Hindutva*, the philosophical basis for Hindu fundamentalism.[83] Its cultural pillar is an organization called Vishva Hindu Parishad, founded in 1964, which has had a political wing since the 1980 establishment of the Bharatiya Janata Party, whose leader, Atal Bihari Vajpayee, was prime minister from 1998-2004.

The assertion of Hindu religious identity provoked unease among some of India's 20 million Sikhs, who worship one God and revere the *Adi Granth*, their holy book.[84] Indian Prime Minister Indira Gandhi was murdered in 1984 by two of her Sikh bodyguards in revenge for sending troops to storm the Sikhs' holiest shrine, the Golden Temple, which had been occupied by militant Sikh separatists. Hundreds of people were killed in the botched government operation.[85]

CURRENT SITUATION
Political Battles

Christian conservatives remain a potent force in American political life, even though they appear to have lost some of their political clout with the election of a liberal, pro-choice president and a decidedly more liberal Congress.

In the 2008 U.S. presidential election, evangelicals were briefly buoyed by the nomination of a Christian conservative, Alaska Gov. Sarah Palin, as the Republican vice presidential candidate. But their hopes of having another evangelical in high office were dashed when Palin and her running mate, Sen. John McCain, R-Ariz., were comfortably beaten by their Democratic rivals in November.

Palin was raised as a Pentecostal and regularly attended the Assemblies of God church in Wasilla, Alaska. In a Republican National Convention speech, she stressed the need to govern with a "servant's heart" — which in the evangelical world means Christian humility.[86]

But as details of her religious and political views were revealed, secular Americans began to question her candidacy. Video footage surfaced of her being blessed by a Kenyan pastor in 2005 who prayed for her to be protected from "every form of witchcraft" and for God to "bring finances her way" and to "use her to turn this nation the other way around."[87] Palin was also videotaped speaking at the same church in June 2008, calling a $30 billion gas pipeline project in Alaska "God's will" and the war in Iraq "a task that is from God."[88]

While Palin ultimately may have hurt the Republican ticket more than helping, the passage on Election Day of referenda banning gay marriage in several states — including California — shows that Christian conservatism remains a significant force. And across the American South and heartland, religious conservatives have pressured state and local governments to pass a variety of "family" and faith-based measures, ranging from

Is Islamic fundamentalism more dangerous than Christian fundamentalism?

YES

Maajid Nawaz
*Director, Quilliam Foundation,
London, England*

Written for *CQ Global Researcher,* February 2009

While not all Muslim fundamentalists are a threat, certain strands of Muslim fundamentalism are more dangerous than Christian fundamentalism. This is simply a truth we must face up to as Muslims. The first stage of healing is to accept and recognize the sickness within. Until such recognition comes, we are lost.

But if Muslim fundamentalism is only a problem in certain contexts, this is not true of political Islam, or Islamism. Often confused with fundamentalism, political Islamism is a modernist project to politicize religion, rooted in the totalitarian political climate of post-World War I Egypt. But this ideology didn't restrict itself to political goals. Instead, its adherents aspired to create a modern, totalitarian state that was illiberal but not necessarily fundamentalist.

In the 1960s, the Muslim Brotherhood — Egypt's largest Islamist group — failed to impose their non-fundamentalist brand of Islam in Egypt. Instead, they fled to religiously ultra-conservative Saudi Arabia. Here they allied with reactionary fundamentalists. It is from this mix of modernist Islamism and fundamentalism that al Qaeda and jihadist terrorism emerged. It was in Saudi Arabia that Osama bin Laden was taught by Muslim Brotherhood exiles. It was from Saudi Arabia that streams of Muslim fundamentalists traveled to Afghanistan and Pakistan where they fell under the spell of the Egyptian Islamist Abdullah Azzam, another inspiration for bin Laden. The root of the present terrorist danger is the alliance between modernist political Islamists and Muslim fundamentalists.

This global jihadist terrorism — modern in its political ideals and tactics yet medieval in both its religious jurisprudence and justification for violence — is more dangerous than Christian fundamentalism. I believe that such terrorism, far from representing the fundamentals of Islam, is actually un-Islamic. However, a Christian may similarly argue that attacking abortion clinics is un-Christian. We both need to acknowledge the role that religion plays in motivating such individuals.

So, having recognized this problem, how can Muslims tackle it? It is not enough for Muslims to merely take a stand against terrorism and the killing of innocent civilians. This is the very least that should be expected of any decent human being. Muslims must also challenge both conservative fundamentalism and the modern Islamist ideology behind jihadist terrorism. Islamism is to blame, alongside Western support for dictatorships, for the situation we face today.

NO

Radwan Masmoudi
*President, Center for the Study of Islam
and Democracy, Washington, D.C.*

Written for *CQ Global Researcher,* February 2009

The term "fundamentalism" can be misleading, because the overwhelming majority of Muslims believe the Koran is the literal word of God and a guide for the individual, the family and society to follow on everything social, political and economic. In a recent Gallup Poll, more than 75 percent of Muslims — from Morocco to Indonesia — said they believe Islamic laws should be either the only source or one of the main sources of laws in their countries. Under a U.S. definition of "fundamentalism," these people would all be considered "fundamentalists."

However, the overwhelming majority of Muslims are peaceful and reject violence and extremism. In the same poll, more than 85 percent of Muslims surveyed said they believe democracy is the best form of government. Thus, they are not interested in imposing their views on others but wish to live according to the teachings of their religion while respecting people of other religions or opinions. Democracy and respect for human rights — including minority rights and women's rights — are essential in any society that respects and practices Islamic values.

It would be a terrible mistake to consider all fundamentalist Muslims a threat to the United States or to mankind. Radical and violent Muslim extremist groups such as al Qaeda and the Taliban represent a tiny minority of all Muslims and a fringe minority of religious (or fundamentalist) Muslims. These extremist groups are a threat both to their own societies and to the West. But they do not represent the majority opinion among religious-based groups that are struggling to build more Islamic societies through peaceful means.

Many Christian fundamentalist groups have resorted to violence, specifically attacks against abortion clinics in the United States. In addition, prominent Christian fundamentalist leaders, such as John Hagee, Pat Robertson and others say Islam is the enemy and have called for the United States to invade Muslim countries like Iraq, Afghanistan and even Iran. These wars have cost the lives of more than 1 million innocent people in these countries and could still cause further deaths and destruction around the world. The devout of all faiths should condemn the killing of innocents and the self-serving labeling of any religion as the "enemy" against which war should be waged. Surely, one — whether Muslim or Christian — can be extremely devout and religious without calling for violence or hoping for Armageddon.

Jewish Settlements Stir Outrage and Support

Left-wing Israelis criticize Israel last December for allowing fundamentalist Jews to build settlements in the Palestinian territories (top). Evangelicals from the U.S.-based Christians United for Israel movement (bottom) support the settlements during a rally in Jerusalem last April. Many analysts say pressure from American fundamentalist Christians led former President George W. Bush, a born-again Christian, to offer unqualified support for Israel and to invade Iraq — policies that have exacerbated U.S.-Muslim relations.

restrictions on access to birth control and abortion to requirements that "intelligent design" be taught in place of or alongside evolution in schools. The laws have triggered ire — and a slew of lawsuits — on the part of groups intent on retaining the Constitution's separation of church and state.[89]

Meanwhile, thousands of conservative Episcopalians in the United States have abandoned their church because of the hierarchy's tolerance of homosexuality and are teaming up with Anglican Protestants in Africa who share their conservative views.[90]

In Latin America, evangelical television preachers are using their fame to launch themselves into politics, notes Dennis Smith, a U.S.-born Presbyterian mission worker who has lived in Guatemala since 1977. He says that in Brazil, Pentecostal preacher Edir Macedo cut a deal with President Luiz Inacio Lula de Silva in which Macedo got to hand-pick the country's vice president. In Guatemala Harold Caballeros, a Pentecostal who preaches that the Mayan Indians there have made a pact with the devil by clinging to their traditional beliefs, is trying to become president, Smith adds.

In Africa, the Somali parliament on Jan. 31 elected a moderate Islamist cleric, Sheik Sharif Ahmed, as the country's new president. The election occurred just as the hard-line Islamist al-Shabaab militia took control of the central Somali town of Baidoa and began imposing its harsh brand of Islamic law there.[91]

Rising Violence

Attacks on Christian minorities in Iraq and India — and efforts to forcibly convert them — have escalated in recent months.

In November militants said to be from the Pakistan-based Lashkar-e-Taiba carried out a meticulously planned attack in Mumbai, India, killing 164 people in a shooting spree that targeted hotels frequented by Western tourists.[92] Ex-Islamist Nawaz says of the group: "I know them well. They want to reconquer India. They see it as being under Hindu occupation now because it was once ruled by Muslim emperors of Turko-Mongol descent. They use the territorial dispute between India and Pakistan over the sovereignty of Kashmir as a pretext for pursuing their global jihad agenda."

Lisa Curtis, a research fellow for South Asia at the Heritage Foundation in Washington, believes that Pakistan is playing a sinister role here. "The Pakistan military's years of support for jihadist groups fighting in Afghanistan and India," she says, is "intensifying linkages between Pakistani homegrown terrorists and al Qaeda."

India's suspicion that forces within the Pakistani government have given Lashkar-e-Taiba a free rein is further straining an already tense relationship between the two nations.

The Lashkar attackers also killed two young Jewish missionaries, Rabbi Gavriel Holtzberg and his wife Rivkah, in an assault on the Chabad center in Mumbai, where they had been based since 2003. While some accuse the Chabad of proselytizing, Rabbi Avi Tawil, who studied with U.S.-born Gavriel Holtzberg for two years in Argentina, insists, "He did not force anyone to accept his philosophy. He was doing social work — working with prisoners for example."

But the Mumbai attacks were not the only violence perpetrated by religious extremists in India last year. Between August and December, members of the paramilitary, right-wing Hindu group Bajrang Dal — using the rallying cry "kill Christians and destroy their institutions" — murdered dozens of Christians, including missionaries and priests, burned 3,000 homes and destroyed more than 130 churches in Orissa state.[93] The attackers were angered at proselytizing by Pentecostal missionaries in the region and tried to force Christians to convert back to Hinduism.[94]

Martha Nussbaum, a professor of law and ethics at the University of Chicago and author of the recent book *The Clash Within: Democracy, Religious Violence and India's Future*, writes that no one should be surprised right-wing Hindus "have embraced ethno-religious cleansing." Since the 1930s, "their movement has insisted that India is for Hindus, and that both Muslims and Christians are foreigners who should have second-class status in the nation."[95]

India's bloodiest religiously based violence in recent years was the slaughter of up to 2,000 Muslim civilians by Hindu mobs in Gujarat state in 2002.[96] A Bajrang Dal leader boasted: "There was this pregnant woman, I slit her open. . . . They shouldn't even be allowed to breed. . . . Whoever they are, women, children, whoever . . . thrash them, slash them, burn the bastards. . . . The idea is, don't keep them alive at all; after that, everything is ours."[97]

In Iraq last fall, in the northern city of Mosul, some 400 Christian families were forced to flee their homes after attacks by Sunni Muslim extremists.[98]

In Nigeria, sectarian violence between Christians and Muslims in the city of Jos spiked again in late November, leaving at least 300 dead in the worst clashes since 2004, when 700 people died. Religious violence in Nigeria tends to break out in the "middle belt" between the Muslim north and the predominantly Christian south.[99]

Then in December Israel launched a massive offensive against the Islamist Hamas government in the Gaza Strip, in response to Hamas' continuous rocket attacks into Israel; at least 1,300 Palestinians died during the 22-day assault. An uneasy truce now exists, but Hamas remains defiant, refusing to accept Israel's right to exist and vowing to fight for the creation of an Islamic Palestinian state in its place.[100]

While most commentators focus on the political dimension of the conflict, Belgian Jewish journalist Anne Eckstein is as concerned about Hamas' religious extremism. "I see nothing in them apart from hatred and death to all who are not Muslims. . . . Jews first but then Christians and everybody else. And those who believe that this is not a war of civilization are very mistaken."

Also in December, troops from, Uganda, southern Sudan and the Democratic Republic of Congo launched a joint offensive to catch Lord's Resistance Army (LRA) leader Kony.[101] The LRA retaliated, massacring hundreds. Kenya-based evangelical Professor Nkansah insists the LRA is "not really religious — no one has ever seen them praying. They are just playing to the Christian communities in Uganda. If they were true Christians, they would not be destroying human life like they are."

Even in areas where religious violence has not broken out, a certain fundamentalist-secular tension exists. In the United Kingdom, for example, a debate has broken out over whether Muslim communities should be allowed to handle family matters — such as divorce and domestic violence cases — in Muslim courts that apply Islamic law. These increasingly common tribunals, despite having no standing under British law, have "become magnets for Muslim women seeking to escape loveless marriages."[102] In Africa, the Tanzanian parliament is having a similar debate, with proponents noting that Kenya, Rwanda and Uganda have had such courts for decades.[103]

In Israel, the majority-secular Jewish population has begun to resent ultra-Orthodox Jewish men who subsist on welfare while immersing themselves in perpetual study of the holy texts. "They claim this is what Jews did in the past, but this is nonsense," says Jerusalem-based journalist Gorenberg, who notes that ultra-Orthodox wives often work outside the home in order to support their very large families. The Haredim are trying to restore ancient Judaism by weaving priestly garments in the traditional way, producing a red heifer using genetic engineering and raising boys in a special compound kept ritually pure for 13 years, says Gorenberg, a fierce critic of fundamentalist Jews.[104]

Many secular Israelis also resent the religious Jews that have settled in the Palestinian territories, arguing they make Muslims hate Israel even more and thus threaten Israel's very security.

OUTLOOK

More of the Same?

Al Qaeda's Egyptian-born chief strategist, Ayman Al-Zawihiri, is very clear about his goal. "The victory of Islam will never take place until a Muslim state is established in the heart of the Islamic world, specifically in the Levant [Eastern Mediterranean], Egypt and the neighboring states of the [Arabian] Peninsula and Iraq."[105]

Former-Islamist Nawaz says such a state would not, as fundamentalists claim, be a return to the past but a modernist creation, having more in common with the totalitarian regimes of 20th-century Europe than with the tolerant Islamic caliphates in the Middle Ages. He thinks Islamists have the greatest chance of seizing power in Egypt and Uzbekistan.

Given the Islamization that she has observed on numerous visits to Uzbekistan, Martha Brill Olcott, a senior associate at the Carnegie Endowment for International Peace in Washington, predicts the country will not remain secular. Because the Muslims there are Sunni, she thinks they will follow an Egyptian or Pakistani model of government.

Georgetown Professor Dallal predicts Iran will remain the world's only theocracy. "I do not think the Iranian model will be replicated," he says. "The religious elite is more institutionalized and entrenched there than elsewhere."

And although young Iranians are more secular than their parents and have been disenchanted with the religious rulers, "We should not assume this is a deep-rooted trend," warns Iranian-born author Hunter. "Look at Arab countries: Forty years ago we thought they were going secular, but not now."

As for Islamist militancy, the signs are mixed. While a Pew survey showed a drop in support for global jihad among Muslims overall, it also found that young Muslims in the United States were more likely to support radical Islam than their parents. Fifteen percent of 18-29-year-olds thought suicide bombing could be justified compared to just 6 percent of those over 30.[106]

And even if, as some analysts suggest, al Qaeda is faltering, other Islamist groups may thrive, such as Hezbollah in Lebanon, Hamas in Gaza and Pakistan's Lashkar-e-Taiba. They attract popular support because they also provide social services, unlike al Qaeda, whose bloody campaigns have alienated most Muslims.[107]

The Israel-Palestine conflict, intractable as ever, will continue to be grist for the Islamist mill. Bin Laden has urged Muslims to "kill the Americans and their allies [and] to liberate the Al-Aqsa Mosque," which is located on the Temple Mount in Jerusalem that Israel has controlled since 1967.[108]

In the Palestinian territories, "Islamist symbols, discourses and practices have become widely disseminated across the factional spectrum," according to Ohio State's Lybarger, but whether it continues depends on the actions of Israel, the United States and other Arab states toward Palestine, he says.[109] Many observers hope President Obama and his newly-appointed Middle East envoy George Mitchell will be able to broker a peace deal, given Obama's aggressive outreach to the Muslim world.

In the United States, the Christian right is likely to remain strong, even as Obama moves to overhaul Bush's faith-based initiatives. Secularists may ask Obama to prohibit groups receiving government funds from discriminating in hiring based on religious beliefs. "Hiring based on religious affiliation is justified," says Stanley Carlson-Thies, director of the Center for Public Justice in Washington, D.C. "Would you ask a senator not to ask about political ideology when selecting staff? A ban would [be] a sweeping change."[110]

Looking farther afield, Baptist minister Land says "by 2025 the majority of Christians . . . will be African, Latin American and Asian. That is where evangelical Christianity is growing fastest." The fastest-growing Christian denominations are in Nigeria, Sudan, Angola, South Africa, India, China and the Philippines, according to the World Christian Database.[111]

But Kenya's Nkansah doubts that Christian-based political parties will emerge in sub-Saharan Africa. "In North Africa almost everyone is Muslim, so it is easier to have Islamic parties. But here, there is more of a mix, and politicians do not want to create unnecessary tensions."

In Guatemala, American Presbyterian missionary Smith says, "Since neither modernity nor democracy has been able to bring security, the rule of law, social tolerance or broad-based economic development" evangelical television preachers will "continue to have great power for the foreseeable future."

Meanwhile, a glimpse of Asia's future might be found in South Korea. "As dusk turns to dark in this capital city," journalist Mosettig wrote, "the skyline glitters with more than the urban lights of office towers and apartment blocks. From the hills that define Seoul's topography and neighborhoods, it is easy to spot lighted electric crosses. They are among the most visible reminders of just how deeply Christianity shapes South Korea."[112]

NOTES

1. Richard A. Oppel Jr. and Pir Zubair Shah, "In Pakistan, Radio Amplifies Terror of Taliban," *The New York Times*, Jan. 24, 2009, www.nytimes.com/2009/01/25/world/asia/25swat.html?_r=1&scp=1&sq=Taliban%20Pakistan&st=cse.

2. "The U.S. Religious Landscape Survey," Pew Forum on Religion and Public Life, Feb. 25, 2008, p. 170, http://religions.pewforum.org.

3. Michelle Goldberg, *Kingdom Coming: The Rise of Christian Nationalism* (2007), p. 7.

4. *Ibid.*, p. 8.

5. Dominionism, Goldberg notes, is derived from a theocratic sect called Christian Reconstructionism, which advocates replacing American civil law with Old Testament biblical law.

6. See World Evangelical Alliance Web site, www.worldevangelicals.org. For background, see David Masci, "Evangelical Christians," *CQ Researcher*, Sept. 14, 2001, pp. 713-736.

7. Quoted in Eliza Griswold, "God's Country," *The Atlantic*, March 2008, www.theatlantic.com/doc/200803/nigeria.

8. Michael Mossetig, "Among Sea of Glittery Crosses, Christianity Makes Its Mark in South Korea," PBS, Nov. 5, 2007, www.pbs.org/newshour/indepth_coverage/asia/koreas/2007/report_11-05.html. For background, see Alan Greenblatt and Tracey Powell,

"Rise of Megachurches," *CQ Researcher*, Sept. 21, 2007, pp. 769-792.

9. Presentation by Wang Zuoan, China's deputy administrator of religious affairs, Sept. 11, 2008, at the Brookings Institution, Washington, D.C.

10. Estimates provided by Samuel Heilman, Sociology Professor and expert on Jewish fundamentalism at City University of New York.

11. "Christians Attacked in Two States of India" World Evangelical Alliance Web site, Dec. 15, 2008, www.worldevangelicals.org/news/view.htm?id=2277.

12. Loren D. Lybarger, *Identity and Religion in Palestine: The Struggle between Islamism and Secularism in the Occupied Territories* (2007), p. 73.

13. See National Counter Terrorism Center's Worldwide Incidents Tracking System, http://wits.nctc.gov.

14. The Shia, who make up 15 percent of the world's 1.4 billion Muslims, believe only the prophet Mohammad's family and descendants should serve as Muslim leaders (imams). Sunnis — who make up the other 85 percent — believe any Muslim can be an imam. Iran is the world's most Shia-dominated country, while there are also significant Shia communities in Iraq, Turkey, Lebanon, Syria, Kuwait, Bahrain, Saudi Arabia, Yemen, Pakistan and Azerbaijan.

15. Lybarger, *op. cit.*

16. "Sharia stoning for Nigeria man," BBC News, May 17, 2007, http://news.bbc.co.uk/2/hi/africa/6666673.stm.

17. For background, see John Felton, "Child Soldiers," *CQ Global Researcher*, July, 2008.

18. Scott Shane, "Global Forecast by American Intelligence Expects Al Qaeda's Appeal to Falter," *The New York Times*, Nov. 20, 2008, www.nytimes.com/2008/11/21/world/21intel.html?_r=1&emc=tnt&tntemail0=y.

19. "Country Reports on Terrorism," Office of the Coordinator for Counterterrorism, U.S. Department of State, April 2008, www.state.gov/documents/organization/105904.pdf.

20. Martin Marty and R. Scott Appleby, eds., *Fundamentalisms Comprehended* (The Fundamentalism Project), 2004, University of Chicago Press.

21. Source: Talk by Egyptian scholar and human rights activist Saad Eddin Ibrahim, at Woodrow Wilson International Center for Scholars, Washington, D.C., Sept. 8, 2008.

22. For background, see Brian Beary, "Future of Turkey," *CQ Global Researcher*, December 2007.

23. Raja Kamal, "Israel's fundamentalist Jews are multiplying," *The Japan Times*, Aug. 21, 2008, http://search.japantimes.co.jp/cgi-bin/eo20080821a1.html.

24. *Ibid.*

25. Mike Madden, "Sundown on Colorado fundamentalists," *Salon.com*, Nov. 2, 2008, www.salon.com/news/feature/2008/11/03/newlifechurch/index.html?source=rss&aim=/news/feature.

26. Susan Jacoby, "Religion remains fundamental to US politics," *The Times* (London), Oct. 31, 2008, www.timesonline.co.uk/tol/comment/columnists/guest_contributors/article5050685.ece.

27. "Human Security Brief 2007," Human Security Report Project, Simon Fraser University, Canada, May 21, 2008, www.humansecuritybrief.info.

28. "Unfavorable views of Jews and Muslims on the Increase in Europe," Pew Research Center, Sept. 17, 2008, p. 4, http://pewglobal.org/reports/pdf/262.pdf.

29. *Ibid.*

30. Andrew MacIntyre and Douglas E. Ramage, "Seeing Indonesia as a normal country: Implications for Australia," Australian Strategic Policy Institute, May 2008, www.aspi.org.au/publications/publication_details.aspx?ContentID=169&pubtype=5.

31. Michael Sullivan, "Megachurch Symbolizes Indonesia's Tolerance," National Public Radio, Oct. 19, 2008, www.npr.org/templates/story/story.php?storyId=95847081.

32. Comments from Pew Forum on Religion and Public Life discussion, "Between Relativism and Fundamentalism: Is There a Middle Ground?" March 4, 2008, Washington, D.C., http://pewforum.org/events/?EventID=172.

33. Sarah Glazer, "Radical Islam in Europe," *CQ Global Researcher*, November 2007.

34. Sayyid Qutb, *Milestones*, *SIME* (Studies in Islam and the Middle East) *Journal*, 2005, p. 125, http://majalla.org/books/2005/qutb-nilestone.pdf.

35. Lybarger, *op. cit.*

36. See Goldberg, *op. cit.*, p. 8.

37. *Ibid.*, p. 208.

38. "Human Security Brief 2007," *op. cit.*, p. 19.

39. Osama Bin Laden, "Text of Fatwa Urging Jihad Against Americans," Feb. 23, 1998, in John J. Donohue and John L. Esposito, *Islam in Transition: Muslim Perspectives* (2007), pp. 430-432.

40. "Tanzania: Muslim paper says war on terror guise to fight Islam," BBC Worldwide Monitoring, Aug. 24, 2008 (translation from Swahili of article in Tanzanian weekly Islamic newspaper *An-Nuur*, Aug. 15, 2008).

41. Barbara Crossette, "The World: (Mid) East Meets (Far) East; A Challenge to Asia's Own Style of Islam," *The New York Times*, Dec. 30, 2001.

42. Pew Global Attitudes Project, "Islamic Extremism: Common Concern for Muslim and Western Publics," July 14, 2005, p. 25, http://pewglobal.org/reports/pdf/248.pdf.

43. Griswold, *op. cit.*

44. Zaki Chehab, *Inside Hamas — The Untold Story of the Militant Islamic Movement* (2007), p. 104.

45. Gabriel Almond, Scott Appleby and Emmanuel Sivan, *Strong Religion: The Rise of Fundamentalisms Around the World* (The Fundamentalism Project), The University of Chicago Press, 2003, p. 110.

46. Rachid Ghannouchi, "The Participation of Islamists in a Non-Islamic Government," in Donohue and Esposito, *op. cit.*, pp. 271-278.

47. Jason Burke, "Don't believe myths about sharia law," *The Guardian* (United Kingdom), Feb. 10, 2008, www.guardian.co.uk/world/2008/feb/10/religion.law1. For background, see Robert Kiener, "Crisis in Pakistan" *CQ Global Researcher*, December 2008, pp. 321-348.

48. Oppel and Shah, *op. cit.*

49. Brenda E. Brasher, *Encyclopedia of Fundamentalism* (2001), p. 397.

50. *Ibid.*, pp. 202-204.

51. *Ibid.*, pp. 465-467.

52. *Ibid.*, p. 186.

53. For background, see the following *CQ Researchers*: Kenneth Jost, "Religion and Politics," Oct. 14, 1994,

pp. 889-912; and David Masci, "Religion and Politics," July 30, 2004, pp. 637-660.

54. For background, see Rachel S. Cox, "Home Schooling Debate," *CQ Researcher*, Jan. 17, 2003, pp. 25-48.

55. David Holthouse, "Casting Stones: An Army of radical Christian Reconstructionists is preparing a campaign to convert conservative fundamentalist churches," Southern Law Poverty Center, winter 2005, www.splcenter.org/intel/intelreport/article.jsp?aid=591.

56. Brasher, *op. cit.*, pp. 407-409.

57. *Ibid.*, p. 86.

58. *Ibid.*

59. Adrian Wooldridge, "The Theocons: Secular America Under Siege," *International Herald Tribune*, Sept. 26, 2006, www.iht.com/articles/2006/09/25/opinion/booktue.php.

60. See Paul Harris, "Bush says God chose him to lead his nation," *The Guardian*, Nov. 2, 2003, www.guardian.co.uk/world/2003/nov/02/usa.religion; and Melissa Rogers and E. J. Dionne Jr., "Serving People in Need, Safeguarding Religious Freedom: Recommendations for the New Administration on Partnerships with Faith-Based Organizations," The Brookings Institution, December 2008, www.brookings.edu/papers/2008/12_religion_dionne.aspx. For background, see Sarah Glazer, "Faith-based Initiatives," *CQ Researcher*, May 4, 2001, pp. 377-400.

61. James Carroll, "Religious comfort for bin Laden," *The Boston Globe*, Sept. 15, 2008, www.boston.com/news/nation/articles/2008/09/15/religious_comfort_for_bin_laden.

62. For background, see Dan Eggen and Paul Kane, "Goodling Says She 'Crossed the Line'; Ex-Justice Aide Criticizes Gonzales While Admitting to Basing Hires on Politics," *The Washington Post*, May 24, 2007, p. A1.

63. Brasher, *op. cit.*, p. 154.

64. Almond, Appleby and Sivan, *op. cit.*, p. 171.

65. Cedric Mayson, "Religious Fundamentalism in South Africa," African National Congress Commission for Religious Affairs, January 2007, http://thebrenthurstfoundation.co.za/Files/terror_talks/Religious%20Fundamentalism%20in%20SA.pdf.

66. Rob Crilly, "Lord's Resistance Army uses truce to rearm and spread its gospel of fear," *The Times* (London), Dec. 16, 2008, www.timesonline.co.uk/tol/news/world/africa/article5348890.ece.

67. *Ibid.*

68. Brasher, *op. cit.*, p. 37.

69. For background, see Peter Katel, "Global Jihad," *CQ Researcher*, Oct. 14, 2005, pp. 857-880.

70. Almond, Appleby and Sivan, *op. cit.*, pp. 177-79.

71. Brasher, *op. cit.*, p. 37.

72. "Human Security Brief 2007," *op. cit.*

73. For background, see Glazer, "Radical Islam in Europe," *op. cit.*

74. "World Christian Database," Center for the Study of Global Christianity, Gordon-Conwell Theological Seminary, www.worldchristiandatabase.org/wcd/home.asp.

75. "Muslim Americans: Middle Class and Mostly Mainstream," Pew Forum on Religion and Public Life, May 22, 2007, p. 4, http://pewforum.org/surveys/muslim-american.

76. Chehab, *op. cit.*, pp. 134-150.

77. Brasher, *op. cit.*, p. 255.

78. "World Christian Database," *op. cit.*

79. Brasher, *op. cit.*, p. 255.

80. *Ibid.*, p. 204.

81. See American Friends of Lubavitch Washington, D.C., www.afldc.org.

82. Brasher, *op. cit.*, p. 222.

83. Almond, Appleby and Sivan, *op. cit.*, pp. 136-139.

84. *Ibid.*, pp. 157-159.

85. *Ibid.*

86. John L. Allen Jr., "McCain's choice a nod not only to women, but post-denominationalists," *National Catholic Reporter*, Aug. 30, 2008, http://ncrcafe.org/node/2073.

87. Garance Burke, "Palin once blessed to be free from witchcraft," The Associated Press, Sept. 25, 2008, http://abcnews.go.com/Politics/wireStory?id=5881256. Video footage at www.youtube.com/watch?v=QIOD5X68lIs.

88. Alexander Schwabe, "Sarah Palin's Religion: God and the Vice-Presidential Candidate," *Spiegel* online, Sept. 10, 2008, www.spiegel.de/international/world/0,1518,577440,00.html. Video footage at www.youtube.com/watch?v=QG1vPYbRB7k.

89. For background see the following *CQ Researchers*: Marcia Clemmitt, "Intelligent Design," July 29, 2005, pp. 637-660; Kenneth Jost and Kathy Koch, "Abortion Showdowns," Sept. 22, 2006, pp. 769-792; Kenneth Jost, "Abortion Debates," March 21, 2003, pp. 249-272; and Marcia Clemmitt, "Birth-control Debate," June 24, 2005, pp. 565-588.

90. See Karla Adam, "Gay Bishop Dispute Dominates Conference; Anglican Event Ends With Leader's Plea," *The Washington Post*, Aug. 4, 2008, p. A8.

91. Jeffrey Gettleman and Mohammed Ibrahim, "Somalis cheer the selection of a moderate Islamist cleric as President," *The New York Times*, Feb. 1, 2009, www.nytimes.com/2009/02/01/world/africa/01somalia.html.

92. Ramola Talwar Badam, "Official: India received intel on Mumbai attacks," The Associated Press, *Denver Post*, Dec. 1, 2008, www.denverpost.com/business/ci_11111305.

93. Somini Sengupta, "Hindu Threat to Christians: Convert or Flee," *The New York Times*, Oct. 12, 2008, www.nytimes.com/2008/10/13/world/asia/13india.html?pagewanted=1&_r=1&sq=Christians percent20India&st=cse&scp=1.

94. "Indian Christians Petition PM for Peace in Orissa at Christmas," World Evangelical Alliance Web site, Dec. 14, 2008, www.worldevangelicals.org/news/view.htm?id=2276.

95. Martha Nussbaum, "Terrorism in India has many faces," *Los Angeles Times*, Nov. 30, 2008, p. A35.

96. For background, see David Masci, "Emerging India," *CQ Researcher*, April 19, 2002, pp. 329-360.

97. Quoted in Nussbaum, *op. cit.*

98. "Iraq: Christians trickling back to their homes in Mosul," IRIN (humanitarian news and analysis service of the U.N. Office for the Coordination of Humanitarian Affairs), Nov. 6, 2008, www.irinnews.org/Report.aspx?ReportId=81317.

99. Ahmed Saka, "Death toll over 300 in Nigerian sectarian violence, The Associated Press, Nov. 29, 2008," www.denverpost.com/breakingnews/ci_11101598.

100. Gilad Shalit, "Hamas rejects Israel's Gaza cease-fire conditions," *Haaretz*, Jan. 28, 2009, www.haaretz.com/hasen/spages/1059593.html.

101. Scott Baldauf, "Africans join forces to fight the LRA," *The Christian Science Monitor*, Dec. 16, 2008, www.csmonitor.com/2008/1217/p06s01-woaf.html.

102. Elaine Sciolino, "Britain Grapples With Role for Islamic Justice," *The New York Times*, Nov. 18, 2008, www.nytimes.com/2008/11/19/world/europe/19shariah.html?_r=1&emc=tnt&tntemail0=y.

103. "Tanzania: Islamic Courts Debate Splits Legislators," *The Citizen* (newsletter, source: Africa News), Aug. 14, 2008.

104. Gershom Gorenberg, "The Temple Institute of Doom, or Hegel Unzipped," *South Jerusalem* (Blog), July 8, 2008, http://southjerusalem.com/2008/07/the-temple-institute-of-doom-or-hegel-unzipped.

105. See Katel, *op. cit.*, p. 859.

106. "Muslim Americans: Middle Class and Mostly Mainstream," *op. cit.*

107. Scott Shane, "Global Forecast by American Intelligence Expects Al Qaeda's Appeal to Falter," *The New York Times*, Nov. 20, 2008, www.nytimes.com/2008/11/21/world/21intel.html?_r1&emc=tnt&tntemail0=y.

108. Bin Laden, *op. cit.*

109. Lybarger, *op. cit.*, p. 244.

110. Carlson-Thies was speaking at a discussion on faith-based initiatives organized by the Brookings Institution in Washington, D.C. on Dec. 5, 2008.

111. See 'fastest growing denominations' category in "World Christian Database," *op. cit.*

112. Michael Mosettig, "Among Sea of Glittery Crosses, Christianity Makes its Mark in South Korea," Nov. 5, 2007, Public Broadcasting Service, www.pbs.org/newshour/indepth_coverage/asia/koreas/2007/report_11-05.html.

BIBLIOGRAPHY

Books

Almond, Gabriel A., Scott Appleby and Emmanuel Sivan, *Strong Religion: The Rise of Fundamentalisms Around the World, University of Chicago Press,* **2003.**
Three history professors synthesize the findings of a five-volume project that looks at 75 forms of religious fundamentalism around the world.

Brasher, Brenda E., ed., *Encyclopedia of Fundamentalism, Routledge,* **2001.**
Academics provide an A-Z on Christian fundamentalism — from its origins in the United States to its spread to other countries and religions.

Donohue, John J., and John L. Esposito, *Islam in Transition: Muslim Perspectives, Oxford University Press,* **2007.**
Essays by Muslim thinkers address key questions, such as the role of women in Islam, the relationship between Islam and democracy and the clash between Islam and the West.

Lybarger, Loren D., *Identity and Religion in Palestine: The Struggle between Islamism and Secularism in the Occupied Territories, Princeton University Press,* **2007.**
A U.S. sociologist who spent several years in the Palestinian territories explores how groups promoting fundamentalist Islam have gradually eclipsed secular nationalism as the dominant political force.

Thomas, Pradip Ninan, *Strong Religion, Zealous Media: Christian Fundamentalism and Communication in India, SAGE Publications,* **2008.**
An associate professor of journalism at the University of Queensland, Australia, examines the influence of U.S televangelists in India and the battle for cultural power between Hindu, Muslim and Christian fundamentalists. SAGE is the publisher of *CQ Global Researcher.*

Articles

"The Palestinians: Split by geography and by politics," *The Economist,* **Feb. 23, 2008, www.economist.com/ world/mideast-africa/displaystory.cfm?story_ id=10740648.**
The secular organization Fatah controls the West Bank while the Islamist group Hamas is in charge in Gaza.

Crilly, Rob, "Lord's Resistance Army uses truce to rearm and spread its gospel of fear," *The Times* (London), **Dec. 16, 2008, www.timesonline.co.uk/ tol/news/world/africa/article 5348890.ece.**
A violent military campaign led by Ugandan Christian fundamentalists threatens to destabilize the neighboring region.

Griswold, Eliza, "God's Country," *The Atlantic,* **March 2008, pp. 40-56, www.theatlantic.com/ doc/200803/nigeria.**
An author recounts her visit to Nigeria, a deeply religious country where Christian and Muslim clerics compete to grow their flocks, and religious tensions often spill over into violence.

Tavernise, Sabrina, "Independent, Tajiks Revel in Their Faith," *The New York Times,* **Jan. 3, 2009, www.nytimes.com/2009/01/04/world/asia/04tajik .html?emc=tnt&tntemail0=y.**
The Central Asian republic has become increasingly Islamic since its independence from the Soviet Union, with strong influence from Saudi Arabia.

Traynor, Ian, "Denmark's political provocateur: Feminist, socialist, Muslim?" *The Guardian,* **May 16, 2008, www .guardian.co.uk/world/2007/may/16/religion.uk.**
The controversial Danish politician Asmaa Abdol-Hamid, a devout Muslim, hopes to become the first person elected to the Danish parliament to wear the Islamic headscarf.

Reports and Studies

"Islamic Extremism: Common Concern for Muslim and Western Publics," *The Pew Global Attitudes Project,* **July 14, 2005, http://pewglobal.org/reports/pdf/248.pdf.**
A U.S.-based research center surveys public opinion in 17 countries on why Islamic extremism is growing.

MacIntyre, Andrew and Douglas E. Ramage, "Seeing Indonesia as a normal country: Implications for Australia," *Australian Strategic Policy Institute,* **May 2008, www.aspi.org.au/publications/publication_ details.aspx?ContentID=169&pubtype=5.**
Two Australian academics argue that claims of rampant Islamic fundamentalism in Indonesia — the world's most populous Muslim country — are exaggerated.

Mayson, Cedric, "Religious Fundamentalism in South Africa," *African National Congress, Commission for Religious Affairs*, January 2007, http://the-brenthurstfoundation.co.za/Files/terror_talks/Religious%20Fundamentalism%20in%20SA.pdf.
A South African activist blames growing fundamentalism in South Africa on U.S. Christian fundamentalists.

Olcott, Martha Brill and Diora Ziyaeva, "Islam in Uzbekistan: Religious Education and State Ideology," *Carnegie Endowment for International Peace*, July 2008, www.carnegieendowment.org/publications/index.cfm?fa=view&id=21980&prog=zru.
Two academics chart the growth of Islam in the Central Asian republic.

For More Information

Association of Evangelicals in Africa, www.aeafrica.org. A continent-wide coalition of 33 national evangelical alliances and 34 mission agencies that aims to "mobilize and unite" evangelicals in Africa for a "total transformation of our communities."

European Jewish Community Centre, 109 Rue Froissart, 1040 Brussels, Belgium; (32) 2-233-1828; www.ejcc.eu. Office of the Chabad Jewish missionary movement's delegation to the European Union.

Evangelical Graduate School of Theology, N.E.G.S.T., P.O. Box 24686, Karen 00502, Nairobi, Kenya; (254) 020-3002415; www.negst.edu. An Evangelical Christian institution devoted to the study of religion in Africa.

Forum 18 News Service, Postboks 6603, Rodeløkka, N-0502 Oslo, Norway; www.forum18.org. News agency reporting on government-sponsored repression of religion in Central Asia.

Organisation of the Islamic Conference, P.O. Box 178, Jeddah 21411, Saudi Arabia; (966) 690-0001; www.oic-oci.org. Intergovernmental organization with 57 member states, which promotes the interests of the Muslim world.

The Oxford Centre for Hindu Studies, 15 Magdalen St., Oxford OX1 3AE, United Kingdom; (44) (0)1865-304-300; www.ochs.org.uk. Experts in Hindu culture, religion, languages, literature, philosophy, history, arts and society.

Pew Forum on Religion and Public Life, 1615 L St., N.W., Suite 700, Washington, DC 20036-5610; (202) 202-419-4550; http://pewforum.org. Publishes surveys on religiosity, including fundamentalist beliefs, conducted around the world.

World Christian Database, BRILL, P.O. Box 9000, 2300 PA Leiden, The Netherlands; (31) (0)71-53-53-566; www.worldchristiandatabase.org. Provides detailed statistical data on numbers of believers, by religious affiliation; linked to U.S.-based Center for the Study of Global Christianity, Gordon-Conwell Theological Seminary.

World Evangelical Alliance, Suite 1153, 13351 Commerce Parkway, Richmond, BC V6V 2X7 Canada; (1) 604-214-8620; www.worldevangelicals.org. Network for evangelical Christian churches around the world.

Worldwide Incidents Tracking System, National Counter Terrorism Center, University of Maryland, College Park, MD 20742; (301) 405-1000; http://wits.nctc.gov. Provides detailed statistics on religiously inspired terrorist attacks across the world from 2004-2008.

The Obama Presidency

12

Can Barack Obama Deliver the Change He Promises?

Kenneth Jost and the *CQ Researcher* Staff

The largest crowd in Washington history cheers President Barack Obama after his swearing in on Jan. 20, 2009. An estimated 1.8 million high-spirited, flag-waving people gathered at the Capitol and National Mall, but thousands more were turned away by police due to overcrowding.

From *CQ Researcher*,
January 30, 2009.

They came to Washington in numbers unprecedented and with enthusiasm unbounded to bear witness and be a part of history: the inauguration of Barack Hussein Obama on Jan. 20, 2009, as the 44th president of the United States and the first African-American ever to serve as the nation's chief executive.

After taking the oath of office from Chief Justice John G. Roberts Jr., Obama looked out at the estimated 1.8 million people massed at the Capitol and National Mall and delivered an inaugural address nearly as bracing as the subfreezing temperatures.

With hardly the hint of a smile, Obama, 47, outlined the challenges confronting him as the fifth-youngest president in U.S. history. The nation is at war, he noted, the economy "badly weakened" and the public beset with "a sapping of confidence."

"Today I say to you that the challenges we face are real," Obama continued in his 18-minute speech. "They are serious and they are many. They will not be met easily or in a short span of time. But know this, America — they will be met."[1] (*See economy sidebar, p. 286; foreign policy sidebar, p. 292.*)

The crowd received Obama's sobering message with flag-waving exuberance and a unity of spirit unseen in Washington for decades. Despite Democrat Obama's less-than-landslide 7 percentage-point victory over John McCain on Nov. 4, hardly any sign of political dissent or partisan opposition surfaced on Inauguration Day or during the weekend of celebration that preceded it. (*See maps, p. 278; poll, p. 280.*)

"It's life-changing for everyone," said Rhonda Gittens, a University of Florida journalism student, "because of who he is,

Obama Victory Changed Electoral Map

Barack Obama won nine traditionally Republican states in the November 2008 election that George W. Bush had won easily in 2004, and his electoral and popular vote totals were significantly higher than Bush's. In 2004, Bush won with 50.7 percent of the vote to John Kerry's 48.3 percent. By comparison Obama garnered 52.9 percent to Sen. John McCain's 45.7. In the nation's new political map, the Democrats dominate the landscape, with the Republicans clustered in the South, the Plains and the Mountain states.

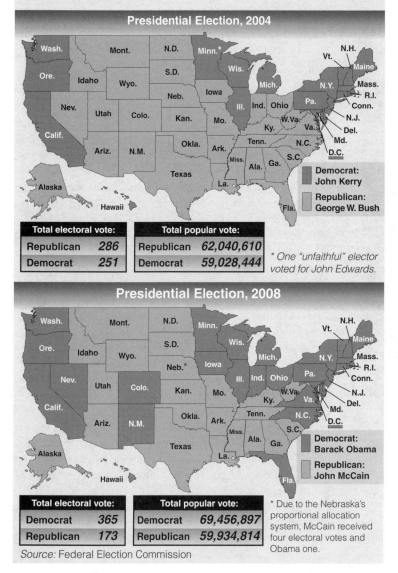

Presidential Election, 2004

Democrat: John Kerry
Republican: George W. Bush

Total electoral vote:	
Republican	286
Democrat	251

Total popular vote:	
Republican	62,040,610
Democrat	59,028,444

* One "unfaithful" elector voted for John Edwards.

Presidential Election, 2008

Democrat: Barack Obama
Republican: John McCain

Total electoral vote:	
Democrat	365
Republican	173

Total popular vote:	
Democrat	69,456,897
Republican	59,934,814

* Due to the Nebraska's proportional allocation system, McCain received four electoral votes and Obama one.

Source: Federal Election Commission

because of how he represents everyone." Gittens traveled to Washington with some 50 other members of the school's black student union.

The inaugural crowd included tens of thousands clustered on side streets after the U.S. Park Police determined the mall had reached capacity. The crowd was bigger than for any previous inauguration — at least three times larger than when the outgoing president, George W. Bush, had first taken the oath of office eight years earlier. The total number also exceeded independent estimates cited for any of Washington's protest marches or state occasions in the past.*

The spectators came from all over the country and from many foreign lands. "He's bringing change here," said Clayton Preira, a young Brazilian accompanying three fellow students on a two-month visit to the United States. "He's bringing change all over the world." The spectators were of all ages, but overall the crowd seemed disproportionately young. "He really speaks to young people," said Christian McLaren, a white University of Florida student.

Most obviously and most significantly, the crowd was racially and ethnically diverse — just like the new first family. Obama himself is the son of a

---

* Crowd estimates for President Obama's inauguration ranged from 1.2 million to 1.8 million. Commonly cited estimates for other Washington events include: March on Washington for Jobs and Freedom, 1963, 250,000; President John F. Kennedy's funeral, 1963, 800,000; inauguration of President Lyndon B. Johnson, 1965, 1.2 million; Peace Moratorium, 1969, 250,000; Million Man March, 1995, 400,000-800,000; March for Life, 1998, 225,000; March for Women's Lives, 2004, 500,000-800,000.

black Kenyan father and a white Kansan mother. His wife Michelle, he often remarks, carries in her the blood of slaves and of slave owners. Among those behind the first lady on the dais were Obama's half-sister, Maya Soetoro-Ng, whose father was Indonesian, and her husband, Konrad Ng, a Chinese-American. Some of Obama's relatives from Kenya came as well, wearing colorful African garb.

The vast numbers of black Americans often gave the event the air of an old-time church revival. In quieter moments, many struggled to find the words to convey the significance, both historic and personal. "It hasn't sunk in yet," Marcus Collier, a photographer from New York City, remarked several hours later.

David Moses, a health-care supervisor in New York City, carried with him a picture of his late father, who had encouraged him and his brother to join the anti-segregation sit-ins of the early 1960s in their native South Carolina. "It's the culmination of a long struggle," Moses said, "that still has a long way to go."

Shannon Simmons, who had not yet been born when Congress passed major civil rights legislation in the 1960s, brought her 12-year-old daughter from their home in New Orleans. "It's historic," said Simmons, who made monthly contributions to the Obama campaign. "It's about race, but it's more than that. I believe he can bring about change." (*See sidebar, p. 282.*)

For black Americans, old and young alike, the inauguration embodied the lesson that Obama himself had often articulated — that no door need be viewed as closed to any American, regardless of race. For Obama himself, the inauguration climaxed a quest that took him from the Illinois legislature to the White House in only 12 years.

To win the presidency, Obama had to defy political oddsmakers by defeating then-Sen. Hillary Rodham Clinton, the former first lady, for the Democratic nomination and then beating McCain, the veteran Arizona senator and Vietnam War hero. Obama campaigned hard against the Bush administration's record, blaming Bush, among other things, for mismanaging the U.S. economy as well as the wars in Iraq and Afghanistan.

After a nod to Bush's record of service and help during the transition, Obama hinted at some of those criticisms in his address. "The nation cannot prosper long when it favors only the prosperous," he declared, referencing tax cuts enacted in Bush's first year in office that Obama had called for repealing.

On national defense, "we reject the false choice between our safety and our ideals," Obama continued. The Bush administration had come under fierce attack from civil liberties and human rights advocates for aggressive detention and interrogation policies adopted after the Sept. 11, 2001, terrorist attacks on the United States. (*See "At Issue," p. 302.*)

Despite the attacks, Obama also sounded conservative notes throughout the speech, blaming economic woes in part on a "collective failure to make hard choices" and calling for "a new era of responsibility." Republicans in the audience were pleased. "He wasn't pointing fingers just toward Bush," said Rhonda Hamlin, a social worker from Alexandria, Va. "He was pointing fingers toward all of us."

With the inauguration behind him, Obama went quickly to work. Within hours, the administration moved to institute a 120-day moratorium on legal proceedings against the approximately 245 detainees still being held at the Guantánamo Bay Naval Base in Cuba. Obama had repeatedly pledged during the campaign to close the prison; two days later he signed a second decree, ordering that the camp be closed within one year.

Then on his first full day as president, Obama on Jan. 21 issued stringent ethics rules for administration officials and conferred separately with his top economic and military advisers to begin mapping plans to try to lift the U.S. economy out of its yearlong recession and bring successful conclusions to the conflicts in Iraq and Afghanistan.

By then, the Inauguration Day truce in partisan conflict was beginning to break down. House Republicans pointed to a Congressional Budget Office study questioning the likely impact of the Democrats' $825-billion economic stimulus package, weighted toward spending instead of tax cuts. "The money that they're going to throw out the door, at the end of the day, is not going to work," said Rep. Devin Nunes, R-Calif., a member of the tax-writing House Ways and Means Committee. (*See "At Issue," p. 303.*)

The partisan division raised questions whether Democratic leaders could stick to the promised schedule of getting a stimulus plan to Obama's desk for his signature by the time of the Presidents' Day congressional recess in mid-February. More broadly, the Republicans' stance presaged continuing difficulties for Obama as he turned to other ambitious agenda items, including his repeated pledge to overhaul the nation's health-care system. (*See sidebar, p. 296.*)

Public Gives Obama Highest Rating

Barack Obama began his presidency with 79 percent of Americans having a favorable impression of him — higher than the five preceding presidents. George W. Bush entered office with a 62 percent favorability rating; he left with a 33 percent approval rating, lowest of post-World War II presidents except Harry S. Truman and Richard M. Nixon.

Do you have a favorable impression of . . . ?

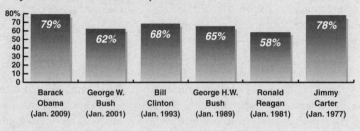

Source: The Washington Post, Jan. 18, 2009

Obama included health care in his inaugural litany of challenges, along with education, climate change and technology. For now, those initiatives lie in the future. In the immediate days after his euphoric inauguration, here are some of the major questions being debated:

Is President Obama on the right track in fixing the U.S. economy?

As president-elect, Obama spent his first full week in Washington in early January first warning of trillion-dollar federal budget deficits for years to come and then making urgent appeals for public support for a close to trillion-dollar stimulus to get the economy moving.

Members of Congress from both parties and advocates and economic experts of all persuasions agree on the need for a good-sized federal recovery program for the seriously ailing U.S. economy. And most agree on a prescription that combines spending increases and tax cuts. But there is sharp disagreement as to the particulars between tax-cutting conservatives and pump-priming liberals, with deficit hawks worried that both of the prescribed remedies could get out of hand.

With the plan's price tag then being estimated somewhere around $800 billion, Obama made his first sustained appeal for public support in a somber, half-hour address on Jan. 8 at George Mason University in Fairfax,

Va., outside Washington. Any delay, he warned, could risk double-digit unemployment. He outlined plans to "rebuild America" ranging from alternative energy facilities and new school classrooms to computerized medical records, but he insisted the plan would not entail "a slew of new government programs." He reiterated his campaign promise of a "$1,000 tax cut for 95 percent of working-class families" but made no mention of business tax cuts being included as sweeteners for Republican lawmakers.

Within days, Obama's plan was taking flack from left and right in the blogosphere. Writing on the liberal HuffingtonPost.com, Robert Kuttner, co-editor of *American Prospect* magazine, denounced the spending plan as too small and the business tax cuts as "huge concessions" in a misguided effort at "post-partisanship." From the right, columnist Neal Boortz accused Obama on the conservative TownHall.com of using the economic crisis as "cover for increased government spending that he's been promising since the day he announced his candidacy."

Allen Schick, a professor of economics at the University of Maryland in College Park and formerly an economics specialist with the Congressional Research Service, sees weaknesses with both components of the Obama plan. "We really have no model to deal with the question of what's the right number" for the stimulus, he says. "And we're not even sure that the stimulus will do the job, especially if a lot of the spending is wasteful."

As for the tax cuts, Schick calls them "harebrained, more intended to look good and buy support than to actually get the economy moving." In particular, he criticized a proposed $3,000 jobs credit for employers. "We know from the past that employers don't hire people for just a few shekels," he says. Eventually, the jobs credit was dropped, but the package still includes business tax breaks such as a $16 billion provision to allow businesses to use 2008 and 2009 losses to offset profits for the previous five years instead of two.

Conservatives favor tax cuts, but not the middle-class tax cut that Obama is proposing. "A well-designed tax cut is the only effective short-term stimulus," says J. D. Foster, a senior fellow at the Heritage Foundation. But Foster, who worked in the Office of Management and Budget in the Bush administration, calls either for extending or making permanent Bush's across-the-board rate cuts, which primarily benefited upper-income taxpayers.

From the opposite side, Chad Stone, chief economist with the liberal Center on Budget Policy and Priorities, endorses Obama's approach. "Tax cuts should be focused on people of low and moderate means, who are much more likely to spend the extra money they get," he says.

Academic economists, however, caution that tax cuts may not deliver a lot of bang for the buck in terms of short-term stimulus. Studies indicate that taxpayers pocketed at least one-third of the $500 tax rebate the government disbursed to counteract the 2001 recession.

Advocates and observers on both sides warn that the spending side of the package may also be less effective than hoped if political forces play too large a role in shaping it. "If it goes to pork, if it goes to green jobs that may sound good in the short term but may not have a market response or a market for them, then it's a waste," Paul Gigot, editorial page editor of *The Wall Street Journal*, said on NBC's "Meet the Press" on Jan. 11.

"If the stuff that gets added is not very effective as stimulus or the things that are good get pulled out, that would not be good," says Stone.

For its part, the budget-restraint advocacy group Concord Coalition sees political forces as driving up the total cost of the package — in spending and tax cuts alike — with no regard for the long-term impact. "Nothing is ever taken off the table," says Diane Lim Rogers, the coalition's chief economist.

Rogers complains of "political pressure to come up with tax cuts even though economists are having trouble figuring out whether they're going to do any good." At the same time, she says spending has to be designed "as thoughtfully as possible, not in a way that the federal government ends up literally just throwing money out the door."

A range of experts also call for renewed efforts to solve the mortgage and foreclosure crisis, saying that homeowners are not going to start spending again without confidence-restoring steps. Indeed, Federal Reserve Chairman Ben Bernanke pointedly told a conference in December that steps to reduce foreclosures "should be high on the agenda" in any economic recovery plan.[2]

Despite questions and concerns about the details, however, support for strong action is all but universal. "We have no choice," said Mark Zandi, chief economist of Moody's Economy.com and a former adviser to the McCain campaign, also on "Meet the Press." "If we don't do something like this — a stimulus package, a foreclosure mitigation plan — the economy is going to slide away."

Is President Obama on the right track in Iraq and Afghanistan?

At the start of his presidential campaign in February 2007, candidate Obama was unflinchingly calling for withdrawing all U.S. combat forces from Iraq within 16 months after taking office. But his tone began changing as he neared the Democratic nomination in summer 2008. And in his first extended broadcast interview after the election, President-elect Obama said on NBC's "Meet the Press" on Dec. 7 only that he would summon military advisers on his first day in office and direct them to prepare a plan for "a responsible drawdown."

Obama also did nothing to knock down host Tom Brokaw's forecast of a "residual force" of 35,000 to 50,000 U.S. troops in Iraq through the end of his term. "I'm not going to speculate on the numbers," Obama said, but he went on to promise "a large enough force in the region" to protect U.S. personnel and to "ferret out any terrorist activity." In addition, Obama voiced disappointment with developments in Afghanistan and said that "additional troops" and "more effective diplomacy" would be needed to achieve U.S. goals there.

Many foreign policy observers are viewing Obama's late campaign and post-election stances as a salutary shift from ideology to pragmatism. "It seems very clear that he will not fulfill his initial pledge to withdraw all U.S. forces from Iraq in 16 months — which is only wise," says Thomas Donnelly, a resident fellow on defense and national security issues at the American Enterprise Institute (AEI).

"I personally have been very impressed with [Obama's] thinking and his way of assembling a national security team," says Kenneth Pollack, director of the Brookings Institution's Saban Center for Middle East Policy. "This is not a man who plays by the traditional American political rules."

First Black President Made Race a Non-Issue

Obama's personal attributes swept voters' doubts aside.

Barack Obama took the oath of office the day after this year's Martin Luther King holiday, and he accepted the Democratic presidential nomination last August on the 45th anniversary of King's celebrated "I Have a Dream" speech.

For millions of Americans, Obama's election as the nation's first African-American president seemed to fulfill the promise of King's "dream" of a nation in which citizens "will not be judged by the color of their skin, but by the content of their character."

"Obviously, for an African-American to win the presidency, given the history of this country . . . is a remarkable thing," Obama said after the election. "If you think about grandparents who are alive today who grew up under Jim Crow, that's a big leap."[1]

While Obama clearly benefited from the sacrifices of the civil rights generation — to which he has paid homage — his politics are different from the veterans of that movement. Older black politicians such as the Rev. Jesse Jackson seemed to base their candidacies mainly on issues of particular concern to African-Americans. But black politicians of Obama's generation, such as Massachusetts Gov. Deval Patrick and Newark Mayor Cory Booker (both Democrats), have run on issues of broader concern — in Obama's case, first on the war in Iraq and later on the economic meltdown.

"The successful ones start from the outside by appealing to white voters first, and work back toward their base of black voters," said broadcast journalist Gwen Ifill, author of the new book *The Breakthrough: Politics and Race in the Age of Obama.*[2]

Black voters initially were reluctant to support Obama — polls throughout 2007 showed Sen. Hillary Rodham Clinton with a big lead among African-Americans — but he picked up their support as it became clear he was the first black candidate with a realistic hope of winning the White House. Clinton's support among blacks dropped markedly in the wake of remarks by former President Bill Clinton that many found demeaning.

But many white Democratic voters remained reluctant to support Obama, particularly in Appalachia. Exit polling

during the Pennsylvania primary, for example, showed that 16 percent of whites had considered race in making their pick, with half of those saying they would not support Obama in the fall.[3]

Obama also was bedeviled by videotaped remarks of his pastor, the Rev. Jeremiah Wright, which were incendiary and deemed unpatriotic. But Obama responded with a widely hailed speech on race in March 2008 in which he acknowledged both the grievances of working-class whites and the continuing legacy of economic disadvantages among blacks. Obama said his own life story "has seared into my genetic makeup the idea that this nation is more than the sum of its parts — that out of many, we are truly one."[4]

As the general election campaign got under way, it was clear that race would continue to be a factor. One June poll showed that 30 percent of Americans admit prejudice.[5] And, despite Obama's lead, there was debate throughout the campaign about the so-called Bradley effect — the suggestion that people will lie to pollsters about their true intentions when it comes to black candidates.*

But neither Obama nor Arizona Sen. John McCain, his Republican rival, made explicit pleas based on race, with McCain refusing to air ads featuring Wright. As the campaign wore on, no one forgot that Obama is black — but most doubters put that fact aside in favor of more pressing concerns.

"For a long time, I couldn't ignore the fact that he was black. I'm not proud of that," Joe Sinitski, a 48-year-old Pennsylvania voter, told *The New York Times*. "I was raised to think that there aren't good black people out there."[6] But Sinitski ended up voting for Obama, along with many other whites won over by Obama's personal attributes or convinced that issues such as the economy trumped race.

Exit polls showed that Obama prevailed among those who considered race a significant factor, 53 to 46 percent.[7] "In difficult economic times, people find the price of prejudice is just

---

* The Bradley effect refers to Tom Bradley, an African-American who lost the 1982 race for governor in California despite being ahead in voter polls going into the election.

a little too high," said outgoing North Carolina Gov. Mike Easley, a Democrat.[8]

"The Bradley effect really was not a significant factor, despite much concern, fear and hyperventilation about it leading up to the election," says Scott Keeter, a pollster with the Pew Research Center. "Race was a consideration to people, but what it wasn't, invariably, was a negative consideration for white voters. It was a positive consideration for many white voters who saw Obama as a candidate who could help the country toward racial reconciliation."

Obama carried more white voters than former Vice President Al Gore or Sen. John Kerry of Massachusetts, the two previous Democratic nominees. Still, he could not have prevailed without black and Hispanic voters, particularly in the three Southern states he carried. In Virginia — a state that had voted Republican since 1964 — Obama lost by 21 points among white voters, according to exit polls.

His victory clearly did not bring racial enmity to its end. In December, Chip Saltsman, a candidate for the Republican Party chairmanship, sent potential supporters a CD containing the song "Barack the Magic Negro," a parody popularized by right-wing talk show host Rush Limbaugh during the campaign. And, when Senate Democrats initially balked in January at seating Roland Burris as Obama's replacement, Rep. Bobby Rush, D-Ill, played the race card, warning them not to "hang or lynch the appointee," comparing the move to Southern governors who sought to block desegregation.[9]

But still polls suggest that most Americans believe Obama's presidency will be a boon for race relations. A *USA Today*/Gallup Poll taken the day after the November election showed that two-thirds predicted black-white relations "will eventually be worked out" — by far the highest total in the poll's history.[10]

In the future, white males may no longer be the default inhabitants of America's most powerful position. The present generation and those in the future are likely to grow up thinking it's a normal state of affairs for the country to be led

Michelle Obama holds the Bible used to swear in President Abraham Lincoln as Barack Obama takes the oath of office from Supreme Court Chief Justice John G. Roberts Jr.

AFP/Getty Images/Tim Sloan

by a black president. "For a lot of African-Americans, it already has made them feel better and more positive about the country and American society," says David Bositis, an expert on black voting at the Joint Center for Political and Economic Studies.

"When you ask my kids what they want to be when they grow up, they always say they want to work at McDonald's or Wal-Mart," said Joslyn Reddick, principal at a predominantly black school in Selma, Ala., a city from which King led an historic march for voting rights in 1965.

"Now they will see that an African-American has achieved the highest station in the United States," Reddick said. "They can see for themselves that dreams can come true."[11]

— Alan Greenblatt,
staff writer, *Governing* magazine

[1] Bryan Monroe, "The Audacity of Victory," *Ebony*, January 2009, p. 16.

[2] Sam Fulwood III, "The New Face of America," *Politico.com*, Jan. 13, 2009.

[3] Alan Greenblatt, "Changing U.S. Electorate," *CQ Researcher*, May 30, 2008, p. 459.

[4] The Obama speech, "A More Perfect Union," is at www.youtube.com/watch?v=pWe7wTVbLUU. The text of the March 18, 2008, speech, "A More Perfect Union," is found in *Change We Can Believe In: Barack Obama's Plan to Renew America's Promise* (2008), pp. 215-232.

[5] Jon Cohen and Jennifer Agiesta, "3 in 10 Americans Admit to Race Bias," *The Washington Post*, June 22, 2008, p. A1.

[6] Michael Sokolove," The Transformation," *The New York Times*, Nov. 9, 2008, p. WK1.

[7] John B. Judis, "Did Race Really Matter?" *Los Angeles Times*, Nov. 9, 2008, p. 34.

[8] Rachel L. Swarns, "Vaulting the Racial Divide, Obama Persuaded Americans to Follow," *The New York Times*, Nov. 5, 2008, p. 7.

[9] Clarence Page, "Hiding Behind Black Voters," *Chicago Tribune*, Jan. 4, 2009, p. 24.

[10] Susan Page, "Hopes Are High for Race Relations," *USA Today*, Nov. 7, 2008, p. 1A.

[11] Dahleen Glanton and Howard Witte, "Many Marvel at a Black President," *Chicago Tribune*, Nov. 5, 2008, p. 6.

Cabinet Includes Stars, Superstars and Surprises

President Obama made his Cabinet selections in record time, and his appointees run the gamut of race, ethnic origin, gender, age and even party affiliation. Those in top posts include Sen. Hillary Rodham Clinton at State and Robert Gates continuing at Defense. Besides Gates, one other Republican was chosen: Transportation's Ray LaHood. New Mexico Gov. Bill Richardson's withdrawal left the Commerce post unfilled along with the director of Drug Control Policy. Cabinet-level appointees include four women, two Asian-Americans, two Hispanics and two African-Americans.

Name, Age Department	Date of Nomination	Date of Confirmation	Previous Positions
Hillary Rodham Clinton, 61, State	Dec. 1	Jan. 21	New York U.S. senator (2001-09); first lady (1993-2001); Arkansas first lady (1979-81, 1983-92)
Timothy Geithner, 47, Treasury	Nov. 24	Jan. 26	President, Federal Reserve Bank of New York (2003-09); under secretary, Treasury (1998-2001)
Robert Gates, 65, Defense*	Dec. 1	Dec. 6, 2006 *	Defense secretary (2006-present); director, CIA (1991-93); deputy national security adviser (1989-91)
Eric Holder, 57, Attorney General	Dec. 1		Deputy attorney general (1997-2001); U.S. attorney (1993-97); judge, D.C. Superior Court (1988-93)
Ken Salazar, 53, Interior	Dec. 17	Jan. 20	Colorado U.S. senator (2005-09); Colorado attorney general (1999-2005)
Tom Vilsack, 58, Agriculture	Dec. 17	Jan. 20	Iowa governor (1999-2007); Iowa state senator (1992-99)
Hilda Solis, 51, Labor	Dec. 19		California U.S. representative (2001-09); California state senator (1995-2001)
Tom Daschle, 61, Health & Human Services	Dec. 11		South Dakota U.S. senator (1987-2005); Senate majority leader (2001, 2001-03); South Dakota U.S. representative (1979-87)
Shaun Donovan, 42, Housing and Urban Development	Dec. 13	Jan. 22	Commissioner, New York City Dept. of Housing Preservation and Development (2004-08); deputy assistant secretary, HUD (2000-01)

Obama invited speculation about a shift toward the center by selecting Clinton and Robert Gates as the two Cabinet members on his national security team along with a retired Marine general, James Jones, as national security adviser. (*See chart, at left.*) Clinton had voted for the Iraq War in late 2002, though she echoed Obama during the campaign in calling for troop withdrawals. As Bush's secretary of Defense, Gates had overseen the "surge" in U.S. forces during 2007.

"This is a group of people who are very sober, very intelligent, fully aware of the importance of Iraq to America's security interests and of the fragility of the situation there," says Pollack.

Some anti-war activists were voicing concern about Obama's seeming shift within days of his election. "Obama has very successfully branded himself as anti-war, but the fact remains that he's willing to keep a residual force in Iraq indefinitely, [and] he wants to escalate in Afghanistan," said Matthis Chiroux of Iraq Veterans Against the War. "My hope is that he starts bringing home the troops from Iraq immediately, but I think those of us in the anti-war movement could find ourselves disappointed."[3]

Since then, however, criticism of Obama's emerging policies has been virtually non-existent from the anti-war and Democratic Party left. "He seems to be accelerating the withdrawal, which is terrific," says Robert Borosage, co-director of the Campaign for America's Future. Borosage is "concerned" about the residual force in Iraq because of the risk that U.S. troops will become involved in "internecine battles." But he adds, "That's what he's promised, and I think he'll fulfill his promise."

Donnelly and Pollack, however, both view a continuing U.S. role in Iraq as vital. "There's good progress, but a long way to go," says Donnelly. "A huge American role is going to be needed through the four years of the Obama administration." Pollack agrees. "Iraq is far from solved. Whether we like it or not, Iraq is a vital interest for the United States of America."

In his campaign and since, Obama has treated Afghanistan as more important to U.S. interests and harshly criticized the Bush administration for — in his view — ignoring the conflict there. Afghanistan "had had a huge rhetorical place in the Obama campaign," says Donnelly. "The idea being that Afghanistan was the good war, the more important war, and that Iraq was a dead end strategically."

P. J. Crowley, a senior fellow at the liberal think tank Center for American Progress, calls Obama's focus on Afghanistan "correct" but emphasizes the need for a multipronged effort to stabilize and reform the country's U.S.-backed government. "Returning our weight of effort [to Afghanistan] is a right approach," says Crowley, who was spokesman for the National Security Council under President Bill Clinton.

"More troops may help in a narrow sense," Crowley continues, "but I don't think anyone suggests that more troops are the long-term solution in Afghanistan. The insertion of U.S. forces is logical in the short- to mid-term, but it has to be part of a broader strategy."

But Pollack questions the value of any additional U.S. troops at all. "The problems of Afghanistan are not principally military; they are principally political and diplomatic," he says. "Unless this new national security team can create a military mission that is of value to what is ultimately a diplomatic problem, it's going to be tough to justify to the country the commitment of those additional troops."

Name, Age, Department	Date of Nomination	Date of Confirmation	Previous Positions
Ray LaHood, 63, Transportation	Dec. 19	Jan. 22	Illinois U.S. representative (1995-2009); state representative (1982-83)
Steven Chu, 60, Energy	Dec. 15	Jan. 20	Director, Lawrence Berkeley National Laboratory, Dept. of Energy (2004-09); professor, UC-Berkeley (2004-present); Nobel Prize winner, physics (1997)
Arne Duncan, 44, Education	Dec. 16	Jan. 20	C.E.O, Chicago Public Schools (2001-09)
Eric Shinseki, 66, Veterans Affairs	Dec. 7	Jan. 20	Chief of staff, Army (1999-2003)
Janet Napolitano, 51, Homeland Security	Dec. 1	Jan. 20	Arizona governor (2003-09); attorney general (1999-2002)
Rahm Emmanuel, 49, Chief of Staff	Nov. 6	NA	Illinois U.S. representative (2003-09); senior adviser to the president (1993-98)
Lisa Jackson, 46, Environmental Protection Agency	Dec. 15	Jan. 22	Chief of staff, governor of New Jersey (2008-09); commissioner, New Jersey Dept. of Environmental Protection (2006-2008)
Peter Orszag, 40, Office of Management and Budget	Nov. 25	Jan. 20	Director, Congressional Budget Office (2007-08); adviser, National Economic Council (1997-98)
Susan Rice, 44, Ambassador to the United Nations	Dec. 1	Jan. 22	Assistant secretary, State (1997-2001); National Security Council (1993-97)
Ron Kirk, 54, Trade Representative	Dec. 19		Mayor of Dallas (1995-2002)

Department heads are listed in order of succession under Presidential Succession Act; nondepartment heads were given Cabinet-level status.

* Gates was confirmed when first nominated by President George W. Bush and did not have to be re-confirmed.

Compiled by Vyomika Jairam; all photos by Getty Images

Bleak Economy Getting Bleaker

Economists widely agree a stimulus plan is needed.

When Barack Obama took office on Jan. 20, he inherited the most battered U.S. economy since World War II — and one of the shakiest to confront a new president in American history.

And the view from the Oval Office is likely to get bleaker before the gloom begins to lift.

"There are very serious questions on the financial side and apprehension among many parties that there may be more bad news to come," says Kent Hughes, director of the Program on Science, Technology, America and the Global Economy at the Woodrow Wilson Center for Scholars.

Already, Obama has stepped into the worst unemployment picture in 16 years, with the jobless rate at 7.2 percent and 11.1 million people out of work. The economy lost 1.9 million jobs during the last four months of 2008 — 524,000 in December alone.[1]

Economists worry that rising unemployment in manufacturing, construction, retailing and other sectors foreshadows an even more dismal future, at the very least in the short term. Dean Baker, co-director for the Center for Economic and Policy Research, a liberal think tank in Washington, says he expects another million or so jobs to disappear through February, then the pace of job loss to slow if Congress acts to stimulate the economy.

Obama must figure out not only how to get people back to work but also how to restore their confidence in the economy. A punishing credit crisis and cascade of grim news from Wall Street has led consumers to stop spending on everything from restaurant meals to houses and autos.[2]

Home sales have plunged in recent months, foreclosures are hitting record levels and a study by PMI Mortgage Insurance Co. estimates that half of the nation's 50-largest Metropolitan Statistical Areas have an "elevated or high probability" of experiencing lower home prices by the end of the third quarter of 2010 compared to the same quarter of 2008.[3]

Retail sales, a key indicator of consumer confidence, fell in December 2008 for the sixth month in a row, according to the Commerce Department.[4] The International Council of Shopping Centers said chain-store sales in December posted their biggest year-to-year decline since researchers began tracking figures in 1970.[5]

Rebecca Blank, a senior fellow at the Brookings Institution and former member of President Bill Clinton's Council of Economic Advisers, says the unemployment numbers "suggest the economy is still on the way down," and the decline in holiday sales is "surely going to lead to some bankruptcies and belt tightening in the retail sector."

Indeed, such trouble is already occurring. The shopping centers group estimated that 148,000 retail stores closed last year and that more than 73,000 will be shuttered in the first half of 2009.[6] Among the latest examples: Bankrupt electronics chain Circuit City said in January that it was closing its remaining 567 stores, putting some 30,000 employees out of work.

To revive the economy, the new administration — most visibly Obama himself — is urging Congress to quickly approve a stimulus package that could approach $900 billion. Much of the money would likely go toward tax cuts and public infrastructure projects, though how, exactly, the government would allocate it remains a matter of intense political debate.

One thing seems certain, though: The cost of a stimulus package, added to the hundreds of billions of dollars already spent to shore up the nation's flagging financial system, will add to the bulging federal deficit.

"The thing you know for sure is that a stimulus is going to add to the debt, which is [now] quite frightening, and it's going to make it worse," says June O'Neill, an economics professor at the City University of New York's Baruch College

Borosage also worries about an increased U.S. military presence in Afghanistan. "A permanent occupation of Afghanistan is a recipe for defeat," he says.

All of the experts stress that U.S. policy in Afghanistan now plays a secondary part in the fight with the al Qaeda terrorist group, which carried out the 9/11 attacks in the United States. "There is no al Qaeda in Afghanistan," says Donnelly. "Al Qaeda has now reconstituted itself in the tribal areas of northwest Pakistan."

Donnelly questions Afghanistan's importance to U.S. interests altogether but ultimately supports continued U.S. involvement. "The only thing worse than being engaged in Afghanistan," he says, "is turning our backs on it."

and a former director of the Congressional Budget Office (CBO) during the Clinton administration.

In January the CBO projected a $1.2 trillion deficit for the fiscal year. A stimulus plan would add even more pressure on Obama to get federal spending under control. "My own economic and budget team projects that, unless we take decisive action, even after our economy pulls out of its slide, trillion-dollar deficits will be a reality for years to come," Obama said.[7]

The battered economy that confronts President Obama includes record foreclosure rates and plummeting home values. Above, a foreclosed home in Nevada, the state with the nation's highest foreclosure rate.

of the financial markets, more closely resembles the Great Depression than any other recession since then.

Most postwar recessions "were the result of the Fed raising rates," says Baker. "That meant we knew how to reverse it. This one, there's not an easy answer to. We're not going to see [another] Great Depression — not double-digit unemployment for a decade." But in terms of the severity of the problem, Baker adds, the Great Depression is the "closest match" to what confronts the new administration.

— Thomas J. Billitteri

Still, a wide spectrum of economists — including conservatives who typically look askance at government spending — agree that a stimulus plan is necessary.

Martin Feldstein, a Harvard University economist and former chair of the Council of Economic Advisers in the Reagan administration, told a House committee in January that stopping the economic slide and restoring "sustainable growth" requires fixing the housing crisis and adopting a "fiscal stimulus of reduced taxes and increased government spending."[8]

Feldstein pointed out that past recessions started after the Federal Reserve raised short-term interest rates to fight inflation. Once inflation was under control, the Fed cut rates, which spurred a recovery. But the current recession is different, Feldstein said: It wasn't caused by the Fed tightening up on fiscal policy, and thus rate cuts haven't succeeded in reviving the economy.

"Because of the dysfunctional credit markets and the collapse of housing demand, monetary policy has had no traction in its attempt to lift the economy," he said.

That poses an especially daunting challenge for Obama.

Baker of the Center for Economic and Policy Research says that the current crisis, occurring amid a broad collapse

[1] Bureau of Labor Statistics, "Employment Situation Summary," Jan. 9, 2009, www.bls.gov/news.release/empsit.nr0.htm.

[2] For coverage of the economic crisis, see the following *CQ Researcher* reports: Thomas J. Billitteri, "Financial Bailout," Oct. 24, 2008, pp. 865-888; Kenneth Jost, "Financial Crisis," May 9, 2008, pp. 409-432; Marcia Clemmitt, "Regulating Credit Cards," Oct. 10, 2008, pp. 817-840; and Marcia Clemmitt, "The National Debt," Nov. 14, 2008, pp. 937-960.

[3] News release, "PMI Winter 2009 Risk Index Indicates Broader Risk Spreading Across Nation's Housing Markets," PMI Mortgage Insurance Co., Jan. 14, 2009.

[4] Bob Willis, "U.S. Economy: Retail Sales Decline for a Sixth Month," Bloomberg, Jan. 14, 2009, www.bloomberg.com.

[5] V. Dion Haynes and Howard Schneider, "A Brutal December for Retailers," *The Washington Post*, Jan. 9, 2009, p. 2D.

[6] *Ibid.*

[7] Quoted in David Stout and Edmund L. Andrews, "$1.2 Trillion Deficit Forecast as Obama Weighs Options," *The New York Times*, Jan. 8, 2009, www.nytimes.com/2009/01/08/business/economy/08deficit.html?scp=2&sq=deficit&st=cse.

[8] Martin Feldstein, "The Economic Stimulus and Sustained Economic Growth," statement to the House Democratic Steering and Policy Committee, Jan. 7, 2009, www.nber.org/feldstein/Economic StimulusandEconomicGrowthStatement.pdf.

Is President Obama on the right track in winning support for his programs in Congress?

As president of Harvard University, Lawrence Summers clashed so often and so sharply with faculty and others that he was forced out after only five years in office. But when Summers went to Capitol Hill as President-elect Obama's

designee to be top White House economic adviser, the normally self-assured economist told lawmakers that he and other administration officials plan to be all ears.

"All of us have been instructed that when it comes to Congress, to listen and not just talk," Summers told House Democrats in a Jan. 9 meeting to discuss Obama's economic recovery plan.[4]

AFP/Getty Images/Mark Ralston

State Department staffers greet new Secretary of State Hillary Rodham Clinton on her first day of work, Jan. 22, 2009.

Within days after the new Congress was sworn in on Jan. 6, however, lawmakers on both sides of the political aisle were, in fact, taking pot shots at Obama's plan. Republicans were calling for hearings after the plan was unveiled — a move seen as jeopardizing Obama's goal of signing a stimulus bill into law before Congress' mid-February recess. Meanwhile, some Democratic lawmakers were questioning the business tax cuts being considered for the package, calling them examples of what they considered the discredited philosophy of "trickle-down economics."

Despite the criticisms, Obama was upbeat about his relations with Congress in an interview broadcast on ABC's "This Weekend" on Jan. 11. "One of the things that we're trying to set a tone of is that, you know, Congress is a co-equal branch of government," Obama told host George Stephanopoulos. "We're not trying to jam anything down people's throats."

Veteran Congress-watchers in Washington are giving Obama high marks in his dealings with Capitol Hill so far, while also praising Congress for asserting its own constitutional prerogatives.

"Obama is off to a very good start with Congress, and, just as importantly, Congress is off to a good start with him," says Thomas Mann, a senior fellow at the Brookings Institution. "No more [status as a] potted plant for the first branch or an inflated sense of presidential authority by the second, but instead a serious engagement between the players at the opposite ends of Pennsylvania Avenue."

Obama is "in good shape," says Stephen Hess, a senior fellow emeritus at Brookings who began his Washington career as a White House staffer under President Dwight D. Eisenhower in the 1950s. Hess credits Obama in particular with seeking to consult with Republican as well as Democratic lawmakers.

"He was very shrewd after talking with Democrats to talk with Republicans," says Hess, who also teaches at George Washington University. "He has given the opposition the sense that he's open, he's listening. He's reached out to them when he doesn't need them — which of course is the right time to reach out to them."

Norman Ornstein, a resident scholar at the American Enterprise Institute, similarly credits Obama with having gone "further in consulting members of the opposition party than any president I can remember." Writing in the Capitol Hill newspaper *Roll Call*, Ornstein also said Obama is well aware of lawmakers' "issues and sensitivities." For example, Ornstein noted the president-elect's personal apology to Senate Intelligence Committee Chair Dianne Feinstein, D-Calif., for failing to give her advance word in early January of the planned nomination of Leon Panetta to head the Central Intelligence Agency.[5]

The lapse of protocol on the Panetta nomination — which Feinstein later promised to support — may well have been the only avoidable misstep by the Obama team in its dealings with Congress. Criticisms of the economic recovery program as it took shape could hardly have been avoided. And Republican senators naturally looked for ways to find fault with some of Obama's Cabinet nominees — such as their criticism of Attorney General-designate Eric Holder for his role in President Clinton's pardon of fugitive financier Marc Rich and Treasury Secretary-designate Timothy Geithner for his late payment of tens of thousands of dollars in federal income taxes.

A prominent, retired GOP congressman, however, says Obama is doing well so far and predicts the economic crisis may give him a longer than usual pass with lawmakers from both parties. "He has the advantage of a honeymoon, and perhaps the second advantage of the economic conditions of the country, which I think will help the Congress to gather around his program," says Bill Frenzel, a guest scholar at the Brookings Institution and a Minnesota congressman for two decades before his retirement in 1991.

Big-Name Policy 'Czars' Head for West Wing

Appointments may signal decline in Cabinet's influence.

President Barack Obama has tapped several high-profile Washington insiders to fill new and existing senior White House positions, indicating the new administration is shifting policy making from the Cabinet to the influential White House West Wing.

The new so-called policy "czars" include former Sen. Tom Daschle, D-S.D., at the Office of Health Reform (he is also Health and Human Services secretary); former assistant Treasury secretary Nancy Killefer, leading efforts to cut government waste as the nation's first chief performance officer; former Environmental Protection Agency Administrator Carol Browner as the new coordinator of energy and climate policy; and former New York City Council member Adolfo Carrion Jr., who is expected to head the Office of Urban Affairs.

"We're going to have so many czars," said Thomas J. Donohue, president of the U.S. Chamber of Commerce. "It's going to be a lot of fun, seeing the czars and the regulators and the czars and the Cabinet secretaries debate."[1]

In another major West Wing appointment, former Treasury secretary and Harvard President Lawrence Summers becomes director of the existing National Economic Council. In the weeks leading up to the inauguration, analysts noted that Summers, and not then-Treasury secretary-designate Timothy Geithner, was leading then-President-elect Obama's efforts to draft a new financial stimulus package.

But Paul Light, an expert on governance at New York University, questions the role the new "czars" will play. "It's a symbolic gesture of the priority assigned to an issue, and I emphasize the word symbolic," he said. "There've been so many czars over the last 50 years, and they've all been failures. Nobody takes them seriously anymore."[2]

— Vyomika Jairam

---

[1] Michael D. Shear and Ceci Connolly, "Obama Assembles Powerful West Wing; Influential Advisers May Compete With Cabinet," *The Washington Post*, Jan. 14, 2009, p. A1.

[2] Laura Meckler " 'Czars' Ascend at White House," *The Wall Street Journal*, Dec. 15, 2005, p. A6.

"We're talking about both Republicans and Democrats," Frenzel continues. "Democrats are going to want to be independent, and Republicans are going to want to take whacks at him when they can. But I think there is a mood of wanting to help the president when they can for a while."

Ornstein and Hess caution, however, that new presidents cannot expect the honeymoon to last very long. Ornstein writes that Obama's hoped-for supermajority support in Congress "may be doable on stimulus" and "perhaps even on health care." But he says an era of "post-partisan politics" will require "some serious steps" by party leaders and rank-and-file members.

For his part, Hess says Obama may eventually begin to disappoint some within his own party — but not yet. "Democrats will for a while cut him a great deal of slack," Hess explains. "Reason No. 1, he's not George W. Bush. Reason No. 2, they're going to get some of what they want. And reason No. 3, some of those folks have become wiser about the way politics is played in this town."

BACKGROUND
'A Mutt, Like Me'

Barack Obama's inauguration as president represents a 21st-century version of the American dream: the election of a native-born citizen, both black and white, with roots in Kansas and Kenya. Abandoned by his father and later living apart from his mother, Obama was nurtured in his formative years by doting white grandparents and educated in elite schools before turning to community organizing in inner-city Chicago and then to a political career that moved from the Illinois statehouse to the White House in barely 12 years.[6]

Barack Hussein Obama was born in Honolulu on Aug. 4, 1961, to parents he later described in his memoir *Dreams from My Father* as a "white as vanilla" American mother and a "black as pitch" Kenyan father. Barack Obama Sr. and Stanley Ann Dunham married, more or less secretly, after having met as students at the University of Hawaii. Stanley Ann's "moderately liberal" parents

CHRONOLOGY: 1961-2006

1960s-1970s *Obama born to biracial, binational couple; begins education in Indonesia after mother's remarriage, then returns to Hawaii.*

1961 Barack Hussein Obama born on Aug. 4, 1961, in Honolulu; parents Stanley Ann Dunham and Barack Obama Sr. meet as students at University of Hawaii; father leaves family behind two years later for graduate studies at Harvard, return to native Kenya.

1967-1971 Obama's mother remarries, family moves to Indonesia; Obama attends a secular public elementary school with a predominantly Muslim student body until mother decides he should return to Hawaii for schooling.

1971-1979 "Barry" Obama lives with grandparents Stanley and Madelyn Dunham; graduates with honors from Punahou School, one of three black students at the elite private school; enrolls in Occidental College in Los Angeles but transfers later to Columbia University in New York City.

1980s-1990s *Works as community organizer in Chicago, gets law degree, enters politics.*

1983 Obama graduates with degree in political science from Columbia University; floods civil rights organizations with job applications.

1985-1988 Works on housing, employment issues as community organizer in Far South Side neighborhood in Chicago.

Summer 1988 Visits Kenya for first time.

1988-1991 Enrolls in Harvard Law School in fall 1988; graduates in 1991 after serving as president of *Harvard Law Review* — the first African-American to hold that position.

1992-1995 Returns to Chicago; marries Michelle Robinson in 1992; runs voter registration project; works as lawyer, lecturer at University of Chicago Law School.

1995 *Dreams from My Father* is published; mother dies just after publication (Nov. 7, 1995).

1996 Elected to Illinois legislature as senator representing Chicago's Hyde Park area; serves for eight years.

2000-2006 *Enters national political stage as U.S. senator, Democratic keynoter.*

2000 Loses badly in Democratic primary for U.S. House seat held by Rep. Bobby Rush.

2002 Opposes then-imminent war in Iraq.

2004 Gains Democratic nomination for U.S. Senate from Illinois. . . . Wins wide praise for keynote address to Democratic National Convention. . . . Elected U.S. senator from Illinois: third African-American to serve in Senate since Reconstruction.

2005-2006 Earns reputation as hard worker in Senate; compiles liberal voting record; manages Democrats' initiative on ethics reform. . . . *Audacity of Hope* is published (October 2006). . . . Deflects intense speculation about possible presidential bid.

accepted the union. In Kenya — where Barack Sr. already had a wife and child — the family did not. The marriage lasted only two years; Barack left his wife and child behind to go to graduate school at Harvard. Stanley Ann filed for divorce, citing standard legal grounds.

His mother's second marriage, to an Indonesian student, Lolo Soetoro, took young Barry, as he was then called, to his Muslim stepfather's native country at the age of 6. Lolo worked as a geologist in post-colonial Indonesia; his mother taught English. They had a child, Obama's half-sister, Maya. (Maya Soetoro-Ng now teaches high school

history in Honolulu.) Barry attended a predominantly Muslim school that would be falsely depicted as an Islamist madrassa during the 2008 campaign. His mother, meanwhile, taught her son about the civil rights struggles in America and eventually sent him back to Hawaii for schooling. The marriage ended later, a victim of cultural and personality differences.

Barry returned to live with grandparents Stanley and Madelyn Dunham — "Gramps" and "Toot" (her nickname came from the Hawaiian word for grandmother). They provided him the stable, supportive home life that he had

CHRONOLOGY: 2007-PRESENT

2007 *Obama enters presidential race as underdog to New York Sen. Hillary Rodham Clinton; nearly matches Clinton in "money primary" in advance of Iowa caucuses.*

Feb. 10, 2007 Obama announces candidacy for Democratic nomination for president at rally in Springfield, Ill., three weeks after Clinton, former first lady, joined race; Democratic field eventually includes eight candidates.

March-December 2007 Democratic candidates engage in 17 debates, with no knockout punches; Obama closes gap with Clinton in polls, fundraising.

2008 *Obama gains Democratic nomination after drawn-out contest with Clinton; beats Republican Sen. John McCain as economic issues take center stage.*

January-February Obama scores upset in Iowa caucuses (Jan. 3); Clinton wins New Hampshire primary (Jan. 9); field narrows to two candidates by end of January.

March-April Clinton wins big-state primaries, including Ohio (March 4) and Pennsylvania (April 22); Obama edges ahead in delegates.

May-June Obama gains irreversible lead after Indiana, North Carolina primaries (May 6); clinches nomination after final primaries (June 2).

July Obama goes to Iraq, reaffirms 16-month pullout timetable; speaks at big rally in Berlin, Germany.

August Obama picks Delaware Sen. Joseph R. Biden as running mate; accepts nomination with speech promising

Iraq withdrawal, domestic initiatives; McCain chooses Alaska Gov. Sarah Palin as running mate.

September-October Obama holds his own in three debates with McCain (Sept. 26, Oct. 7, Oct. 15); McCain challenge to go to Washington to push financial bailout plan ends with advantage to Obama.

Nov. 4 Obama victory is signaled with victories in "red states" in East, Midwest; networks declare him winner as polls close in West (11 p.m., Eastern time).

November-December Obama completes Cabinet selections; works on economic recovery plan; vacations in Hawaii.

2009 *Obama inaugurated before largest crowd in Washington history.*

Jan. 5-19 Obama, in Washington, starts public campaign for economic recovery plan. . . . Congress reconvenes with Democrats holding 256-178 majority in House with one vacancy, 57-41 majority in Senate with two seats vacant. . . . More high-level nominations; Commerce post in limbo after Bill Richardson withdraws because of ethics investigation in New Mexico.

Jan. 20 Obama is inaugurated as 44th president; uses inaugural address to detail "serious" challenges at home, abroad; promises that challenges "will be met." . . . President moves quickly over next week to reverse some Bush administration policies; lobbies Congress on economic stimulus package, but Republicans continue to push for less spending, more tax cuts.

---

somewhat lacked so far. He gained admission to the prestigious Punahou School as one of only three black students. His father visited once — Barack's only time spent with him after the divorce — and spoke to one of his son's classes about life in Africa. Obama's mother came back to Hawaii for studies in anthropology, but when she returned to Indonesia for field work Barack chose to stay in Hawaii.

At Punahou, Obama excelled as a student and played with the state championship basketball team his senior year. He graduated in 1979 and enrolled at Occidental

College in Los Angeles. Two years later, he transferred to Columbia University in New York. By now, Obama was well aware of racial issues in the United States — and his ambiguous place in the story. "I learned to slip back and forth between my black and white worlds," he wrote in *Dreams from My Father.* More recently, as president-elect, Obama referred self-deprecatingly to his background. In describing the kind of puppy he would have preferred to get for his two young daughters, but for Malia's allergies, Obama said, "A mutt, like me."

Myriad Global Problems Confront Obama

Two wars, the Middle East and terrorism top the list.

President Barack Obama faces immense foreign-policy challenges — two wars and a turbulent global scene that includes continuing conflict in the Middle East — all against the backdrop of a global economic crisis.

Tens of thousands of U.S. troops are at war in Iraq and Afghanistan. Israel, America's closest Mideast ally, has just suspended a devastating military offensive in the Gaza Strip that could restart at any time. And Islamist terrorism remains a constant threat, with al Qaeda leader Osama bin Laden still at large.[1]

Obama divided his early days in office between wartime matters, the latest Mideast crisis and the economic meltdown. By all indications, he will be walking a tightrope between domestic and international affairs for the forseeable future.

"A president in these circumstances is going to want to do everything possible to ensure that the transformative and ambitious and very difficult projects of domestic policy that have been designated as the priority for this new administration are not inhibited or disrupted by early failures, in counterterrorism or foreign policy," Steve Coll, president and CEO of the New America Foundation, a nonpartisan think tank, told a pre-inauguration conference on security issues.

Obama's inaugural address restated his commitment to withdraw U.S. forces from Iraq, which is more peaceful after more than five years of war but still violent and torn by political intrigue.[2]

In Afghanistan, however, escalating warfare is tied to another source of U.S. worries: Pakistan. Concern escalated in late November following coordinated terrorist attacks on hotels and other sites in Mumbai — India's financial and cultural capital — which were traced to a jihadist group in Pakistan with deep ties to that country's intelligence agency.[3] Some 175 people were killed and 200 wounded.

The group, Lashkar-e-Taiba, also has at least some operational link to al Qaeda and bin Laden, who is believed to be hiding in Pakistan's northern tribal region, bordering Afghanistan. Another al Qaeda ally, the Taliban guerrillas who are fighting the Afghan government and U.S.

and NATO troops in Afghanistan, use Pakistan as a headquarters.[4]

"Moreover," a government commission on weapons of mass destruction and terrorism said in December, "given Pakistan's tense relationship with India, its buildup of nuclear weapons is exacerbating the prospect of a dangerous nuclear arms race in South Asia that could lead to a nuclear conflict."[5]

The other daunting foreign-policy issue facing the new Obama administration — conflict between Israel and the Palestinians — offers slender prospects for peace. "Two states living side by side in peace and security — right now that stands about as much chance as Bozo the Clown becoming president of the United States," says Aaron David Miller, a former Mideast peace adviser to six secretaries of State.

The biggest obstacle, Miller says, is the "broken and dysfunctional" state of the Palestinian national movement. Fatah, the secular party that runs the West Bank, has a negotiating relationship with Israel. Hamas, the elected Islamist party and militia that initially seized power in an anti-Fatah coup in Gaza in 2007, deems Israel illegitimate. Hamas sponsored or tolerated rocket fire into Israel from Gaza but halted rocketing at the beginning of a cease-fire that began in June 2008. But Israel accused Hamas of building up its arsenal and retaliated by limiting the flow of goods into the region. In December, Hamas announced it wouldn't renew the already shaky truce, blaming the Israeli embargo and military moves. From then on, Hamas stepped up rocketing.

Israel's recent 22-day anti-Hamas offensive in Gaza cost some 1,300 Palestinian lives. The Palestinians estimated the civilian death toll at 40 percent to 70 percent of the fatalities; Israel put the toll at about 25 percent of the total. Israeli fatalities totaled 13, including three civilians.[6]

The scale of Israel's Gaza offensive is renewing calls for the U.S. government to change its relationship to Israel. "The days of America's exclusive ties to Israel may be coming to an end," Miller wrote in *Newsweek* in January. Obama, however, reaffirmed his support for Israel in his Jan. 26 interview with the Arabic-language network Al Arabiya.[7]

Those interests also would require devising a response to what the United States believes is a nuclear arms development project by Iran, which supports Hamas politically and financially — a sign, for some, of how all Middle Eastern issues are interconnected.

"One of the great mistakes we have made has been to believe we can compartmentalize these different policies, that we can somehow separate what is happening between Israel and the Palestinians from what's happening in Iraq and what's happening in Iran and what's happening in Egypt and Saudi Arabia and everywhere else in the Middle East," said Kenneth M. Pollack, a senior fellow at the Brookings Institution and former CIA analyst of the region. "Linkage is a reality."[8]

Another set of connections ties past U.S. support for NATO membership by Ukraine and Georgia to chilled U.S. relations with Russia, which views the potential presence of Western military allies — and U.S. missiles — on its borders as hostile.

Despite the Cold War echoes of that dispute, some foreign-affairs experts argue that Obama actually confronts a less perilous international panorama than some of his recent predecessors. "We don't have the Cold War and World War II," says Michael Mandelbaum, director of the foreign policy program at Johns Hopkins University's School of Advanced International Studies. "Those were existential threats. What the incoming president faces are annoying and troublesome, but not existential threats."

That picture could change if jihadist radicals took over nuclear-armed Pakistan. For now, Mandelbaum argues the biggest international and domestic dangers are one and the same — the economic meltdown.

But success for the huge spending package that Obama wants will require participation by China, America's major creditor. "China has been lending us money by buying our bonds," Mandelbaum says. "That huge stimulus package is not going to work unless we get some cooperation from the Chinese."

Palestinians in Gaza search the rubble of their homes for usable items after an Israeli air strike on Jan. 5, 2009.

Getty Images

In short, the American way of life very much depends on China, Mandelbaum says: "For what Americans care about, for what matters in the world, the issue of where and how we borrow money for the stimulus and where and how we rebalance the economy dwarfs Gaza in importance, and is more important than Iraq and Afghanistan."

— Peter Katel

[1] For coverage of the Iraq and Afghanistan wars, the Middle East and Islamic fundamentalism, see the following *CQ Researcher* reports: Peter Katel, "Cost of the Iraq War," April 25, 2008, pp. 361-384; Peter Katel, "New Strategy in Iraq," Feb. 23, 2007, pp. 169-192; and Peter Katel, "Middle East Tension," Oct. 27, 2006, pp. 889-912. Also see the following *CQ Global Researcher* reports: Roland Flamini, "Afghanistan on the Brink," June 2007, pp. 125-150; Robert Kiener, "Crisis in Pakistan," December 2008, pp. 321-348; and Sarah Glazer, "Radical Islam in Europe," November 2007, pp. 265-294.

[2] Alissa J. Rubin, "Iraq Unsettled by Political Power Plays," *The New York Times*, Dec. 25, 2008, www.nytimes.com/2008/12/26/world/middleeast/26baghdad.html; and Alissa J. Rubin, "Bombs Kill 5 in Baghdad, but Officials Avoid Harm," *The New York Times*, Jan. 20, 2009, www.nytimes.com/2009/01/21/world/middleeast/21iraq.html.

[3] Jane Perlez and Somini Sengupta, "Mumbai Attack is Test for Pakistan on Curbing Militants," *The New York Times*, Dec. 3, 2008, www.nytimes.com/2008/12/04/world/asia/04pstan.html?scp=5&sq=MumbaiLashkar ISI&st=cse.

[4] For a summary and analysis, see K. Alan Kronstadt and Kenneth Katzman, "Islamist Militancy in the Pakistan-Afghanistan Border Region and U.S. Policy," Congressional Research Service, Nov. 21, 2008, http://fpc.state.gov/documents/organization/113202.pdf.

[5] See "World at Risk," Commission on the Prevention of Weapons of Mass Destruction Proliferation and Terrorism, December 2008, p. xxiii.

[6] See Steven Erlanger, "Weighing Crimes and Ethics in the Fog of Urban Warfare," *The New York Times*, Jan. 16, 2009, www.nytimes.com/2009/01/17/world/middleeast/17israel.html?scp=1&sq=Gaza civiliandeathpercent&st=cse; Amy Teibel, "Last Israeli troops leave Gaza, completing pullout," The Associated Press, Jan. 21, 2009, http://news.yahoo.com/s/ap/ml_israel_palestinians.

[7] Aaron David Miller, "If Obama Is Serious, He should get tough with Israel," *Newsweek*, Jan. 3, 2009, www.newsweek.com/id/177716.

[8] Quoted in Adam Graham-Silverman, "Conflict in Gaza Strip Presents Immediate Challenge for New President," *CQ Today*, Jan. 20, 2009.

AFP/Getty Images/Timothy A. Clary

Barack Obama's riveting, highly personal keynote address at the 2004 Democratic National Convention made him an overnight star and presidential contender.

Graduating from Columbia in 1983 with a degree in political science, Obama decided to take on the so-called Reagan revolution by becoming a community organizer — aiming, as he wrote, to bring about "change . . . from a mobilized grass roots." Obama flooded civil rights organizations to no avail until he was hired in 1985 by Gerald Kellman, a white organizer looking for an African-American to help with community development and mobilization in a Far South Side section of Chicago. Obama's three years in Chicago brought him face to face with the gritty realities of urban life and the disillusionment of the disadvantaged. He later described the time as "the best education I ever had."[7]

Obama enrolled in Harvard Law School in 1988.[8] He wrote nothing about the decision in his memoir and has said little about it elsewhere. Before going, he visited Kenya, where his father had died in an automobile accident six years earlier. Obama described enjoying the meeting with his extended family while acutely conscious of the cultural gap. At Harvard, he excelled as a student, played pick-up basketball and had only a limited social life after meeting his future wife, Michelle Robinson, a lawyer he had met while working for a Chicago law firm as a summer associate. His election in 1990 as president of the *Harvard Law Review* — as a compromise between conservative and liberal factions — marked the first time an African-American had held the prestigious position.

His barrier-breaking gained enough attention to get Obama an invitation from a literary agent, Jane Dystel, to write a book.[9] Obama planned to write about race relations, but in the three years of writing it turned into more of a personal memoir. Obama has said he was unmindful of political consequences in the writing and that he rejected a suggestion from one of his editors to delete references to drug use while in college. The book garnered respectable reviews — and the audio version won a Grammy — but no more than middling sales. Obama's mother read page proofs and lived just long enough to see it published. She died of ovarian cancer in November 1995.[10]

Red, Blue and Purple

Obama needed only 10 years to rise from the back benches of the Illinois legislature to a front seat on the national political stage. His political ambition misled him only once: in a failed run for the U.S. House. But he succeeded in other endeavors on the strength of hard work, personal intelligence, political acumen and earnest efforts to bridge the differences of race, class and partisan affiliation.

Obama entered politics in 1995 as the chosen successor of a one-term state senator, Alice Palmer. But he turned on his mentor when she sought re-election after all, following a losing bid in a special election for a U.S. House seat. Obama successfully challenged signatures on Palmer's nominating petitions and had her disqualified (and the other candidates too) to win the Democratic nomination unopposed and eventual election.

As a Democrat in a Republican-controlled legislature and a liberal with no connection to his party's organization, Obama worked to develop personal ties — some formed in a weekly poker game. Among his accomplishments: ethics legislation, a state earned-income tax credit and a measure, backed by law enforcement, to require videotaped interrogations in all capital cases.[11]

After four years in office, Obama decided in 2000 to mount a primary challenge to the popular and much better known Democratic congressman, Bobby Rush. The race was foolhardy from the outset. But — as Obama recounts in his second book, *The Audacity of Hope* — he suffered a grave embarrassment when he failed to return from a family vacation in Hawaii in time to vote on a major gun control bill in a specially called legislative session. Rush won handily.[12] In the 2008 presidential campaign, Obama's absence on the gun control vote was cited along with many other instances when he voted "present" as evidence of risk-averse gamesmanship on his part — a depiction vigorously disputed by the campaign.

His ambition unquenched, Obama began deciding by fall 2002 to run for the U.S. Senate seat then held by Republican Peter Fitzgerald, a vulnerable incumbent who eventually decided not to seek re-election. In October, at the invitation of a peace activist group, he delivered to an anti-war rally in Chicago his now famous speech opposing the then-imminent U.S. war in Iraq. Obama formally entered the Senate race in 2003 as the underdog to multimillionaire Blair Hull and state Comptroller Dan Hynes. But Hull's candidacy collapsed after allegations of abuse against his ex-wife. Hynes ran a lackluster campaign, while Obama waged a determined, disciplined drive that netted him nearly 53 percent of the vote in a seven-way race.[13]

Obama's debut on the national stage came in July 2004 after the presumptive Democratic presidential nominee, Massachusetts Sen. John Kerry, picked him to deliver the keynote address at the party's convention. Obama drafted the speech himself, according to biographer David Mendell. The night before, he told a friend, "My speech is pretty good." It was better than that. Obama wove his personal story together with verbal images of working-class America to lead up to the passage — rebroadcast thousands of times since — envisioning a unified nation instead of the "pundits' " image of monochromatic "Red States" and "Blue States." The speech "electrified the convention hall," *The Washington Post* reported the next day, and made Obama a rising star to be watched.[14]

By the time of the speech, political fortune had already shone on Obama back in Illinois. Divorce files of his Republican opponent in the Senate race, Jack Ryan, made public in June, showed that Ryan had pressured his wife to go with him to sex clubs and have sex in front of others. Ryan, a multimillionaire businessman, resisted pressure to withdraw for more than a month. Once Ryan bowed out — three days after Obama's speech — GOP leaders had to scramble for an opponent. They eventually lured Alan Keyes, a conservative African-American from Maryland, to be the sacrificial lamb in the race. Obama won with a record-setting 70 percent of the vote to take his seat in January 2005 as only the third African-American to serve in the U.S. Senate since Reconstruction.

Obama entered the Senate with the presidency on his mind but also the recognition that he must succeed first in a club with low tolerance for celebrity without substance. A profile in Congressional Quarterly's *Politics in America* published with his presidential campaign under way in 2007 credited Obama with "a reputation as a hard worker, a good listener and a quick study."[15]

With Democrats in the majority, Obama was designated in 2007 to spearhead the party's work on ethics reform — a role that prompted an icy exchange with his future opponent, Sen. McCain, who had expected to work with Democrats on a bipartisan approach. The eventual package included a ban on senators' discounted trips on corporate jets, but not — as Obama had pushed for — outside enforcement of ethics rules.

Obama had more success working with other Republicans, including Oklahoma's Tom Coburn (Internet access to government databases) and Indiana's Richard Lugar (international destruction of conventional weapons). Overall, however, his voting record was solidly liberal and reliably party-line. In the 2008 race, the McCain campaign repeatedly tried to debunk Obama's image of post-partisanship by challenging him to cite a significant example of departing from Democratic Party positions.

'Yes, We Can'

Obama won the Democratic nomination for president in a come-from-behind victory over frontrunner Hillary Clinton on the strength of fundraising prowess, message control and a pre-convention strategy focused on amassing delegates in caucus as well as primary states. He took an even bigger financial advantage into the general election but pulled away from McCain only after the nation's dire economic news in October drove the undecideds decisively toward the candidate promising "change we can believe in."[16]

Despite intense speculation and Obama's evident interest, he decided to run only after heart-to-heart talks with Michelle while vacationing in Hawaii in December 2006. Michelle's reluctance stemmed from the effects on the family and fear for Obama's personal safety. In the end, she agreed — with one stipulation: Obama had to give up smoking. That promise remains a work in progress. In his post-election appearance on NBC's "Meet the Press" on Dec. 7, Obama promised only that, "you will not see any violations" of the White House's no-smoking rule while he is president.

Daschle Appointment Shows Commitment to Health-Care Reforms

But a vote on a specific plan may be delayed until next year.

"The flaws in our health system are pervasive and corrosive. They threaten our health and economic security," said former Sen. Tom Daschle, D-S.D., President Obama's nominee for secretary of Health and Human Services (HHS), at his initial confirmation hearing before the Senate Health, Education, Labor, and Pensions (HELP) Committee on Jan. 8.[1]

Throughout his campaign, Obama promised to make good-quality health care accessible to all Americans. Many observers see his choice of Daschle — who recently coauthored a book laying out a plan for universal insurance coverage — to lead both HHS and a new White House Office of Health Policy as a sign of the new president's commitment to health-care reform, which he has called the key to economic security.[2] "I talk to hardworking Americans every day who worry about paying their medical bills and getting and keeping health insurance for their families," Obama said.[3]

In the final presidential debate on Oct. 15, 2008, Obama laid out the essence of his health overhaul. "If you've got health insurance through your employer, you can keep your health insurance," he said. "If you don't have health insurance, then what we're going to do is to provide you the option of buying into the same kind of federal pool [of private insurance plans] that [Republican presidential nominee] Sen. McCain and I enjoy as federal employees, which will give you high-quality care, choice of doctors at lower costs, because so many people are part of this insured group," Obama said.[4]

In addition, Obama's plan would:

- require insurance companies to accept all applicants, including those with already diagnosed illnesses — or "preexisting conditions" — that insurers often decline to cover;
- create a federally regulated national "health insurance exchange" where people could buy coverage from a range of approved private insurers and possibly from a public insurance program as well;
- provide subsidies to help lower-income people buy coverage;
- require all children to have health insurance; and
- require employers except small businesses to either provide "meaningful" coverage to workers or pay a percentage of payroll toward the costs of a public plan.[5]

Points of potential controversy include whether all Americans should be required to buy health coverage.

During the presidential primary campaign, Obama sparred with fellow Democratic candidate Sen. Hillary Rodham Clinton, D-N.Y., who called for a mandate on individuals to buy insurance. Obama disagreed, saying, "my belief is that if we make it affordable, if we provide subsidies to those who can't afford it, they will buy it," and that only children's coverage should be required.[6]

But many analysts, including Daschle, point out that unless coverage is required many people will buy it only after they become sick, making it impossible for health insurance to perform its main task — spreading the costs of care among as many people as possible, not just among those who happen to be sick at a given time.

"The only way we can achieve universal coverage is to require everybody to either purchase private insurance or enroll in a public program," Daschle wrote.[7]

Obama entered the race with a speech to an outdoor rally on a cold Feb. 10, 2007, in Springfield, Ill. After acknowledging the "audacity" of his campaign, Obama laid out a platform of reshaping the economy, tackling the health-care crisis and ending the war in Iraq. He started well behind Clinton in the polls and in organization. In the early debates — with eight candidates in all — Obama himself rated his performance as "uneven," according to *Newsweek*'s post-election account.[17] By December, however, Obama had pulled ahead of Clinton

If Obama ends up authorizing a new government-run insurance plan to compete with private insurers for enrollees, as most Democrats favor, the plan could face tough opposition from Republicans.

"Forcing private plans to compete with federal programs, with their price controls and ability to shift costs to taxpayers, will inevitably doom true competition and could ultimately lead to a single-payer, government-run healthcare program," said Sen. Michael Enzi, R-Wyo., the top Republican on the HELP Committee. "Any new insurance coverage must be delivered through private health-insurance plans."[8]

Congressional Democrats stand ready to work with the Obama administration to move health-care reform quickly. Two very influential senators, HELP Committee Chairman Sen. Edward Kennedy, D-Mass., and Finance Committee Chairman Sen. Max Baucus, D-Mont., were already crafting health-reform legislation last year and are expected to begin a strong push for legislation soon. But the press of other business and the time-consuming process of gathering support for a specific plan will put off a vote until the end of this year or the beginning of 2010, predicted Rep. Pete Stark, D-Calif., chairman of the House Ways and Means Health Subcommittee. "I don't think we'll do it in the first 100 days," said Stark.[9]

Ironically, the struggling economy, which leaves many more Americans worried about their jobs and therefore their health coverage, may have opened the door for reform by giving business owners, doctors and others a greater stake in getting more people covered, said Henry Aaron, a senior fellow in economic studies at the centrist Brookings Institution. "Before the economic collapse . . . the odds of national reform were nil," but the nation's economic stress makes it somewhat more likely, especially since Congress has been spending large amounts of money on other industries, Aaron said.[10]

Nevertheless, Aaron and some other analysts say the climate for health-care reform may not be much different from that in 1993 when the tide quickly turned against the Clinton administration's attempt at providing universal health care.

The times are "similar," and despite the desire of many for reform, the details will be painful and will spark push-back, Stuart Butler, vice president of the conservative Heritage Foundation, told PBS' "NewsHour." "When you say, 'We've got to make the system efficient by reducing unnecessary costs' . . . that means people's jobs and . . . doctors are going to rebel against that."[11]

— Marcia Clemmitt

[1] Quoted in "Daschle: Health Care Flaws Threaten Economic Security," CNNPolitics.com, Jan. 8, 2009, www.cnn.com/2009/POLITICS/01/08/daschle.confirmation.

[2] For background see the following *CQ Researcher* reports by Marcia Clemmitt: "Universal Coverage," March 30, 2007, pp. 265-288, and "Rising Health Costs," April 7, 2006, pp. 289-312.

[3] Barack Obama, "Modern Health Care for All Americans," *The New England Journal of Medicine*, Oct. 9, 2008, p. 1537.

[4] Quoted in "In Weak Economy, Obama May Face Obstacles to Health Care Reform," PBS "NewsHour," Nov. 20, 2008, www.pbs.org.

[5] "2008 Presidential Candidate Health Care Proposals: Side-by-Side Summary," health08.org, Kaiser Family Foundation, www.health08.org.

[6] Quoted in Jacob Goldstein, "Clinton and Obama Spar Over Insurance Mandates," *The Wall Street Journal* Health Blog, Feb. 1, 2008, http://blogs.wsj.com.

[7] Quoted in Teddy Davis, "Obama and Daschle at Odds on Individual Mandates," ABC News blogs, Dec. 11, 2008, http://blogs.abcnews.com.

[8] "Enzi Asks Obama Health Cabinet Nominee Daschle Not to Doom Health-Care Competition," press statement, office of Sen. Mike Enzi, Jan. 8, 2009, http://enzi.senate.gov.

[9] Quoted in Jeffrey Young, "Rep. Stark: No Health Reform Vote in Early '09," *The Hill*, Dec. 17, 2008, http://thehill.com.

[10] Quoted in Ben Weyl, "Experts Predict a Health Overhaul Despite Troubled Economy," *CQ Healthbeat*, Dec. 9, 2008.

[11] "In Weak Economy, Obama May Face Obstacles to Health Care Reform," *op. cit.*

in some New Hampshire polling and was in a virtual dead-heat in the all-important "money primary."

The Iowa caucuses on Jan. 3, 2007, gave Obama an unexpected win with about 38 percent of the vote and left only two other viable candidates standing: former North Carolina Sen. John Edwards, who came in second; and Clinton, who finished a disappointing third. Five days later, however, Clinton regained her stride with a 3-percentage-point victory over Obama in the first-in-the-nation New Hampshire primary. Edwards' third-place

Vice President Biden Brings Foreign-Policy Savvy

"I want to be the last guy in the room on every important decision."

The inauguration of Joseph R. Biden Jr. as the 47th vice president of the United States caps a journey almost as improbable as Barack Obama's. During seven terms as a U.S. senator from Delaware, Biden has never lived in Washington, instead commuting daily by train from Wilmington. In 1972, at age 29, he became the sixth-youngest senator ever elected, leading many to believe the White House was in his future.

But after two failed presidential campaigns — in 1988 and in the last election — Biden seemed fated to remain a Senate lifer.

Along the way he rose to become chairman of the Judiciary Committee and gained national prominence while leading the confirmation hearings of Supreme Court nominees Robert Bork and Clarence Thomas. He had also served twice as chairman of the Foreign Relations Committee.

Obama's limited time in the Senate and lack of international experience led to increased speculation that he would select Biden as his running mate to bridge the gap. "[Joe Biden is] a leader who sees clearly the challenges facing America in a changing world, with our security and standing set back by eight years of failed foreign policy," Obama said in introducing Biden as his selection on Aug. 23, 2008.

But the new president has yet to clarify the specific role Biden will play in the new administration. The appointment of Hillary Rodham Clinton as secretary of State all but ensures that Biden, despite his impressive résumé, will not be the point man on foreign policy as initially expected.

Nor does anyone expect him to emulate former Vice President Dick Cheney's muscular role. Upon taking office in 2001, Cheney demanded — and President George W. Bush approved — a mandate to give him access to "every table and every meeting," expressing his voice in "whatever area the vice president feels he wants to be active in," recalls former White House Chief of Staff Joshua B. Bolten.[1]

Cheney's push to expand presidential war-making authority is arguably his most lasting legacy, but he also served as a gatekeeper for Supreme Court nominees, editor of tax proposals and arbiter of budget appeals.

While most vice presidents arrive eager to expand the influence of their position, Biden faces the unusual conundrum of figuring out how to scale it back. "The only value of power is the effect, the efficacy of its use," he told *The New York Times*. "And all the power Cheney had did not result in effective outcomes." But without any direct constitutional authority in the executive branch, Biden does not want to return to the days when vice presidents were neither

finish kept him in the race, but he dropped out on Jan. 30 after finishing third in primaries in Florida and his birth state of South Carolina.

The one-on-one between Obama and Clinton continued through May. Clinton bested Obama in a series of supposedly "critical" late-season primaries — notably, Ohio and Pennsylvania — even as Obama pulled ahead in delegates thanks to caucus state victories and also-ran proportional-representation winnings from the primaries. He turned the most serious threat to his campaign — his relationship with the sometimes fiery black minister, Jeremiah Wright — into a plus of sorts with a stirring speech on racial justice delivered in Philadelphia on March

18. With Clinton's "electability" arguments unavailing, Obama mathematically clinched the nomination on June 3 as the two split final primaries in Montana and South Dakota. Clinton withdrew four days later, promising to work hard for Obama's election.

With nearly three months before the convention, Obama went to Iraq and Europe to burnish his national security and foreign policy credentials. His 16-month timetable for withdrawal now essentially matched the Iraqi government's own position — weakening a Republican line of attack. An address to a huge and adoring crowd in Berlin underscored Obama's promise to raise U.S. standing in the world. The McCain campaign countered

seen nor heard. "I don't think the measure is whether or not I accrete the vestiges of power; it matters whether or not the president listens to me."[2]

And although he says he doesn't seek to wield as much influence as Cheney, many don't expect the loquacious Biden to follow Al Gore either, who in 1992 was assigned a defined portfolio by President Bill Clinton to work on environmental and technology matters. "I think his fundamental role is as a trusted counselor," said Obama senior adviser David Axelrod. "I think that when Obama selected him, he selected him to be a counselor and an adviser on a broad range of issues."[3]

And that's exactly how Biden — who at first balked at accepting the position — wants it. "I don't want to have a portfolio," Biden says. "I don't want to be the guy who handles U.S.-Russian relations or the guy who reinvents government."

"I want to be the last guy in the room on every important decision."

"It's irrelevant what the outside world perceives. What is relevant is whether or not I'm value-added," Biden contends. And very few debate his credentials for the position.

"I'm the most experienced vice president since anybody. Anybody ever serve 36 years as a United States senator?" he asks.[4]

But in all likelihood Biden's first move to Washington will surely be his last.

At age 66, he says he has no plans to pursue the presidency, or return to the Senate for that matter, in 2016 — the last full year of a possible second term for Obama. That suggests he'll truly serve Obama's ambitions rather than his own.

"This is in all probability, and hopefully, a worthy capstone in my career," he said.

— Darrell Dela Rosa

Newly sworn in Vice President Joseph R. Biden, his wife, Jill, and son Beau greet crowds during the Inaugural Parade.

[1] Barton Gellman and Jo Becker, " 'A Different Understanding With the President,' " *The Washington Post*, June 24, 2007, blog.washingtonpost .com/cheney/chapters/chapter_1.

[2] Peter Baker, "Biden Outlines Plans to Do More With Less Power," *The New York Times*, Jan. 14, 2009, www.nytimes.com/2009/01/15/us/ politics/15biden.html?_r=1.

[3] Helene Cooper, "For Biden, No Portfolio but the Role of a Counselor," *The New York Times*, Nov. 25, 2008, www.nytimes.com/2008/11/26/ us/politics/26biden.html.

[4] Baker, *op. cit.*

with an ad mocking Obama's celebrity status. On the eve of the convention, Obama picked Biden as his running mate. The selection won praise as sound, if safe. The four-day convention in Denver (Aug. 25-28) went off without a hitch. Obama's acceptance speech drew generally high marks, but some criticism for its length and predictable domestic-policy prescriptions.

McCain countered the next day by picking Alaska Gov. Sarah Palin as his running mate. The surprise selection energized the GOP base but raised questions among observers and voters about his judgment. For the rest of the campaign, the McCain camp tried but failed to find an Obama weak spot. Obama had already survived personal attacks about ties to Rev. Wright, indicted Chicago developer Tony Rezko and one-time radical William Ayers. He had also fended off attacks for breaking his pledge to limit campaign spending by taking public funds. Improved ground conditions in Iraq shifted the contest from national security — McCain's strength — to the economy: Democratic turf. Obama held his own in three debates and used his financial advantage — he raised a record $742 million in all — to engage McCain not only in battleground states but also in supposedly safe GOP states.

By Election Day, the outcome was hardly in doubt. Any remaining uncertainty vanished when Virginia, Republican since 1968, went to Obama early in the evening. By 9:30, one blog had declared Obama the winner. The networks waited until the polls closed on the West Coast — 11 p.m. in the East — to declare Obama to be the 44th president of the United States. In Chicago's Grant Park, tens of thousands of supporters chanted "Yes, we can," as Obama strode on stage.

"If there is anyone out there," Obama began, "who still doubts that America is a place where all things are possible; who still wonders if the dream of our founders is alive in our time; who still questions the power of our democracy, tonight is your answer."[18]

A Team of Centrists?

President-elect Obama began the 76 days between election and inauguration by hitting nearly pitch-perfect notes in his dealings with official Washington — including President Bush and members of Congress — and with the public at large. Beginning with his first post-election session with reporters, Obama sounded both somber but hopeful in confronting what he continually referred to as the worst economic crisis in generations. He completed his selection of Cabinet appointees in record time before taking an end-of-December vacation with his family in Hawaii. Some discordant notes were sounded as Inauguration Day neared in January. But on the eve of the inauguration, polls showed Obama entering the Oval Office with unprecedented levels of personal popularity and hopeful support. (*See graph, p. 280.*)

Acknowledging the severity of the economic crisis, Obama started the announcement of Cabinet-level appointments on Nov. 24 by introducing an economic team that included New York Federal Reserve Bank President Timothy Geithner to be secretary of the Treasury. Geithner had been deeply involved in the Fed's moves in the financial bailout. Obama also named Summers, who had served as deputy undersecretary of the Treasury in the Clinton administration, as special White House assistant for economic policy.

A week later, Obama introduced a national security team that included Hillary Clinton as secretary of State and Gates as holdover Pentagon chief. Clinton accepted the post only after weighing the offer against continuing in the Senate with possibly enhanced visibility and influence. In addition, the appointment required former President Clinton to disclose donors to his post-presidential foundation to try to reduce potential conflicts of interest with his wife's new role.

Along with Gates, Obama also introduced Gen. Jones, a retired Marine commandant and former North Atlantic Treaty Organization supreme commander, as his national security adviser. He also said that he would nominate Holder, a former deputy attorney general, for attorney general; Gov. Janet Napolitano of Arizona for secretary of Homeland Security; and Susan E. Rice, a former assistant secretary of State, for ambassador to the United Nations with Cabinet rank. Holder was in line to be the first African-American to head the Justice Department.

Other Cabinet nominations followed in rapid succession: New Mexico Gov. Bill Richardson, like Clinton one of the contenders for the Democratic nomination, for Commerce; Gen. Eric Shinseki, a critic of Iraq War policies, for Veterans Affairs; and former Senate Democratic Leader Tom Daschle of South Dakota, for Health and Human Services and a new White House office as health reform czar.

Obama picked Shaun Donovan, commissioner of New York City's housing department, for Housing and Urban Development; outgoing Illinois Rep. Ray LaHood, a Republican, for Transportation; and Chicago public schools Commissioner Arne Duncan, a reformer with good relations with Chicago teacher unions, for Education. Steven Chu, a Nobel Prize-winning scientist and an advocate of measures to reduce global warming, was picked for Energy. Sen. Kenneth Salazar, a Colorado Democrat with a moderate record on environmental and land use issues, was tapped for Interior. Former Iowa Gov. Tom Vilsack, who had supported Clinton for the nomination, was chosen for Agriculture. And Rep. Hilda Solis, a California Democrat and daughter of a union family, was designated for Labor.

As Obama prepared to leave for Hawaii, some supporters were griping about the moderate cast of his selections. "We just hoped the political diversity would have been stronger," Tim Carpenter, executive director of Progressive Democrats of America, told Politico.com. But official Washington appeared to be giving him top marks. *The Washington Post* described the future Cabinet as dominated by "practical-minded centrists who have straddled big policy debates rather than staking out the strongest pro-reform positions."[19]

Obama arrived in Washington on Jan. 4 to enroll daughters Malia, 10, and Natasha ("Sasha"), 7, in the private Sidwell Friends School and begin two hectic work weeks before a long weekend of pre-inaugural events. By then, problems had begun to arise, including a corruption scandal over the selection of Obama's successor in the Senate; the withdrawal of one of his Cabinet nominees; and questions about several of his nominees for top posts.

The Senate seat controversy stemmed from a federal investigation of Illinois Gov. Rod Blagojevich that included tape-recorded comments by the Democratic chief executive that were widely depicted as attempting to sell the appointment for political contributions or other favors. In charging Blagojevich with corruption, U.S. Attorney Patrick Fitzgerald specifically cleared Obama of any involvement. But Obama had been forced to answer questions on the issue from Hawaii and had lined up with Senate Democratic Leader Harry Reid in promising not to seat any Blagojevich appointee. When Blagojevich went ahead and appointed former state Comptroller Roland Burris, an African-American, Reid initially resisted but eventually bowed to the fait accompli and welcomed Burris to the Senate.

Richardson had withdrawn from the Commerce post on Jan. 3 after citing a federal probe into a possible "pay for play" scandal in New Mexico.

Two other Cabinet nominees faced critical questions as Senate confirmation hearings got under way. Treasury Secretary-designate Geithner was disclosed to have failed to pay Social Security and Medicare taxes for several years and to have paid back taxes and interest only after being audited. Attorney General-designate Holder faced questions about his role in recommending that President Clinton pardon fugitive financier Marc Rich and in submitting a pardon application for members of the radical Puerto Rican independence movement FALN. Both seemed headed toward confirmation, however.

CURRENT SITUATION

Moving Quickly

Beginning with his first hours in office, President Obama is moving quickly to put his stamp on government policies by fulfilling campaign promises on such issues as government ethics, secrecy and counterterrorism. Along with the flurry of domestic actions, Obama opened initiatives on the diplomatic front by promising an active U.S. role to promote peace in the Middle East and naming high-level special envoys for the Israeli-Palestinian dispute and the strategically important region of South Asia, including Afghanistan and Pakistan.

In the biggest news of his first days in office, Obama on Jan. 22 signed executive orders to close the Guantánamo prison camp within one year and to prohibit the use of "enhanced" interrogation techniques such as waterboarding by CIA agents or any other U.S. personnel. Human rights groups hailed the actions. "Today is the beginning of the end of that sorry chapter in our nation's history," said Elisa Massimino, executive director and CEO of Human Rights First.

Some Republican lawmakers, however, questioned the moves. "How does it make sense," House GOP Whip Eric Cantor asked, "to close down the Guantánamo facility before there is a clear plan to deal with the terrorists inside its walls?"

An earlier directive, signed late in the day on Jan. 20, ordered Defense Secretary Gates to halt for 120 days any of the military commission proceedings against the remaining 245 detainees at Guantánamo. Separately, Obama directed a review of the case against Ali Saleh Kahlah al-Marri, a U.S. resident and the only person designated as an enemy combatant being held in the U.S.

The ethics and information directives signed on Jan. 21 followed Obama's campaign pledges to limit the "revolving door" between government jobs and lobbyist work and to make government more transparent and accountable.

The new ethics rules bar any executive branch appointees from seeking lobbying jobs during Obama's administration. They also ban gifts from lobbyists to anyone in the administration. Good-government groups praised the new policies as the strictest ethics rules ever adopted. Fred Wertheimer, president of the open-government group Democracy 21, called them "a major step in setting a new tone and attitude for Washington."

On information policy, Obama superseded a Bush administration directive promising legal support for agencies seeking to resist disclosure of government records under the Freedom of Information Act. Instead, Obama called on all agencies to release information whenever possible. "For a long time now, there's been too much secrecy in this city," Obama said at a swearing-in ceremony for senior White House staff.

Obama also signed an executive order aimed at greater openness for presidential records following the

Should Congress and the president create a commission to investigate the Bush administration's counterterrorism policies?

YES Frederick A. O. Schwarz Jr.
Chief Counsel, Brennan Center for Justice, New York University School of Law; co-author, Unchecked and Unbalanced: Presidential Power in a Time of Terror (New Press, 2008)

Written for *CQ Researcher*, January 2009

In his inaugural address, President Obama rejected "as false the choice between our safety and our ideals." Throughout our history, seeking safety in times of crisis has often made it tempting to ignore the wise restraints that make us free and to rush into actions that do not serve the nation's long-term interests. (The Alien and Sedition Acts at the dawn of the republic and the herding of Japanese citizens into concentration camps early in World War II are among many historic examples.) After 9/11 we again overreacted to crisis, this time by descending into practices including torture, extraordinary rendition, warrantless wiretapping and indefinite detention. Each breached American values and thus made America less safe.

Our new president is taking steps to reject these actions. And some say this is all that is needed because we need to look forward. Others clamor for criminal prosecutions because to hold our heads high wrongdoers should be held to account.

But, to me, neither of these positions is right. Prosecution is not likely to be productive, and could well be unfair. At the same time, failure to learn more about how we went wrong poses two dangers: First, if we blind our eyes to the truth, we increase the risk of repetition when the next crisis comes.

Second, clearly and fairly assessing and reporting what went wrong — and right — in our reactions to 9/11 will honor America's commitment to openness and the rule of law. Committing ourselves to a full exploration is consistent with the ethos the new president articulated on his first day in office: "The way to make government responsible is to hold it accountable. And the way to make government accountable is to make it transparent."

For these two reasons, I have recommended that the president and Congress appoint an independent, nonpartisan commission to investigate national counterterrorism policies. This is the best way to achieve accountability and an understanding of how to design an effective counterterrorism policy that comports with fundamental values.

Shortly after his reelection in 1864, President Abraham Lincoln nicely articulated the necessity of learning from the past without seeking punishment: "Let us study the incidents of [recent history], as philosophy to learn wisdom from, and none of them as wrongs to be revenged."

NO David B. Rivkin Jr. and Lee A. Casey
Washington attorneys who served in the Justice Department under Presidents Reagan and George H. W. Bush

Written for *CQ Researcher*, January 2009

A special commission would be both unnecessary and harmful. First, multiple congressional inquiries have already aired and analyzed all of the Bush administration's key legal and policy decisions. Indeed, whether through disclosures, leaks, media and/or congressional investigations, both the process and substance of the administration's war-related decisions have been publicized to an unprecedented extent. If any further inquiry into these policies is necessary, the normal congressional and executive branch investigatory tools are always available, including additional hearings.

Second, a special commission would be fundamentally unfair, beginning — as it would — with the proposition that the Bush policies represent systematic wrongdoing. The Bush policies were based upon well-established case law and reasonable legal extrapolation from the available authorities. Simply because the Supreme Court ultimately decided to change the legal landscape does not mean the Bush administration ignored the law; it did not. Moreover, although there have been many problems and certainly some abuses over the past seven years — Abu Ghraib being a case in point — these have been remarkably rare when compared with past armed conflicts and/or counterterrorism campaigns like the one Britain conducted in Northern Ireland.

A commission would also inevitably involve attacks on career officials in the intelligence community and the departments of Justice and Defense, not merely Bush political appointees. When combined with past investigations, the commission's work would inevitably burden, distract and demoralize the nation's intelligence capabilities. The end result would be the extension of a bureaucratic culture that already favors excessive caution and inaction among our key intelligence and law enforcement officials — the very developments, acknowledged by the 9/11 Commission, as contributing mightily to the analytical, legal and policy failures of 9/11.

Finally, a commission would warp our constitutional fabric and harm civil liberties. While many commissions have operated throughout American history, they have not focused on potential prosecutions. Such a private or quasi-governmental commission would not be constrained by the legal and constitutional limits on Congress and the executive branch, thus raising a host of important constitutional questions.

That the commission's supporters — so determined to vindicate the rights of enemy combatant detainees — seem untroubled by these issues is both ironic and terribly sad.

Will Obama's economic stimulus revive the U.S. economy?

YES
Dean Baker
Co-director, Center for Economic and Policy Research

Written for *CQ Researcher*, January 2009

President Obama's stimulus proposal is a very good start toward rescuing the economy. In assessing the plan, it is vitally important to recognize the seriousness of the downturn. The economy lost an average of more than 500,000 jobs a month in the last three months of 2008. In fact, the actual job loss could have been over 600,000 a month due to the way in which the Labor Department counts jobs in new firms that are not in its survey.

The recent announcements of job loss suggest that the rate of job loss may have accelerated even further. It is possible that we are now losing jobs at the rate of 700,000 a month. This is important, because people must understand the urgency of acting as quickly as possible.

With this in mind, the package being debated does a good job of getting money into the economy quickly. According to the projections of the Congressional Budget Office (CBO), 62 percent of the spending in the package will reach the economy before the end of 2010, with most of the rest coming in 2011. This money will be giving the economy a boost when we need it most.

At this point, there is considerable research on the impact of tax cuts, and the evidence suggests that they do not have nearly as much impact on the economy, primarily because a large portion of any tax cut is saved. According to Martin Feldstein, President Reagan's chief economist, just 10 percent of the tax cuts sent out last spring were spent. The rest was saved. Increased savings can be beneficial to household balance sheets, but savings will not boost the economy right now.

There will also be long-term benefits from President Obama's package. For example, the CBO projected we would save more than $90 billion on medical expenses over the next decade by computerizing medical records, which will be financed through the stimulus. In addition, weatherizing homes and offices and modernizing the electrical grid will substantially reduce our future energy use.

The Obama administration projects that this package will generate close to 4 million jobs, and several independent analysts have arrived at similar numbers. This will not bring the economy back to full employment, but it is still a huge improvement over doing nothing.

The cost of this bill sounds large, but it is important to remember that the need is large. If we were to just do nothing, the economy would continue to spiral downward, with the unemployment rate reaching double-digit levels in the near future.

NO
J. D. Foster
Norman B. Ture Senior Fellow in the Economics of Fiscal Policy, The Heritage Foundation

Written for *CQ Researcher*, January 2009

President Barack Obama promises to create 3.5 million new jobs by the end of 2010, and that vow provides a clear measure by which to judge whether his policies work.

U.S. employment stood at about 113 million people in December 2008, so the Obama jobs pledge will be met if 116.5 million people are working by the end of 2010. Reaching this goal will require effective stimulus policies — and the only fiscal policy that can come close to reaching the goal is to cut marginal tax rates.

Obama's target for jobs creation was chosen carefully. Employment peaked at about 115.8 million jobs in November 2007. Obama's jobs pledge at that time was to create 2.5 million jobs, for a total of 116.5 million private sector jobs.

The November 2008 jobs report showed a half-million jobs lost, so his job-creating target rose by a half-million, affirming the 116.5 million target. Then last month's jobs report showed another half-million jobs lost, and the president raised the target again to its current 3.5 million total.

To stimulate the economy, Obama and congressional Democrats have focused on massive new spending programs. However, the federal budget deficit is likely to exceed $2.5 trillion over the next two years even before any stimulus is added. If deficit spending were truly stimulative, the economy would be at risk of overheating by now, not sliding deeper into recession.

Additional deficit spending won't be any more effective than the first $2 trillion, because government spending doesn't create additional demand in the economy. Deficit spending must be financed by borrowing, so while government spending increases demand, government borrowing reduces demand. Worse, since the government's likely to borrow between $3 trillion and $4 trillion over the next two years, the enormous waves of government debt will likely drive interest rates up. That would only prolong the recession and weaken the recovery.

An effective fiscal stimulus would defer the massive 2011 tax hike (higher tax rates on dividends and capital gains are scheduled to kick in), and also cut individual and corporate tax rates further to reduce the impediments to starting new businesses, hiring, working and investing.

To meet his goal, President Obama should junk his ideology and the wasteful spending that goes with it and focus on cutting marginal tax rates. That's the only way to hit his jobs creation target.

congressionally established five-year waiting period after any president leaves office. The order supersedes a Bush administration directive in 2001 by giving the incumbent president, not a former president, decision-making authority on whether to invoke executive privilege to prevent release of the former president's records.

On foreign policy, Obama on his first full day in office turned to the fragile cease-fire in Gaza by placing calls to four Mideast leaders: Egyptian President Hosni Mubarak, Israeli Prime Minister Ehud Olmert, Jordanian King Abdullah and Palestinian Authority President Mahmoud Abbas. Obama offered U.S. assistance to try to solidify the ceasefire that had been adopted over the Jan. 17-18 weekend by Israel and Hamas, the ruling party in Gaza.

Israel had begun an offensive against Hamas on Dec. 27 in an effort to halt cross-border rocket attacks into Israel by Hamas supporters. During the transition, Obama had limited himself to a brief statement regretting the loss of life on both sides. White House press secretary Robert Gibbs said Obama used the calls from the Oval Office to pledge U.S. support for consolidating the cease-fire by preventing the smuggling of arms into Hamas from neighboring Egypt. He also promised U.S. support for "a major reconstruction effort for Palestinians in Gaza," Gibbs said.

The next day, Obama took a 10-block ride to the State Department for Hillary Clinton's welcome ceremony as secretary following her 94-2 Senate confirmation on Jan. 21. As part of the event, Clinton announced the appointment of special envoys George Mitchell for the Middle East and Richard Holbrooke for Afghanistan and Pakistan.

In his remarks, Obama renewed support for a two-state solution: Israel and a Palestinian state "living side by side in peace and security." He also promised to refocus U.S. attention on what he called the "perilous" situation in Afghanistan, where he said violence had increased dramatically and a "deadly insurgency" had taken root.

Returning to domestic issues, Obama on Jan. 23 signed — as expected — an order to lift the so-called Mexico City policy prohibiting U.S. aid to any nongovernmental organizations abroad that provide abortion counseling or services. The memorandum instructed Secretary of State Clinton to lift what Obama called the "unwarranted" restrictions. The policy was first put in place by President Ronald Reagan in 1984, rescinded by President Clinton in 1993 and then reinstituted by President Bush in 2001.

After the weekend, Obama reversed another of Bush's policies on Jan. 26 by directing Environmental Protection Agency Administrator Lisa Jackson to reconsider the request by the state of California to adopt automobile emission standards stricter than those set under federal law. In a reversal of past practice, the Bush administration EPA had denied California's waiver request in December 2007. On the same day, Obama instructed Transportation Secretary Ray LaHood to tighten fuel efficiency standards for cars and light trucks beginning with 2011 model cars.

Working With Congress

President Obama is pressing Congress for quick action on an economic stimulus plan even as bipartisan support for a proposal remains elusive. Meanwhile, the new administration is struggling to find ways to make the financial bailout approved before Obama took office more effective in aiding distressed homeowners and unfreezing credit markets.

House Democrats moved ahead with an $825-billion stimulus package after the tax and spending elements won approval in separate, party-line votes by the House Ways and Means Committee on Jan. 22 and the House Appropriations Committee the day before. The full House was scheduled to vote on the package on Jan. 28 after deadline for this issue, but approval was assured given the Democrats' 256-178 majority in the chamber.

Obama used his first weekly address as president on Jan. 24 — now not only broadcast on radio but also posted online as video on YouTube and the White House Web site — to depict his American Recovery and Reinvestment Plan as critical to get the country out of an "unprecedented" economic crisis. The plan, he said, would "jump-start job creation as well as long-term economic growth." Without it, he warned, unemployment could reach double digits, economic output could fall $1 trillion short of capacity and many young Americans could be forced to forgo college or job training.

Without mentioning the tax and spending plan's minimum total cost, Obama detailed a long list of infrastructure improvements to be accomplished in energy, health care, education and transportation. He mentioned a $2,500 college tax credit but did not note other items in the $225 billion in tax breaks included in the plan — either his long-advocated $1,000 tax break for working families or the various business tax cuts added as sweeteners for Republicans.

Republicans, however, remained unconvinced. Replying to Obama's address, House Minority Leader John Boehner called the plan "chock-full of government programs and projects, most of which won't provide immediate relief to our ailing economy." On "Meet the Press" the next day, the Ohio lawmaker again called for more by way of tax cuts, criticized the job-creating potential of Obama's plan and warned of opposition from most House Republicans.

Appearing on another of the Sunday talk shows, McCain told "Fox News Sunday" host Chris Wallace, "I am opposed to most of the provisions in the bill. As it stands now, I would not support it."

On a second front, the principal members of Obama's economic team are assuring Congress of major changes to come in the second stage of the $700-billion financial rescue plan approved last fall. During confirmation hearings, Treasury Secretary-designate Geithner promised the Senate Finance Committee on Jan. 21 to expect "much more substantial action" to address the problem of troubled banks that has chilled both consumer and corporate credit markets since fall 2008.

Geithner's comments on the financial bailout were overshadowed by sharp questions from Republican senators about the nominee's tax problems while working for the International Monetary Fund. For several years, Geithner failed to pay Social Security and Medicare taxes, which the IMF — as an international institution — does not withhold from employees' pay as domestic employers do. Geithner repeatedly apologized for the mistake and pointed to his payment of back taxes plus interest totaling more than $40,000. In the end, the committee voted 18-5 to recommend confirmation; the full Senate followed suit on Jan. 26 in a 60-34 vote.*

On the bailout, Geithner said he would increase the transparency and accountability of the program once he assumed the virtually unfettered responsibility for dispensing the remaining $350 billion. He acknowledged criticisms that so far the program has benefited large financial institutions but done little for small businesses. He also promised to restrict dividends by companies that receive government help.

---

* Attorney General-designate Holder, Obama's other controversial Cabinet nominee, was expected to be confirmed by the full Senate on Jan. 29 or 30, after deadline for this issue, following the Senate Judiciary Committee's 17-2 vote on Jan. 28 to recommend confirmation.

With many banks still holding billions in troubled assets on their balance sheets, speculation is increasing in Washington and in financial circles about dramatic action by the government. Possible moves include the creation of a government-run "bad bank" to buy distressed assets from financial institutions or even outright nationalization of one or more banks.

"People continue to be surprised by the poor condition of the banks," says Dean Baker, co-director of the Center for Economic and Policy Research, a liberal think tank in Washington. "Whatever plans they may have made a month ago might be seen as inadequate given the severity of the problem of the banking system."

With the stimulus package on the front burner, however, Obama went to Capitol Hill on Jan. 27 for separate meetings to lobby House and Senate Republicans to support the measure. The closed-door session with the full House GOP conference lasted an hour — slightly longer than scheduled, causing the president to be late for the start of the meeting on the other side of the Capitol with Republican senators.

In between meetings, Obama challenged GOP lawmakers to try to minimize partisan differences. "I don't expect 100 percent agreement from my Republican colleagues, but I do hope we can put politics aside," he said.

For their part, House Republican leaders expressed appreciation for the president's visit and his expressed willingness to compromise. But some renewed their opposition to the proposal in its current form. Rep. Tom Price of Georgia, chairman of the conservative House Republican Study Committee, said the proposal "remains rooted in a liberal, big-government ideology."

Obama's meeting with GOP senators came on the same day that the Senate Finance and Appropriations committees were marking up their versions of the stimulus package. The Senate was expected to vote on the proposal over the weekend, giving the two chambers two weeks to iron out their differences if the bill was to reach Obama's desk before the Presidents' Day recess.

OUTLOOK

Peril and Promise

One week after taking office, President Obama is getting high marks from experts on the presidency for carefully stage-managing his first policy initiatives while discreetly moving to set realistic expectations for the months ahead.

"He's started out quite impressively," says Fred Greenstein, professor of politics emeritus at Princeton University in New Jersey and the dean of American scholars on the U.S. presidency. "So far, it's been a striking rollout week."

Other experts agree. "The Obama administration has met expectations for the first week," says Meena Bose, chair of the Peter S. Kalikow Center for the Study of the American Presidency at Hofstra University in Hempstead, N.Y. "There's been virtually no drama, which is an indication of how he intends to run his administration."

"The indications are all positive," says Bruce Buchanan, a professor of political science at the University of Texas in Austin and author of several books on the presidency. Like the others, Buchanan says Obama is holding on to popular support while striving either to win over or to neutralize Republicans on Capitol Hill.

The wider world outside Washington, however, is giving Obama no honeymoon in office. The U.S. economy is continuing to lag, while violence and unrest continue to simmer in three global hot spots: Gaza, Iraq and Afghanistan.

On the economy, Obama has initiated a daily briefing from senior adviser Summers in addition to the daily briefing on foreign policy and national security issues. "Frankly," Obama told congressional leaders on Jan. 23, "the news has not been good." The day before, the Commerce Department had reported that new-home construction fell to its slowest pace since reporting on monthly rates began in 1959. On the same day, new claims for unemployment benefits matched the highest level seen in a quarter-century.[20]

Meanwhile, leading U.S. policy makers were giving downbeat assessments of events in Afghanistan and Iraq. In testimony to the Senate Armed Services Committee, Defense chief Gates warned on Jan. 27 to expect "a long and difficult fight" in Afghanistan. A few days earlier, the outgoing U.S. ambassador to Iraq, Ryan Crocker, warned that what he called "a precipitous withdrawal" could jeopardize the country's stability and revive al Qaeda in Iraq. And special envoy Mitchell left Washington for the Mideast on Jan. 26, just as the fragile cease-fire between Hamas and Israel was jeopardized by the death of an Israeli soldier from a roadside bomb and an Israeli air strike in retaliation.

Obama continues to work at the problems with the same kind of message control that served him well in the election. After reaping a full day's worth of mostly favorable news coverage on the Guantánamo issue, the administration began directing laser-like attention to the economy

from Jan. 22 on. For example, the repeal of the Bush administration's ban on funding international groups that perform abortions was announced late on Friday, Jan. 23 — a dead zone for news coverage.

On foreign policy, Obama emphasized the Mitchell and Holbrooke appointments by personally going to the State Department for the announcements. And he underscored the inaugural's outreach to Muslims by granting his first formal television interview as president to the Arabic satellite television network Al Arabiya. Obama called for a new partnership with the Muslim world "based on mutual respect and mutual interest." One of his main tasks, he told the Dubai-based network in an interview aired on Jan. 27, is to communicate that "the Americans are not your enemy."[21]

Obama and his senior aides are also signaling to supporters that some of their agenda items will have to wait. In a pre-inauguration interview with *The Washington Post*, for example, he reiterated his support for a labor-backed bill to make it easier to unionize workers but downgraded it to a post-stimulus agenda item. Similarly, press secretary-designate Gibbs repeated Obama's support for repealing the military's "don't ask, don't tell" policy on homosexuals on the transition's Web site on Jan. 13, but the next day expanded on the answer: "Not everything will get done in the beginning," Gibbs said.[22]

Greenstein and Bose view Obama's inaugural address — which many observers faulted for rhetorical flatness — as a conscious, initial step to lower expectations about the pace of the promised "change we can believe in." Greenstein calls it a "get-down-to-work" address. Obama himself again evoked the inaugural's theme of determination in the face of adversity when he spoke to congressional leaders immediately following the address.

"What's happening today is not about me," Obama said at the joint congressional luncheon on Inauguration Day. "It is about the American people. They understand that we have arrived at a moment of great challenge for our nation, a time of peril, but also extraordinary promise."

"President Obama has done everything he can to tamp down this sense that he somehow walks on water," says Bose. "He has done everything he can to show that he is a man of substance.

"We have to recognize that these challenges aren't going to be met overnight and that we have to have confidence that we're going to meet them," she continues. "Now the question is, 'Can he govern? Can he show results?'"

NOTES

1. The text and video of the inaugural address are available on the redesigned White House Web site: www.whitehouse.gov. Some crowd reaction from Christopher O'Brien of CQ Press' College Division.

2. Quoted in Clea Benson, "An Economy in Foreclosure," *CQ Weekly*, Jan. 12, 2009.

3. Quoted in Aamer Madhani, "Will Obama Stick to Timetable?" *Chicago Tribune*, Nov. 6, 2008, p. 11.

4. Quoted in Shailagh Murray and Paul Kane, "Democratic Congress Shows It Will Not Bow to Obama," *The Washington Post*, Jan. 11, 2009, p. A5.

5. Norman Ornstein, "First Steps Toward 'Post-Partisanship' Show Promise," *Roll Call*, Jan. 14, 2009.

6. For a compact, continuously updated biography, see Barack Obama, www.biography.com. Background also drawn from Barack Obama, *Dreams from My Father: A Story of Race and Inheritance* (2004 ed.; originally published 1995). See also David Mendell, *Obama: From Promise to Power* (2007).

7. Quoted in Serge Kovaleski, "Obama's Organizing Years: Guiding Others and Finding Himself," *The New York Times*, July 7, 2008, p. A1.

8. Background drawn from Jody Kantor, "In Law School, Obama Found Political Voice," *The New York Times*, Jan. 28, 2007, sec. 1, p. 1.

9. Background drawn from Janny Scott, "The Story of Obama, Written by Obama," *The New York Times*, May 18, 2008, p. A1.

10. For a story on his mother's influence on Obama, see Amanda Ripley, "A Mother's Story," *Time*, April 21, 2008, p. 36.

11. See David Jackson and Ray Long, "Showing his bare knuckles: In first campaign, Obama revealed hard-edged, uncompromising side in eliminating party rivals," *Chicago Tribune*, April 4, 2007, p. 1; Rick Pearson and Ray Long, "Careful steps, looking ahead: After arriving in Springfield, Barack Obama proved cautious, but it was clear to many he had ambitions beyond the state Senate," *ibid.*, May 3, 2007, p. 1.

12. See Barack Obama, *The Audacity of Hope: Thoughts on Reclaiming the American Dream* (2006), pp. 105-107.

13. See David Mendell, "Obama routs Democratic foes; Ryan tops crowded GOP field," *Chicago Tribune*, March 17, 2004, p. 1.

14. For the full text of the 2,165-word speech, see http://obamaspeeches.com/002-Keynote-Address-at-the-2004-Democratic-National-Convention-Obama-Speech.htm. For Mendell's account, see *Obama, op. cit.*, pp. 272-285. Obama's conversation with Martin Nesbitt may have been reported first in David Bernstein, "The Speech," *Chicago Magazine*, July 2007; the anecdote is briefly repeated in Evan Thomas, *"A Long Time Coming": The Inspiring, Combative 2008 Campaign and the Historic Election of Barack Obama* (2009), p. 6. For the Post's account, see David S. Broder, "Democrats Focus on Healing Divisions," July 28, 2004, p. A1.

15. *CQ's Politics in America 2008* (110th Congress), www.cnn.com/video/#/video/world/2007/01/22/vause.obama.school.cnn.

16. Some background from Thomas, *op. cit.*

17. *Ibid.*, p. 9.

18. Many versions of the speech are posted on YouTube, including a posting of CNN's coverage.

19. Carpenter was quoted in Carrie Budoff Brown and Nia-Milaka Henderson, "Cabinet: Middle-of-the-roaders' dream?" *Politico*, Dec. 19, 2008; Alec MacGillis, "For Obama Cabinet, a Team of Moderates," *The Washington Post*, Dec. 20, 2008, p. A1.

20. See Kelly Evans, "Home Construction at Record Slow Pace," *The Wall Street Journal*, Jan. 23, 2009, p. A3.

21. See Paul Schemm, "Obama tells Arabic network US 'is not your enemy,'" The Associated Press, Jan. 27, 2009.

22. Obama quoted in Dan Eggen and Michael D. Shear, "The Effort to Roll Back Bush Policies Continues," *The Washington Post*, Jan. 27, 2009, p. A4; Gibbs quoted in, "Obama aide: Ending 'don't ask, don't tell' must wait," CNN.com, Jan. 15, 2009.

BIBLIOGRAPHY

Books by Barack Obama

Dreams from My Father: A Story of Race and Inheritance (Three Rivers Press, 2004; originally published by Times Books, 1995) is a literate, insightful memoir written in the three years after Obama's graduation from Harvard Law School. The three parts chronicle his "origins" from his birth through college, his three years as a community organizer in Chicago and his two-month pre-law school visit to his father's homeland, Kenya.

The Audacity of Hope: Thoughts on Reclaiming the American Dream (Crown, 2006) is a political manifesto written as Obama considered but had not definitively decided on a presidential campaign. The book opens with a critique of the "bitter partisanship" of current politics and an examination of "common values" that could underline "a new political consensus." Later chapters specifically focus on issues of faith and of race. Includes index.

Change We Can Believe In: Barack Obama's Plan to Renew America's Promise (Three Rivers Press, 2008), which includes a foreword by Obama, outlines steps for "reviving our economy," "investing in our prosperity," "rebuilding America's leadership" and "perfecting our union." Also includes texts of seven speeches from his declaration of candidacy on Feb. 7, 2007, to his July 24, 2008, address in Berlin.

Books About Barack Obama

The only objective, full-length biography is *Obama: From Promise to Power* (Amistad/Harper Collins, 2007) by David Mendell, the *Chicago Tribune* political reporter who began covering Obama in his first race for the U.S. Senate. An updated version was published in 2008 under the title *Obama: The Promise of Change.*

Two critical biographies appeared during the 2008 campaign: David Freddoso, *The Case Against Barack Obama: The Unlikely Rise and Unexamined Agenda of the Media's Favorite Candidate* (Regnery, 2008); and Jerome Corsi, *The Obama Nation: Leftist Politics and the Cult of Personality* (Threshold, 2008). Freddoso, a writer with National Review Online, wrote what one reviewer called a "fact-based critique" depicting Obama as "a fake reformer and a real liberal." Corsi, a conservative author and columnist best known for his book *Unfit for Command* attacking Democratic presidential nominee John Kerry in 2004, came under fierce criticism from the Obama campaign and independent observers for undocumented allegations about Obama's background.

Two post-election books chronicle the 2008 campaign. Evan Thomas, *"A Long Time Coming": The Inspiring, Combative 2008 Campaign and the Historic Election of Barack Obama* (Public Affairs, 2009) is the seventh in *Newsweek*'s quadrennial titles documenting presidential campaigns on the basis of reporting by a team of correspondents, with some reporting specifically not for publication until after the election. Chuck Todd and Sheldon Gawiser, *How Barack Obama Won: A State-by-State Guide to the Historic 2008 Presidential Election* (Vintage, 2009) gives an analytical overview of the campaign and election with detailed voting analyses of every state. A third title, *Obama: The Historic Journey*, is due for publication Feb. 16 by *The New York Times* and Callaway; the author is Jill Abramson, the *Times*' managing editor, in collaboration with the newspaper's reporters and editors.

Other books include John K. Wilson, *Barack Obama: The Improbable Quest* (Paradigm, 2008), an admiring analysis of Obama's political views and philosophy by a lawyer who recalls having been a student in Obama's class on racism and the law at the University of Chicago Law School; Paul Street, *Barack Obama and the Future of American Politics* (Paradigm, 2009), a critical depiction of Obama as a "power-conciliating centrist"; and Jabiri Asim, *What Obama Means: For Our Culture, Our Politics, Our Future* (Morrow, 2009) a depiction of Obama as creating a new style of racial politics — less confrontational than in the past but equally committed to social justice and more productive of results.

Articles

Purdum, Todd, "Raising Obama," *Vanity Fair*, March 2008.

The magazine's national editor, formerly a *New York Times* reporter, provided an insightful portrait of Obama midway through the 2008 primary season.

Von Drehle, David, "Person of the Year: Barack Obama: Why History Can't Wait," *Time*, Dec. 29, 2008.
Time's selection of Obama as person of the year includes an in-depth interview of the president-elect by Managing Editor Richard Stengel, Editor-at-large von Drehle and Time Inc. Editor-in-chief John Huey. The full text is at time.com/obamainterview.

On the Web

The Obama administration unveiled a redesigned White House Web site (www.whitehouse.gov) at 12:01 p.m. on Jan. 20, 2009 — even before President-elect Obama took the oath of office. The "Briefing Room" includes presidential announcements as well as a "Blog" sometimes being updated several times a day. "The Agenda" incorporates Obama's campaign positions, subject by subject. The site includes video of the president's speeches, including the inaugural address as well as the weekly presidential address — previously broadcast only on radio.

For More Information

American Enterprise Institute for Public Policy Research, 1150 17th St., N.W., Washington, DC 20036; (202) 862-5800; www.aei.org. Conservative think tank researching issues on government, economics, politics and social welfare.

Campaign for America's Future, 1825 K St., N.W., Suite 400, Washington, DC 20006; (202) 955-5665; www.ourfuture.org. Advocates progressive policies.

Center for American Progress, 1333 H St., N.W., 10th Floor, Washington, DC 20005; (202) 682-1611; www.americanprogress.org. Left-leaning think tank promoting a government that ensures opportunity for all Americans.

Center for Economic and Policy Research, 1611 Connecticut Ave., N.W., Suite 400, Washington, DC 20009; (202) 293-5380; www.cepr.net. Promotes open debate on key economic and social issues.

Center on Budget and Policy Priorities, 820 First St., N.E., Suite 510, Washington, DC 20002; (202) 408-1080; www.cbpp.org. Policy organization working on issues that affect low- and moderate-income families and individuals.

Concord Coalition, 1011 Arlington Blvd., Suite 300, Arlington, VA 22209; (703) 894-6222; www.concordcoalition.org. Nonpartisan, grassroots organization promoting responsible fiscal policy and spending.

Heritage Foundation, 214 Massachusetts Ave., N.E., Washington, DC 20002; (202) 546-4400; www.heritage.org. Conservative think tank promoting policies based on free enterprise, limited government and individual freedom.

13

HPV Vaccine

Should It Be Mandatory for School Girls?

Nellie Bristol

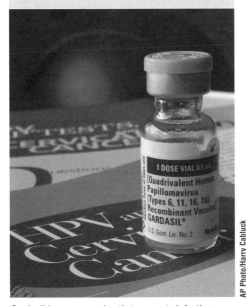

Gardasil is a new vaccine that prevents infections from HPV, a sexually transmitted disease that causes cervical cancer. Many state lawmakers want to make inoculations mandatory for school attendance, but conservative groups say requiring the expensive shots would encourage inappropriate sexual activity and override parental autonomy.

From *CQ Researcher*,
May 11, 2007.

As far as Tamika Felder was concerned, it was, "Look out, world, here I come." With visions of meeting celebrities and making her mark in the world, she moved from her native South Carolina to Washington, D.C., primed and ready to fulfill her ambition to become a TV producer. By age 24, she was an associate producer for a political cable network and had interviewed Hillary Rodham Clinton and Vice President Al Gore. Life was "exhilarating."

But in 2001, a boil under her arm forced her to see a physician. After treating the condition, he recommended an overdue, full physical, including a routine Pap smear to screen for cervical cancer. Always healthy except for the inevitable concerns about her plus-size frame, Felder was stunned when her doctor told her she had advanced cervical cancer. At age 25, Felder had a radical hysterectomy.

To help sort through her emotional turmoil, Felder began researching the disease. She discovered cervical cancer is caused almost exclusively by the extremely common human papillomavirus (HPV), which infects 80 percent of women by their 50s.[1] She got angry.

"How could this be so common yet I've never heard of it?" she asks. Her ire intensified when she discovered how it was transmitted. "My best friend's husband said, 'I think that's caused by an STD.' I remember getting really angry with him — how dare you say something like that to me — and then finding out he was right."

Genital HPV is thought to be the most common sexually transmitted disease in the United States. According to the Centers for

311

HPV Legislation Introduced in Half of Country

Legislation requiring HPV vaccinations for young girls has been passed or is under consideration in 24 states and Washington, D.C.

Legislation introduced

No legislation pending

Source: "State by State: HPV Vaccination Legislation," PBS, Feb. 23, 2007, www.pbs.org.

Disease Control and Prevention (CDC), about 6.2 million women and men are newly infected every year.[2] While some infections cause genital warts, many cases cause no symptoms, and 90 percent clear up spontaneously within two years. But 10 percent of cases turn into persistent infections, and certain HPV strains are considered "known human carcinogens" that can cause cervical and other cancers in women and certain cancers in men.[3] In 2007, an estimated 11,100 American women will be diagnosed with cervical cancer and 3,700 will die from it.[4] While cervical cancer is declining in industrialized countries like the United States, some 390,000 new cases are diagnosed each year in developing countries where Pap smears and early diagnosis are not readily available. (*See sidebar, p. 322.*)

But public health experts hope those numbers will be slashed by a highly effective new vaccine that prevents four of the most problematic strains of genital HPV. With an impressive 90-100 percent effectiveness record and no major safety incidents, Merck & Co.'s Gardasil was given expedited approval by the Food and Drug Administration (FDA), which licensed it on June 8, 2006. The vaccine targets cancer-causing HPV strains 16 and 18, which cause 70 percent of cervical cancer,

and types 6 and 11, which cause 90 percent of genital warts.[5]

When Merck presented its data on the vaccine to the CDC's Advisory Committee on Immunization Practices (ACIP) in June 2006, the panel broke out in applause, recalls William Schaffner, chairman of preventive medicine at Vanderbilt University's School of Medicine. (Schaffner attends ACIP meetings as a non-voting member representing the National Foundation for Infectious Disease.) The vaccine is "a hugely impressive advance for women's health," says Schaffner, who also is a member of Merck's data-safety monitoring board.

The CDC panel recommended the vaccine for routine use in 11- to 12-year-old girls. The shots can be started as young as 9, and the panel suggests "catch-up" vaccinations for females ages 13-26. The endorsement became official CDC policy on March 12, 2007. The agency's non-binding vaccine recommendations are the basis for immunization policies nationwide. Most states adopt the CDC's recommended immunization schedule as the list of shots children must get before they can attend — or continue attending — public school; most private schools adopt the same schedule.

But an early push by Merck and some state lawmakers to require preteen girls to receive the vaccine in order to continue attending school created a national stir. Parents and others resisted the mandates, proposed in 24 states, citing uncertainty about the inoculation's long-term safety and discomfort with immunizing preteen girls against a sexually transmitted disease (STD).

Conservative groups joined the opposition, saying the vaccine would encourage inappropriate sexual activity and override parental autonomy. Others questioned whether an STD vaccine should be required for school attendance as are vaccines for more broadly communicable illnesses such as measles and whooping cough.

"This, to me, is clearly an issue between a child and a parent, especially when we know that this vaccine is for a disease that is shown to only occur through sexual activity," said California Republican state Sen. George Runner. "I have certain values and issues which I deal with my daughter on. And it seems to send an inconsistent message about sexual activity."[6]

Also at issue is Gardasil's cost. At $120 for each dose of a three-dose regimen, it is one of the most expensive vaccines ever made. While insurers are approving reimbursement, physicians complain they are not paying enough to cover the costs of administering the shots.[7]

Stories about intense lobbying by Merck further soured the public. Press reports discussed Merck's financial support of Women in Government, a group for female state legislators whose members are in the forefront of vaccine-mandating bills.[8] In Texas, where the first mandate was enacted via executive order by Republican Gov. Rick Perry, reports noted that the governor's former top aide is now a Merck lobbyist.[9] The order was later rescinded by the state legislature.[10]

In New Mexico, Democratic Gov. Bill Richardson vetoed a bill to require the shot after initially promising to sign it.[11] So far, only Virginia and the District of Columbia have mandated the vaccine for girls entering sixth grade, but with a delayed effective date and a generous opt-out provision for parents uncomfortable with the vaccine.[12]

Public health and vaccine experts, while excited about the vaccine's potential to reduce cervical cancer, complained the move to require the immunization occurred more quickly than for most new shots. Many say the effort was premature and bypassed established state public health structures.

"Mandates are usually carefully considered as part of a strategy to gain additional control of an infection that you don't have good control of with the first vaccine strategies," says Joseph Bocchini, chairman of the American Academy of Pediatrics' (AAP) Infectious Disease Committee. In addition, before a vaccine is mandated, health experts say the public needs to be

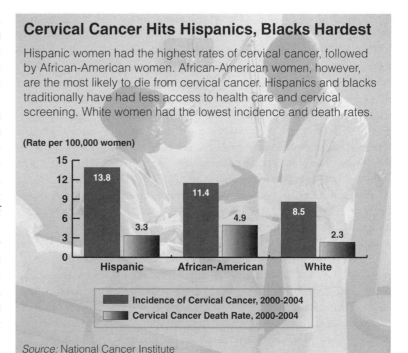

Cervical Cancer Hits Hispanics, Blacks Hardest

Hispanic women had the highest rates of cervical cancer, followed by African-American women. African-American women, however, are the most likely to die from cervical cancer. Hispanics and blacks traditionally have had less access to health care and cervical screening. White women had the lowest incidence and death rates.

(Rate per 100,000 women)

	Incidence of Cervical Cancer, 2000-2004	Cervical Cancer Death Rate, 2000-2004
Hispanic	13.8	3.3
African-American	11.4	4.9
White	8.5	2.3

Source: National Cancer Institute

educated about the disease, and processes need to be in place to ensure adequate payment and availability.

Partly in anticipation of the Gardasil controversy, the Association of Immunization Managers (AIM) developed a position statement addressing immunization mandates. The group cautioned last June that mandated vaccinations should be carefully considered and used sparingly in order to maintain public support.[13]

AIM Executive Director Claire Hannan says drug manufacturers' lobbying of state legislatures has become "a trend that we're concerned about." AIM is particularly worried that controversial immunizations might be mandated with liberal opt-out provisions that could spill over into other vaccines vital for public safety.

Despite the controversy, many physicians say the vaccine is popular with their patients.[14] A Merck spokeswoman said the company had distributed 5 million doses from June 2006 through this March, valued at $600 million. Public health and vaccine experts expect the noise surrounding Gardasil to die down, but the controversy is likely a harbinger of vaccine debates to come. Vaccines, now a minimally profitable business eschewed by most U.S. drug companies, could become big business as immunizations are developed for other diseases, including STDs.[15]

Meanwhile, vaccine-safety advocates are concerned about adding more injections to the already full schedule of childhood immunizations. "The question becomes, 'How many vaccines are we going to require our children to get to go school?'" asks Barbara Loe Fisher, co-founder and president of the National Vaccine Information Center. "Somebody at some point has got to give evidence that we are not harming an ever-increasing minority of children by giving too many vaccines."

The CDC currently recommends children receive 48 doses of 14 vaccines by age 6 and 53 doses of 15 vaccines by age 12. With addition of the HPV vaccine, girls would receive 56 doses of 16 vaccines by age 12.[16]

"The HPV vaccine has caused an awakening that is probably 25 years in the making," says Fisher. "You are not going to see another vaccine introduced in this country for children that does not undergo the same scrutiny."

But for Felder and others who have suffered the life-altering effects of invasive cervical cancer, the debate is superfluous. "When people politicize things that ought not be politicized, it always astounds me," said Texas cervical cancer survivor Cheryl Swope Lieck.[17]

"I'm for the vaccine being mandatory, going through what I went through," Felder says. "If you walked a day in my shoes, you would never want this to happen to anybody else."

As experts, lawmakers and parents study the new HPV vaccine, here are some of the questions being asked:

Is the new HPV vaccine safe and effective?

Public health and vaccine experts are confident about the safety of Gardasil and thrilled about its potential to reduce a terrible cancer for women.

"This vaccine represents an important medical breakthrough," says Anne Schuchat, director of the CDC's National Center for Immunization and Respiratory Diseases. "As a result, these vaccine recommendations address a major health problem for women and represent a significant advance in women's health. It has been tested in thousands of women around the world and has been found to be safe and effective in providing protection against the two types of HPV that cause most cervical cancers."

Vaccine expert Paul Offit of Children's Hospital in Philadelphia agrees: "Not only is it phenomenally safe,

it's phenomenally effective," says Offit, who has patent holdings with Merck on a rotavirus vaccine.

"It's not a perfect vaccine, but, wow, it's a fabulous vaccine," says Vanderbilt's Schaffner.

Despite the widespread confidence, some experts say long-term safety can only be judged with time.

"For any vaccine, you never know its complete safety pattern when the vaccine is licensed. It's not until you get into the hundreds of thousands and millions [of doses] that you begin to look at more rare [negative] events," says Walter Orenstein, associate director of the Emory Vaccine Center at Emory University's School of Medicine. "The safety information we have is very encouraging, so I have no problem recommending the vaccine. But certainly we always have to have our eyes open and look to see if there are rare events that we weren't able to detect in the early trials."

Arthur Allen — a Washington, D.C., journalist and author of the 2007 book *Vaccine: The Controversial Story of Medicine's Greatest Lifesaver* — says that while Gardasil's clinical trials were good-sized, they were not large enough to find an unusual side effect "that might be serious enough for people to not want to take the vaccine."

Clinical-trial enrollees who received the Gardasil formulation now on the market have been followed, at most, for slightly more than five years. Safety data on the vaccine are available from seven studies, covering 11,778 individuals ages 9-26 who received the vaccine and 9,686 who received placebos. According to the CDC, detailed safety data are available for 5,088 vaccine recipients. The most commonly reported immediate adverse reaction was pain at the injection site. Almost 84 percent of Gardasil recipients reported such pain, compared to 75 percent of those receiving a placebo containing aluminum and 49 percent of those receiving a saline solution. A small number of individuals reported severe pain (2.8 percent) while even smaller numbers cited severe swelling and redness.[18]

Systemic reactions were reported in similar numbers by those receiving vaccines and those who received placebos, according to the CDC. The most common reported reaction was fever, with 13 percent of vaccine recipients and 11 percent of placebo recipients reporting mild to moderate fevers.[19]

To enhance immune response, Gardasil contains aluminum, which has been used in many vaccines. But

Fisher questions its safety, saying aluminum has been shown to cause problems like swelling and chronic joint pain.

Fisher also worries about Gardasil's effect on younger girls. She says the clinical studies included only 1,200 girls under 16. "That is just not enough information," she says. "What is the scientific evidence that it is safe and effective to use in the age group for which it has been recommended? If you look at it on balance, it has not been proven."

Diane Harper, director of the Gynecological Cancer Prevention Researcher Group at Dartmouth Medical School in New Hampshire, agrees with Fisher. "It is silly to mandate vaccination of 11-to-12-year-old girls. There is not enough evidence gathered on side effects to know that safety is not an issue," she said, calling the inoculations of preadolescent girls "a great big public health experiment."[20] (Harper is a lead researcher for both Merck and GlaxoSmithKline HPV vaccines.)

Fisher also says there have been an increasing number of cases of fainting associated with the vaccine. "There's something unusual, we feel, going on," she says.

While noting that adolescent girls — the main age group receiving the shots — are generally more prone to fainting, the CDC concedes that it nonetheless has tracked increased fainting episodes associated with Gardasil. Recent data show comparable rates of dizziness and injection-site pain compared to other vaccines, but also an increase in syncope, or loss of consciousness associated with a drop in blood pressure. HPV vaccine recipients reported syncope in 9 percent of cases vs. 5 percent for a diphtheria, tuberculosis and pertussis (DTaP) booster shot and 6 percent for Hepatitis B inoculations.[21] The effect is noted in CDC documents, and agency recommendations suggest vaccine providers observe patients for 15 minutes after inoculations.[22]

Jennifer Allen, a spokeswoman for Merck, said safety and efficacy were assumed for the 9-15 age group even though the studies were done primarily on older girls and women — a common practice in immunology called bridging. Studies of younger girls were not feasible, Allen notes, because of the need to perform pelvic examinations to determine cervical condition. She also said it would have been too long to wait for study purposes until a 9-year-old becomes sexually active and is potentially exposed to HPV. Some version of the vaccine has been tested without serious incident in 25,000 individuals over a 10-year period, she added.

Factors Associated with Female HPV Infection

The Centers for Disease Control and Prevention has identified four major risk factors for HPV acquisition in women:

- **Under 25 years of age**
- **Increasing number of sex partners**
- **First sexual intercourse at 16 years or younger**
- **Male partner has had multiple sex partners**

Source: "Human Papillomavirus: HPV Information for Clinicians," Centers for Disease Control and Prevention, April 2007

Time also will show whether the vaccine remains effective over the long term. While antibody levels remain adequate for inoculations given so far, booster shots may eventually be needed. Some experts worry that by giving the initial inoculation at such a young age, the protection could wear off just when women are most at risk for HPV infection. Vaccine supporters are keeping a close eye on the antibody levels of clinical-trial participants.

New research published in *The New England Journal of Medicine* on May 10 suggests vaccine effectiveness may be more modest in the prevention of HPV infections from all strains. The results could mean that other oncogenic (cancer-causing) types of HPV are filling the "biological niche" left behind after elimination of the two primary HPV strains, according to physicians George Sawaya and Karen Smith-McCune of the Department of Obstetrics, Gynecology and Reproductive Sciences at the University of California, San Francisco. "At least 15 oncogenic HPV types have been identified, so targeting only two types may not have had a great effect on overall rates of preinvasive lesions," they write.[23]

While saying delaying vaccination could mean missing out on protection for many women, Sawaya and Smith-McCune note the vaccine still poses many

Should Boys Get HPV Vaccine, Too?

Early studies show it's safe, effective

Since the human papillomavirus (HPV) is transmitted between males and females, some health experts think the virus won't be sufficiently contained until boys receive HPV vaccinations, too.

Early studies have shown Merck's new vaccine, Gardasil, to be both effective at producing immunity in boys and safe. It has been licensed for males in the 25 nations of the European Union and in Australia. In the United States, however, the Food and Drug Administration (FDA) also requires proof that the vaccine actually prevents infection, and that has not been proven in males. Results from ongoing studies are expected by the end of 2008.

"If you think about the broader population and population dynamics, it stands to reason that you're more efficient at reducing infection that both a male and female can transmit if they are both protected with a vaccine," says Anna Giuliano, a University of Florida researcher conducting studies for Merck on HPV and males.

Studies have identified HPV DNA on male genitalia, anal mucus and in the oral cavity. Prevalence is from 2.3 percent to 34.8 percent for types known to cause cancer. Peak incidence was found in men between ages 30 and 39. Penile HPV prevalence rose with increasing numbers of sexual partners. Overall, HPV prevalence appeared to be lower in men than women, suggesting penile tissue could be less receptive to high-risk HPV.[1]

Nonetheless, vaccinating boys and men could reduce rates of genital warts and cancers more likely to affect males. In particular, it could prevent anal cancer, which occurs at a rate of about 35 per 100,000 for men with histories of receptive anal intercourse.[2] The rate could be double that for men with

immune-system-weakening HIV infection. Studies show that 90 percent of 14,500 cases of anal cancer in developed countries are attributable to cancer-causing HPV.[3] In the United States in 2003, of the 4,167 new cases of anal cancer, 90 percent were attributable to HPV. The virus also is suspected in a few of the 29,627 cases of oral cavity and pharynx cancer and in 40 percent of 1,059 cases of penile cancer.[4]

"The thought of vaccinating males has been around for a while," Giuliano says. "At some point, a virgin boy is going to obtain HPV from a girl." Males and females also develop warts at the same rate, she notes.

According to early trials with boys, "immune response is at least as high as or perhaps even higher than what we see in girls for the same age, so we're really hopeful that we're going to see efficacy in men in the adult trials," Giuliano says. Another hopeful sign for male efficacy is that in female trials the vaccine worked against lesions of the vulva. "The skin of the vulva is very similar to the skin of the penis, so the hope is that given those similarities, the vaccine will also work in preventing warts and other lesions in the penile area," she explains.

---

[1] Ann N. Burchell, *et al.*, "Epidemiology and Transmission Dynamics of Genital HPV Infection," in *Vaccine: HPV Vaccines and Screening in the Prevention of Cervical Cancer* (2006).

[2] David Tuller, "New Vaccine for Cervical Cancer Could Prove Useful in Men, Too," *The New York Times*, Jan. 30, 2007.

[3] Maxwell D. Parkin, *et al.*, "The Burden of HPV-Related Cancers," in *Vaccine: HPV Vaccines and Screening in the Prevention of Cervical Cancer* (2006).

[4] "Quadrivalent Human Papillomavirus Vaccine," *Morbidity and Mortality Weekly Report*, Centers for Disease Control and Prevention, March 12, 2007.

---

unanswered questions. "[A] cautious approach may be warranted in light of important unanswered questions about overall vaccine effectiveness, duration of protection and adverse effects that may emerge over time," they write.

Should the HPV vaccine be mandatory?

Vaccines are considered one of the most effective public health interventions of modern times. Mandates ensure their wide application, protecting public health in the

case of easily communicable diseases, like measles or mumps. They also ensure that all children will get protection.

"For all the childhood diseases for which there are vaccines and mandates, disparities, for all intents and purposes, have been eliminated," says Vanderbilt's Schaffner. "It is the most color-blind health-delivery program that we have in the United States."

But the HPV vaccine is controversial, and the vaccine doesn't have a long-term safety record. "A vaccine has to be out and used voluntarily a much longer time before it's ethically justified to mandate it," said *Vaccine* author Allen.

Further, HPV can only be spread through intimate contact, a fact that makes some people uncomfortable with mandating a vaccine, particularly for preteens.

"Bioethicists, who generally hold the values of patient autonomy and informed consent to be preeminent, tend to be skeptical about compulsory-vaccination laws. Not surprisingly, some have expressed wariness about or opposition to mandating HPV vaccination," writes James Colgrove, a research scientist at the Mailman School of Public Health at Columbia University. "Because HPV is not casually transmissible, they argue there is a less-compelling rationale for requiring protection against it than against measles or pertussis, for instance; in the absence of potential harm to a third party, such law may be considered unacceptably paternalistic."[24]

That attitude was reflected in a debate over the vaccine in Washington.

"The D.C. government is strongly urging your daughter to be immunized against . . . the human papillomavirus, or HPV, which is sexually transmitted," wrote *Washington Post* columnist Courtland Milloy. "After all, your daughter is 11 and probably black, so the assumption is she'll be having unprotected sex in no time — but don't take offense."[25]

The Family Research Council, a conservative Christian lobbying group based in Washington, D.C., opposes HPV mandates because they "infringe on the right of parents to make decisions regarding their children's medical care." In addition, they say, because genital HPV is sexually transmitted rather than being spread by casual contact, there is insufficient public health justification for requiring the vaccine for school attendance. Moreover, the group says, forcing girls to get an HPV vaccine may "lead them to believe that the vaccine is the

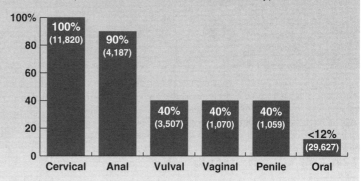

HPV Causes All Cervical Cancer in U.S.

Human papillomavirus (HPV) caused 100 percent of the 11,820 cases of cervical cancer reported in the United States in 2003. By comparison, HPV caused 90 percent of the 4,187 cases of anal cancer and less than 12 percent of the nearly 30,000 cases of oral cavity and pharynx cancer.

Percentage of Cancer Cases Caused by HPV in U.S., 2003
(total number of cases of each variety)

Source: "Quadrivalent Human Papillomavirus Vaccine: Recommendations of the Advisory Committee on Immunization Practices (ACIP)," Morbidity and Mortality Weekly Report, Centers for Disease Control and Prevention, March 12, 2007

only available way to reduce the risk of cervical cancer, which is untrue." Abstinence and monogamy with an uninfected partner is the surest way to prevent HPV, the group insists.[26]

Moreover, Fisher of the National Vaccine Information Center questions the need to mandate a vaccine for a disease that is on the wane anyway. Overall rates of invasive cervical cancer declined 17 percent between 1998 and 2002 in the United States, largely as a result of successful screening programs.[27] Higher rates of cancer persist, however, for Hispanic and African-American women, who traditionally have had less access to health care and cervical screening.[28] (*See graph, p. 313.*)

Requiring the shots is the only way to guarantee access to inoculations for girls at highest risk of developing cervical cancer, say mandate supporters. "The goal was really to ensure that all girls and women have access to this life-saving technology regardless of their socioeconomic status," says Indiana state Sen. Connie Lawson, R-Danville, board chair of Women in Government. "This is a cancer that can be eliminated as long as we

Kim Skeen holds her 2-month-old daughter as she tells California lawmakers on March 13, 2007, why she opposes a bill that would mandate vaccinations for young girls against the cervical cancer-causing human papillomavirus. Karen Holgate, left, of the California Family Council, also opposes the bill; Dr. Jenny Biller, an obstetrician and gynecologist at the University of California-Davis Medical Center, right, supports the measure.

continue the appropriate screening and testing and then the immunization."

Some girls only receive health care as a result of mandates, says Lawson. "Schools are the one place where all girls can be reached because many of these girls do not have a regular physician."

"We have a social and moral obligation to protect our daughters," said mandate sponsor Mary Cheh (D), a Washington, D.C., City Council member.[29]

"Vaccinating against HPV will save lives — period," agreed her cosponsor David Catania (I).

Providing the vaccine at younger ages helps ensure that females will receive it at the optimal time — before they begin sexual activity. Gardasil appears to have no effect on existing infections. Most HPV infections occur within two to five years after sexual activity begins, with peak infection rates occurring between ages 14 and 24.[30]

Moreover, mandating the vaccine makes it more likely the shots will be paid for, experts say. Government programs that pay for vaccines for low-income and uninsured patients only cover payments for children, leaving uninsured adults to face the high vaccination costs themselves.[31]

Gov. Perry believes the vaccine is an essential step for eradicating cervical cancer and is disappointed the Texas

legislature decided to delay that goal by rescinding his mandate order, a spokeswoman said. Democratic Texas state Rep. Jessica Farrar agrees. "Why would you not want to prevent cervical cancer?" she said.[32]

Did the quick push for HPV mandates harm public health?

In a statement praised by vaccine experts, the Association of Immunization Managers (AIM) said earlier this year that vaccine mandates should be "used sparingly, approached cautiously and considered only after an appropriate vaccine implementation period."

During that period, AIM says, "It is critical to ensure that the necessary elements are in place to support a school/child-care requirement." These include physician support for the vaccine, public acceptance and adequate vaccine supply. Also necessary are data to ensure vaccine safety. "Inappropriate application of mandates risks loss of support for immunization programs and reversal of policy and program gains," AIM added.[33]

Many public health experts say the move to mandate Garadasil has been premature. For most vaccines, the time between licensure and when they are required by states for school attendance is a matter of months, if not years. The chicken-pox vaccine, for example, was licensed in 1995. The first jurisdictions to make it mandatory did so in 1997.[34] The CDC recommended that infants receive a Hepatitis B vaccine in 1991. The first state to mandate the Hepatitis B vaccine for newborns was Massachusetts in 1993.[35]

But for the HPV vaccine, state legislators began proposing mandates weeks before the CDC had officially recommended it. The first mandate bill was introduced in Michigan in September 2006; the CDC recommendation became official this March.[36]

"I don't think with any vaccine you start out with a mandate," says Emory's Orenstein.

There are several reasons to take a go-slow approach with vaccine mandates, experts say. First, the public needs to be educated. This is especially true of the HPV vaccine, which targets an infection many people have never heard of. While millions of women get Pap smears, many don't know the test looks for an infection they may have acquired months previously that may cause cancer 10 years later.

In addition, it takes time for physicians to accept new vaccines and to set up appropriate systems for their administration and purchase. Insurers also need to work through their payment policies.

"When you think about mandates, you've got to make sure the system's in place, that the vaccine's available and there are places the kids can get the vaccine," says AAP's Bocchini.

Experts also are concerned that the controversy around the HPV vaccine could lead to expanding exemption policies for all shots, which could threaten immunization levels for other diseases. Parental concerns have led some states considering HPV mandates to propose more liberal exemption policies for the shot. In Virginia, for example, the HPV requirement was passed with its own unique opt-out clause. Under the provision, parents would not have to submit their exemption decision in writing, as is required for other mandatory immunizations.

"While I believe that this vaccine shows great promise for preventing cancer, I believe the decision to administer this vaccine should be made by the parents," said Democratic Virginia Gov. Tim Kaine.[37]

Special exemptions for certain vaccines "can create a slippery slope, where at the next legislative session somebody says, 'Well, why should the exemption be different for HPV — it should be [the same] for all vaccines,'" says Daniel Salmon, an associate professor in the Department of Epidemiology and Health Policy Research at the University of Florida. "I'm concerned the push for HPV is going to result in loosening the exemption law, and that could have a very profound impact" on vaccination rates. (Salmon sits on a Merck vaccine-policy board.)

States can soften vaccine requirements either by expanding the range of reasons accepted for claiming exemption and/or making the opt-out process easier. In some states, documentation is required from a physician or religious leader for approval of medical or religious exemptions. Others, like California, simply require a parent's signature on a form. "In such a case, the parent, by signing, may simply be pursuing the path of least resistance," writes Salmon. "It is certainly easier to sign your name than to make an appointment with a doctor or at a clinic, arrange to take the child to the clinic and spend time (and perhaps money) to have a child immunized."[38]

The 19 states with the toughest exemption requirements have fewer exemptions than states with less-demanding requirements, Salmon's research shows. Only 16 states reported denying exemption requests. "Children

with exemptions to school-immunization requirements have had higher rates of vaccine-preventable diseases and contributed to outbreaks of such diseases," he writes. "The risks of measles and pertussis in school-age children in the USA with non-medical exemption has been reported to be 22-35 times and 5.9 times higher, respectively, than those in vaccinated children."[39]

According to the Association of Immunization Managers, most states allow religious exemptions from childhood immunization, all states honor medical exemptions and 21 states allow philosophic or personal exemptions. "Opening existing law or regulation for this purpose risks having an effective mandate weakened, altered in an undesirable way or even revoked entirely," the group said.[40]

In response to criticism that mandates were pushed too quickly, Merck spokeswoman Allen counters that everything about the vaccine was fast — from initial FDA approval to the CDC's recommendations. The move to mandate was in keeping with that same spirit, she says. Nonetheless, the company discontinued its Gardasil lobbying campaign in February, largely in response to public reaction. "There was clearly a lot of focus on Merck's involvement," she says.

As to the controversy's effects on public health, Indiana state Rep. Lawson says, "It's unfortunate that the conversation has been diverted from this life-saving technology to some of the other issues that have arisen."

BACKGROUND

Sex and Cervical Cancer

The connection between cervical cancer and sex was first suspected in Italy in the 1840s. Physicians in Florence noticed that prostitutes and married women suffered the disease but not nuns, except for those who had sex before they entered the convent. Even more curious, physicians found surprisingly high rates of cervical cancer among second wives of men whose first wives died of the disease.[41]

In the 1920s, the Pap smear screen for cervical cancer was first developed by American physician George Papanicolaou, and by 1941 its efficacy had been proved. Papanicolaou began his research using a nasal speculum on female guinea pigs. He began studying human cervical scrapings in 1920 and soon was able to identify

CHRONOLOGY

1840s *Physicians in Florence, Italy, suspect link between sex and cervical cancer when they notice prostitutes and married women get the disease but not nuns.*

1920s-1930s *Pap smear research begins, later becoming one of the most effective cancer-screening technologies. American physician George Papanicoloau's research with vaginal smears shows detection of abnormal cervical cells.*

1940s-1950s *Improvements to Pap smear and further evidence of efficacy documented. Test starts to become more widely used.*

1941 Efficacy of Pap smears argued, first research published in the *American Journal of Obstetrics and Gynecology.*

1943 Pap smear enters into general use.

1980s-1990s *Links between HPV and cervical cancer established. Research begins to pinpoint the strains responsible for cancer. Vaccine development begins. First clinical trials conducted.*

1983 Compelling link between HPV and cervical cancer posited by German virologist Harald zur Hausen.

1993 Development of a vaccine against human papillomavirus begins.

1997 Trials begin for Merck's Gardasil, GSK's (GlaxoSmithKline) Cervarix.

1999 HPV proposed as a necessary factor for development of cervical cancer.

2000-Present *Vaccine development shows promise. Vaccine-approval efforts begin; legislative and regulatory processes are launched.*

2001 Gardasil Phase III studies start.

2002 Gates Foundation funds efforts at Seattle-based PATH (Program for Appropriate Technology in Health) to develop rapid HPV testing technology for cervical-cancer prevention in developing countries

2003 Women in Government begins cervical-cancer reduction campaign; Cervarix Phase III studies begin.

May 2005 Gates Foundation gives grants to PATH, World Health Organization to accelerate access to HPV vaccines in developing countries.

June 2005 Gates Foundation gives grant to University of Colorado, Denver, and Harvard to develop inexpensive HPV vaccines.

December 2005 Merck submits application for Gardasil to Food and Drug Administration (FDA) for fast-track approval.

June 2006 Gardasil licensed by FDA. . . . Association of Immunization Managers develops position statement on mandating vaccines. . . . Advisory Council on Immunization Practices (ACIP) recommends routine vaccination of females ages 11-12 and "catch-up" vaccinations of 13-26-year-olds.

September 2006 Michigan becomes first state to introduce HPV vaccine-mandate legislation. . . . Gardasil approved for sale and marketing in European Union.

November 2006 New Hampshire becomes first state to offer HPV vaccine free to all girls. . . . Merck launches national advertising campaign for Gardasil.

January 2007 Press reports link Merck, lawmakers.

February 2007 Gov. Rick Perry, R-Texas, orders HPV vaccine for sixth-grade girls. . . . Merck suspends state lobbying campaign. . . . Rep. Phil Gingrey, R-Ga., reintroduces Parental Right to Decide Protection Act.

March 2007 Centers for Disease Control and Prevention recommends HPV vaccine. . . . GSK files license application with FDA for Cervarix. . . . New Mexico legislature passes mandate bills.

April 2007 Virginia and Washington, D.C., require HPV vaccine for sixth-grade girls. . . . New Mexico governor vetoes mandate bill. . . . Texas legislature approves bill to rescind governor's executive order.

malignant cells.[42] Pap testing was improved in 1947 by J. Ernest Ayre, a Canadian gynecologist, who developed a way to directly test cervical tissue.[43]

But it wasn't until the 1980s that a compelling link between HPV and cervical cancer was posited by German virologist Harald zur Hausen. In 1983 he concluded that HPV 16 was present in 50 percent of cervical cancer cells.[44] "He and his colleagues really made the breakthrough," says National Cancer Institute (NCI) Principal Investigator John Schiller, one of the inventors named on a series of U.S. government patents on HPV vaccines licensed to Merck and GlaxoSmithKline.

The NCI and others built on zur Hausen's work. "The data became stronger and stronger," Schiller says. "Almost all the tumors have one type or another of these HPVs in them, and if you turn off the [HPV] genes, the tumor cells die."

Epidemiologists tried to confirm the results by looking for HPV DNA in pre-malignant lesions, but tests weren't sensitive enough. The subsequent development of the sophisticated polymerase chain-reaction method of reproducing DNA for study allowed more accurate tests.

"Suddenly the correlation was huge," Schiller says. "There are almost no cervical cancers that don't have HPV in them. If you never had an HPV infection, you'd have almost zero chance of getting cervical cancer."

Using these discoveries as a base, vaccine development began in 1993. It proved to be a challenge.

The sheer number of papillomaviruses, named for the papilla, or bud, created by the tumors, posed a major obstacle.[45] There are 100 strains known to infect humans, 40 of which are passed through sexual contact. Vaccines would have to target several strains to be effective, and cross immunization appeared to be minimal. Also, the virus does not grow well in cell cultures. Even if it did, live, attenuated (weakened) vaccines could contain cancer-causing viruses making them inappropriate for use as a preventive measure in healthy individuals.

Researchers eventually developed test vaccines using virus-like particles (VLPs) that produce a non-infectious shell of the virus that doesn't contain viral DNA.[46] These reverse Trojan horses fool the body into developing an immune response. VLPs proved effective in papillomavirus infections found in cottontail rabbits, cows and dogs. It was time to turn to humans.

Merck began clinical trials on a vaccine in the 1990s, according to company spokeswoman Allen. Initial trials

for Gardasil began in 1997. Phase III studies began in 2001. In December 2005, the company submitted its first license application to the FDA and requested fast-track approval. To qualify for such expedited review, a product must target a serious or life-threatening disease and address an unmet medical need. The FDA granted the request, and a process that normally takes 10 to 12 months took six.

In June 2006, the CDC's Advisory Committee on Immunization Practices recommended the vaccine be included on the current CDC childhood immunization schedule; the CDC concurred on March 12.[47]

Lobbying Missteps

The CDC develops official government recommendations regarding vaccines, but individual states decide whether or not to require those immunizations for school attendance. In the case of the HPV vaccine, state legislators — anxious to be in the forefront of an impressive anti-cancer technology — and following one of the most intense lobbying efforts by a vaccine manufacturer — rushed to sponsor HPV-vaccine-related legislation. Two bills were introduced in 2006. By April 2007, legislators in at least 39 states had introduced bills that would require, fund or educate the public about the HPV vaccine. At least 24 states and the District of Columbia considered legislation to make the vaccine mandatory.[48]

The quick move to legislate alienated some state public health officials. "School and child-care requirements for any vaccine must be pursued through existing state processes," said the Association of Immunization Managers. "Legislators, advocates, consumer groups, manufacturers and others interested in pursuing school and health-care immunization requirements should first contact the state health agency program." Most states, it continued, have an existing process for considering vaccine mandates designed to "ensure a thorough evaluation of all the relevant issues."[49]

Vanderbilt's Schaffner says it is unwise for pharmaceutical manufacturers to "kind of do an end-run around public health departments and go directly to legislators," because it "always causes confusion." And if the efforts are successful, he adds, it also makes the public health structure "really grumpy."

Many attribute the push to well-meaning lawmakers who may not fully understand the intricacies of

New Vaccine Could Aid Developing Countries

But high cost may limit distribution

Cervical cancer remains a major killer of women globally, even though the disease has declined dramatically in developed countries with the advent of reliable Pap smear testing.

Cervical cancer is the second-most-common cancer among women worldwide. There were an estimated 493,000 new cases in 2002 and 274,000 deaths. Eighty-three percent of the cases occurred in developing countries, where cervical cancer accounts for 15 percent of female cancers. The highest incidence rates are in sub-Saharan Africa, Melanesia, Latin America and the Caribbean. High rates also are found in south-central and southeast Asia.[1]

New HPV vaccines could have their biggest impact in these high-risk areas. But the poverty that limits Pap smear screening in the developing world — causing the disparities in global cervical cancer rates — will also affect vaccine distribution.

"In the industrialized world, while the cost will be higher, this vaccine will be affordable and welcomed by most health workers, adolescents and parents," writes American researcher Mark Kane.[2] "In the poorer countries in the developing world, however, it has taken two decades for new vaccines to become available in the public sector."

Besides the difficulty of financing the expensive HPV vaccine — which costs $360 for the three shots — many countries have a shortage of health facilities and workers. Moreover, the vaccine must be kept constantly cool to prevent spoilage. Reliable refrigeration is unavailable in many rural areas without electrical power or in urban areas that experience power outages. In addition, the vaccine is best administered to adolescent girls three times over a six-month period — logistically not an easy task in many countries.

The vaccine's connection to a sexually transmitted disease (STD) also could cause even more problems in other countries than it has here. "The age at which girls initiate sexual activity varies considerably between different countries and cultures," notes researcher Thomas Wright. "The average reported age of first sexual intercourse varies in [European Union] countries from 15 years for women in the Czech Republic to 20+ years in Italy. In Portugal, only 25 percent of 18-year-old women have been sexually active, whereas in Iceland 72 percent have been."[3]

In developing countries, the age for female sexual initiation is lower in Latin America, with a median age of 15, and higher in Africa and Asia, with a median age of 18-20, according to a World Health Organization report.[4]

To overcome some of these obstacles, researchers at the National Cancer Institute, in academia and in

immunization policy. Others say connections with Merck tainted the effort and call into question the legislators' motivation. Indeed, a member of Women in Government reportedly quit the group over concern about the Merck connection.[50]

Indiana state Sen. Lawson, the group's chairman, acknowledges Merck gives the group "unrestricted grants" similar to those they receive from other corporate sponsors. But the funds did not influence members' decisions to sponsor mandate legislation, she insists. The group's campaign to reduce cervical cancer started in 2003. Although they were aware of clinical trials for Gardasil, she says, "When we first began on this issue, we had no idea the vaccine was going to be approved."

Some analysts attribute Merck's hastiness to financial interest. Since GlaxoSmithKline could have its own HPV vaccine on the market by next year, Merck wants to capture as much of the market as it can while it's the only game in town, they say.

While most experts think the controversy ultimately will die down and the vaccine will be widely used, the missteps could delay its acceptance.

"They've muddied the waters with their clumsiness," says Peter Lurie, of Public Citizen's Health Research Group. "There's no question [Merck has been] aggressive about this." Drugmakers "have too much influence [over legislators], and in this case they overplayed it. The merits of the vaccine would have sold themselves."

Bocchini of the American Academy of Pediatrics agrees. "Lobbying for mandates was wrong because it raised these other issues that ended up being very controversial and could potentially hurt the vaccine program," he says. "The private industry are working to develop a vaccine that doesn't need to be kept cold — perhaps an inhaled form, according to Principal Investigator John Schiller of the National Cancer Institute.[5] In the meantime, global health groups are studying the best methods for disseminating the vaccine in developing countries. Toward that end, PATH, a nonprofit U.S. health organization, is conducting research in India, Peru, Uganda and Vietnam under a five-year $27.8 million grant from the Bill & Melinda Gates Foundation.[6]

The Seattle-based foundation also is funding research at the University of Colorado and Harvard on inexpensive HPV vaccines.[7]

Schiller says independent production in industrializing countries like India may be the only way to reduce the price to an affordable level and ensure the vaccine's worldwide distribution.

"If the only women who get it are women who are destined to be Pap screened for cervical cancer anyway, it's not going to have that big an impact on public health, especially for the cost it's going to be," he says. "None of us are going to be satisfied if the only people who get the vaccine are women who already are most protected from cervical cancer because we have screening — that's not why we developed the vaccine."

Bill and Melinda Gates are funding research into the best way to supply HPV vaccine in developing countries.

[1] Schiller is one of several inventors named on a series of U.S. government patents on HPV vaccines licensed to Merck and GlaxoSmithKline (GSK).

[2] Maxwell Parkin, *et al.*, "The Burden of HPV-Related Cancers," in *Vaccine: HPV Vaccines and Screening in the Prevention of Cervical Cancer* (2006).

[3] Mark Kane, *et al.*, "HPV Vaccine Use in the Developing World," in *ibid*.

[4] Thomas Wright, *et al.*, "HPV Vaccines and Screening in the Prevention of Cervical Cancer; Conclusions from a 2006 Workshop of International Experts," in *ibid*.

[5] World Health Organization, "Sexual Relations Among Young People in Developing Countries," 2001.

[6] PATH press release, June 5, 2006.

[7] Bill & Melinda Gates Foundation, http://gatesfoundation.org.

decision for mandates really should be a public health issue. It was an error to go to the legislature."

Vaccine Payments

When states mandate a vaccine, they must also address funding issues, including whether to require government insurers to cover the vaccine. Funding aid may be key for Gardasil, considered one of the most expensive vaccines ever. The retail cost of the three-shot inoculation regime is $120 per dose, or $360. The federal Vaccines for Children fund negotiated a discount and pays $96 per dose, according to Merck. While experts attribute most of the high cost to company efforts to recover development costs, they also say Gardasil is highly purified and expensive to make.

The CDC recommendations cleared Gardasil's inclusion in the Vaccines for Children fund, which pays for 41 percent of childhood immunizations — those for children who are Medicaid-eligible or have insufficient insurance coverage or are covered by American Indian or Alaska Native health programs.[51] The federal government also helps pay for immunizations through the Immunization Grant Program, Medicaid and the State Children's Health Insurance Program.

Most private- and public-sector insurers will cover them, according to the Kaiser Family Foundation, "but policies about who will be covered and what amount will be paid are still being determined." Most girls in the target age group have private insurance, according to the foundation, but 12 percent of girls ages 9 to 18 — and 29 percent of women 19 to 26 — are uninsured.[52] Uninsured adults have no public coverage for vaccines, but Merck has established a vaccine assistance program for low-income uninsured women ages 19 to 26.

The HPV vaccine was added to the National Vaccine Injury Compensation Program (VICP) on Feb. 1, 2007. Established by Congress in 1986, the program limits the liability of manufacturers of mandated vaccines and health practitioners who administer them. It requires the parents of a child allegedly harmed by a childhood vaccine to first seek redress from this federal compensation system before filing an action in a regular court.

To be included in the program, a vaccine must be a routine childhood immunization recommended by the Centers for Disease Control. It also must be approved by Congress as a taxable vaccine for purposes of the Vaccine Injury Compensation Trust Fund. Even though inclusion in the program is limited to childhood vaccines, individuals of all ages can file a claim with the fund, a spokesman says. VICP is a no-fault alternative to the traditional tort system for resolving vaccine injury claims. It provides compensation to people found to be injured by eligible vaccines.

Attacking STDs

Gardasil is the second vaccine approved for an STD associated with cancer. The first was the Hepatitis B vaccine, licensed in 1986 and recommended for use in infants in 1991. Although the disease is primarily transmitted by sex or intravenous drug use (through the use of dirty needles), the District of Columbia and 47 states now require the vaccine for children.[53] Introduction of the Hepatitis B vaccine, however, didn't provoke the same resistance as the HPV vaccine, even though it is given at younger ages.

Public health experts say that how a virus is transmitted is less an issue for most parents than a vaccine's safety and efficacy. According to the American Academy of Pediatrics' Bocchini, a recent survey shows that for most parents, "whether it's a sexually transmitted disease is not as important as whether it's a significant infection." Once parents are educated about HPV and cervical cancer, he says, "the vast majority is going to choose to give their daughter this vaccine, and young women will choose to take the vaccine."

Optimism is high about the HPV vaccine's potential to reduce cervical-cancer rates, but regular Pap screening is still recommended for sexually active women and all women over 21, according to the American Cancer Society

and The American College of Obstetricians and Gynecologists.[54] Testing should begin within three years after the onset of sexual activity. Women should have a test every year until age 30. Women over 30 or with three normal tests may be screened every two to three years.[55]

The Pap smear has been one of the most effective cancer-screening devices ever. Cervical cancer rates drop by up to 75 percent in countries with solid Pap testing programs, studies show. The single most predictive factor for development of cervical cancer is lack of regular Pap smears.[56]

While 82 percent of women report having a Pap smear in the last three years, those numbers are lower for women with less than a high-school education, women without health insurance and certain racial/ethnic populations including Hispanics and Asians, according to the Center for Disease Control and Prevention.[57]

"We now have a vaccine against a cancer that, ironically, we probably have the best public-health intervention for already," the National Cancer Institute's Schiller says referring to the Pap smear. But he adds, the vaccine and screening "should be considered in tandem."

The widely used screening tests detect abnormal cervical cells that can lead to cancer. Of the 55 million Pap tests performed each year in the United States, about 6 percent are abnormal and require medical follow-up at a cost estimated at $3 billion annually.[58] Follow-up can include additional Pap testing or further examination using a colposcopy and possible removal of abnormal cells. Cells can be excised or destroyed using electrical current, freezing or laser therapy.[59]

CURRENT SITUATION

State Action

The Michigan Senate introduced the first bill to mandate the vaccine — for girls entering sixth grade — in September 2006, according to the National Conference of State Legislatures (NCSL). The bill was not enacted. Since then, legislators in at least 39 states and the District of Columbia have introduced measures that would require, fund or educate the public about the vaccine. So far this year, bills have been introduced in 24 states and the District that would make the vaccine mandatory.[60]

In addition, some state health departments are offering HPV vaccine-financing programs. New Hampshire, for example, announced in late 2006 it will provide the vaccine at no cost to all girls under 18.[61]

Washington, D.C., and Virginia were the first jurisdictions to mandate the vaccine. The District of Columbia has the highest rate of cervical cancer in the country — 13.3 cases per 100,000 women compared to the U.S. rate of 8.1 per 100,000. The city council approved a measure on April 19 requiring girls entering the sixth grade to receive the inoculation starting in fall 2009.[62] Parents can choose to exempt their children for any reason. Virginia, which has one of the nation's lowest rates of cervical cancer — 6.3 cases per 100,000 women — adopted a mandate on April 11, which also has a liberal exemption provision.

"Because the human papillomavirus is not communicable in a school setting, a parent or guardian, at the parent or guardian's sole discretion, may elect for their child not to receive the human papillomavirus vaccine, after having reviewed materials describing the link between the human papillomavirus and cervical cancer . . ." the Virginia law says.[63] The law, which won't be implemented until Oct. 1, 2008, requires girls to receive all three doses of the vaccine, the first before they enter sixth grade.[64]

Texas was the first state to enact a mandate — through an executive order signed by Gov. Perry on Feb. 2, 2007. It required all girls entering the sixth grade to receive the HPV vaccine. Shots for low-income girls were to be subsidized by the state and parents were allowed to waive the requirement.[65]

But two months later, the Texas legislature rescinded the mandate. The action prohibits schools from mandating the HPV vaccine but requires distribution of information about the inoculation to parents or legal guardians at the appropriate time in the immunization schedule. Texas has the second-highest incidence of cervical cancer in the country, 10.1 cases per 100,000 women, and also the second-highest cervical cancer death rate, 2.9 deaths per 100,000. The U.S. rate is 2.4 deaths per 100,000.[66]

New Mexico lawmakers also enacted an HPV vaccine requirement for girls between 9 and 14, with a parental opt-out. After initially saying he would support the bill, Gov. Richardson — who is seeking the Democratic nomination for president — vetoed it. "While everyone recognizes the benefits of this vaccine, there is insufficient time to educate parents, schools and health-care providers," he said.[67]

Richardson instead signed a bill that requires insurance plans in the state to cover the FDA-approved HPV

Texas cancer survivor Cheryl Swope Lieck, with her daughters Allison, 12, and Cameron, 8, supports Gov. Rick Perry's executive order requiring sixth-grade girls to be vaccinated against the virus that causes most cases of cervical cancer. In April, the legislature rescinded the order.

inoculations for girls 9 to 14, at existing deductibles and coinsurance. The states also enacted a bill to create a panel to study cervical-cancer disparities among the states. New Mexico's cervical-cancer incidence rate is 6.9 cases per 100,000 women and its mortality rate is 1.5 per 100,000.

As of April 27, 2007, a total of 93 bills related to HPV vaccine had been introduced in state legislatures. Ten measures call for funding the immunizations. Two have passed state legislatures, in Colorado and in South Dakota. The Colorado law allocates 4 percent of state tobacco-settlement money to a cervical cancer-immunization fund. South Dakota gives the Department of Health $9.2 million to offer the HPV vaccine to women 11 to 18.

An additional 13 bills were introduced that would fund insurance coverage of the vaccine. New Mexico's was signed into law. Education related to HPV and cervical cancer was the issue in another 23 bills. Utah passed legislation establishing a campaign to raise awareness of the causes, prevention and risks of cervical cancer. Indiana, North Dakota and Washington state all authorized HPV education programs.

Two legislatures, Maryland and New Mexico, passed bills to establish a task force or advisory board related to HPV and cervical cancer. Maryland and

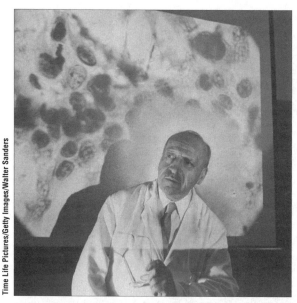

Greek-born American physician George N. Papanicolaou began developing his famed Pap smear test for cervical cancer in the 1920s, using female guinea pigs. By 1941 its efficacy had been proved. Slides of cancer cells are projected in the background.

California saw mandate legislation withdrawn after initial introduction.[68]

Congressional Action

On the federal level, Rep. Phil Gingrey, a Georgia Republican who is also an obstetrician/gynecologist, introduced a bill this year designed to discourage states from mandating the HPV vaccine. With 27 cosponsors, the Parental Right to Decide Protection Act would prohibit use of federal funds by any state or locality to implement an HPV vaccine mandate.

"Federal funds should not be used to implement a mandatory vaccine program for a disease that does not threaten the public health of schoolchildren in the course of casual, daily interaction between classmates and inserts the government into the lives of children, parents and physicians," the bill says.

The legislation would not prohibit federal funds from supporting voluntary vaccine programs, and any state with an optional program could use Medicaid and education money to provide vaccinations. A spokeswoman for Gingrey said the congressman is hoping to attach the bill to fiscal 2008 Health and Human Services Department funding legislation.

While welcoming the development of an HPV vaccine, Gingrey said, parents should oversee its administration. "As an ob-gyn physician, I understand the importance of protecting Americans from sexually transmitted diseases, and I applaud the development of an HPV vaccine," he said in a statement. "States should require vaccinations for communicable disease, like measles and the mumps. But you can't catch HPV if an infected schoolmate coughs on you or shares your juice box at lunch. Whether or not girls get vaccinated against HPV is a decision for parents and physicians, not state governments."[69]

FDA Action, Clinical Trials

The Food and Drug Administration is considering whether to license Cervarix, the HPV vaccine made by the British company GlaxoSmithKline (GSK), which submitted its application on March 29. Cervarix targets HPV 16 and 18, the two strains most prevalent in cervical cancer tumors. GSK is conducting worldwide clinical trials involving nearly 30,000 females ages 10 to 55.[70] If licensure is granted, an action that could take 10 to 12 months, GSK will then present its data to the ACIP.

GSK's vaccine differs from Merck's because it targets only two HPV strains and uses a different adjuvant, or substance that improves and extends the effects of a vaccine. GSK claims that its adjuvant, composed of aluminum salt and monophosphoryl lipid A, is more effective than Merck's, which is aluminum.

In its recommendations for Gardasil, the CDC called for further studies in several areas, including how long the protection lasts, surveillance of rates of cervical cancer precursors and genital warts and vaccination safety.[71]

More than 80 HPV-related clinical trials are ongoing, — 40 for Gardasil and 22 for Cervarix — according to a government registry.[72] They range from studies of the effects of green-tea extract in preventing cervical cancer in HPV-positive women to the effects of male circumcision on transmission of STDs, including HPV. Several studies compare Gardasil and Cervarix. Others look at safety and effectiveness and the effects of the use of HPV vaccines with other immunizations, including those for diphtheria, tetanus and pertussis.[73]

HPV vaccine safety also is being monitored through the government's Vaccine Adverse Events Reporting System (VAERS) and CDC's Vaccine Safety Datalink. VAERS collects data submitted by pharmaceutical manufacturers,

AT ISSUE

Will the HPV vaccine promote promiscuity?

YES Janice Shaw Crouse, Ph.D.
Director and Senior Fellow, Beverly LaHaye Institute, Concerned Women for America

Written for *CQ Researcher*, May 2007

The short answer is, "Yes, of course, there will be people who will use the HPV vaccine as protection from their promiscuity." Inevitably, there are people who abuse and/or misuse technological and scientific advances. While such misuse is regrettable, the HPV vaccine is a major step forward in protecting women from cervical cancer.

While there is nothing inherently wrong with the vaccine, Concerned Women for America opposes mandating the vaccine. As with any new development, misinformation is widespread, and many questions remain.

If the HPV vaccine is viewed as a magic bullet that will provide absolute protection against a major sexually transmitted disease, some parents will shirk the responsibility to teach their daughters about making good decisions. While, the physical aspect of sexual activity and possible health consequences are very important concerns, the emotional and psychological aspects are also significant.

If parents are only concerned about health consequences, they might fail to address the other substantive concerns about their adolescent's potential early sexual activity. We warn parents against the false security that they might feel from the vaccine. More important, we encourage parents to use the vaccine as a springboard for appropriate discussions with their teens about the benefits of abstinence and the freedom that comes from waiting to have sex until after marriage.

There is still much that we don't know about the vaccine's limitations. We know that there have been some negative reactions. There are reports of headaches and dizziness; some users have temporarily lost vision, others have lost consciousness and some have had seizures.

Some parents are concerned about the increased number of vaccines that are recommended for children today. In addition, parents need to recognize that cancer screening through annual Pap smears has been highly successful in identifying cervical conditions that might lead to cancer.

The availability of the HPV vaccine brings another dimension to parental responsibility toward their preteens and teenagers. The potential for using the supposed safety of the vaccine as an excuse for promiscuity is certainly possible. Parental involvement in preparing their teens for good decision-making will be much more important in a climate where the HPV vaccine is available. No parent wants to see a teen become promiscuous; certainly, conscientious parents will want to warn their children that early sexual activity is harmful, even if there is protection against cervical cancer.

NO Arthur L. Caplan
Chairman, Department of Medical Ethics, University of Pennsylvania

Written for *CQ Researcher*, May 2007

When ideology is allowed to trump the facts, women die. Recently, a major study showed that abstinence-only sex education had absolutely no impact on the sexual behavior of the students taking these classes. Harry Wilson, associate commissioner of the Family and Youth Services Bureau in the Department of Health and Human Services, offered this assessment of the finding: "These interventions are not like vaccines. You can't expect one dose to be protective all throughout the youth's high school career."

Put aside the fact that the study was the latest of many studies of state programs showing that abstinence-only sex education has no impact whatsoever on sexual behavior and pregnancy prevention. Ignore the fact that the kids in the most recent study had a minimum of 50 hours of abstinence-only sex ed.

What Commissioner Wilson failed to grasp was that his blind allegiance to abstinence-only sex-education in the face of a mountain of data that it does not work is going to kill American girls and women. And it is precisely because Wilson, the Bush administration and Congress are not using a vaccine but rather continuing to rely on a blind faith in abstinence and chastity that those girls and women will die.

The new cervical cancer vaccine made by Merck, Gardasil, is 100 percent effective in preventing infection with the two strains of papillomavirus that cause 70 percent of the cervical cancer cases in the United States. The virus is transmitted by sexual intercourse and other forms of sexual contact. The Food and Drug Administration and the Centers for Disease Control and Prevention's Advisory Committee on Immunization have signed off on the HPV vaccine, noting that it is incredibly safe and effective. So why isn't every young woman in America lining up to get this vaccine? Wilson and those who continue to believe that abstinence and chastity until marriage are sufficient answers to the danger of cervical cancer are directly to blame.

Last year nearly 4,000 women died in the United States from cervical cancer. Many of them were poor, had a job with no health insurance or lacked access to good gynecological care. Ten thousand other women were told they needed surgery and chemotherapy to battle cervical cancer. Every one of those cases was preventable.

There is no evidence that just telling young boys and girls not to have sex will stop them from doing so or that offering a vaccine will trigger sexual activity. But there is irrefutable evidence that giving young women the HPV vaccine will save thousands of lives and prevent a great deal of suffering.

> "The HPV vaccine has really re-energized people's thinking about STD vaccines. Before this vaccine, no STD vaccine has worked at all or very, very poorly"
>
> — *John Schiller,*
> *Principal Investigator*
> *National Cancer Institute*

health-care providers, state immunization programs and vaccine recipients or their parents.

OUTLOOK

Years Away?

Every year in the U.S., 3.5 million to 5 million Pap tests show abnormal results. Follow-up on those tests, including repeat Paps, other tests, cell destruction or removal and — in the worse cases — cancer treatment costs $3 billion annually. The HPV vaccine will help reduce both those costs and the angst associated with cervical abnormalities.

"All that downstream activity will be reduced ideally by 70 percent," said Emory University's Schaffner. "Now, in the real world, it will be somewhat less than 70 percent, but that's okay. We'll reduce it by a whole lot."

Of course, that scenario assumes near universal immunization with the new vaccine, which may be years off. But vaccine experts expect the current debate to die down and acceptance to increase.

"I don't think it will be a long, drawn-out debate," says Johns Hopkins University bioethicist James Hodge. He predicts mandates to be enacted by a few key states, followed by others, although maybe with a more liberal opt-out policy than is available for other vaccines.

The National Cancer Institute's Schiller agrees. "There's going to be a burst of activity with the public, and then it's going to be like Hepatitis B and become routine," he says. Physician acceptance will be key, he added. "If the physician recommends this vaccine, parents won't even think twice."

Even when the vaccine becomes widespread, empirical reductions in cervical cancer will take 10 to 30 years to materialize, experts say.

Experts expect at least one more HPV vaccine, GSK's Cervarix, to be approved next year. But being less comprehensive, the GSK vaccine may cost less, potentially forcing Merck to lower the price for Gardasil. Research

also could continue to address other strains of HPV, but that process could take years.

Despite the controversy, successful licensure of an HPV vaccine is having a positive effect on other STD-vaccine activities. "The HPV vaccine has really re-energized people's thinking about STD vaccines," Schiller said. "Before this vaccine, no STD vaccine has worked at all or very, very poorly."

The success has created "a sea change in the way people think about the potential of STD vaccines," he says. "Now people really think they can work." Although still years away, progress is being made for vaccines addressing both chlamydia and herpes simplex. National Institutes of Health researcher Harlan Caldwell says his work on a chlamydia vaccine is moving to non-human primate trials and then, he hopes, on to Phase I trials for humans.

"We're mid-river, if you will, but we can see the other side, and we're very encouraged by that," he says. "I can't tell you one year, two years, but it's going the right way." Chlamydia — the most frequently reported sexually transmitted bacterial infection in the United States — can cause pelvic inflammatory disease, which can lead to infertility.

The controversy over HPV's sexually transmitted nature has not deterred Caldwell from his work in the basic science needed to develop a chlamydia vaccine, he says. "If I'm successful, then at some point in time it will be used for public health reasons and to develop vaccines to prevent chlamydia STDs and, hopefully, blinding trachoma. That's my charge, and I leave the rest of it to others," he says.

Herpes researcher David Knipe of Harvard says a biotechnology company is now developing a herpes-simplex product for use in Phase I clinical trials. He expects trials to begin in about two years. Herpes is an incurable, chronic infection that causes genital ulcers and makes sufferers more prone to HIV infection. About 145 million Americans have herpes.

The success of Gardasil and other recently licensed immunizations, including those for meningococcal disease and rotavirus, also could reinvigorate drug company interest in vaccines. Vaccine development had been abandoned by many companies over the years as a result of low profit margins and liability concerns.

"Vaccines used to be a marginal thing, but now with these new vaccines — these are blockbuster drugs," Schiller says. "Because of that, I think companies are looking again at getting back into the vaccine world."

NOTES

1. Centers for Disease Control and Prevention (CDC), press release, June 29, 2006.

2. CDC, "Quadrivalent Human Papillomavirus Recommendations of the Advisory Committee on Immunization Practices (ACIP)," *Morbidity and Morality Weekly Report*, March 12, 2007.

3. Department of Health and Human Services, Public Health Service, National Toxicology Program, 11th Report on Carcinogens, January, 2005.

4. CDC, *op. cit.*, March 12, 2007.

5. *Ibid.*

6. Kate Folmar, "Sexual Overtone Raised in Bid to Protect Girls," *San Jose Mercury News*, Jan. 21, 2007.

7. Sandra Boodman, "Who Gets Stuck?" *The Washington Post*, May 1, 2007.

8. John Carreyrou, "Questions on Efficacy Cloud a Cancer Vaccine," *The Wall Street Journal*, April 16, 2007.

9. Liz Peterson, "States Asked to Require HPV Vaccine for Schoolgirls," The Associated Press Jan. 30, 2007.

10. National Conference of State Legislatures, "HPV Vaccine Update," April 27, 2007.

11. "Actions Taken on HPV Vaccine Proposals in N.M., Utah, Washington, D.C.," www.KaiserNetwork.org.

12. National Conference of State Legislatures, *op. cit.*

13. Association of Immunization Managers "Position Statement: School and Child Care Immunization Requirements," June 2006.

14. Boodman, *op. cit.*

15. Jonathan Rockoff, "New Life Seen For Vaccine Industry," *Chicago Tribune*, June 6, 2006.

16. For background, see Kathy Koch, "Vaccine Controversies," *CQ Researcher*, Aug. 25, 2000, pp. 641-672.

17. Jamie Stengle, "Cancer Survivors, Doctors Say Politics Clouds HPV Vaccine Debate," The Associated Press, March 5, 2007.

18. CDC, *op. cit.*, March 12, 2007.

19. *Ibid.*

20. Dianne Finch, "HPV Vaccine Not Fully Tested on Young Girls Says Dartmouth Researcher," National Public Radio April 20, 2007.

21. CDC, *op. cit.*, March 12, 2007.

22. *Ibid.*

23. George F Sawaya and Karen Smith-McCune, "HPV Vaccination-Answers, More Questions," *The New England Journal of Medicine*, May 10, 2007.

24. James Colgrove, "The Ethics and Politics of Compulsory HPV Vaccination," *The New England Journal of Medicine*, December 2006.

25. Courtland Milloy, "District's HPV Proposal Tinged with Ugly Assumptions," *The Washington Post*, Jan. 10, 2007.

26. Moira Gaul, "The HPV Vaccine and School Mandates: Questions and Answers," Family Research Council, February 2007; www.frc.org/get.cfm?i=IF07B01&v=PRINT.

27. Mona Saraiya, "Cervical Cancer Incidence in a Prevaccine Era in the United States, 1998-2002," *Obstetrics and Gynecology*, Vol. 109, No. 2, pp. 360-370.

28. CDC, *op. cit.*, March 12, 2007.

29. Nikita Stewart, "Council Votes for Girls' Vaccination," *The Washington Post*, April 20, 2006.

30. CDC, "Human Papillomavirus: HPV Information for Clinicians," November 2006.

31. Kaiser Family Foundation, "HPV Vaccine: Implementation and Financing Policy," January 2007.

32. Quoted in Corrie MacLaggan and Selby W. Gardner, "Bill to Halt HPV Mandate Heads to Governor's Desk," *Austin American-Statesman*, April 26, 2007.

33. Association of Immunization Managers, "Position Statement: School and Child Care Immunization Requirements," June 2006.

34. Immunization Action Coalition, "Varicella Prevention Mandates," www.immunize.org.

35. Immunization Action Coalition, "Hepatitis B Prevention Mandates," www.immunize.org

36. National Conference of State Legislatures, *op. cit.*

37. Gov. Tim Kaine, press release, March 26, 2006.

38. Daniel Salmon, "Mandatory Immunization Laws and the Role of Medical, Religious and Philosophical Exemptions," www.vaccinesafety.edu.

39. Daniel Salmon, *et al.*, "Compulsory Vaccination and Conscientious or Philosophical Exemptions: Past, Present and Future," *The Lancet*, Feb. 4, 2006.

40. Association of Immunization Managers, *op. cit.*

41. Donald G. McNeil. "How a Vaccine Search Ended in Triumph," *The New York Times*, Aug. 29, 2006.

42. Stylianos P. Michalas, "The Pap test: George N. Papanicolaou (1883-1962) A Screening Test for the Prevention of Cancer of the Uterine Cervix," *European Journal of Obstetrics & Gynecology and Reproductive Biology*, (90) 2000.

43. Sharon Brown, "Pap Smears Can Prevent Cervical Cancer," *US Pharmacist,* September 2002.

44. Peter McIntyre, "Finding the Viral Link: the Story of Harald zur Hausen," *Cancer World*, July-August 2005.

45. McNeil, *op. cit.*

46. John Schiller and Philip Davies, "Delivering on the Promise: HPV Vaccines and Cervical Cancer," *Nature Review Microbiology*, April 2004.

47. CDC, *op. cit.*, March 12, 2007.

48. National Conference of State Legislatures, *op. cit.*

49. Association of Immunization Managers, *op. cit.*

50. Carreyrou, *op. cit.*

51. Kaiser Family Foundation, *op. cit.*

52. *Ibid.*

53. Immunization Action Coalition, Hepatitis B Prevention Mandates, *op. cit.* Also see "Critics Blame Hepatitis Vaccine for Injuries; But Health Officials Say It's Safe," in Koch, *op. cit.*, pp. 646-47.

54. CDC, *op. cit.*, March 12, 2007.

55. *Ibid.*

56. National Cancer Institute, "Fact Sheet, The Pap Test: Questions and Answers," March 2007.

57. CDC, *op. cit.*, March 12, 2007.

58. Kaiser Family Foundation, *op. cit.*

59. CDC, *op. cit.*, November 2006.

60. National Conference of State Legislatures, *op. cit.*

61. *Ibid.*

62. Stewart, *op. cit.*

63. National Conference of State Legislatures, *op. cit.*

64. *Ibid.*

65. *Ibid.*

66. Kaiser Family Foundation, statehealthfacts.org.

67. Daniel Vock, "Quick Cancer Mandate Raises Health Concerns," www.Stateline.org.

68. National Conference of State Legislatures, *op. cit.*

69. Rep. Phil Gingrey, press release, Feb. 16, 2007.

70. GlaxoSmithKline, press release, March 29, 2007.

71. CDC, *op. cit.*, March 12, 2007.

72. http://clinicaltrials.gov.

73. *Ibid.*

BIBLIOGRAPHY

Books

Allen, Arthur, *Vaccine: The Controversial Story of Medicine's Greatest Lifesaver*, W.W. Norton, 2007.
Journalist Allen documents the amazing and sordid history of vaccines and explores controversies past and present.

Henderson, Gregory S., MD, PhD, and Batya Swift Yasgur, MA, MSA, *Women at Risk: The HPV Epidemic and Your Cervical Health*, Penguin Putnam, 2002.
A specialist in cancers of the reproductive system discusses HPV and its related disorders.

Link, Kurt, MD, *The Vaccine Controversy: The History, Use and Safety of Vaccinations*, Praeger, 2005.
A retired internist discusses common vaccines as well as the immune system and infection and vaccine history.

Palefsky, Joel, MD, with Jody Handley, *What Your Doctor May Not Tell you About HPV and Abnormal Pap Smears*, Warner Books, 2002.
Palefsky, a professor of medicine at the University of California-San Francisco who is a leading expert on HPV, gives a comprehensive overview of HPV and its effects.

Articles

Bosh, F.X., *et al.*, *Vaccine: HPV Vaccines and Screening in the Prevention of Cervical Cancer*, Elsevier, 2006.

Experts from around the world contributed to this collection of 30 articles on scientific aspects of HPV and vaccines.

Carreyrou, John, "Questions on Efficacy Cloud a Cancer Vaccine," *The Wall Street Journal*, April 16, 2007.
The author quotes researchers skeptical about Gardasil's efficacy and details Merck's lobbying efforts.

McNeil, Donald G., "How a Vaccine Search Ended in Triumph," *The New York Times*, Aug. 29, 2006.
McNeil describes the discovery of HPV's connection to cancer and the vaccine-development process.

Orenstein, Walter, *et al.*, "Immunizations in the United States: Success, Structure, and Stress," *Health Affairs*, May/June 2005.
Orenstein discusses the successes and challenges of the U.S. immunization system.

Peterson, Liz Austin, "States Asked to Require HPV Vaccine for Schoolgirls," *The Associated Press*, Jan. 30, 2007.
Peterson describes Merck's efforts to encourage state adoption of Gardasil mandates.

Rawson, Kate, "The HPV Race: Merck Breaks Ground; GSK Tries to Make Up Ground," *RPM Report*, December 2005.
Rawson outlines business issues and development processes for Merck's Gardasil and GlaxoSmithKline's Cervarix.

Vilos, George A., MD, "After Office Hours: The History of the Papanicolaou Smear and the Odyssey of George and Andromache Papanicolaou," *Obstetrics and Gynecology*, March 1998.
This biography of Pap smear inventor Papanicolaou and his wife details how he discovered that abnormal cervical cells can be detected through vaginal smears.

Walboomers, Jan, *et al.*, "Human Papillomavirus is a Necessary Cause of Invasive Cervical Cancer Worldwide," *Journal of Pathology*, 1999.
The article reports that nearly all cervical cancers are likely caused by HPV.

zur Hausen, Harald, "Viruses in Human Cancers," *Science*, Nov. 22, 1991.
The German virologist credited with connecting HPV to cervical cancer discusses other virus-cancer links.

Reports and Studies

"Human Papillomavirus: HPV Information for Clinicians," *Centers for Disease Control and Prevention*, November 2006.
The CDC discusses transmission, prevention, detection and clinical management of HPV infection.

"Quadrivalent Human Papillomavirus Vaccine: Recommendations of the Advisory Committee on Immunization Practices," *Morbidity and Mortality Weekly Report*, Centers for Disease Control and Prevention, March 12, 2007.
The CDC's official recommendation regarding HPV vaccine includes detailed information on HPV transmission and the vaccine itself.

***Association of Immunization Managers, Position Statement: School and Child Care Immunization Requirements*, June 2006.**
Group of state immununization program managers' statement on the process for mandating vaccinations.

Dunne, Eileen, MD, MPH, *et al.*, "Prevalence of HPV Infection Among Females in the United States," *Journal of the American Medical Association*, Feb. 28, 2007.
Dunne and her colleagues provide the latest data on prevalence of HPV in the United States, broken down by age and strains.

For More Information

American Cancer Society, 1599 Clifton Rd., Atlanta, GA 30329; 1-800-227-2345; www.cancer.org. Nationwide, community-based health organization dedicated to eliminating cancer as a major health problem through research, advocacy, education and service.

Centers for Disease Control and Prevention, 1600 Clifton Rd., Atlanta, GA 30333; (404) 639-3311; www.cdc.gov. Agency of the Department of Health and Human Services promoting health and the quality of life by preventing and controlling disease, injury and disability.

Concerned Women for America, 1015 15th St., N.W., Suite 1100, Washington, DC 20005; (202) 488-7000; www.cwfa.org. Coalition of conservative women that promotes the incorporation of biblical values and family traditions into public policy.

Family Research Council, 801 G St., N.W., Washington, DC 20001; (202) 393-2100; www.frc.org. Christian organization formulating public policy that upholds the institutions of marriage and the family; opposes mandating HPV vaccine for young girls.

National Vaccine Information Center, 204 Mill St., Suite B1, Vienna, VA 22180; (703) 938-0342; www.909shot .com. Nonprofit educational organization advocating safety and informed-consent protections in the mass-vaccination system.

Women in Government, 2600 Virginia Ave., N.W., Suite 709, Washington, DC 20037; (202) 333-0825; www .womeningovernment.org. An association of female state legislators that seeks solutions to complex public-policy issues.

14

Declining Birthrates

Will the Trend Worsen Global Economic Woes?

Sarah Glazer

Sonja Hackethal, a management consultant in Frankfurt, Germany, says a new-child cash bonus from the government influenced her decision to have more children; her twins are due in February. Germany is among many countries with birthrates far below the level needed to prevent population decline. Europe's principal problem — aging populations supported by shrinking proportions of young workers — also threatens the economies of Japan, China and, to a lesser extent, the United States.

From *CQ Researcher*, November 21, 2008.

Courtesy Sonja Hackethal

Sonja Hackethal, a management consultant and mother of two from Frankfurt, is just the kind of university-educated woman the German government is counting on.

Almost half the women in Germany with university degrees remain childless (compared to about 30 percent of all German women). Moreover, the country has one of the lowest birthrates in Europe — far below the level needed to prevent population decline.

In a bid to raise the birthrate among highly paid career women like Hackethal, the government last year started offering to pay women two-thirds of their salary — paid monthly for up to a year — if they stayed home with their child.[1]

The 37-year-old Hackethal, now pregnant with twins due in February, says the cash bonus influenced her decision to get pregnant again because it allows her to take more time off from her job than she did for her second child.

"When I started out, I thought my career was so important that I would have no children at all because it was so impossible," she says, alluding to Germany's notorious scarcity of child-care facilities, its rigid practice of sending German children home for lunch until they reach university (schools have no cafeterias) and German executives' inflexibility about rescheduling evening meetings.

In her view, the new policy not only helps financially but also lends a social stamp of approval to returning to her job after a long maternity leave — this in a country where working mothers are tagged disparagingly as "Raven mothers," after the bird that abandons its young.

Worldwide Fertility Rate Is Declining

The world's fertility rate, or the average number of children born to women, has been dropping steadily since the 1950s. It is now at about 2.6, or above the 2.1 level needed to prevent declines in the population. After 2040, however, the rate is expected to drop below the replacement rate.

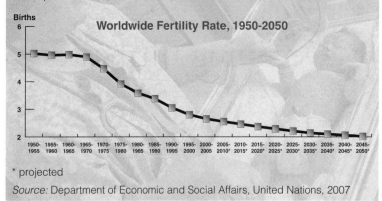

Worldwide Fertility Rate, 1950-2050

* projected

Source: Department of Economic and Social Affairs, United Nations, 2007

Last year the birthrate in Germany ticked up slightly — from 1.3 births per woman to 1.4 — and the government quickly attributed the rise to its new policy.[2]

But experts are more skeptical. Wolfgang Lutz, a demographer at the Institute for Applied Systems Analysis in Vienna, Austria, says cash bonuses may accelerate parents' decision to have a child but generally do not influence the number of children born overall. If Germany has the same reaction as Austria, which started baby bonuses in 2002, the birthrate uptick will be small and short-lived. Austria's rate already has slumped back.

"I would certainly be very cautious about interpreting this as a real change in trends," Lutz says.

Germany is a microcosm of many of the issues that Europe as a whole is facing as birthrates have plunged below the level needed to prevent declining population. Europe's principal problem — growing aging populations supported by shrinking proportions of young workers — also threatens the economies of Japan, China, and, to a lesser extent, the United States. In addition to labor shortages and slowed economic growth, many experts predict that an aging population's mounting social security and health costs will impose crushing taxes on young workers.

In the 1970s, when the average fertility rate in Europe was 2.5 children per woman, the press was full of predictions of overpopulation.[3] Now, not a single country in the 27-member European Union has a fertility rate above 2.1 children per woman — the magic number demographers consider necessary for Europe's population to replace itself.[4]

"If we don't do something, we're in big trouble," says Jonathan Grant, president of the think tank RAND Europe. Low birthrates threaten future living standards in Europe and could put "unbearable strain" on the welfare state, according to a RAND report.[5] "We've been very poor at explaining why low fertility matters," he says. "It matters for our kids. This is an investment in the future."

While the coming retirement of American baby boomers has engendered similar public handwringing about the burden it will put on Social Security, U.S. problems pale next to Europe's. Western Europe has far more generous pensions, earlier retirement ages and much lower birthrates.

The United States is exceptional among major industrialized Western nations: Its 2.1 birthrate produces enough children to replace the population as older citizens die.[6] In Europe, almost all countries are below the replacement level.[7] In Japan the rate is only 1.3 children per woman.[8] (*See graph, p. 339.*)

The ratio of workers to retirees currently stands at about 3 to 1 in the United States and is projected to fall to around 2 to 1 by 2030. While that is a dramatic shift, the ratio has already fallen to roughly 2 to 1 in Italy, Sweden and the United Kingdom, for example, and to less than 2 to 1 in Spain.[9] By 2030, Italy is expected to have a mere 0.7 workers for each retiree — or more people collecting benefits than paying taxes. Canada and Japan are not far behind.[10]

With much of the world now entering an economic downturn, many of these demographic trends could be exacerbated as young couples put off having children until their economic future is more secure. In addition, governments' ability to pay the rising health and pension costs of millions of retiring baby boomers could become even more tenuous.

Germany is not alone in trying to increase its birthrate with sweeteners to prospective mothers and fathers. Similar policies have been instituted in Japan, South Korea and Australia, where low birthrates are viewed as a crisis. But, as Japan's decades-long low fertility suggests, cash benefits may not work without a fundamental restructuring of society towards the belief that women can be good mothers and workers at the same time.

Support for that theory comes from the observation that countries that traditionally expect women to stay home once they have children, such as Italy and Japan, continue to have extremely low birthrates. When those attitudes are combined with economic insecurity and a lack of trust in government, as in Eastern Europe, they spell the lowest birthrates in Europe. Countries like Russia or Poland that remain at 1.3 for many years can expect to halve their current population in less than 45 years.[11]

Surprisingly, some countries in Western Europe appear to be recovering fertility, and demographers are even projecting rising populations for them. Official forecasts now project France, Sweden and Britain will gain population by 2050.[12] Italy and Spain have already grown in population since their nadir in the mid-1990s, thanks largely to immigration, according to Francesco Billari, a professor of demography at Bocconi University in Milan. (*See sidebar, p. 346.*)

The recovery also is being spurred by older women having more children — "almost sufficient to compensate for the sharply reduced birthrates of younger women," according to David Coleman, a demographer at Oxford University in England.[13]

Countries with comprehensive policies making it easier to combine work and family, such as France and Sweden, have seen birthrates rising to among the highest in Europe. France almost reached replacement this year (2.0 births per woman), although experts debate how much of the credit is due to its pro-natalist policies.

Japan's Upside-Down Age Pyramid

Japan's rapidly aging population has altered the nation's once classic age pyramid (left) — with young people forming the base and older people at the tip — to a coffin-shaped structure with a growing population of old people (right). By 2050, nearly 40 percent of Japan's citizens will be 65 and over, compared to 22 percent in 2007 and just 5 percent in 1950. Those 15 to 64 will account for 52 percent of the population in 2050, down from 65 percent in 2007.

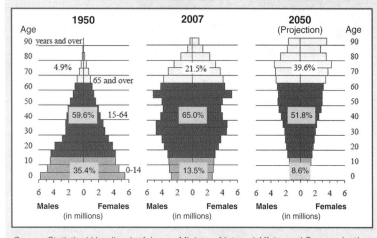

Japan's Changing Age Distribution

Source: Statistical Handbook of Japan, Ministry of Internal Affairs and Communication

The United States and Britain, which also have high rates, don't offer the free child care and long maternity leaves of France, but their job markets are more flexible when it comes to part-time work and re-entering the labor market.

For many years, young couples in Europe have reported wanting more children — an average of two — than they actually have, a desire presumably hindered by finances or society.

But the most recent surveys from German-speaking countries are finding the desired number of children averages less than two. This new trend could lead to a downward spiral in fertility as social attitudes shift toward seeing one-child or childless families as normal, predicts Lutz.[14]

In those countries where fertility has plummeted, Lutz says, "we may be raising a generation that just considers children much less important than in their parents' generation." Does that mean the evolutionary drive to perpetuate our species could vanish?

Getty Images/Sean Gallup

Cem Ozdemir, recently elected to lead the Green Party in Germany, reflects immigrants' increasingly important roles in many nations struggling with low birthrates and aging populations. The son of Turkish immigrants, Ozdemir is the first person with a foreign background to hold a top political post in Germany.

"The species won't die out so quickly because there's still lots of population growth in Africa and other parts of the world," he says. "But there may be many fewer Germans."

That may be one reason why encouraging fertility raises a red flag, especially to young Germans, reminiscent of Third Reich policies to encourage breeding of blond, blue-eyed Aryans. Some question whether such policies are a jingoistic, racist reaction against a rising tide of immigrants, even though immigrants may offer such countries their best hope for providing enough young workers.

Declining populations could be a boon for the environment by reducing humans' carbon footprint on the Earth, say experts like Stanford University biologist Paul Ehrlich, author of *The Population Bomb.* But as China's billion-plus consumers continue to buy more cars and live more like Americans, their increasing consumption — and that of others in the Third World — could outweigh any benefits.

Besides, world population is expected to grow at least until mid-century before seeing any declines. Even if the world's average family size fell immediately to just slightly more than two children, population still would gain several billion before it stopped growing. That's because of the built-in momentum generated by the unprecedented number of youths entering reproductive years, according to Population Action International.[15]

As experts study the latest population trends, these are some of the issues being debated in the political and academic arenas:

Will today's low birthrates cause economic problems?

Imagine a country whose population is older than Florida's is today — a country with twice as many retirees but only 18 percent more workers to support them. Large numbers of impoverished elderly residents languish in substandard nursing homes, many without sons or daughters to visit them. With fewer buyers, houses stand empty. Furniture makers, construction firms and auto dealers find themselves with shrinking sales and profits.

That's the kind of scenario for the United States of 2030 painted by economists worried about a future where millions of American baby boomers will be retiring while the number of tax-paying workers per retiree is shrinking. The nation will be "plunging headlong toward Third World status" unless there are major changes in pension policy and health care, Boston University economist Lawrence J. Kotlikoff has warned.[16]

In his book *Fewer,* American Enterprise Institute Senior Fellow Ben J. Wattenberg paints an even more dire picture for countries with far lower birthrates than the United States, which still produces enough children to replace its population.[17]

"I find it unlikely that nations facing fewer births, fewer customers and an aging problem will show substantial [economic] growth rates," he writes.[18]

For years, demographers have hailed declining birthrates as an economic boon for poor countries like India, because it means fewer mouths to feed. But recently, demographers have started focusing their concern on the rich countries of the world, where couples are having too few children to replace themselves. Sixty-two countries, representing 42 percent of the world's population, have birthrates below population replacement levels.[19]

Workers' collective tax contributions to "pay as you go" social security systems like that in the United States and Europe will be dwindling just as the costs of health, pensions and social security will be rising for a swelling population of retirees, who are living ever longer. The result: unbearable fiscal strains on the government, the economy and workers.

Europe, which now has four workers for every retiree, is expected to see a decline to two workers per retiree by 2050, according to Grant of RAND Europe. The impact could be enormous: Approximately 40-60 percent of public spending is sensitive to the age structure, largely through spending on health and pensions.[20]

Compared to the $1 out of every $7 in payroll now distributed in the United States to Social Security and Medicare, workers could see $4 out of every $10 going to support those benefits, Phillip Longman, a senior fellow at the New America Foundation, a progressive think tank in Washington, D.C., warns in his book *The Empty Cradle*.[21]

Just as possible, retirees could see their benefits reduced because fewer workers are contributing. "What could you buy with your Social Security check or your IRA if everyone else in your generation had simply forgotten to have children or had failed to invest in them?" Longman asks.[22]

Fewer workers also mean less demand for houses, goods and services, which can lead to slower growth. If trends in declining birthrates and increasing life expectancy continue, the European economy would be some 20 percent smaller in 2050 than if current growth rates continued, according to a RAND Europe analysis.[23]

"The economic consequences are pretty challenging," says Grant, the report's coauthor, adding that this grim analysis was made before the recent global economic downturn.

"Economists may be able to construct models of how economies could grow amid a shrinking population, but in the real world it has never happened," Longman has argued. A nation's gross domestic product is the product of its labor force times average output per worker. If, for example, Italy's working-age population plunges by 41 percent by 2050, each worker's output would have to increase by at least that amount to keep Italy's economic growth rate from falling below zero, Longman estimates.[24]

Yet when countries first begin to age, they can actually benefit economically because fewer children are putting demands on government and family budgets for housing, feeding and educating them. Japan, now one of the world's oldest nations, with a fifth of its population over 65, is one of the countries that benefited during its birthrate decline in the 1980s.

Indeed, as much as 30 percent of the economic boom of the "Asian tigers" — Hong Kong, South Korea, Singapore and Taiwan — was due to this so-called "demographic dividend," according to Leiwen Jiang, senior demographer at Population Action International.

Partly for this reason, not everyone buys a gloom-and-doom scenario. "Nobody is denying the importance of very low fertility rates," especially when it comes to the question of how countries will pay for rising pension and health obligations, says Michael S. Teitelbaum, vice president of the Alfred P. Sloan Foundation, which supports research in science, technology and economics. But he adds, "It doesn't automatically mean prosperity will plummet," especially if productivity per capita rises.

Labor scarcity could even "provide an incentive to increase efficiency and productivity," Stanford biologist Ehrlich has argued.[25]

"Much of Western Europe has had low fertility for a long time," Teitelbaum notes. "I haven't noticed any signs of economic prosperity deteriorating in those societies" — at least prior to the current economic crisis, he observes.

China is an example of a country that is enjoying an economic boom, even with a rapidly declining fertility rate, because productivity and output have been increasing at such a fast pace.

But that kind of benefit lasts only about one generation in most rapidly aging countries, most demographers agree. As workers start to retire, their swelling numbers cost far more in pensions and health care than any savings the government reaps at first — such as by educating a shrinking proportion of children.

In the long term, however, the consequences of declining fertility are likely to be far more burdensome for a country like China, which got old before it got rich, than for Europe and the United States, which got rich before they got old, as Longman puts it. China's population is aging at one of the fastest rates ever recorded thanks to its one-child policy. The percentage of elderly in China is projected to triple from 8 percent to 24 percent in 2050, and the ratio of working-age adults to the elderly is projected to decline drastically from 9 to 1 to 2.5 to 1.[26]

"We say we can't afford it, but the U.S. will have it easy compared to China," says Robert Retherford, coordinator of population and health studies at the East-West Center in Honolulu. For one thing, he says, the U.S. fertility rate has declined much more gradually. For another, the United States is a rich country.

Already, health-care costs in China are rising faster than economic growth and individual earnings. That's been compounded by the health-care system's shift to a market-based system in the early 1980s and an increase in chronic diseases.[27]

In China, where per capita income is a fraction of an American's, "they can't afford to buy the same benefits," notes Retherford. "At least we in the U.S. and Europe are rich countries. You take away from younger people to give to the old, but there's still a lot to go around."

Even for rich countries, some experts say the most pessimistic forecasts could be avoided if governments raise retirement ages or take in more immigrants as workers. (*See sidebar, p. 346.*)

Some are skeptical that pensioners accustomed to their perks will ever let this happen politically. But according to demographer Tomas Sobotka at the Vienna Institute of Demography, the retirement age has already been raised in the Czech Republic, and he thinks Western European countries will soon follow suit.

Older workers, however, don't provide the kind of innovation that nations need to grow economically, argues Longman, pointing to the economic stagnation in aging Japan. "Why is there so little ingenuity in Japan? Where's the energy and entrepreneurism that would get them out of this?" he asks. "It's not there because they don't have a youngish population to produce it."

In a surprise development, the latest population figures coming out of Western Europe show that some of the first European countries to experience very low fertility levels, like Spain and Italy, have actually increased in population since their nadir in the mid-'90s, rather than declining in size as demographers had predicted. A leading reason is a surprisingly high number of immigrants. Since many are young, they've also helped boost fertility rates and slow the aging process.[28]

Taking note of these trends, Sobotka says, "It's incorrect to talk about Europe as a continent that is facing a gloomy future because of low birthrates." He predicts that the rising participation of women in the labor force together with continuing immigration will help swell the ranks of young workers and that fertility rates will continue to rise in Scandinavia and France.

Are falling birthrates good for the environment?

"The battle to feed all of humanity is over," and has been lost, proclaimed Ehrlich in his 1968 bestseller, *The Population Bomb.* "In the 1970s the world will undergo famines — hundreds of millions of people are going to starve to death in spite of any crash programs embarked upon now."[29]

Ehrlich's prediction turned out to be wrong. He did not foresee that improvements in agricultural technology growing out of the "Green Revolution" would produce soaring yields per acre. By 2002, a bushel of soybeans in the United States required 36 percent less land than in 1970.[30]

Today, Ehrlich continues to worry about overpopulation and its effect on the environment. He recently wrote that he saw population shrinkage "as a hugely positive trend."[31]

Asked about demographers' worries over declining birthrates, he responds, "The demographers don't have a clue about how serious the environmental problems are — particularly in countries that consume a lot — in destroying our life-support systems."

Expanding populations are putting rising demands on food, energy and materials, straining ecosystems and resources in developing countries, Ehrlich warns. He cites increasingly common shortages of fresh water in the developing world and mounting evidence of global warming. Given current technology, Ehrlich estimates that the ideal population for the Earth is about 2 billion — less than a one-third of today's population — or approximately the world's population in the 1930s.[32]

Ehrlich is not alone in his concerns, as countless campaigns by environmental groups attest. Optimum Population Trust, a green think tank in Manchester, England, that campaigns for population limits, supports British Home Minister Phil Woolas' recent pledge to try to cap Britain's growing population at 70 million, partly through immigration restrictions.

Even though Britain's birthrate remains below the replacement level, this goal strikes a chord with many Britons. By 2060, Britain will be the EU's largest country, growing from its current 61 million to 77 million, according to recent projections by Eurostat, the European Union's statistical service.[33]

"We're saying you have to focus on environmental problems first because without that we're all finished," says Rosamund McDougall, the group's policy director. She expresses a common English sentiment that the country is feeling a bit cramped. England has the second-highest population density in Europe.

"You can spend an hour in a traffic jam now trying to get a half-hour out of London," she observes. "People in this country are only too aware of congestion, of the new housing proposed in our greenbelts" — green areas around cities that are supposed to be protected from development but that some politicians have proposed using to ease housing shortages. "It's making people very angry; they feel the government isn't listening. There are very few places where you can be silent or see the stars without light pollution. It's tragic."

Even with birthrates continuing to decline, world population is projected to grow at least until the middle of the century, reaching about 9.2 billion people — a third more than now — by 2050, according to UN projections.[34]

For example, even if India continues its slide from six children per woman in the 1950s to three today, the country will add another half-billion people over the next 50 years, notes John Bongaarts, vice president and distinguished scholar at the Population Council, an international, nongovernmental organization headquartered in New York.

With rising populations in Africa and Southeast Asia, "Billions more will be added to the planet," he notes. "They all want our lifestyle, with three-bedroom houses. Eventually they will get there, and it will have a disastrous effect on food consumption."

So the world still has plenty to worry about, Ehrlich says. By 2050 many developing countries are likely to be richer, with higher-consuming populations, placing even more strain on the planet.

"With the 6.7 billion people we have now, we use about 1.4 times the productive capacity of the planet," he says. "If you jack everyone up to U.S. standards of living, then you need more planets" — about two more Earths if the 5 billion-plus people in developing nations match the consumption patterns of the 1.2 billion in industrialized nations, he estimates.

Fertility Rates Increasing in Some Countries

In most developed countries, the fertility rate — or the average number of children born to women — is below the population-replacement level, but it nonetheless has risen over the past decade in several low-fertility nations, including Italy and Spain. Eight countries, including Russia, Bulgaria and Spain, have seen rates rebound 20 percent from the low point of the mid-1990s. The United States is almost the only large, Western, industrialized nation whose rate is above the replacement level; Great Britain and France are close behind. In some countries, the increases largely can be explained by rising immigration and the higher birthrates of foreign-born women.

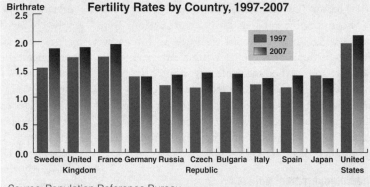

Fertility Rates by Country, 1997-2007

Source: Population Reference Bureau

Paradoxically, declining birthrates often mean more single and childless households, which use more energy per capita than a family of four sharing the same washing machine, notes demographer Jiang of Population Action International. He points out that population is just one factor in calculating energy consumption and environmental effects.

Changes in economic activity also affect energy consumption. Population aging in the United States would contribute to a decline in carbon emissions of 40 percent by 2085, Jiang has calculated. Partly, he says, this is because "you will have less people working" and using energy in their offices and factories.

But affluence can also be good for the environment, because rich countries can use their wealth for pollution control, the American Enterprise Institute's Wattenberg argues in his book *Fewer.* "Notwithstanding major population growth, the environment is getting better" in the United States, he writes.[35] Between 1976 and 2002, ambient air pollution levels for lead dropped by 97 percent, carbon monoxide by 72 percent and sulfur dioxides by 33 percent.[36]

During the Cold War, the free-market nations of Western Europe were vastly more affluent than those in communist Eastern Europe, but pollution of every sort — air, water and ground — was far more extensive in Eastern Europe, Wattenberg points out. Pollution levels in affluent South Korea are also much lower than in impoverished North Korea.

As for concerns that developing countries like China will put unbearable burdens on agriculture as their per capita meat consumption rises, Wattenberg argues that the agricultural efficiencies of the Green Revolution "have barely begun" in some parts of the world.[37]

Skeptics point out that Ehrlich's predictions were dead wrong, so why should we listen to him now? "There's no doubt Paul Ehrlich was too early" in his predictions about feeding the world, says Bongaarts.

But today Bongaarts shares many of Ehrlich's worries. Recent spikes in world food prices are a harbinger of things to come, he fears. "I'm 99 percent convinced that we will have many more problems in the future with environmental constraints."

Should government policies to increase birthrates be implemented?

Last year Australian Treasurer Peter Costello proudly announced that the country's offer of a one-time cash bonus of more than $2,500 to parents for every new baby had led to the highest number of births since 1971.[38]

He had introduced the policy in 2004, urging Australian couples to have "one for the husband, one for the wife and one for country."[39] Costello credited the policy with starting a mini-baby boom, boosting the fertility rate to 1.8 children per woman from 1.7 in 2002.[40]

Similar bonuses have been tried in Spain, Singapore, Germany and Scandinavia. Yet studies generally find bonuses have temporary effects — lasting only about one to three years. They have a short-term effect on fertility rates because they speed up, but do not increase, births that couples have already decided upon, researchers find.

"So far we haven't found clear evidence of one single policy having an effect on fertility," says demographer Billari at Bocconi University in Milan.

Nevertheless, Billari is among the experts who believe that what seems to be working is a combination of comprehensive policies that make it easier for parents to combine working with having children, as in France and Scandinavia.

The fact that France's birthrate has been among the highest in Europe for several years (and almost hit replacement last year) has some experts convinced that its raft of policies deserves some of the credit. These include a 16-week paid maternity leave, financial assistance for child care and nannies, a guarantee of the same job after a three-year paid leave, free full-day kindergarten for ages 2-and-a-half to 6 and a monthly stipend for each child until age 20.

"France is the only convincing case where it has worked," says demographer Lutz at the Institute for Applied Systems Analysis in Vienna. In his view, France meets several criteria for success. Parents trust that the government will help them out. (France has had pro-natalist policies since the 19th century.) The programs are comprehensive. And the tax incentives are enormous. A family divides its income by the number of family members and pays tax only on the resulting fraction.

As a result, "even a millionaire with five children pays practically no tax in France," Lutz points out. "This is a massive redistribution from those who have no children to those who do. No other country could even dream of it."

Indeed that's probably why South Korea has failed to push its birthrate above 1.3, even with the offer of $1,000 bonuses. They're not big enough, Lutz has told the government.

In addition, baby bonuses need to be long-term over many years, and people need to be able to count on them, Lutz says. In 2006, after then-Russian President Vladimir Putin called the low birthrate "Russia's gravest problem," the government promised families a whopping bonus of 11,000 euros (then about $9,600) for every child born after the first. Looking at the latest statistics, Carl Haub, a demographer at the Population Reference Bureau in Washington, suggested Russia as a candidate for policies that have worked; its birthrate rose from 1.3 to 1.4 this year.[41]

But Lutz is doubtful. "People don't trust that the government will live up to its policies because there's so much political instability."

The former Soviet countries were another place where pro-natalist policies worked, according to Lutz, because people felt they could rely on them. In communist East

Germany, for example when a woman became pregnant she automatically received an apartment — a strong incentive in a society plagued with housing shortages. Fertility rose to almost two children per woman. But after communism fell in 1989, along with its pro-natalist policies, fertility plummeted in Russia along with the rest of the Soviet-bloc countries.

Many demographers share Lutz's view that France is a success story.

Yet Nicholas Eberstadt, a scholar in political economy at the American Enterprise Institute, remains skeptical. He thinks France's current high birthrate could just as easily be the result of high immigrant birthrates that get averaged with low birthrates for French women. But since France doesn't collect birth statistics by ethnicity it's hard to know for sure.

Governments face two insurmountable hurdles with such policies, according to Eberstadt: changing couples' desired family size and paying enough to make it worthwhile for increasingly well-paid women to forgo their income in order to have children — what economists call the "opportunity cost."

That amount could cost much more than the national pension expenditures that governments already are worried about, he estimates.

"If women are 50 percent of the labor force, why don't you think that program might be 50 percent of gross domestic product?" he asks. "In the final analysis it's hard to see how you can bribe parents into having more children than they'd have in the first place."

Some mothers in Germany have made a similar calculation in response to the government's 1,800-euro monthly cash baby bonus — the top rate for highly paid women. "For me, it doesn't make any rational sense to have a child for this 1,800 euros per month when you compare it to the overall cost of educating a child. Because it costs 100,000 to 200,000 euros to send a child to private school or university," says Nathalie Fetzer-Hoernig, 37, a management consultant from Frankfurt. Although the birth of her third daughter qualified her for the bonus, she says the decision to have another child was made before the new bonus policy was announced.

Even a comprehensive package of work-family incentives won't have any effect if the culture still expects mothers to stay home and forgo a career, Japan's experience suggests.

Japan has failed to raise its birthrate above 1.3 even after more than 10 years of emulating French policies like subsidized, affordable child care. That's because Japan's culture and labor market make it almost impossible for a woman who wishes to have a career to raise children too, concluded a study by Patricia Boling, a professor of political science and women's studies at Purdue University in West Lafayette, Ind.[42]

Japan's dual labor market tends to confine mothers to low-paid, part-time jobs, which carry such low status that they virtually expel women from highly paid careers. Career-track women in good jobs are expected to work 10-15 hours a day and are penalized if they leave work early to pick up children. Nor can they count on husbands for help with child care: The combination of long hours, long commutes and work-related evening entertainment means husbands are rarely home.

By contrast, French women can count on a 35-hour workweek and help with housework from their husbands, according to Boling. And, they are not seen as abandoning the nest when they return to work.

Sweden's generous paid paternity and maternity leaves are often credited for the country's relatively high fertility rates. But the New America Foundation's Longman remains skeptical. "Do you know when the highest number of dads takes off?" he asks. "It just happens to coincide with the opening of elk season."

Are these incentives just a way for governments to push up the birthrate for native-born mothers to stave off a rising tide of immigrants? Maybe. Italian President Silvio Berlusconi offered 1,000-euro payments only for children of mothers born in the European Union — an embarrassment that the subsequent government reversed. "This was really put out as an anti-immigrant message," says Billari, who criticized the policy on discrimination grounds.

Some suspect Germany's baby bonus aimed at college-educated women has a similar purpose since its sliding income scale rewards a middle-class German woman more than an uneducated, low-income Turkish woman.

That's a major problem with pro-natalist policies, Eberstadt says, because "the bribe that will induce a college-educated woman to have an extra child may be much bigger than the bribe that would induce a high-school dropout. It has to be unfair, skewed toward the wealthy, and that's really difficult for a democratic society to stomach."

BACKGROUND

Birthrates and Wealth

In 1798, the English political economist Thomas R. Malthus published his famous "Essay on the Principle of Population," predicting that population increases would outgrow the world's ability to feed itself.

To some extent, his observation is borne out historically in the continually shortening period over which the world population has been doubling.[43] But Malthus did not foresee that the human race would devise methods to feed itself through improvements in agricultural technology.

From the 16th to the 18th centuries, European countries experienced repeated subsistence crises — plagues, famines and wars — reducing population drastically and making underpopulation the principal concern in many European countries. In contrast to Malthus, some economists who had lived through such mass casualties in Germany and Spain believed growing population was the most important cause of wealth.

To this day, the debate "has never been fully resolved," according to economist Sylvester J. Schieber. It is still reflected in the arguments of environmentalists who fear an overburdened planet versus economists who fear that declining birthrates and fewer workers mean slower economic growth.[44]

After the mid-18th century, increases in life expectancy coupled with high birthrates contributed to a new trend — population growth. From the Industrial Revolution on, improved hygiene, sanitation and public health would lead to better standards of living and increasing longevity. (Similar improvements are continuing in the developing world today and are expected to continue to drive up population levels at least until the middle of the 21st century.)

During the 19th century, fertility began declining in many developed countries, spurring anxiety about maintaining the "purity" of a nation's native stock. Following its crushing defeat in the Franco-Prussian War, French patriots began to blame France's decline in power on the fact that France had a lower fertility rate than Germany.

France's explicitly pro-natalist policies date from this period.

Anxiety and Eugenics

Indeed, worries about population decline "seem to crop up at predictable moments when a dominant political or economic power begins to feel unsure of its mastery and uncertain about the future," writes the Sloan Foundation's Teitelbaum, coauthor of *The Fear of Population Decline*.[45] Demography offers a convenient explanation for national problems, he finds.

Concerns about dropping birthrates among the "right" kind of people inspired eugenics movements in the 20th century in the United States and Europe, as governments feared the loss of native populations. By the 1920s and '30s they would have unsavory and eventually horrifying results.

By 1900, the U.S. birthrate had fallen 40 percent from 100 years before — most sharply among the white Protestant middle class, graduates of elite colleges and women graduates. President Theodore Roosevelt harangued upper- and middle-class women to have more children. In his 1906 State of the Union address, he scolded them for "willful sterility" — a problem he feared was leading to "race suicide." These sentiments were the drivers of a "positive eugenics" movement to encourage breeding by genetically superior individuals.[46]

But "negative eugenics" to discourage breeding of the inferior wasn't far behind. Concerns about the "overfertility of the mentally and physically defective" inspired Margaret Sanger in her crusade for birth control, which she thought would take pressure off educated women to breed. Sterilization for the insane and blemished would permit "decent and responsible citizens" to have only the number of children they could "decently rear" rather than to enter "a cradle competition with the irresponsible," she told Vassar College women in 1926.[47]

In the 1920s, amid concern about falling birthrates, 12 states passed laws calling for the sterilization of the "feeble-minded," homeless and paupers.[48]

Then came Adolf Hitler and the Nazis, who "plundered" such pro-natalist and eugenics ideas in the service of racial purity and anti-Semitism, Columbia University historian Matthew Connelly writes in his recent history of world population control, *Fatal Misconception.* It is this history — and the horrific memory of Nazi eugenic matchmaking

C H R O N O L O G Y

19th Century *People live longer due to improved hygiene and public health; fertility begins declining in many developed countries.*

1871 France encourages higher birthrates after suffering crushing defeat in Franco-Prussian War.

1900s-1920s *U.S. birthrate falls, especially among educated middle class, spurring a "positive eugenics" movement to raise birthrates among the fittest and "negative" movement to prevent reproduction of the "unfit"; 12 states pass laws to sterilize homeless, paupers and "feeble-minded." U.S. shuts off immigration in 1920s.*

1900 U.S. birthrate is 40 percent below level 100 years earlier.

1906 President Theodore Roosevelt urges upper- and middle-class women to have children to avoid "race suicide."

1930s-1950s *Hitler comes to power, adopts eugenics ideas to promote breeding camps for Aryans, extermination for Jews and others considered "unfit." Cheap housing, federal credit and rising incomes spur postwar baby boom.*

1945 After World War II, GI's come home ready to start families.

1946 Rising birthrates signal beginning of baby boom.

1957 U.S. birthrate peaks at 3.6, highest level since 1898.

1960s *Overpopulation concerns increase as the world grows at unprecedented rate.*

1964 Last year of baby boom births.

1968 Paul Ehrlich's bestselling *Population Bomb* predicts widespread famine by the 1970s.

1970s-1990s *Social changes in U.S. and Europe — including sexual revolution and more women in the labor force — contribute to later marriage and motherhood and declining birthrates. Almost half of European countries introduce policies to boost birthrates. Germany, Sweden and France shut off immigration amid concerns about declining native fertility.*

1971 U.S. women who want four children are in the minority, compared to a majority in 1966.

1972 Birthrate of white Americans drops below replacement level for first time — to 1.9 children per woman.

1978 Fully half of American women are in labor force, up from 34 percent in 1958.

1989 Iron Curtain falls. Birthrates fall to new lows in former communist countries as pro-natalist policies disappear and economies crash.

2000s *World birthrate declines, but population is projected to grow at least until mid-century. Countries with declining birthrates offer new-baby bonuses. Europe continues to grow, largely because of immigration.*

2000 Majority of U.S. mothers with young children are working.

2002 Birthrates in 17 countries in Europe plunge to 1.3 children per woman or less.

2004 Australia introduces baby bonus.

2007 World fertility falls to 2.6 children per woman, half the rate in the 1950s. . . . Germany introduces baby bonus aimed at educated women.

2008 Report shows Europe's population is increasing, due largely to immigration. . . . U.S. Census Bureau announces ethnic minorities will be in the majority in 2042. . . . In Germany, Cem Ozdemir, the son of Turkish immigrants, is the first child of immigrants elected to a national political post, symbolizing the assimilation of new immigrants into European society.

Helping Retirees Find Meaningful Second Careers

Putting 77 million baby boomers back to work could aid economy.

Anne Nolan was approaching her mid-50s when her company was sold, and she suddenly found herself jobless. She couldn't afford to retire, but with almost 30 years in corporate management she couldn't stomach the thought of yet another corporate job.

As a child of the '60s, Nolan had long wanted to work with the homeless — but she figured that winning the lottery was the only way she could afford it. Then one day on her way to buy yet another lottery ticket, Nolan realized that the management skills she had honed in the business world could help her realize her dream — even without hitting the jackpot.

"It suddenly became clear — I wanted to do something that would make me feel passionate," she says.

Indeed, when Nolan went for a job interview at the agency now known as Crossroads Rhode Island, the state's largest provider of services to the homeless, she wept. "I was so touched by the people and the humanity and the pain of the place," recalls Nolan, an ebullient 60-year-old. Now, seven years later, she heads the organization.[1]

She has never regretted her decision to work there, even though she took a pay cut — to half her corporate salary — and had to refinance her house several times to get by. Nolan plans to keep working until she's 70. Meanwhile, when Mondays approach, she says, "I never experience the Sunday night blues."

If more baby boomers follow Nolan's path and keep working, some experts think the United States could avoid the financial problems expected when the nation's growing population of retired baby boomers is set to receive Social Security, but fewer and fewer workers are paying into the system.

The country's 77 million baby boomers — born between 1946 and 1964 — started turning 60 in 2006 and promise to grow over the next two decades into the largest retirement bulge ever. If current employment patterns continue, the ratio of workers to retirees will plunge from 3.3 to 1 to 2.2 to 1 by 2030, according to Social Security actuaries.

Anne Nolan found her life's work, running a service organization for the homeless in Rhode Island, after losing her job several years ago.

That drop could mean not only lower living standards for heavily taxed workers but also threats to Social Security and Medicare financing, according to the Urban Institute, a Washington think tank.[2]

Yet baby boomers are healthier and live longer than their parents, and they come from a generation that may be more interested in doing something for society than playing golf once they reach retirement age. A recent nationwide survey by Civic Ventures, a nonprofit group in San Francisco, finds that 5-8 million people between ages 44 and 70 are already engaged in socially meaningful second careers. Of the rest in this big cohort, half are interested in moving into fields with social impact like education, health care, government and the nonprofit sector, the survey found.[3]

Civic Ventures President Marc Freedman says the idea of socially meaningful "encore careers" for boomers germinated from the group's experience running Experience Corps, which places over-55s in inner-city schools as part-time tutors and mentors. Like the Peace Corps, it was designed to offer a paid stint of no more than two years. But

and concentration camps for Jews and others the Nazis considered unfit — that continues to underlie much of the discomfort governments face when proposing explicit policies to increase their native-born birthrates.

"Hitler had breeding camps for blond, blue-eyed Germanic offspring. These very extreme developments make it still difficult in Germany and the whole German-speaking world to discuss whether governments should do

after 10 years, "We discovered nobody ever left," Freedman says, and many used the experience to launch second careers, often in teaching. That got Freedman thinking about the millions of baby boomers about to retire.

"If people are living longer, the real question is what kind of work will they do?" he says. Will it be clerking at Wal-Mart? Or will it be work that, in Freedman's words, "uses their experience, that's satisfying, connects them to other people" and contributes to society. Civic Venture's survey "showed there are millions of people — not just Bill Gates or Al Gore — who are having these significant second acts," he observes.

The recent economic downturn is likely to force older people to work longer, especially as nest eggs lose value. During the 12 months ending Sept. 30, retirement accounts had lost $1.6 trillion, the Urban Institute reported.[4]

ReServe, a nonprofit in New York City that matches retired professionals to part-time public service work with nonprofits and government, is already seeing a change in how retirees view the promised stipend of $10 per hour, once considered largely symbolic. "A growing proportion actually need it," as retirement funds dry up, says Scott Kariya, who recruits for the organization.

One question is how open employers — even nonprofits — will be to hiring older workers, especially if they want an employee with up-to-date technology skills or someone adept at on-the-job training.[5] "If you want someone who is instant-messaging all day, don't come to us," says Kariya; what ReServe promises is "maturity and experience."

Yet it's not just baby boomers who seem eager to work into their later years — if Nathan M. Fuchs is any indication. Now 77, Fuchs retired last year from his lifelong career as senior trial counsel to the Securities and Exchange Commission and signed up with ReServe without taking a break. As he explains it, "I couldn't contemplate doing nothing every day." He describes himself as "vigorous," and his sonorous voice could still carry across a courtroom.

He now works three days a week with the Brooklyn district attorney making presentations to senior citizens about identity-theft scams that target them. "This fits into the skills I learned as a trial lawyer — to speak to people and have an understanding of the legal issues," Fuch says modestly. "I enjoy the fact that I can help somebody, because the money lost to them would be so catastrophic."

Nathan M. Fuchs, 77, a retired attorney, teaches senior citizens about identity theft. "I couldn't contemplate doing nothing every day," he says.

At the younger end of the spectrum, Kariya, a trim, youthful-looking 52-year-old, exemplifies the kind of successful young baby boomer who may soon join the growing army of retirees. After 23 years as a corporate headhunter, he was financially secure enough to retire two years ago, happy to leave his 60-hour workweek behind. He joined ReServe for a stipend and is now a paid program officer there responsible for recruiting new civic-minded organizations as employers.

Kariya describes his former career, where he was paid on commission, as "all about the money." At ReServe, by contrast, he feels he is doing "really meaningful work. There are so many people with extremely good skills. It's a waste not to use them."

---

[1] www.crossroadsri.org.

[2] Gordon B. T. Mermin, *et al.*, "Will Employers Want Aging Boomers?" Urban Institute, July 2008, p. 7, www.urban.org and www.retirementpolicy .org.

[3] "Encore Career Survey," The MetLife Foundation/Civic Ventures, 2008, www.civicventures.org/publications/surveys/encore_career_survey/Encore_ Survey.pdf.

[4] Richard W. Johnson, *et al.*, "How is the Economic Turmoil Affecting Older Americans?" Urban Institute, October 2008.

[5] Mermin, *et al.*, *op. cit.*

---

anything to raise the birthrate," says demographer Lutz at the Institute for Applied Systems Analysis in Vienna.

Such a fear is justified, according to Connelly. "If history is any guide, projections of population decline may lead to increasingly coercive policies, even crash programs," he warns.[49]

A common government response, historically, has been anti-immigration legislation. Amid concerns about declining

Can Immigrants Make Up for Low Birthrates?

In some surprising cases the answer is yes.

A tourist in Italy quickly becomes aware that the woman cleaning the hotel room is Romanian, the man moving the pool chairs is Albanian and the handbag hawkers on the sidewalks are African.

If declining birthrates in countries like Italy won't provide enough workers to support retirees, could the missing births be replaced by immigrants?

"Probably not," has been the conventional wisdom among many experts ever since a U.N. report tried to answer the question in 2000. To keep the ratio of workers to retirees constant in Europe, 700 million immigrants would have to be allowed into the continent by 2050, and immigrants would have to make up 75 percent of the European Union (EU) population, the report calculated. That would mean 70 percent of Italy's inhabitants would be immigrants by mid-century.[1]

Such a high immigration rate "seems out of reach," the U.N. report concluded, both for practical reasons and because there would be too much political opposition to such an incursion of foreigners.[2]

But more recent data coming out of Europe suggest the U.N. forecast may have been overly pessimistic — at least for some parts of Europe. Immigration was the "main driver" in last year's 2.9 million increase in Europe's population, according to the EU's statistical service, Eurostat. In fact, immigration accounted for 76 percent of the increase. Out of 48 countries in Europe, 21 saw increases mostly or entirely due to immigration last year.[3]

Those increases were reflected in a slight uptick in the birth rate in Spain last year. The increase was due to foreign-born mothers, although a new cash bonus of $3,500 also may have helped. However, the Spanish birth bonus is small compared to incentives offered by other European countries.

In Italy, more than half a million new immigrants were counted last year, slightly more than the number of births for the year. Contrary to earlier predictions that the populations of Spain and Italy would be in decline by now, they have actually increased since their nadir in the mid-1990s — largely due to migration.

"This is an absolutely unexpected and unprecedented migration impact on population," says demographer Tomas Sobotka at the Vienna Institute of Demography. These trends could eliminate many of the economic problems people worry about, he observes. "Migration can stabilize the working

African immigrants seeking to enter Italy illegally were intercepted by the Italian coast guard recently and brought to the island of Lampedusa.

population and to some extent contribute to higher fertility and slow down aging a little because migrants are young."

Whether Italians will tolerate a continuing flow of immigrants at these levels is another question. A recent spate of violence against immigrants from Africa and China has spurred a national discussion in the country about racism.[4] It also has spurred new government attempts to restrict migration.

"Italians haven't yet accepted the idea that there is a real demographic revolution going on and that the composition of the population is going to be different," says Francesco Billari, a professor of demography at Bocconi University in Milan.

The concept of citizenship in Italy is tied to one's blood, Billari points out. Unlike the United States, a person born in Italy of foreign parents does not automatically become a citizen. As the reactions to the recent election of Sen. Barack Obama as U.S. president showed, many immigrants of color in Europe doubt they would ever be permitted to rise that high in the political process in Europe.[5]

Italian racial preferences came to light in an embarrassing way in the case of Italian Premier Silvio Berlusconi's "baby bonus," when immigrant parents were accidentally sent checks for their offspring and then were asked to return the money. The government hadn't intended to promote those births; it restricted the bonus to children of EU-born parents.[6]

More recently, Berlusconi's government has pledged to imprison illegal immigrants for up to four years and to continue destruction of Roma (gypsy) encampments. But some of those efforts have been blocked by the EU's "free movement directive," which gives EU citizens the same rights as locals.[7] The surprising number of foreigners may reflect government tolerance for some illegal immigrants because they are needed as workers, suggests Sobotka. He estimates that although about half of those who immigrate to Spain do so illegally, many are legalized by the government within a year or two.

Some experts remain skeptical that immigration can solve Europe's low birthrate problem. According to Nicholas Eberstadt, a scholar in political economy at the American Enterprise Institute, many migrants who arrive in Europe are uneducated and lack the skills that contribute to the sort of innovation needed by a dynamic economy. Simultaneously, many educated Europeans have been moving abroad to places like the United States.

Those leaving Europe are "native born, young and disproportionately well-educated — exactly the kinds of productive citizens any headhunter would like to be cherry-picking," he says.

In Japan, a country considered especially hostile to immigration, births rose slightly in 2007 due solely to a rise in immigrant births, according to demographer Carl Haub of the Population Reference Bureau. That's surprising, because with few exceptions, Japan bars foreigners from becoming citizens. If Japan's population continues at its low 1.3 birthrate for many years, a large majority of Japanese would have to be foreign-born after only two generations in order for Japan to maintain its current population.[8]

In Great Britain, immigrants' higher-than-replacement birthrate has contributed to a rising birthrate for the nation as a whole. Last year, mothers in Britain had an average of 1.9 children — the highest rate since 1980.[9] But the government doesn't seem to be rejoicing. Home Office Minister Phil Woolas recently pledged to cap Britain's rising population of 61 million before it reaches 70 million, saying the government's new Australia-style immigration points system would make it harder for immigrants — especially the uneducated — to enter.[10]

Forecasts of doom for other low-fertility countries may have been overstated as well. Eight countries — Sweden, Russia, Bulgaria, Spain, Estonia, Latvia, the Czech Republic and Ukraine — have seen their birthrates rebound about 20 percent from the low point of the mid-1990s, notes Haub, suggesting low fertility rates may have bottomed out on the continent. "In Europe the trend is up," he observes but cautions, "these rates are still very low."

Eberstadt says the outlook for Russia's continuing population decline is particularly grim because it also has a high rate of death at early ages, largely from chronic diseases like heart disease. If Russia had the same death rate as France since the end of communism, it would have 18 million more people — a World War II scale of casualties. Eberstadt explains the gap as the "difference between a health explosion in Western Europe and a health nightmare in Russia; I'm not sure it's going to be easy to turn around."

The current economic downturn is likely to change some of these trends in unexpected directions. With fewer jobs to offer, will Western Europe be as appealing to immigrants? Will younger people put off having children even longer?

In Zaragoza, a bustling industrial city in Spain's fastest-growing region, Francisco Braulio, 31, works for one of the parts makers that supply the General Motors plant there, but the GM plant just had layoffs, which led to its temporary shutdown. He said he and his wife have put off talk of having children for now as a cost-saving measure.[11]

Historically, birthrates tend to be extremely sensitive to economic conditions. In Sweden the birthrate fell to its lowest point in the country's history, 1.5, during the deep economic recession and high unemployment in the 1990s.

Iceland, the only country in Western Europe with a replacement birthrate, has been admired for several years as a model of gender equity and supporter of working parents. But as the country faces economic collapse, all that could change.

"I would almost bet a little money that the fertility rate of Iceland will decline in the next few years," says American demographer Michael S. Teitelbaum, coauthor of *The Fear of Population Decline.* "There will be a lot of deferral of childbearing as unemployment rises."

[1] United Nations, *Replacement Migration* (2000).

[2] Quoted in Michael S. Teitelbaum, "Western Experiences with International Migration in the Context of Population Decline," paper for the National Welfare Policy Seminar, Government of Japan, Dec. 16, 2003.

[3] Eurostat, European Commission, "Population in Europe 2007: First Results," 2008, at http://epp.eurostat.ec.europa.eu.

[4] Rachel Donadio, "Italy's Attacks on Migrants Fuel Debate on Racism," *International Herald Tribune,* Oct. 13, 2008.

[5] Steven Erlanger, "Can Europe Produce an Obama?" *International Herald Tribune,* Nov. 12, 2008.

[6] Kathryn Joyce, " 'Missing' ": The 'Right' Babies," *The Nation,* Oct. 28, 2008.

[7] "When Brussels Trumps Rome," *The Economist,* Oct. 25-31, 2008, p. 51.

[8] Fred R. Harris, ed., *The Baby Bust* (2006), p. 42.

[9] Office of National Statistics, U.K., "Fertility, Rise in UK Fertility Continues," Aug. 21, 2008, www.statistics.gov.uk.

[10] Alan Travis, "Tougher Rules Aimed at Curbing Population Rise," *The Guardian,* Oct. 25, 2008.

[11] Nelson D. Schwartz, "Europeans Feel Pain of Financial Gridlock," *International Herald Tribune,* Nov. 4, 2008, p. 1.

native fertility in the 1920s, the United States shut off immigration. Germany, Sweden and France would do the same in the 1970s.[50]

Boom and Bust

In the 1930s and '40s, low fertility rates in Europe and the United States led to dire projections about the future, only to be proven wrong by the unexpected rise in fertility rates in the late 1940s and '50s that became known as the baby boom.[51] Between 1946 and 1964, fertility rose sharply in many developed countries, starting with the postwar baby boom.

In 1945, millions of GI's returning from World War II brought with them a pent-up desire for families. They were greeted by the new availability of federal credit for middle- and lower-class home ownership and the introduction of mass-produced tract housing. The cheap housing and rising incomes made having children economically attractive, and Americans responded by marrying younger and beginning families earlier — boosting overall fertility.

The number of children produced by American women went from a low average of 2.1 children in 1936 to 3.6 in 1957 — an extraordinarily high rate for a modern, industrial society and a rate not seen in the United States since 1898.[52]

The American baby boom was an exception in the long history of falling fertility, writes historian Herbert S. Klein. After just two decades, Americans "were back again to marrying later, producing children later and producing fewer children."[53]

The 1970s and '80s ushered in an era of sweeping social changes, among them the sexual revolution, made possible by increasing contraception. Feminism helped to change the aspirations of women from the housewifery of the 1950s, bringing with it increasing education, growing entry into the labor force and later marriage and motherhood.

A sea change in attitudes could be seen by the early 1970s in the United States. By 1971, women who wanted four children were a minority, down from the majority in 1966. While in 1958 only 34 percent of women had been in the labor force, by 1978 half were, increasing every year to reach over 60 percent by the beginning of the 21st century. By 2000, the majority of mothers with young children were working.[54]

Each succeeding generation after the 1960s reduced its fertility so much that by the last decades of the 20th century the birthrates of native-born, white Americans had become similar to those of their peers in Northern Europe. By 1972, the birthrate of this group had dropped below replacement level for the first time — to just 1.9 children per woman — and to this day has not returned to the replacement level of 2.1 births per woman.

Mexican-Americans continue to make up for the declining fertility rates of whites and almost every other ethnic group, both because they are the single, most important part of the Hispanic population and because they have the highest birthrates of any group in the country.[55]

During the post-1964 era, the family began to lose importance as the dominating factor in determining fertility. The rate of divorce rose along with the number of single-parent families and the number of children born outside marriage — and not just among the poor. By 2000, single-parent families made up 27 percent of households with children.[56]

Population Explosion

Unprecedented growth rates in world population — from 1 percent a year in the 1940s to more than 2 percent during the '60s — led to worries about a population explosion.

Warnings about an overpopulated future were issued by experts in the United States, little realizing the boom was nearing an end. In 1968, Stanford biologist Ehrlich's best-seller *The Population Bomb* predicted worldwide famines in the 1970s and '80s due to overpopulation. Another expert predicted humans would outweigh the Earth in less than 1,200 years.[57]

Contrary to the predictions, fertility in developed countries began falling again in the 1970s, and the rate of expansion in the world population fell to 1.85 percent annually. By the 1990s, world population was growing by only 1.4 percent per year, demonstrating the post-World War II baby boom had definitively reached an end.[58]

Europe also began to see unprecedented, low birthrates. In 1989, after the fall of the Iron Curtain, many family-friendly policies were discontinued in Eastern and Central Europe, which, coupled with economic insecurity, led to plunging birthrates.

Western Europe, too, saw dropping rates, but they had more to do with increasing levels of education and delayed childbearing. In Southern European countries,

AT ISSUE

Will low fertility rates in Europe cause economic problems?

YES Phillip Longman
*Schwartz Senior Fellow, New America
Foundation; author,* The Empty Cradle

Written for *CQ Researcher*, November 2008

Between 1750 and 1850, the population of Western Europe doubled. The booming population spurred the discovery of more efficient means of producing and distributing food, energy and other scarce goods, while also producing more bright minds to work on such challenges. Eventually, rising productivity allowed the economy to provide for additional increases in population, even as each person consumed more. Love it or hate it, this is the system that built the modern world.

Can Europe sustain economic growth in the face of a shrinking workforce today? Certainly it can, within limits. But a nation's gross domestic product is literally the sum of its workforce multiplied by average output per worker. Thus, a decline in the workforce implies a decline in an economy's growth potential. When the number of workers falls, economic growth can only occur through compensating increases in productivity.

The European Commission projects that Europe's potential growth rate over the next 50 years will fall by 40 percent due to the shrinking workforce. Europe's economic growth will depend entirely on getting more out of each remaining worker, many of them unskilled, recent immigrants or aging native-born workers forced to delay retirement.

To remain affluent, Europe will require a thorough re-engineering — not just of its formal economy but of the family as well. To cite a few of the design challenges:

- How does industry maintain economies of scale when the number of consumers is perpetually falling?
- What sustains the value of your house, or your retirement portfolio, when there are ever fewer younger people to whom you might sell your accumulated assets?
- How does a population dominated by middle-aged and elderly people bear the risks necessary for entrepreneurism and technological dynamism?
- How can young people afford to raise and educate children if the cost and duration of education continue to rise, even as more of their wages go for pensions and health care?
- How can a government meet its obligations to the swelling ranks of the elderly when there are fewer and fewer workers to tax?
- How can nations with shrinking populations defend themselves militarily against more youthful competitors?

If you know the answers to those questions, you can safely bet on Europe's vibrant future.

NO Tomas Sobotka
*Research Scientist, Vienna
Institute of Demography, Austria*

Written for *CQ Researcher*, November 2008

Low fertility is unlikely to pose a serious challenge to prosperity in most parts of Europe. Eastern Europe is a main exception, where low fertility rates, emigration and generally poor health promise a prolonged population decline.

In Western Europe (except Germany) and the Nordic countries, low fertility rates are either barely supporting population stability or they are bolstered by substantial immigration, especially in Southern and Central Europe.

Moreover, Europe may see a modest increase in future fertility rates, as women who have postponed childbearing eventually give birth. Most Europeans aspire to have two children, and governments, worried about the possible consequences of low birthrates, have become increasingly receptive to supporting these aspirations.

Nevertheless, governments will have to make bold adjustments to their aging societies. Aging in Europe is largely driven by ever increasing life expectancy and can be only partly moderated by higher fertility or immigration. Only an unrealistically high rise in fertility rates could temporarily halt this trend, but this would create a large baby boom generation that would inevitably translate after many decades to a still larger bulge in the proportion of elderly.

Many inevitable accommodations to population aging may be more important than higher fertility rates, such as:

- Increasing retirement age and allowing flexibility in retirement. Mandatory and early retirement age puts many experienced and productive workers out of work. An effective retirement age of less than 60 years, as in France, is unsustainable when life expectancy exceeds 80.
- Increasing women's participation in work. If Southern Europe can catch up with the Nordic countries in women's labor participation, the shrinkage of the labor force can be mitigated.
- Making the labor market more flexible, to increase employment rates among young adults.
- Supporting healthy and active aging. This is where Europe has a clear advantage over the United States, where obesity, health disparities and costly health care affect older workers' productivity.

The threat of low fertility rates should not be exaggerated. Many European societies have successfully accommodated low fertility in the past. Europe's future prosperity and quality of life will be driven more by the quality of its health care, education and flexible adjustments to aging than by higher fertility and population growth.

high unemployment for young people was another factor. In the early 1990s, Italy and Spain were the first in Europe to plunge to fertility levels as low as 1.3. By 2002 there were 17 countries in this category in Southern, Central and Eastern Europe.

Today, world fertility has fallen to 2.6 children per woman in 2007, about half its level in the 1970s. The decline has been driven mostly by developing countries, where the education of women has been the most important influence in producing smaller families. Even so, the least-developed countries retain fertility rates at 5.0, presaging continued global population growth at least to the middle of this century, according to U.N. projections.

CURRENT SITUATION

Social Changes

Demographers have been surprised to find a link between higher birthrates and unmarried couples living together. In Iceland — the only Western European nation with a birthrate at replacement — more than half of births occur outside marriage. In Scandinavia, which has among the highest birthrates in Europe, 40-50 percent of births are outside marriage.

According to one explanation, at least in wealthy societies cohabitation is a sign of gender equity. Sequential relationships may also lead to more children. "Assume a child is something that cements the relationship between partners, but you have only one child," says demographer Billari of Bocconi University in Milan. "So how will you ever get two children in your life? You have two partnerships."

By contrast, in countries like Italy or Spain, which have a strong emphasis on traditional marriage and look disapprovingly on cohabitation and divorce, people postpone marriage until they are far older. That means they start having children much closer to the end of their childbearing years — and thus have fewer.

Another aspect of Spanish and Italian society that leads to fewer children is the tendency of men (and women, too) to live with their parents until age 25 or 30 — a trend that has won Italian men the moniker of "big babies." Universities do not have dormitories, and rent is expensive, so it makes economic sense for young people to live at home. Moreover, says Billari,

Southern Europe has had a history of massive emigration during its hardest economic eras. "So leaving the parents has been associated for a long time with a sad moment of migration. Now, when the economic situation is much better, people postpone this sad moment of leaving your parents."

Ironically, the economic downturn may keep young Italians living at home and eating mom's pasta a few years longer, further depressing birthrates.

Policy Changes

Western Europe's population of 30-to-45-year-olds is expected to shrink by 18 percent by 2030, and the cohort of 15-to-24-year-olds by 40 percent.

The ensuing shrinkage in workers could be reversed and Western Europe's labor force could actually grow if retirement ages were raised from the 50s to the higher retirement ages — around the mid-60s — common in other developed countries, the American Enterprise Institute's Eberstadt has calculated.[59]

Improving health and life expectancy means retirees can count on 20 years of retirement for men and 25 for women in some countries. Italy, France, Belgium and Luxembourg all penalize continuing work into one's late 50s through forgone pension payments and additional taxes.[60]

Raising the retirement age might be a simpler way to ease the coming financial woes of low birthrate countries than boosting birthrates, Eberstadt argues. Other observers are skeptical that French workers will ever give up their treasured pension benefits, as last year's crippling nine-day transit strike in Paris showed. The strike was a response to President Nicolas Sarkozy's proposal to make transit workers retire two-and-a-half years later.[61]

The other approach is to try to boost fertility rates. Between 1976 and 2003, as the majority of European and other countries came to view their birthrates as too low, almost half tried some kind of policy aimed at persuading couples to have more babies.[62] Baby bonuses have been offered in Australia, Spain, Singapore, Germany and Scandinavia.[63]

But even if these policies lead to dramatic increases in birthrates, the annual number of births would continue to decline. That's because decades of low fertility in the 1960s and '70s have produced progressively smaller cohorts of females who grow up to be mothers, resulting in a downward spiral known as "negative population momentum."

In some countries, efforts to raise the birthrate through cash incentives are already being questioned as too expensive for the little they gain. "The baby bonus is just not an effective expenditure. We pay for every birth, and most parents would have a first child anyway, irrespective of the bonus," Australian Professor Ross Guest of Griffith University said recently.[64]

Japan has tried virtually every one of the incentives offered by Western Europe's leading fertility countries — including bonuses, paid maternity leave and child care — with the addition of subsidized dating services.

But most Japanese women with two kids would probably prefer a six-year maternity leave in keeping with the cultural expectation that women should stay home with their children until age 3, several experts recently suggested. That might work, but it would be very costly. The problem, not just in Japan but in other developed countries in economic recession, is that it is hard to make society more family-friendly without jeopardizing efforts to jump-start the economy in the direction of more efficiency.[65]

OUTLOOK

Downward Spiral

"If your parents never had children, chances are . . . neither will you," comedian Dick Cavett quipped.[66]

If, as surveys in some countries show, young European couples no longer desire to have enough children to replace themselves, they could be raising a generation in which childless and one-family couples are the norm, leading to a downward spiral of ever-decreasing births.

To pessimists, the tendency of educated modern women to have children at later and later ages, and therefore ever fewer of them, could spell the end of some countries as we know them.

Many of today's leaders echo French Premier Georges Clemenceau's despairing warning as that country debated the Treaty of Versailles that "if France turns her back on large families . . . France will be lost because there will be no more Frenchmen."[67]

But population forecasts, especially 35-40 years out, have been notoriously wrong, as the failure to forecast the

postwar baby boom showed. So many social and technological changes could occur in the meantime that the future is essentially "unknowable," in the words of the Sloan Foundation's Teitelbaum.[68]

One futuristic possibility: Women will leapfrog the traditional biological age limit for childbearing through the use of new reproductive technologies. For example, although in vitro fertilization (IVF) is currently so expensive that it is out of reach of all but the well-to-do, Denmark already pays for it through its health insurance, and more than 4 percent of its babies are born this way. According to the RAND Europe think tank, subsidizing IVF could be more cost-effective than increasing baby bonuses.[69]

Could the natural limit be pushed much beyond age 40? Perhaps. Egg banking — freezing healthy female eggs for future use — could "stop the biological clock entirely," according to RAND Europe's Grant.[70]

In the future, concerns about the loss of nations could also abate because notions of what constitutes nationhood and even race are changing surprisingly quickly. Recent predictions by the U.S. Census Bureau that non-Hispanic whites will constitute a minority in their own country by 2042 seemed to raise surprisingly little alarm in the United States.[71]

That may be because most people who describe their origin as Hispanic-American also identify themselves as white. They won't necessarily strike Americans as a recognizable "minority" by the time they constitute a hefty portion of the majority. After all, only a century ago, Irish Catholics, Italians, Eastern European Jews and even some Germans were not considered white by other Americans.[72]

While Europe has been more resistant to absorbing foreigners than the melting pot of America, it too has redefined people it once considered minorities as native stock. Up to a decade ago, Southern Italians were derided as too dark and too short to be authentic Italians.[73]

Just this month, a son of Turkish immigrants became the first leader with an immigrant background to be elected to a top political post in Germany. News accounts noted that Cem Ozdemir, elected to lead the Green Party, spoke German with an accent that reflected the southwestern German region where he grew up.[74]

The fear that there will be "no more Germans" could evaporate if Germans come to see second- and third-generation Turks, with their perfect German accents, as Germans themselves.

NOTES

1. The period in which the sum is paid is extended up to 14 months if the father also takes leave from work.

2. "German Family Minister Calls New Parental Leave Law a Success," *Deutche Welle*, Dec. 15, 2007, www.dw-world.de/dw/article/0,2144,3005391,00.html.

3. Jonathan Grant and Stijn Hoorens, "The New Pronatalism? The Policy Consequences of Population Ageing," *Public Policy Research*, March-May 2006.

4. A fertility rate of 2.1 is considered the replacement rate for Western Europe. The number can be higher for developing nations with higher rates of infant mortality and premature death, according to Population Action International.

5. Grant and Hoorens, *op. cit.*

6. Iceland and New Zealand also had birthrates of 2.1 in 2007.

7. Only Turkey and Azerbaijan have fertility rates higher than 2.1. See Eurostat, European Commission, "Population in Europe: 2007: First Results," 2008.

8. Population Reference Bureau, "Fertility Rates for Low Birth-Rate Countries, 1995 to Most Recent Year," Oct. 22, 2008 at www.prb.org.

9. See "Boomer Retirement Roundtable," Round 2, *Atlantic Unbound*, January 2008, www.theatlantic.com/doc/200801u/boomer-retirement-round-2.

10. "Getting Greyer — and Poorer Too?" *Economist*, May 30, 2005.

11. Francesco C. Billari, "Lowest-Low Fertility in Europe: Exploring the Causes and Finding Some Surprises," *Japanese Journal of Population*, March 2008, pp. 2-16, p. 3.

12. "Suddenly, the old world looks younger," *Economist*, June 14, 2007.

13. *Ibid.*

14. *Ibid.* One study showed the number of children Germans want is 1.75 on average; another found one-fifth of German women are satisfied with having no children.

15. Population Action International, Fact Sheet, "Toward 7 Billion: Why the World Population is Still Growing," May 2, 2005, www.populationaction.org.

16. Lawrence J. Kotlikoff and Scott Burns, *The Coming Generational Storm* (2004), pp. xvii-xviii.

17. Ben J. Wattenberg, *Fewer* (2004).

18. *Ibid.*, p. 213.

19. Calculation by Population Action International based on 2005 data, the most recent available, from United Nations, "World Population Prospects: the 2006 Revision."

20. Grant and Hoorens, *op. cit.*, pp. 113-125.

21. Phillip Longman, *The Empty Cradle* (2004), p. 142.

22. *Ibid.*, p. 140.

23. Grant and Hoorens, *op. cit.*, p. 28.

24. Phillip Longman, "The Global Baby Bust," *Foreign Affairs*, June 1, 2004.

25. Paul Ehrlich, "Enough Already," *New Scientist*, Sept. 30, 2006, pp. 47-50.

26. Toshiko Kaneda, "China's Concern Over Population Aging and Health," June 2006, Population Reference Bureau at www.prb.org.

27. *Ibid.*

28. Carl Haub, "Tracking Trends in Low Fertility Countries: An Uptick in Europe?" Population Reference Bureau, September 2008, www.prb.org; Eurostat, *op. cit.*

29. Quoted in Longman, *op. cit.*, p. 133.

30. *Ibid.*, p. 134. For background, see N. A. Haverstock and R. C. Schroeder, "Green Revolution," in *Editorial Research Reports*, March 25, 1970, available in *CQ Researcher Plus Archive*, www.cqpress.com.

31. Paul and Anne Ehrlich, "Enough Already," *New Scientist*, Sept. 30 2006, pp. 47-50.

32. *Ibid.* For background, see Marcia Clemmitt, "Climate Change," *CQ Researcher*, Jan. 27, 2006, pp. 73-96; and Colin Woodard, "Curbing Climate Change, *CQ Global Researcher*, February 2007, pp. 25-48.

33. "Multiplying and Arriving," *The Economist*, Aug. 28, 2008.

34. "World Population Prospects: The 2006 Revision: Highlights," United Nations, 2007, p. ix.

35. Wattenberg, *op. cit.*, p. 214.

36. *Ibid.*, p. 154.

37. *Ibid.*

38. The payment of 4,187 Australian dollars ($2,675 US) rose to 5,000 Australian dollars ($3,195 US) in July.

39. Stephen Lum and Lauren Wilson, "Time to Rethink Baby Boom," *The Australian*, March 14, 2008, www.theaustraliannews.com.

40. "Spain Adopts Baby Bonus Scheme," Reuters, July 5, 2007.

41. Haub, *op. cit.*

42. Patricia Boling, "Demography, Culture and Policy: Understanding Japan's Low Fertility," *Population and Development Review*, June 2008, pp. 307-326.

43. The world's population more than doubled in the last half-century. See Population Action International, "Toward 7 Billion."

44. Steven A. Nyce and Sylvester J. Schieber, *The Economic Implications of Aging Societies* (2006), p. 10.

45. Michael S. Teitelbaum, *et al.*, "Demography is Not Destiny," *Foreign Affairs*, Sept/October 2004, www.foreignaffairs.org.

46. Longman, *The Empty Cradle, op. cit.*, p. 159.

47. *Ibid.*, p. 161.

48. *Ibid.*, pp. 161-162.

49. Matthew Connelly, *Fatal Misconception* (2008), p. 382.

50. Longman, *The Empty Cradle, op. cit.*, p. 23.

51. Teitelbaum, *op. cit.*

52. Herbert S. Klein, "The U.S. Baby Bust in Historical Perspective," in Harris, *op. cit.*, p. 129.

53. *Ibid.*

54. *Ibid.*, pp. 130-132.

55. *Ibid.*, p. 144.

56. *Ibid.*, p. 132.

57. Nyce and Schieber, *op. cit.*, pp. 10-11.

58. *Ibid.*

59. Nicholas Eberstadt and Hans Groth, "Europe's Coming Demographic Challenge: Unlocking the Value of Health," American Enterprise Institute, press release, Dec. 3, 2007.

60. *Ibid.*

61. Molly Moore, "French Transit Workers Back on Job," *The Washington Post*, Nov. 24, 2007.

62. Fred R. Harris, ed., *The Baby Bust* (2006), p. 99.

63. "Spain Adopts Baby Bonus Scheme," *op. cit.*

64. Stephen Lunn and Lauren Wilson, "Time to Rethink Baby Bonus," *The Australian*, March 14, 2008, www.theaustraliannews.com.au.

65. Harris, *op. cit.*, p. 47.

66. Dick Cavett, www.brainyquote.com/quotes/quotes/d/dickcavett123484.html.

67. Michael S. Teitelbaum, "Western Experiences with International Migration in the Context of Population Decline," Paper for the National Welfare Policy Seminar, Government of Japan, Dec. 16, 2003.

68. *Ibid.*

69. "Should ART Be Part of a Population Policy Mix?" 2006, RAND Europe, www.rand.org/randEurope.

70. Jonathan Grant, *et al.*, "A Second Reproductive Revolution," RAND Review, summer 2008, www.rand.org/publications/randreview.

71. Sam Roberts, "A Nation of None and All of the Above," *The New York Times*, Aug. 17, 2008 at www.nytimes.com.

72. *Ibid.*

73. Kathryn Joyce, "Missing: The Right Babies," *The Nation*, Feb. 14, 2008, www.thenation.com.

74. Judy Dempsey, "German Greens Pick Son of Turks as Leader," *International Herald Tribune*, Nov. 17, 2008, p. 3.

BIBLIOGRAPHY

Books

Connelly, Matthew, *Fatal Misconception: The Struggle to Control World Population, Belknap Press of Harvard University Press*, 2008.
A Columbia University historian raises some disturbing questions about rich countries' efforts to limit immigration and fertility abroad while raising fertility at home.

Freedman, Marc, *Encore: Finding Work that Matters in the Second Half of Life, Public Affairs*, 2007.
The CEO of a nonprofit focusing on careers after retirement argues that "encore careers" can help prevent a disastrous financial burden imposed by retiring baby boomers.

Harris, Fred R., *et al.*, *The Baby Bust: Who Will Do the Work? Who Will Pay the Taxes? Rowman & Littlefield,* 2006.
Essays by distinguished demographers examine government responses to declining fertility in Europe, Japan and the United States.

Longman, Phillip, *The Empty Cradle: How Falling Birthrates Threaten World Prosperity and What to Do About It, Basic Books,* 2004.
A senior fellow at the New America Foundation warns that religious fundamentalists could account for an increasing share of the population if the secular middle class continues its low birthrate.

Wattenberg, Ben J., *Fewer: How the New Demography of Depopulation Will Shape Our Future, Ivan R. Dee,* 2004.
Declining birth rates promise big economic problems in the United States but monumental ones in Europe and Japan, argues a senior fellow at the American Enterprise Institute.

Articles

"After the Boom," *The Atlantic,* January 2008, www.theatlantic.com/doc/200801u/boomer-retirement-roundtable.
Atlantic editors discuss the repercussions of impending baby boomer retirements.

"Multiplying and Arriving," *The Economist,* Aug. 28, 2008, www.economist.com/research/articlesBySubject/displaystory.cfm?subjectid=894664&story_id=12010087.
Fueled by immigration and relatively high birthrates, Great Britain is projected to be the European Union's most populous country by 2060.

"Suddenly, the Old World Looks Younger," *The Economist,* June 14, 2007, www.economist.com/world/europe/displaystory.cfm?story_id=9334869.
The long-term decline in European birthrates is starting to bottom out — and is even rising — in some countries in part due to new government policies.

Ehrlich, Paul, and Anne, "Enough Already," *New Scientist,* Sept. 30, 2006, pp. 47-50, www.newscientist.com.

Stanford biologist Ehrlich, author of the 1968 bestseller *Population Bomb,* and his wife and coauthor argue that declining birthrates and population are good for the environment.

Joyce, Kathryn, "Missing: The 'Right' Babies," *The Nation,* Feb. 14, 2008.
Christian pro-family groups have seized on the issue of declining birthrates to promote a right-wing agenda against abortion, birth control and women's rights.

Miezkowski, Katherine, "Do We Need Population Control?" *Salon.com,* Sept. 17, 2008, www.salon.com/env/feature/2008/09/17/population_control/index.html.
Scientists and historians discuss whether overpopulation is still a problem.

Roberts, Sam, "A Nation of None and All of the Above," *The New York Times,* Aug. 17, 2008, p. WK6.
Although minority ethnic groups will form the majority of the American population by 2042, more and more ethnic minorities identify themselves as "white."

Reports and Studies

"Population in Europe 2007: First Results," *Eurostat,* Sept. 12, 2008, epp.eurostat.ec.europa.eu/portal/page?_pageid=2173,45972494&_dad=portal&_schema=PORTAL &m.
A report from the European Union's statistical service shows that immigration is keeping Europe's population growing despite low birthrates.

"World Population Prospects: The 2006 Revision," *U.N. Population Division,* 2007, www.un.org/esa/population/publications/wpp2006/wpp2006.htm.
The United Nations agency provides the most commonly quoted projections for the world's future population growth and birthrates.

Haub, Carl, "Tracking Trends in Low Fertility Countries: An Uptick in Europe?" *Population Reference Bureau,* September 2008, www.prb.org/Articles/2008/tfrtrendsept08.aspx.
Several low-fertility countries may have started to recover from low birthrates, based on 2007 rates recorded by the Population Reference Bureau.

For More Information

Center for Conservation Biology, Department of Biological Sciences, Stanford University, Stanford, CA 94305; (650) 723-3171; www.stanford.edu/group/CCB. Research arm of Stanford University's Department of Biological Sciences focusing on managing the planet's life-support systems.

Encore, 114 Sansome St., Suite 850, San Francisco, CA 94104; (415) 430-0141; www.encore.org. Nonprofit organization dedicated to helping retirees do socially meaningful work after retirement or as they approach retirement age.

International Institute for Applied Systems Analysis, Schlossplatz 1, A-2361, Laxenburg, Austria; +43 2236 8070; www.iiasa.ac.at. International research organization providing demographic data about Europe and other parts of the world.

Optimum Population Trust, 12 Meadowgate, Urmston, Manchester, England M41 9LB; +07976 370 221; www .optimumpopulation.org. A "green" think tank campaigning for capping population growth in Great Britain in order to protect the environment.

Population Action International, 1300 19th St., N.W., Suite 200, Washington, DC 20036; (202) 557-3400; www.populationaction.org. Private, nonprofit organization advocating family planning and sexual and reproductive rights and conducting research related to those fields.

Population Reference Bureau, 1875 Connecticut Ave., N.W., Suite 520, Washington, DC 20009; (202) 483-1100; www.prb.org. Nonprofit organization providing up-to-date statistics on world population and birthrates.

RAND Europe, Milton Road, Cambridge, England CB4 1YG; +44 1223 353 329; www.rand.org/randeurope. Independent research institute producing reports on declining birthrates and immigration in Europe.

United Nations Population Division, 777 United Nations Plaza, 5th Floor, New York, NY 10017; (212) 687-3366; www.population.org. U.N. division providing statistics and projections on world population.

Urban Institute, 2100 M St., N.W., Washington, DC 20037; (202) 833-7200; www.urban.org. Nonpartisan economic and social policy research group analyzing the economic impact of delaying retirement in the United States and the obstacles older people face in continuing to work.

15

Rapid Urbanization

Can Cities Cope With Rampant Growth?

Jennifer Weeks

Children scavenge for recyclables amid rubbish in the Dharavi slum in Mumbai, India. About a billion people worldwide live in slums — where sewer, water and garbage-collection services are often nonexistent. If impoverished rural residents continue streaming into cities at current rates, the world's slum population is expected to double to 2 billion within the next two decades, according to the United Nations.

From *CQ Researcher*, April 2009.

India's most infamous slum lives up to its reputation. Located in the middle of vast Mumbai, Dharavi is home to as many as 1 million people densely packed into thousands of tiny shacks fashioned from scrap metal, plastic sheeting and other scrounged materials. Narrow, muddy alleys crisscross the 600-acre site, open sewers carry human waste and vacant lots serve as garbage dumps. There is electricity, but running water is available for only an hour or so a day. Amid the squalor, barefoot children sing for money, beg from drivers in nearby traffic or work in garment and leather shops, recycling operations and other lightly regulated businesses.

Moviegoers around the globe got a glimpse of life inside Dharavi in last year's phenomenally popular Oscar-winning film "Slumdog Millionaire," about plucky Jamal Malik, a fictional Dharavi teenager who improbably wins a TV quiz-show jackpot. The no-holds-barred portrayal of slum life may have been shocking to affluent Westerners, but Dharavi is only one of Asia's innumerable slums. In fact, about a billion people worldwide live in urban slums — the ugly underbelly of the rapid and haphazard urbanization that has occurred in many parts of the world in recent decades. And if soaring urban growth rates continue unabated, the world's slum population is expected to double to 2 billion by 2030, according to the U.N.[1]

But all city dwellers don't live in slums. Indeed, other fast-growing cities presented cheerier faces to the world last year, from Dubai's glittering luxury skyscrapers to Beijing's breathtaking, high-tech pre-Olympic cultural spectacle.

World Will Have 26 Megacities by 2025

The number of megacities — urban areas with at least 10 million residents — will increase from 19 to 26 worldwide by the year 2025, according to the United Nations. The seven new megacities will be in Asia and sub-Saharan Africa. Most megacities are in coastal areas, making them highly vulnerable to massive loss of life and property damage caused by rising sea levels that experts predict will result from climate change in the 21st century.

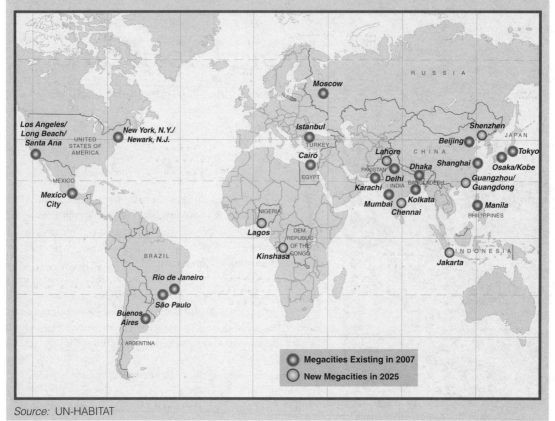

Source: UN-HABITAT

Today, 3.3 billion people live in cities — half the world's population — and urbanites are projected to total nearly 5 billion (out of 8.1 billion) worldwide by 2030.[2] About 95 percent of that growth is occurring in the developing world, especially in Africa and Asia.[3]

These regions are going through the same threefold evolution that transformed Europe and North America over a 200-year period between 1750 and 1950: the industrialization of agriculture, followed by rural migration to cities and declining population growth as life expectancy improves. But today's developing countries are modernizing much faster — typically in less than 100 years — and their cities are expanding at dizzying rates: On average, 5 million people in developing countries move to cities every month. As urban areas struggle to absorb this growth, the new residents often end up crowded into already teeming slums. For instance, 62 percent of city dwellers in sub-Saharan Africa live in slums, 43 percent in southern Asia, 37 percent in East Asia and 27 percent in Latin America and the Caribbean, according to UN-HABITAT, the United Nations agency for human settlements.[4]

UN-HABITAT defines a slum as an urban area without at least one of the following features:

- Durable housing,
- Adequate living space (no more than three people per room),
- Access to clean drinking water,
- Access to improved sanitation (toilets or latrines that separate human waste from contact with water sources), or
- Secure property rights.[5]

But all slums are not the same. Some lack only one basic necessity, while others lack several. And conditions can be harsh in non-slum neighborhoods as well. Thus, experts say, policies should focus on specific local problems in order to make a difference in the lives of poor city dwellers.[6]

Cities "are potent instruments for national economic and social development. They attract investment and create wealth," said HABITAT Executive Director Anna Tibaijuka last April. But, she warned, cities also concentrate poverty and deprivation, especially in developing countries. "Rapid and chaotic urbanization is being accompanied by increasing inequalities, which pose enormous challenges to human security and safety."[7]

Today, improving urban life is an important international development priority.[8] One of the eight U.N. Millennium Development Goals (MDGs) — broad objectives intended to end poverty worldwide by 2015 — endorsed by world leaders in 2000 was environmental sustainability. Among other things, it aims to cut in half the portion of the world's people without access to safe drinking water and achieve "significant improvement" in the lives of at least 100 million slum dwellers.[9]

Tokyo Is by Far the World's Biggest City

With more than 35 million residents, Tokyo is nearly twice as big as the next-biggest metropolises. Tokyo is projected to remain the world's largest city in 2025, when there will be seven new megacities — urban areas with at least 10 million residents. Two Indian cities, Mumbai and Delhi, will overtake Mexico City and New York as the world's second- and third-largest cities. The two largest newcomers in 2025 will be in Africa: Kinshasa and Lagos.

Population of Megacities, 2007 and 2025
(in millions)

2007		2025 (projected)	
Tokyo, Japan	35.68	Tokyo, Japan	36.40
New York, NY/Newark, NJ	19.04	Mumbai, India	26.39
Mexico City, Mexico	19.03	Delhi, India	22.50
Mumbai, India	18.98	Dhaka, Bangladesh	22.02
São Paulo, Brazil	18.85	São Paulo, Brazil	21.43
Delhi, India	15.93	Mexico City, Mexico	21.01
Shanghai, China	14.99	New York, NY/Newark, NJ	20.63
Kolkata, India	14.79	Kolkata, India	20.56
Dhaka, Bangladesh	13.49	Shanghai, China	19.41
Buenos Aires, Argentina	12.80	Karachi, Pakistan	19.10
Los Angeles/Long Beach/ Santa Ana (CA)	12.50	Kinshasa, Dem. Rep. Congo	16.76
Karachi, Pakistan	12.13	Lagos, Nigeria	15.80
Cairo, Egypt	11.89	Cairo, Egypt	15.56
Rio de Janeiro, Brazil	11.75	Manila, Philippines	14.81
Osaka/Kobe, Japan	11.29	Beijing, China	14.55
Beijing, China	11.11	Buenos Aires, Argentina	13.77
Manila, Philippines	11.10	Los Angeles/Long Beach/ Santa Ana (CA)	13.67
Moscow, Russia	10.45	Rio de Janeiro, Brazil	13.41
Istanbul, Turkey	10.06	Jakarta, Indonesia	12.36
		Istanbul, Turkey	12.10
		Guangzhou/Guangdong, China	11.84
		Osaka/Kobe, Japan	11.37
		Moscow, Russia	10.53
		Lahore, Pakistan	10.51
		Shenzhen, China	10.20
		Chennai, India	10.13

New megacities in 2025

Source: UN-HABITAT

Global Population Is Shifting to Cities

Half a century ago, less than a third of the world's population lived in cities. By 2005, nearly half inhabited urban areas, and in 2030, at least 60 percent of the world's population will be living in cities, reflecting an unprecedented scale of urban growth in the developing world. This will be particularly notable in Africa and Asia, where the urban population will double between 2000 and 2030.

Worldwide Urban and Rural Populations

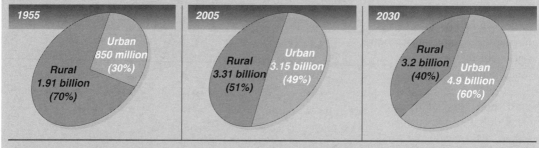

Source: U.N. Department of Economic and Social Affairs; U.N. Population Fund

Delivering even the most basic city services is an enormous challenge in many of the world's 19 megacities — metropolises with more than 10 million residents. And smaller cities with fewer than 1 million inhabitants are growing even faster in both size and number than larger ones.[10]

Many fast-growing cities struggle with choking air pollution, congested traffic, polluted water supplies and inadequate sanitation services. The lack of services can contribute to larger social and economic problems. For example, slum dwellers without permanent housing or access to mass transit have trouble finding and holding jobs. And when poverty becomes entrenched it reinforces the gulf between rich and poor, which can promote crime and social unrest.

"A city is a system of systems. It has biological, social and technical parts, and they all interact," says George Bugliarello, president emeritus of Polytechnic University in New York and foreign secretary of the National Academy of Engineering. "It's what engineers call a complex system because it has features that are more than the sum of its parts. You have to understand how all of the components interact to guide them."

Improving life for the urban poor begins with providing shelter, sanitation and basic social services like health care and education. But more is needed to make cities truly inclusive, such as guaranteeing slum dwellers' property rights so they cannot be ejected from their homes.[11]

Access to information and communications technology (ICT) is also crucial. In some developing countries, ICT has been adopted widely, particularly cell phones, but high-speed Internet access and computer use still lag behind levels in rich nations. Technology advocates say this "digital divide" slows economic growth in developing nations and increases income inequality both within and between countries. Others say the problem has been exaggerated and that there is no critical link between ICTs and poverty reduction.

Managing urban growth and preventing the creation of new slums are keys to both improving the quality of life and better protecting cities from natural disasters. Many large cities are in areas at risk from earthquakes, wildfires or floods. Squatter neighborhoods are often built on flood plains, steep slopes or other vulnerable areas, and poor people usually have fewer resources to escape or relocate.

For example, heavy rains in northern Venezuela in 1999 caused mudslides and debris flows that demolished many hillside shantytowns around the capital city of Caracas, killing some 30,000 people. In 2005 Hurricane Katrina killed more people in New Orleans' lower-income neighborhoods, which were located in a flood plain, than in wealthier neighborhoods of the Louisiana port city that were built on higher ground. As global warming raises sea levels, many of the world's largest cities are expected to be increasingly at risk from flooding.

Paradoxically, economic growth also can pose a risk for some cities. Large cities can be attractive targets for terrorist attacks, especially if they are symbols of national prosperity and modernity, such as New York City, site of the Sept. 11, 2001, attack on the World Trade Center. Last November's coordinated Islamic terrorist attacks in Mumbai followed a similar strategy: Landmark properties frequented by foreigners were targeted in order to draw worldwide media coverage, damage India's economy and send a message that nowhere in India was safe.[12]

Today the global economic recession is creating a new problem for city dwellers: Entry-level jobs are disappearing as trade contracts evaporate and factories shut down. Unable to find other jobs, many recent migrants to cities are returning to rural areas that are ill-prepared to receive them, and laborers who remain in cities have less money to send to families back home.[13]

As national leaders, development experts and city officials debate how to manage urban growth, here are some issues they are considering:

Does urbanization make people better off?

With a billion city dwellers worldwide trapped in slums, why do people keep moving to cities? Demographic experts say that newcomers hope to earn higher incomes and find more opportunities than rural areas can offer.

"Often people are fleeing desperate economic conditions," says David Bloom, a professor of economics and demography at Harvard University's School of Public Health. "And the social attractions of a city — opportunities to meet more people, escape from isolation or in some cases to be anonymous — trump fears about

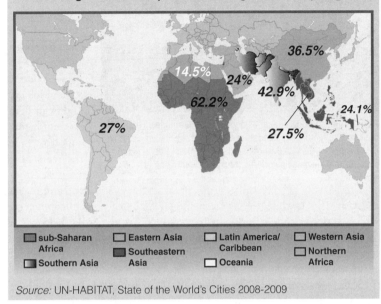

Most African City Dwellers Live in Slums

Most of the world's slum dwellers are in developing countries, with nearly two-thirds of sub-Saharan Africa's city dwellers living in slums.

Percentage of Urban Populations Living in Slums, by Region

36.5%
14.5%
24%
42.9%
62.2%
24.1%
27%
27.5%

☐ sub-Saharan Africa
☐ Southern Asia
☐ Eastern Asia
■ Southeastern Asia
☐ Latin America/ Caribbean
☐ Oceania
☐ Western Asia
☐ Northern Africa

Source: UN-HABITAT, State of the World's Cities 2008-2009

difficult urban conditions. If they have relatives or friends living in cities already, that reduces some of the risk."

When nations attract foreign investment, it creates new jobs. In the 1990s both China and India instituted broad economic reforms designed to encourage foreign investment, paving the way for rapid economic growth. That growth accelerated as information technology advances like the Internet, fiber-optic networks and e-mail made it faster and cheaper to communicate worldwide in real time.[14] As a result, thousands of manufacturing and white-collar jobs were "outsourced" from the United States to India, China and other low-wage countries over the past decade.[15]

These jobs spurred major growth in some cities, especially in areas with educated, English-speaking work forces. The large southern Indian city of Bangalore became a center for information technology — dubbed "India's Silicon Valley." Other cities in India, Singapore and the Philippines now host English-language call centers that manage everything from computer technical support to lost-baggage complaints for airlines. In a

Packed buses in Dhaka take residents in the Bangladeshi capital to their homes in outlying villages on the eve of the Muslim holiday Eid al-Adha — the "Festival of Sacrifice." Rapidly growing cities have trouble keeping up with the transportation needs of residents.

twist on this model, the Chinese city of Dalian — which was controlled by Japan from 1895 through World War II and still has many Japanese speakers — has become a major outsourcing center for Japanese companies.[16]

Some observers say an increasingly networked world allows people to compete for global "knowledge work" from anywhere in the world instead of having to emigrate to developed countries. In his best-seller *The World Is Flat*, author and *New York Times* columnist Thomas Friedman cites Asian call centers as an example of this shift, since educated Indians can work at the centers and prosper at home rather than seeking opportunity abroad. While he acknowledges that millions of people in developing countries are poor, sick and disempowered, Friedman argues that things improve when people move from rural to urban areas.

"[E]xcess labor gets trained and educated, it begins working in services and industry; that leads to innovation and better education and universities, freer markets, economic growth and development, better infrastructure, fewer diseases and slower population growth," Friedman writes. "It is that dynamic that is going on in parts of urban India and urban China today, enabling people to compete on a level playing field and attracting investment dollars by the billions."[17]

But others say it's not always so simple. Educated newcomers may be able to find good jobs, but migrants

without skills or training often end up working in the "informal economy" — activities that are not taxed, regulated or monitored by the government, such as selling goods on the street or collecting garbage for recycling. These jobs are easy to get but come without minimum wages or labor standards, and few workers can get credit to grow their businesses. Members of ethnic minorities and other underprivileged groups, such as lower castes in India, often are stuck with the dirtiest and most dangerous and difficult tasks.[18]

And some countries have experienced urban growth without job growth. Through the late 1980s, many Latin American countries tried to grow their economies by producing manufactured goods at home instead of importing them from abroad.

"Years of government protection insulated these industries from outside competition, so they did not feel pressure to become more productive. Then they went under when economies opened up to trade," says Steven Poelhekke, a researcher with DNB, the national bank of the Netherlands. "In Africa, industrialization has never really taken off. And without job creation governments cannot deliver benefits for new urbanites."[19]

Meanwhile, when cities grow too quickly, competition for land, space, light and services increases faster than government can respond. Real estate prices rise, driving poor residents into squatter neighborhoods, where crowding and pollution spread disease. "When cities get too big, the downsides to city life are bigger than the benefits for vulnerable inhabitants," says Poelhekke.

Broadly, however, urbanization has reduced the total number of people in poverty in recent years. According to a 2007 World Bank study, about three-quarters of the world's poor still live in rural areas. Poor people are urbanizing faster than the population as a whole, so some poverty is shifting to cities. Yet, clearly, many of those new urbanites are finding higher incomes — even if they end up living in city slums — because overall poverty rates (urban plus rural) fall as countries urbanize. While the persistence of urban poverty is a serious concern, the authors concluded, if people moved to the cities faster, overall poverty rates would decline sooner.[20]

Many development advocates say policy makers must accept urbanization as inevitable and strive to make it more beneficial. "We need to stop seeing migration to cities as a problem," says Priya Deshingkar, a researcher at the Overseas Development Institute in Hyderabad, India. "These people were already vulnerable because they can't make a living in rural areas. Countries need to rethink their development strategies. The world is urbanizing, and we have to make more provisions for people moving to urban areas. They can't depend on agriculture alone."

Should governments limit migration to cities?

Many governments have tried to limit urban problems by discouraging migration to cities or regulating the pace of urban growth. Some countries use household registration policies, while others direct aid and economic development funds to rural areas. Political leaders say limiting migration reduces strains on city systems, slows the growth of slums and keeps villages from languishing as their most enterprising residents leave.

China's *hukou* system, for example, requires households to register with the government and classifies individuals as rural or urban residents. Children inherit their *hukou* status from their parents. Established in the 1950s, the system was tightly controlled to limit migration from agricultural areas to cities and to monitor criminals, government critics and other suspect citizens and groups.[21]

In the late 1970s China began privatizing farming and opened its economy to international trade, creating a rural labor surplus and greater demand for city workers. The government offered rural workers temporary residence permits in cities and allowed wealthy, educated citizens to buy urban *hukou* designations. Many rural Chinese also moved to cities without changing their registration. According to recent government estimates, at least 120 million migrant workers have moved to Chinese cities since the early 1980s.[22] Today *hukou* rules are enforced inconsistently in different Chinese cities, where many rural migrants cannot get access to health care, education, affordable housing or other urban services because they are there illegally.

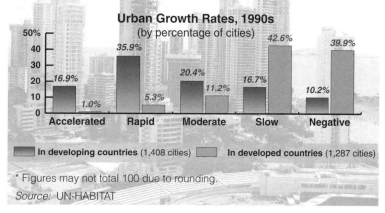

Cities in Developing World Growing Rapidly

More than half the developing world's cities experienced fast annual growth in the 1990s, compared to just 6.3 percent of those in wealthier countries. Conversely, more than 80 percent of cities in the wealthier countries had slow or negative growth, compared to about a quarter of those in developing countries.

Urban Growth Rates, 1990s
(by percentage of cities)

	Accelerated	Rapid	Moderate	Slow	Negative
In developing countries (1,408 cities)	16.9%	35.9%	20.4%	16.7%	10.2%
In developed countries (1,287 cities)	1.0%	5.3%	11.2%	42.6%	39.9%

* Figures may not total 100 due to rounding.
Source: UN-HABITAT

Chinese officials say they must manage growth so all areas of the country will benefit. In a 2007 report to the 17th Communist Party Congress, President Hu Jintao promised to promote "a path of urbanization with Chinese characteristics" that emphasized "balanced development of large, medium-sized and small cities and towns."[23]

But critics say the *hukou* system has created an urban underclass and should be scrapped. When the municipality of Chongqing (which omits an estimated 4.5 million migrant workers from its official population figures) established November 4 as Migrant Workers' Day in 2007, the *Asia Times* commented, "By not changing the [*hukou*] system and instead giving the migrant workers a special holiday, it's a bit like showing starving people menus instead of feeding them."[24]

India and Vietnam also control migration to urban areas by requiring people to register or show local identity cards to access social services. "They're both trying to promote rural development and keep from overburdening urban areas," says Deshingkar at the Overseas Development Institute. "But it doesn't work. People move despite these regulations. It just makes it harder for them, and if they can access services it's at a price."

Many experts say governments should not try to halt rural-to-city migration because when migrant workers send large shares of their wages home to their families in the country it helps reduce rural poverty and inequality. In Dhaka, Bangladesh, for example, remittances from city workers provide up to 80 percent of rural households' budgets, according to the Coalition for the Urban Poor.[25]

Urban growth also helps rural economies by creating larger markets for agricultural products — including high-value products like meat, chicken and fish that people tend to add to their diets as their incomes rise. Cities can promote economic growth in surrounding areas by creating a demand for local farmers' products. For instance, South Africa's Johannesburg Fresh Produce Market offers vendors stalls, overnight storage space, business-skills training and financing; it also requires market agents to buy at least 10 percent of their produce from small, low-income farms.[26]

However, the rootless lifestyle adopted by so-called circular migrants — those who move back and forth between the city and the country — makes people vulnerable, Deshingkar points out. "There are roughly 100 million circular migrants in India now, and they're completely missed by official statistics because the government only counts permanent migrants," she says. "They can't get any insurance or social services, so they carry all the risk themselves."

Beyond the fact that anti-migration policies usually fail, experts say the biggest factor driving population increase in many fast-growing cities is not new residents moving in but "natural increase" — the rate at which people already living there have children. Natural increase accounts for about 60 percent of urban growth worldwide, while 20 percent comes from domestic and international migration and 20 percent results from reclassification of rural areas as urban.[27]

Family-planning programs helped reduce poverty rates in several developing Asian countries — including South Korea, Taiwan, Thailand, Singapore, Indonesia and Malaysia — where having smaller families increased household savings and reduced national education costs.[28] In contrast, artificial birth control is difficult to obtain in the Philippines, where the population is 80 percent Catholic and the government supports only "natural" family planning. Several professors at the University of the Philippines have calculated that if Filipinos had

followed Thailand's example on family planning in the 1970s, the Philippines would have at least 4 million fewer people in poverty and would be exporting rice rather than importing it. Instead, the Philippine government's opposition to family planning "contributed to the country's degeneration into Southeast Asia's basket case," said economist Arsenio Balisacan.[29]

Can we make large cities greener?

Many fast-growing cities are unhealthy places to live because of dirty air, polluted water supplies and sprawling waste dumps. City governments worldwide are increasingly interested in making their cities greener and more sustainable.

Greening cities has many up-front costs but can provide big payoffs. For example, energy-efficient buildings cost less to operate and give cities cachet as centers for advanced technology and design.

Green policies also may help cities achieve broader social goals. When Enrique Peñalosa was elected mayor of Bogotá, Colombia, in 1998, the city was overrun with traffic and crime. Wealthy residents lived in walled-off neighborhoods, while workers were squeezed into shanties on the city's outskirts. Under Peñalosa's rule, the city built hundreds of new parks and a rapid-transit bus system, limited automobile use, banned sidewalk parking and constructed a 14-mile-long street for bicyclists and pedestrians that runs through some of the city's poorest neighborhoods. The underlying goal of the programs: Make Bogotá more people-friendly for poor residents as well as the rich.

"[A]nything that you do in order to increase pedestrian space constructs equality" said Peñalosa, who now consults with city officials in other developing countries. "It's a powerful symbol showing that citizens who walk are equally important to those who have a car."[30] His administration also invested funds that might otherwise have been spent building highways in social services like schools and libraries. Air pollution decreased as more residents shifted to mass transit. Crime rates also fell, partly because more people were out on the streets.[31]

"Mobility and land use may be the most important issues that a mayor can address, because to unlock the economic potential of cities people have to be able to move from one area to another," says Polytechnic University's Bugliarello. "You also have to take care of water supplies and sanitation, because cities concentrate people and pathologies. Appropriate technologies aren't always the most expensive options, especially if cities get together and form markets for them."

For example, bus rapid transit (BRT) systems, which create networks of dedicated lanes for high-speed buses, are much cheaper than subways but faster than conventional buses that move in city traffic. By 2007 some 40 cities worldwide had developed BRT systems, including Bogotá; Jakarta, Indonesia; and Guayaquil, Ecuador. Many others are planned or under construction.[32]

Some developing countries are planning entire green cities with walkable neighborhoods, efficient mass transit and renewable-energy systems. Abu Dhabi, part of the United Arab Emirates on the Persian Gulf, is designing a $20 billion project called Masdar City, which it bills as the world's first carbon-neutral, zero-waste city. Located on the coast next to Abu Dhabi's airport, Masdar City will be a mixed-use community with about 40,000 residents and 50,000 commuters traveling in to work at high-tech companies. Plans call for the city to be car-free and powered mainly by solar energy.[33]

Abu Dhabi wants to become a global hub for clean technologies, according to Khaled Awad, property development director for the Masdar initiative. "It lets us leverage our energy knowledge [from oil and gas production] and our research and development skills and adapt them to new energy markets," he said.

"If we can do it there, we can do it anywhere," said Matthias Schuler, an engineer with the German climate-engineering firm Transsolar and a member of the international Masdar City design and planning team.[34] He points out that average daytime summer temperatures in Abu Dhabi are well over 100 degrees Fahrenheit, and coastal zones are very humid. "You can't find a harsher climate."

In China, meanwhile, green urban design is gaining support as a way to attract foreign investment and demonstrate environmental awareness. But some showpiece projects are falling short of expectations.

Huangbaiyu was supposed to be a sustainable "green village" that would provide new homes for a farming town of more than 1,400 in rural northeast China. But the master plan, produced by a high-profile U.S. green architecture firm, called for 400 densely clustered bungalows without enough yard space for livestock. This meant that villagers would lose their existing income from backyard gardens, sheep flocks and trout ponds. The plan also proposed to use corncobs and stalks to fuel a biogas plant for heat, but villagers needed these crop wastes as winter feed for their goats.

By December 2008 the Chinese builder had constructed 42 houses, but only a few were occupied. The designer blamed the builder for putting up low-quality houses, but others said the plan did not reflect what villagers wanted or needed.[35] Planners "inadvertently designed an ecologically sound plan — from the perspectives of both birds and the green movement — that would devastate the local economy and bankrupt the households whose lives were to be improved," wrote Shannon May, an American graduate student who lived in the old village of Huangbaiyu for two years and wrote her dissertation on the project.[36]

Dongtan, a larger Chinese city designed as a green project with zero-carbon-emission buildings and transit systems, has also been sidetracked. Groundbreaking on the model city of 500,000 on a Manhattan-sized island near Shanghai is more than a year behind schedule. High-rise towers are sprouting up around the site, leading some observers to call the project expensive "greenwashing" — attempting to make lavish development acceptable by tacking on environmentally friendly features.

" 'Zero-emission' city is pure commercial hype," said Dai Xingyi, a professor at Fudan University in Shanghai. "You can't expect some technology to both offer you a luxurious and comfortable life and save energy at the same time. That's just a dream."[37]

Construction is also under way on a new green city southeast of Beijing for 350,000 residents, co-developed by China and Singapore. Tianjin's features include renewable-energy sources, efficient water use and green building standards. Premier Wen Jiabao attended the 2008 groundbreaking.[38]

Although China's green development projects have a mixed record so far, "The government is starting to recognize that it has responsibility for environmental impacts beyond its borders, mainly by promoting renewable energy," says Alastair MacGregor, associate vice president of AECOM, an international design firm with large building projects in China. "Chinese culture is playing catch-up on sustainability."

More than 130 buildings designed to LEED (Leadership in Energy and Environmental Design)

China Aggressively Tackles Air Pollution

"No country in developing Asia takes those challenges more seriously"

China's large cities have some of the world's worst air pollution, thanks to rapid industrial growth, heavy use of coal and growing demand for cars.

The capital, Beijing, lost its 1993 bid to host the 2000 Summer Olympic Games partly because the city was so polluted. A chronic grey haze not only sullied Beijing's international image but also threatened to cause health problems for athletes and impair their performances.

When Beijing was chosen in 2001 to host the 2008 Summer Games, it pledged to put on a "green Olympics," which was widely understood to include clearing the air.

Between 2001 and 2007, however, China's economy grew beyond all predictions, with its gross domestic product expanding by up to 13 percent a year.[1] Beijing's air pollution worsened as new factories, power plants and cars crowded into the city. Winds carried in more pollutants from other burgeoning cities, including nitrogen oxides and sulfur dioxide — which contribute to acid rain and smog — and fine particulates, which can cause or worsen heart and lung problems.

With the Olympic deadline looming, many observers predicted Beijing would not meet its targets even if it relied heavily on authoritarian measures like shutting down factories and limiting auto use.[2] International Olympic Committee President Jacques Rogge said some outdoor endurance sports might have to be postponed if they occurred on high-pollution days — an embarrassing prospect for Chinese leaders.[3]

But China met its promised target, keeping Beijing's daily air pollution index — based on combined measurements of sulfur dioxide, nitrogen dioxide and fine particulates — below 100 during the month the Olympics took place. A 100 index score means air quality will not affect daily activities, compared to a maximum score of 500, when officials warn residents to stay indoors. In fact, during the Olympics in August 2008 Beijing's daily air pollution reached the lowest August measurements since 2000, sometimes even dropping into the 20s.[4]

"No country in Asia has bigger air quality challenges than China, but no country in developing Asia takes those challenges more seriously," says Cornie Huizenga, executive director of the Clean Air Initiative for Asian Cities (CAI-Asia), an international network based in the Philippines and founded by the Asian Development Bank, the World Bank and the U.S. Agency for International Development. "China has taken a whole series of long-term structural measures to address air pollution. The Olympics put a

standards — which measure energy efficiency and healthy indoor working conditions — are planned or under construction in Beijing, Shanghai, Chongqing, Wuhan and other Chinese cities.[39] Chinese investors see LEED buildings as premium products, not as an everyday model, said MacGregor.

Some Chinese cities are developing their own green standards. About half of worldwide new construction between 2008 through 2015 is projected to occur in China, so even greening a modest share of that development would be significant.

"China could end up being a sustainability leader just by virtue of its size," MacGregor predicted.[40]

BACKGROUND

From Farm to Factory

At the beginning of the 19th century only 3 percent of the world's population lived in cities, and only Beijing had more than a million inhabitants.[41] Then new technologies like the steam engine and railroads began to transform society. As the Industrial Revolution unfolded, people streamed from rural areas to manufacturing centers in Europe and the United States seeking a better income and life. This first great wave of urbanization established cities like London, Paris and New York as centers of global commerce.

magnifying glass on Beijing and made them focus there, but its programs are much bigger."

For instance, China continuously monitors air quality in more than 100 cities, requires high-polluting provinces and companies to close small, inefficient emission sources and install pollution-control equipment and has new-car emissions standards roughly equivalent to U.S. and Western European laws.

"For the Olympics China took temporary measures on top of those policies, like closing down large facilities and keeping cars off the roads. All of this plus good weather let Beijing deliver what it promised for the Games," says Huizenga.

Now China is further expanding air pollution regulations. During the Olympics, the Ministry of Environment announced that in 2009 it would start monitoring ultra-fine particle and ozone pollution — persistent problems in many developed countries. And Beijing officials plan to increase spending on public transportation.

Local pollution sources, weather patterns and geography influence air pollution, so China's policies for cleaning up Beijing's air might not work in other large cities. Mexico City, for instance, also has tried to reduce its severe air pollution but is hampered by the city's high altitude (7,200 feet). Car engines burn fuel inefficiently at high altitudes, so they pollute more than at sea level. And while automobiles are the biggest emission sources, scientists also found that leaking liquefied petroleum gas (LPG) — which most Mexican households burn for

cooking and heating — also contributes to Mexico City's air pollution.[5]

"We need better-harmonized air quality monitoring in developing countries before we can compare them," says Huizenga. "But other cities should be able to make progress on a large scale like Beijing. There's a lot of low-hanging fruit, such as switching to cleaner transportation fuels, getting rid of vehicles with [high-polluting] two-stroke engines, managing dust at construction sites and cutting pollution from coal-fired power plants. But to make them work, you also need effective agencies with enough people and money to carry [out] policies."

---

[1] Michael Yang, "China's GDP (2003-2007)," forum.china.org.cn, Nov. 10, 2008; "China Revises 2007 GDP Growth Rate to 13%," Jan. 15, 2009, http://english.dbw.cn.

[2] Edward Russell, "Beijing's 'Green Olympics' Test Run Fizzles," *Asia Times*, Aug. 10, 2007; Jim Yardley, "Beijing's Olympic Quest: Turn Smoggy Sky Blue," *The New York Times*, Dec. 29, 2007; David G. Streets, *et al.*, "Air Quality during the 2008 Beijing Olympic Games," *Atmospheric Environment*, vol. 41 (2007).

[3] "IOC President: Beijing Air Pollution Could Cause Events to Be Delayed During 2008 Olympics," The Associated Press, Aug. 7, 2007.

[4] "Summary: AQ in Beijing During the 2008 Summer Olympics," Clean Air Initiative for Asian Cities, www.cleanairnet.org/caiasia/1412/article-72991.html. Weather conditions are important factors in air pollution levels — for example, summer heat and humidity promote the formation of ground-level ozone, a major ingredient of smog — so to put conditions during the Olympics in context, scientists compared them to readings taken in August of previous years.

[5] Tim Weiner, "Terrific News in Mexico City: Air Is Sometimes Breathable," *The New York Times*, Jan. 5, 2001.

It also spawned horrific slums in factory towns and large cities. Tenement houses became a feature of working-class neighborhoods, with little access to fresh air or clean drinking water. Often whole neighborhoods shared a single water pump or toilet, and trash was usually thrown into the streets.[42]

German social scientist and a co-founder of communist theory Friedrich Engels graphically described urban workers' living conditions in cities like London and Manchester in 1844: "[T]hey are penned in dozens into single rooms. . . . They are given damp dwellings, cellar dens that are not waterproof from below or garrets that leak from above. . . . They are supplied bad, tattered or

rotten clothing, adulterated or indigestible food. . . . Thus are the workers cast out and ignored by the class in power, morally as well as physically and mentally."[43]

Engels and his collaborator Karl Marx later predicted in *The Communist Manifesto* that oppression of the working class would lead to revolution in industrialized countries. Instead, public health movements began to develop in Europe and the United States in mid-century. Seeking to curb repeated cholera and typhoid epidemics, cities began collecting garbage and improving water-supply systems. A new medical specialty, epidemiology (the study of how infections are spread) developed as scientists worked to track and contain illnesses. Cities built

CHRONOLOGY

1700s-1800s *Industrial Revolution spurs rapid urban growth in Europe and the U.S. Expanding slums trigger reforms and public health laws.*

1804 World population reaches 1 billion.

1854 British doctor John Snow discovers the connection between contaminated drinking water and a cholera outbreak in London.

1897 Brazil's first *favela* (shanty town), is established outside Rio de Janeiro.

1900-1960s *Europe and the United States are the most urbanized. Africa and Asia begin gaining independence and struggle to develop healthy economies.*

1906 An earthquake and subsequent fire destroy much of San Francisco, killing more than 3,000 people.

1927 World population reaches 2 billion.

1949 Chinese communists defeat nationalists, establishing the People's Republic of China, which aggressively promotes industrial development.

1960 World population hits 3 billion.

1964 Tokyo becomes first Asian city to host the Olympic Games and soon after that displaces New York as the world's largest city.

1970s-1990s *Urbanization accelerates in Asia and Africa. Many U.S. and European cities shrink as residents move to suburbs.*

1971 East Pakistan secedes from West Pakistan and becomes the independent nation of Bangladesh; populations in Dhaka and other cities grow rapidly.

1974 World population reaches 4 billion.

1979 China initiates broad economic reforms, opens diplomatic and trade relations with the United States and starts to ease limits on migration to cities.

1985 An earthquake in Mexico City kills some 10,000 people and damages water-supply and transit systems.

1987 World population reaches 5 billion.

1991 India institutes sweeping market reforms to attract foreign investors and spur rapid economic growth.

1999 World population reaches 6 billion.

2000s *Most industrialized countries stabilize at 70-80 percent urban. Cities continue to grow in Asia and Africa.*

2000 International community endorses the U.N. Millennium Development Goals designed to end poverty by 2015, including improving the lives of slum dwellers.

2001 Many international companies shift production to China after it joins the World Trade Organization; migration from rural areas accelerates. . . . Terrorists destroy World Trade Center towers in New York City, killing thousands. . . . Taiwan completes Taipei 101, the world's tallest skyscraper (1,671 feet), superseding the Petronas Towers in Kuala Lumpur, Malaysia (1,483 feet).

2005 United Nations condemns Zimbabwe for slum-clearance operations that leave 700,000 people homeless.

2007 The nonprofit group One Laptop Per Child unveils a prototype $100 laptop computer designed for children in developing countries to help close the "digital divide" between cities and rural areas.

2008 More than half of the world's population lives in cities. . . . Beijing hosts Summer Olympic Games. . . . Coordinated terrorist attacks in Mumbai kill nearly 170 people and injure more than 300.

2009 A global recession leaves millions of urban workers jobless, forcing many to return to their home villages.

2030 World's urban population is expected to reach 5 billion, and its slum population could top 2 billion.

2070 About 150 million city dwellers — primarily in India, Bangladesh, China, Vietnam, Thailand, Myanmar and Florida — could be in danger due to climate change, according to a 2008 study.

green spaces like New York's Central Park to provide fresh air and access to nature. To help residents navigate around town, electric streetcars and subway trains were built in underground tunnels or on elevated tracks above the streets.

Many problems persisted, however. Homes and factories burned coal for heat and power, blanketing many large cities in smoky haze. Horse-drawn vehicles remained in wide use until the early-20th century, so urban streets were choked with animal waste. Wealthy city dwellers, seeking havens from the noise, dirt and crowding of inner cities, moved out to cleaner suburban neighborhoods.

Despite harsh conditions, people continued to pour into cities. Economic growth in industrialized countries had ripple effects in developing countries. As wealthier countries imported more and more raw materials, commercial "gateway cities" in developing countries grew as well, including Buenos Aires, Rio de Janeiro and Calcutta (now Kolkata). By 1900, nearly 14 percent of the world's population lived in cities.[44]

End of Empires

Worldwide migration from country to city accelerated in the early-20th century as automation spread and fewer people were needed to grow food. But growth was not uniform. Wars devastated some of Europe's major cities while industrial production swelled others. And when colonial empires dissolved after World War II, many people were displaced in newly independent nations.

Much of the fighting during World War I occurred in fields and trenches, so few of Europe's great cities were seriously damaged. By the late 1930s, however, long-range bombers could attack cities hundreds of miles away. Madrid and Barcelona were bombed during the Spanish Civil War, a prelude to intensive air attacks on London, Vienna, Berlin, Tokyo and elsewhere during World War II. In 1945 the United States dropped atomic bombs on the Japanese cities of Hiroshima and Nagasaki, destroying each. For centuries cities had walled themselves off against outside threats, but now they were vulnerable to air attacks from thousands of miles away.

After 1945, even victorious nations like Britain and France were greatly weakened and unable to manage overseas colonies, where independence movements were underway. As European countries withdrew from their holdings in the Middle East, Asia and Africa over the next 25 years, a wave of countries gained independence, including Indonesia, India, Pakistan, the Philippines, Syria, Vietnam and most of colonial Africa. Wealthy countries began providing aid to the new developing countries, especially in Asia and Latin America. But some nations, especially in Africa, received little focused support.

By mid-century most industrialized countries were heavily urbanized, and their populations were no longer growing rapidly. By 1950 three of the world's largest cities — Shanghai, Buenos Aires and Calcutta — were in developing countries. Populations in developing countries continued to rise through the late 1960s even as those nations struggled to industrialize. Many rural residents moved to cities, seeking work and educational opportunities.

In the 1950s and '60s U.S. urban planners heatedly debated competing approaches to city planning. The top-down, centralized philosophy was espoused by Robert Moses, the hard-charging parks commissioner and head of New York City's highway agency from 1934 to 1968. Moses pushed through numerous bridge, highway, park and slum-clearance projects that remade New York but earned him an image as arrogant and uncaring.[45] His most famous critic, writer and activist Jane Jacobs, advocated preserving dense, mixed-use neighborhoods, like New York's Greenwich Village, and consulting with residents to build support for development plans.[46] Similar controversies would arise later in developing countries.

By the 1960s car-centered growth characterized many of the world's large cities. "Circle over London, Buenos Aires, Chicago, Sydney, in an airplane," wrote American historian Lewis Mumford in 1961. "The original container has completely disappeared: the sharp division between city and country no longer exists." City officials, Mumford argued, only measured improvements in quantities, such as wider streets and bigger parking lots.

"[T]hey would multiply bridges, highways [and] tunnels, making it ever easier to get in and out of the city but constricting the amount of space available within the city for any other purpose than transportation itself," Mumford charged.[47]

Population Boom

In the 1970s and '80s, as populations in developing countries continued to grow and improved agricultural

Cities Need to Plan for Disasters and Attacks

Concentrated populations and wealth magnify impact

Flash floods in 1999 caused landslides in the hills around Caracas, Venezuela, that washed away hundreds of hillside shanties and killed an estimated 30,000 people — more than 10 times the number of victims of the Sept. 11, 2001, terrorist attacks in the United States.

Because cities concentrate populations and wealth, natural disasters in urban areas can kill or displace thousands of people and cause massive damage to property and infrastructure. Many cities are located on earthquake faults, flood plains, fire-prone areas and other locations that make them vulnerable. The impacts are magnified when high-density slums and squatter neighborhoods are built in marginal areas. Political instability or terrorism can also cause widespread destruction.

Protecting cities requires both "hard" investments, such as flood-control systems or earthquake-resistant buildings, and "soft" approaches, such as emergency warning systems and special training for police and emergency-response forces. Cities also can improve their forecasting capacity and train officials to assess different types of risk.[1] Although preventive strategies are expensive, time-consuming and often politically controversial, failing to prepare for outside threats can be far more costly and dangerous.

Global climate change is exacerbating flooding and heat waves, which are special concerns for cities because they absorb more heat than surrounding rural areas and have higher average temperatures — a phenomenon known as the urban heat island effect. According to a study by the Organization for Economic Cooperation and Development (OECD), about 40 million people living in coastal areas around the world in 2005 were exposed to so-called 100-year floods — or major floods likely to occur only once

every 100 years. By the 2070s, the OECD said, the population at risk from such flooding could rise to 150 million as more people move to cities, and climate change causes more frequent and ferocious storms and rising sea levels.

Cities with the greatest population exposure in the 2070 forecast include Kolkata and Mumbai in India, Dhaka (Bangladesh), Guangzhou and Shanghai in China, Ho Chi Minh City and Hai Phong in Vietnam, Bangkok (Thailand), Rangoon (Myanmar) and Miami, Florida. Cities in developed countries tend to be better protected, but there are exceptions. For example, London has about the same amount of flooding protection as Shanghai, according to the OECD.[2]

"All cities need to look at their critical infrastructure systems and try to understand where they're exposed to natural hazards," says Jim Hall, leader of urban research at England's Tyndall Centre for Climate Change Research. For example, he says, London's Underground subway system is vulnerable to flooding and overheating. Fast-growing cities planning for climate change, he adds, might want to control growth in flood-prone areas, improve water systems to ensure supply during droughts or build new parks to help cool urban neighborhoods. "Risks now and in the future depend on what we do to protect cities," says Hall.

In some cities, residents can literally see the ocean rising. Coastal erosion has destroyed 47 homes and more than 400 fields in recent years in Cotonou, the capital city of the West African nation of Benin, according to a local non-profit called Front United Against Coastal Erosion. "The sea was far from us two years ago. But now, here it is. We are scared," said Kofi Ayao, a local fisherman. "If we do not find a solution soon, we may simply drown in our sleep one day."[3]

methods made farmers more productive, people moved to the cities in ever-increasing numbers. Some national economies boomed, notably the so-called Asian tigers — Hong Kong, Singapore, Taiwan and South Korea — by focusing on manufacturing exports for industrialized markets and improving their education systems to create productive work forces.

Indonesia, Malaysia, the Philippines and Thailand — the "tiger cubs" — went through a similar growth phase in the late 1980s and early '90s.

After China and India opened up their economies in the 1980s and '90s, both countries became magnets for foreign investment and created free-trade areas and special economic zones to attract business activity. Cities in

Social violence can arise from within a city or come as an attack from outside. For example, in 2007 up to 600 people were killed when urban riots erupted in Kenya after a disputed national election.[4]

Urban leaders often justify slum-clearance programs by claiming that poor neighborhoods are breeding grounds for unrest. Others say slums are fertile recruiting grounds for terrorist groups. Slums certainly contain many who feel ill-treated, and extreme conditions may spur them into action. Overall, however, experts say most slum dwellers are too busy trying to eke out a living to riot or join terrorist campaigns.

"Poverty alone isn't a sufficient cause [for unrest]," says John Parachini, director of the Intelligence Policy Center at the RAND Corp., a U.S. think tank. "You need a combination of things — people with a profound sense of grievance, impoverishment and leaders who offer the prospect of change. Often the presence of an enemy nearby, such as an occupying foreign power or a rival tribal group or religious sect, helps galvanize people."

Last November's terrorist attacks in Mumbai, in which 10 gunmen took dozens of Indian and foreign hostages and killed at least 164 people, showed an ironic downside of globalization: Wealth, clout and international ties can make cities terrorist targets.

"Mumbai is India's commercial and entertainment center — India's Wall Street, its Hollywood, its Milan. It is a prosperous symbol of modern India," a RAND analysis noted. Mumbai also was accessible from the sea, offered prominent landmark targets (historic hotels frequented by

AP Photo/Pavel Rahman

A Bangladeshi boy helps slum residents cross floodwaters in Dhaka. Rising waters caused by global warming pose a significant potential threat to Dhaka and other low-lying cities worldwide.

foreigners and local elites) and had a heavy media presence that guaranteed international coverage.[5]

But serendipity can also make one city a target over another, says Parachini. "Attackers may know one city better or have family links or contacts there. Those local ties matter for small groups planning a one-time attack," he says.

Developing strong core services, such as police forces and public health systems, can be the first step in strengthening most cities against terrorism, he says, rather than creating specialized units to handle terrorist strikes.

"Basic governance functions like policing maintain order, build confidence in government and can pick up a lot of information about what's going on in neighborhoods," he says. "They make it harder to do bad things."

[1] George Bugliarello, "The Engineering Challenges of Urban Sustainability," *Journal of Urban Technology*, vol. 15, no. 1 (2008), pp. 64-65.

[2] R. J. Nicholls, *et al.*, "Ranking Port Cities with High Exposure and Vulnerability to Climate Extremes: Exposure Estimates," *Environment Working Papers No. 1*, Organization for Economic Cooperation and Development, Nov. 19, 2008, pp. 7-8, www.olis.oecd.org/olis/2007doc .nsf/LinkTo/NT0000588E/$FILE/JT03255617.PDF.

[3] "Rising Tides Threaten to Engulf Parts of Cotonou," U.N. Integrated Regional Information Network, Sept. 2, 2008.

[4] "Chronology: Kenya in Crisis After Elections," Reuters, Dec. 31, 2007; "The Ten Deadliest World Catastrophes 2007," Insurance Information Institute, www.iii.org.

[5] Angel Rabasa, *et al.*, "The Lessons of Mumbai," *RAND Occasional Paper*, January 2009.

those areas expanded, particularly along China's southeast coast where such zones were clustered.

As incomes rose, many Asian cities aspired to global roles: Seoul hosted the 1988 Summer Olympics, and Malaysia built the world's tallest skyscrapers — the Petronas Twin Towers, completed in 1998, only to be superseded by the Taipei 101 building in Taiwan a few years later.

Some Asian countries — including Malaysia, Sri Lanka and Indonesia — implemented programs to improve living standards for the urban poor and helped reduce poverty. However, poverty remained high in Thailand and the Philippines and increased in China and Vietnam.[48]

Cities in South America and Africa also expanded rapidly between 1970 and 2000, although South America

AP Photo

Security officers forcibly remove a woman from her home during land confiscations in Changchun, a city of 7.5 million residents in northeast China, so buildings can be demolished to make way for new construction. Some rapidly urbanizing governments use heavy-handed methods — such as land confiscation, eviction or slum clearance — so redevelopment projects can proceed.

was farther ahead. By 1965 Latin America was already 50 percent urbanized and had three cities with populations over 5 million (Buenos Aires, São Paulo and Rio de Janeiro) — a marker sub-Saharan Africa would not achieve for several decades.[49] Urban growth on both continents followed the "primacy" pattern, in which one city is far more populous and economically and politically powerful than all the others in the nation. The presence of so-called primate cities like Lima (Peru), Caracas (Venezuela) or Lagos (Nigeria) can distort development if the dominant city consumes most public investments and grows to a size that is difficult to govern.

Latin America's growth gradually leveled out in the 1980s: Population increases slowed in major urban centers, and more people moved to small and medium-sized cities.[50] On average the region's economy grew more slowly and unevenly than Asia's, often in boom-and-bust cycles.[51] Benefits accrued mostly to small ruling classes who were hostile to new migrants, and income inequality became deeply entrenched in many Latin American cities.

Africa urbanized quickly after independence in the 1950s and '60s. But from the mid-1970s forward most countries' incomes stagnated or contracted. Such "urbanization without growth" in sub-Saharan Africa created the world's highest rates of urban poverty and income inequality. Corruption and poor management reinforced wealth gaps that dated back to colonial

times. Natural disasters, wars and the spread of HIV/AIDS further undercut poverty-reduction efforts in both rural and urban areas.[52]

New Solutions

As the 21st century began, calls for new antipoverty efforts led to an international conference at which 189 nations endorsed the Millennium Development Goals, designed to end poverty by 2015. Experts also focused on bottom-up strategies that gave poor people resources to help themselves.

An influential proponent of the bottom-up approach, Peruvian economist Hernando de Soto, stirred debate in 2000 with his book *The Mystery of Capital: Why Capitalism Triumphs in the West and Fails Everywhere Else*. Capitalist economies did not fail in developing nations because those countries lacked skills or enterprising spirit, de Soto argued. Rather, the poor in those countries had plenty of assets but no legal rights, so they could not prove ownership or use their assets as capital.

"They have houses but not titles; crops but not deeds; businesses but not statutes of incorporation," de Soto wrote. "It is the unavailability of these essential representations that explains why people who have adapted every other Western invention, from the paper clip to the nuclear reactor, have not been able to produce sufficient capital to make their domestic capitalism work." But, he asserted, urbanization in the developing world had spawned "a huge industrial-commercial revolution" which clearly showed that poor people could contribute to economic development if their countries developed fair and inclusive legal systems.[53]

Not all experts agreed with de Soto, but his argument coincided with growing interest in approaches like microfinance (small-scale loans and credit programs for traditionally neglected customers) that helped poor people build businesses and transition from the "extra-legal" economy into the formal economy. Early microcredit programs in the 1980s and '90s had targeted mainly the rural poor, but donors began expanding into cities around 2000.[54]

The "digital divide" — the gap between rich and poor people's access to information and communications technologies (ICTs) — also began to attract the attention of development experts. During his second term (1997-2001), U.S. President Bill Clinton highlighted the issue as an obstacle to reducing poverty both domestically and at the global level. "To maximize potential, we must turn the digital divide among and within our nations into digital opportunities," Clinton said at the Asia Pacific Economic Cooperation Forum in 2000, urging Asian nations to expand Internet access and train citizens to use computers.[55] The Millennium Development Goals called for making ICTs more widely available in poor countries.

Some ICTs, such as mobile phones, were rapidly adopted in developing countries, which had small or unreliable landline networks. By 2008, industry observers predicted, more than half of the world's population would own a mobile phone, with Africa and the Middle East leading the way.[56]

Internet penetration moved much more slowly. In 2006 some 58 percent of the population in industrial countries used the Internet, compared to 11 percent in developing countries and only 1 percent in the least developed countries. Access to high-speed Internet service was unavailable in many developing regions or was too expensive for most users.[57] Some antipoverty advocates questioned whether ICTs should be a high priority for poor countries, but others said the issue was not whether but when and how to get more of the world's poor wired.

"The more the better, especially broadband," says Polytechnic University's Bugliarello.

While development experts worked to empower the urban poor, building lives in fast-growing cities remained difficult and dangerous in many places. Some governments still pushed approaches like slum clearance, especially when it served other purposes.

Notoriously, in 2005 President Robert Mugabe of Zimbabwe launched a slum-clearance initiative called Operation Murambatsvina, a Shona phrase translated by some as "restore order" and others as "drive out the trash." Thousands of shacks in Zimbabwe's capital, Harare, and other cities across the nation were destroyed, allegedly to crack down on illegal settlements and businesses.

"The current chaotic state of affairs, where small-to-medium enterprises operated outside of the regulatory framework and in undesignated and crime-ridden areas, could not be countenanced much longer," said Mugabe.[58]

But critics said Mugabe was using slum clearance as an excuse to intimidate and displace neighborhoods that supported his opponents. In the end, some 700,000 people were left homeless or jobless by the action, which the United Nations later said violated international law.[59]

Over the next several years Mugabe's government failed to carry out its pledges to build new houses for the displaced families.[60]

CURRENT SITUATION
Economic Shadow

The current global economic recession is casting a dark cloud over worldwide economic development prospects. Capital flows to developing countries have declined sharply, and falling export demand is triggering layoffs and factory shutdowns in countries that produce for Western markets. But experts say even though the overall picture is sobering, many factors will determine how severely the recession affects cities.

In March the World Bank projected that developing countries would face budget shortfalls of $270 billion to $700 billion in 2009 and the world economy would shrink for the first time since World War II. According to the bank, 94 out of 116 developing countries were already experiencing an economic slowdown, and about half of them already had high poverty levels. Urban-based exporters and manufacturers were among the sectors hit hardest by the recession.[61]

These trends, along with an international shortage of investment capital, will make many developing countries increasingly dependent on foreign aid at a time when donor countries are experiencing their own budget crises. As workers shift out of export-oriented sectors in the cities and return to rural areas, poverty may increase, the bank projected.

The recession could mean failure to meet the Millennium Development Goals, especially if donor countries pull back on development aid. The bank urged nations to increase their foreign aid commitments and recommended that national governments:

- Increase government spending where possible to stimulate economies;
- Protect core programs to create social safety nets for the poor;
- Invest in infrastructure such as roads, sewage systems and slum upgrades; and
- Help small- and medium-size businesses get financing to create opportunities for growth and employment.[62]

Getty Images/Daniel Berehulak

AFP/Getty Images/Pal Pillai

Slum Redevelopment Plan Stirs Controversy

Conditions for the 60,000 families living in Mumbai's Dharavi neighborhood (top) — one of Asia's largest slums — are typical for a billion slum dwellers around the globe. Slums often lack paved roads, water-distribution systems, sanitation and garbage collection — spawning cholera, diarrhea and other illnesses. Electric power and telephone service are usually poached from available lines. Mumbai's plans to redevelop Dharavi, located on 600 prime acres in the heart of the city, triggered strong protests from residents, who demanded that their needs be considered before the redevelopment proceeds (bottom). The project has stalled recently due to the global economic crisis.

President Barack Obama's economic stimulus package, signed into law on Feb. 17, takes some of these steps and contains at least $51 billion for programs to help U.S. cities. (Other funds are allocated by states and may provide more aid to cities depending on each state's priority list.) Stimulus programs that benefit cities include $2.8 billion for energy conservation and energy efficiency, $8.4 billion for public transportation investments, $8 billion for high-speed rail and intercity passenger rail service, $1.5 billion for emergency shelter grants, $4 billion for job training and $8.8 billion for modernizing schools.[63]

Governments in developing countries with enough capital may follow suit. At the World Economic Forum in Davos, Switzerland, in January, Chinese Premier Wen Jibao announced a 4 trillion yuan stimulus package (equivalent to about 16 percent of China's GDP over two years), including money for housing, railways and infrastructure and environmental protection. " 'The harsh winter will be gone, and spring is around the corner,' " he said, predicting that China's economy would rebound this year.[64]

But according to government figures released just a few days later, more than 20 million rural migrant workers had already lost their jobs in coastal manufacturing areas and moved back to their home towns.[65] In March the World Bank cut its forecast for China's 2009 economic growth from 7.5 percent to 6.5 percent, although it said China was still doing well compared to many other countries.[66]

In India "circular migration" is becoming more prevalent, according to the Overseas Development Institute's Deshingkar. "Employment is becoming more temporary — employers like to hire temporary workers whom they can hire and fire at will, so the proportion of temporary workers and circular migrants is going up," she says. "In some Indian villages 95 percent of migrants are circular. Permanent migration is too expensive and risky — rents are high, [people are] harassed by the police, slums are razed and they're evicted. Keeping one foot in the village is their social insurance."

Meanwhile, international development aid is likely to decline as donor countries cut spending and focus on their own domestic needs. "By rights the financial crisis shouldn't undercut development funding, because the total amounts given now are tiny compared to the national economic bailouts that are under way or being debated in developed countries," says Harvard economist Bloom. "Politically, however, it may be hard to maintain aid budgets."

At the World Economic Forum billionaire philanthropist Bill Gates urged world leaders and organizations to keep up their commitments to foreign aid despite the global financial crisis. "If we lose sight of our long-term priority to expand opportunity for the world's poor and abandon our commitments and partnerships to reduce inequality, we run the risk of emerging from the current economic downturn in a world with even greater disparities in health and education and fewer opportunities for people to improve their lives," said Gates, whose Bill and Melinda Gates Foundation supports efforts to address both rural and urban poverty in developing nations.[67]

In fact, at a summit meeting in London in early April, leaders of the world's 20 largest economies pledged $1.1 trillion in new aid to help developing countries weather the global recession. Most of the money will be channeled through the International Monetary Fund.

"This is the day the world came together to fight against the global recession," said British Prime Minister Gordon Brown.[68]

Slum Solutions

As slums expand in many cities, debate continues over the best way to alleviate poverty. Large-scale slum-clearance operations have long been controversial in both developed and developing countries: Officials typically call the slums eyesores and public health hazards, but often new homes turn out to be unaffordable for the displaced residents. Today development institutions like the World Bank speak of "urban upgrading" — improving services in slums instead of bulldozing them.[69]

This approach focuses on improving basic infrastructure systems like water distribution, sanitation and electric power; cleaning up environmental hazards and building schools and clinics. The strategy is cheaper than massive demolition and construction projects and provides incentives for residents to invest in improving their own homes, advocates say.[70]

To do so, however, slum dwellers need money. Many do not have the basic prerequisites even to open bank accounts, such as fixed addresses and minimum balances, let alone access to credit. Over the past 10 to 15 years, however, banks have come to recognize slum dwellers as potential customers and have begun creating microcredit programs to help them obtain small loans and credit cards that often start with very low limits. A related concept, micro-insurance, offers low-cost protection in case of illness, accidents and property damage.

Now advocates for the urban poor are working to give slum dwellers more financial power. The advocacy group, Shack/Slum Dwellers International (SDI), for example, has created Urban Poor Funds that help attract direct

Reflecting China's stunningly rapid urbanization, Shanghai's dramatic skyline rises beside the Huangpu River. Shanghai is the world's seventh-largest city today but will drop to ninth-place by 2025, as two south Asian megacities, Dhaka and Kolkata, surpass Shanghai in population.

AP Photo/Zhou Junxiang

investments from banks, government agencies and international donor groups.[71] In 2007 SDI received a $10 million grant from the Gates foundation to create a Global Finance Facility for Federations of the Urban Poor.

The funds will give SDI leverage in negotiating with governments for land, housing and infrastructure, according to Joel Bolnick, an SDI director in Cape Town, South Africa. If a government agency resists, said Bolnick, SDI can reply, " 'If you can't help us here, we'll take the money and put it on the table for a deal in Zambia instead.' "[72]

And UN-HABITAT is working with lenders to promote more mortgage lending to low-income borrowers in developing countries. "Slum dwellers have access to resources and are resources in themselves. To maximize the value of slums for those who live in them and for a city, slums must be upgraded and improved," UN-HABITAT Executive Director Tibaijuka said in mid-2008.[73]

Nevertheless, some governments still push slum clearance. Beijing demolished hundreds of blocks of old city neighborhoods and culturally significant buildings in its preparations to host the 2008 Olympic Games. Some of these "urban corners" (a negative term for high-density neighborhoods with narrow streets) had also been designated for protection as historic areas.[74] Developers posted messages urging residents to take government resettlement fees and move, saying, "Living in the Front Gate's courtyards is ancient history; moving to an apartment makes

Reuters/George Esiri

Two-thirds of sub-Saharan Africa's city dwellers live in slums, like this one in Lagos, Nigeria, which has open sewers and no clean water, electric power or garbage collection. About 95 percent of today's rapid urbanization is occurring in the developing world, primarily in sub-Saharan Africa and Asia.

you a good neighbor," and "Cherish the chance; grab the good fortune; say farewell to dangerous housing."[75]

Beijing's actions were not unique. Other cities hosting international "mega-events" have demolished slums. Like Beijing, Seoul, South Korea, and Santo Domingo in the Dominican Republic were already urbanizing and had slum-clearance programs under way, but as their moments in the spotlight grew nearer, eviction operations accelerated, according to a Yale study. Ultimately, the study concluded, the benefits from hosting big events did not trickle down to poor residents and squatter communities who were "systematically removed or concealed from high-profile areas in order to construct the appearance of development."[76]

Now the debate over slum clearance has arrived in Dharavi. Developers are circling the site, which sits on a square mile of prime real estate near Mumbai's downtown and airport. The local government has accepted a $3 billion redevelopment proposal from Mukesh Mehta, a wealthy architect who made his fortune in Long Island, N.Y., to raze Dharavi's shanties and replace them with high-rise condominiums, shops, parks and offices. Slum dwellers who can prove they have lived in Dharavi since 1995 would receive free 300-square-foot apartments, equivalent to two small rooms, in the new buildings. Other units would be sold at market rates that could reach several thousand dollars per square foot.[77]

Mehta contends his plan will benefit slum residents because they will receive new homes on the same site. "Give me a better solution. Until then you might want to accept this one," he said last summer.[78] But many Dharavi residents say they will not be able to keep small businesses like tanneries, potteries and tailoring shops if they move into modern high-rises, and would rather stay put. (*See "At Issue," p. 377.*)

"I've never been inside a tall building. I prefer a place like this where I can work and live," said Usman Ghani, a potter born and raised in Dharavi who has demonstrated against the redevelopment proposals. He is not optimistic about the future. "The poor and the working class won't be able to stay in Mumbai," he said. "Many years ago, corrupt leaders sold this country to the East India Company. Now they're selling it to multinationals."[79]

OUTLOOK
Going Global

In an urbanizing world, cities will become increasingly important as centers of government, commerce and culture, but some will be more influential than others. Although it doesn't have a precise definition, the term "global city" is used by city-watchers to describe metropolises like New York and London that have a disproportionate impact on world affairs. Many urban leaders around the world aspire to take their cities to that level.

The 2008 *Global Cities Index* — compiled by *Foreign Policy* magazine, the Chicago Council on Global Affairs and the A. T. Kearney management consulting firm — ranks 60 cities on five broad criteria that measure their international influence, including:

- Business activity,
- Human capital (attracting diverse groups of people and talent),
- Information exchange,
- Cultural attractions and experiences, and
- Political engagement (influence on world policy making and dialogue).[80]

The scorecard is topped by Western cities like New York, London and Paris but also includes developing-country cities such as Beijing, Shanghai, Bangkok, Mexico

Will redevelopment of the Dharavi slum improve residents' lives?

YES
Mukesh Mehta
Chairman, MM Project Consultants

Written for *CQ Global Researcher,* April 2009

Slum rehabilitation is a challenge that has moved beyond the realm of charity or meager governmental budgets. It requires a pragmatic and robust financial model and a holistic approach to achieve sustainability.

Dharavi — the largest slum pocket in Mumbai, India, and one of the largest in the world — houses 57,000 families, businesses and industries on 600 acres. Alarmingly, this accounts for only 4 percent of Mumbai's slums, which house about 7.5 million people, or 55 percent of the city's population.

Mumbai's Slum Rehabilitation Authority (SRA) has undertaken the rehabilitation of all the eligible residents and commercial and industrial enterprises in a sustainable manner through the Dharavi Redevelopment Project (DRP), following an extensive consultative process that included Dharavi's slum dwellers. The quality of life for those residents is expected to dramatically improve, and they could integrate into mainstream Mumbai over a period of time. Each family would receive a 300-square-foot home plus adequate workspace, along with excellent infrastructure, such as water supply and roads. A public-private partnership between the real estate developers and the SRA also would provide amenities for improving health, income, knowledge, the environment and socio-cultural activities. The land encroached by the slum dwellers would be used as equity in the partnership.

The primary focus — besides housing and infrastructure — would be on income generation. Dharavi has a vibrant economy of $600 million per annum, despite an appalling working environment. But the redevelopment project would boost the local gross domestic product to more than $3 billion, with the average family income estimated to increase to at least $3,000 per year from the current average of $1,200. To achieve this, a hierarchy of workspaces will be provided, including community spaces equivalent to 6 percent of the built-up area, plus individual workspaces in specialized commercial and industrial complexes for leather goods, earthenware, food products, recycling and other enterprises.

The greatest failure in slum redevelopment has been to treat it purely as a housing problem. Improving the infrastructure to enable the local economy to grow is absolutely essential for sustainable development. We believe this project will treat Dharavi residents as vital human resources and allow them to act as engines for economic growth. Thus, the DRP will act as a torchbearer for the slums of Mumbai as well as the rest of the developing world.

NO
Kalpana Sharma
Author, Rediscovering Dharavi:
Stories from Asia's Largest Slum

Written for *CQ Global Researcher,* April 2009

The controversy over the redevelopment of Dharavi, a slum in India's largest city of Mumbai, centers on the future of the estimated 60,000 families who live and work there.

Dharavi is a slum because its residents do not own the land on which they live. But it is much more than that. The settlement — more than 100 years old — grew up around one of the six fishing villages that coalesced over time to become Bombay, as Mumbai originally was called. People from all parts of India live and work here making terra-cotta pots, leather goods, garments, food items and jewelry and recycling everything from plastic to metal. The annual turnover from this vast spread of informal enterprises, much of it conducted inside people's tiny houses, is an estimated $700 million a year.

The Dharavi Redevelopment Plan — conceived by consultant Mukesh Mehta and being implemented by the Government of Maharashtra state — envisages leveling this energetic and productive part of Mumbai and converting it into a collection of high-rise buildings, where some of the current residents will be given free apartments. The remaining land will be used for high-end commercial and residential buildings.

On paper, the plan looks beautiful. But people in Dharavi are not convinced. They believe the plan has not understood the nature and real value of Dharavi and its residents. It has only considered the value of the land and decided it is too valuable to be wasted on poor people.

Dharavi residents have been left with no choice but to adapt to an unfamiliar lifestyle. If this meant a small adjustment, one could justify it. But the new form of living in a 20-story high-rise will force them to pay more each month, since the maintenance costs of high-rises exceed what residents currently spend on housing. These costs become unbearable when people earn just enough to survive in a big city.

Even worse, this new, imposed lifestyle will kill all the enterprises that flourish today in Dharavi. Currently, people live and work in the same space. In the new housing, this will not be possible.

The alternatives envisaged are spaces appropriate for formal, organized industry. But enterprises in Dharavi are informal and small, working on tiny margins. Such enterprises cannot survive formalization.

The real alternative is to give residents security of tenure and let them redevelop Dharavi. They have ideas. It can happen only if people are valued more than real estate.

In addition to Dubai's glittering, new downtown area filled with towering skyscrapers, the city's manmade, palm-tree-shaped islands of Jumeirah sport hundreds of multi-million-dollar second homes for international jetsetters. Development has skidded to a temporary halt in the Arab city-state, much as it has in some other rapidly urbanizing cities around the globe, due to the global economic downturn.

City and São Paulo. Many of these cities, the authors noted, are taking a different route to global stature than their predecessors followed — a shorter, often state-led path with less public input than citizens of Western democracies expect to have.

"Rulers in closed or formerly closed societies have the power to decide that their capitol is going to be a world-class city, put up private funds and spell out what the city should look like," says Simon O'Rourke, executive director of the Global Chicago Center at the Chicago Council on Global Affairs. "That's not necessarily a bad path, but it's a different path than the routes that New York or London have taken. New global cities can get things done quickly — if the money is there."

Abu Dhabi's Masdar Initiative, for example, remains on track despite the global recession, directors said this spring. The project is part of a strategic plan to make Abu Dhabi a world leader in clean-energy technology. "There is no question of any rollback or slowing down of any of our projects in the renewable-energy sector," said Sultan Ahmed Al Jaber, chief executive officer of the initiative, on March 16.[81] Last year the crown prince of Abu Dhabi created a $15 billion fund for clean-energy investments, which included funds for Masdar City.

Money is the front-burner issue during today's global recession. "Unless a country's overall economic progress is solid, it is very unlikely that a high proportion of city dwellers will see big improvements in their standard of living," says Harvard's Bloom. In the next several years, cities that ride out the global economic slowdown successfully will be best positioned to prosper when world markets recover.

In the longer term, however, creating wealth is not enough, as evidenced by conditions in Abu Dhabi's neighboring emirate, Dubai. Until recently Dubai was a booming city-state with an economy built on real estate, tourism and trade — part of the government's plan to make the city a world-class business and tourism hub. It quickly became a showcase for wealth and rapid urbanization: Dozens of high-rise, luxury apartment buildings and office towers sprouted up seemingly overnight, and man-made islands shaped like palm trees rose from the sea, crowded with multi-million-dollar second homes for jetsetters.

But the global recession has brought development to a halt. The real estate collapse was so sudden that jobless expatriate employees have been fleeing the country, literally abandoning their cars in the Dubai airport parking lot.[82]

Truly global cities are excellent in a variety of ways, says O'Rourke. "To be great, cities have to be places where people want to live and work." They need intellectual and cultural attractions as well as conventional features like parks and efficient mass transit, he says, and, ultimately, they must give residents at least some role in decisionmaking.

"It will be very interesting to see over the next 20 years which cities can increase their global power without opening up locally to more participation," says O'Rourke. "If people don't have a say in how systems are built, they won't use them."

Finally, great cities need creative leaders who can adapt to changing circumstances. Mumbai's recovery after last November's terrorist attacks showed such resilience. Within a week stores and restaurants were open again in neighborhoods that had been raked by gunfire, and international travelers were returning to the city.[83]

The Taj Mahal Palace & Tower was one of the main attack targets. Afterwards, Ratan Tata, grand-nephew of the Indian industrialist who built the five-star hotel, said, "We can be hurt, but we can't be knocked down."[84]

NOTES

1. Ben Sutherland, "Slum Dwellers 'to top 2 billion,' " BBC News, June 20, 2006, http://news.bbc.co.uk/2/hi/in_depth/5099038.stm.

2. United Nations Population Fund, *State of World Population 2007: Unleashing the Potential of Urban Growth* (2007), p. 6.

3. UN-HABITAT, *State of the World's Cities 2008/2009* (2008), p. xi.

4. UN-HABITAT, *op cit.*, p. 90.

5. *Ibid.*, p. 92.

6. *Ibid.*, pp. 90-105.

7. Anna Tibaijuka, "The Challenge of Urbanisation and the Role of UN-HABITAT," lecture at the Warsaw School of Economics, April 18, 2008, p. 2, www.unhabitat.org/downloads/docs/5683_16536_ed_warsaw_version12_1804.pdf.

8. For background see Peter Katel, "Ending Poverty," *CQ Researcher*, Sept. 9, 2005, p. 733-760.

9. For details, see www.endpoverty2015.org. For background, see Peter Behr, "Looming Water Crisis," *CQ Global Researcher*, February 2008, pp. 27-56.

10. Tobias Just, "Megacities: Boundless Growth?" Deutsche Bank Research, March 12, 2008, pp. 4-5.

11. Commission on Legal Empowerment of the Poor, *Making the Law Work for Everyone* (2008), pp. 5-9, www.undp.org/legalempowerment/report/Making_the_Law_Work_for_Everyone.pdf.

12. Angel Rabasa, *et al.*, "The Lessons of Mumbai," *RAND Occasional Paper*, 2009, pp. 1-2, www.rand.org/pubs/occasional_papers/2009/RAND_OP249.pdf.

13. Wieland Wagner, "As Orders Dry Up, Factory Workers Head Home," *Der Spiegel*, Jan. 8, 2009, www.spiegel.de/international/world/0,1 518,600188,00.html; Malcolm Beith, "Reverse Migration Rocks Mexico," *Foreign Policy.com*, February 2009, www.foreignpolicy.com/story/cms.php?story_id=4731; Anthony Faiola, "A Global Retreat As Economies Dry Up," *The Washington Post*, March 5, 2009, www.washington-post.com/wp-dyn/content/story/2009/03/04/ST2009030404264.html.

14. For background, see David Masci, "Emerging India, *CQ Researcher*, April 19, 2002, pp. 329-360; and Peter Katel, "Emerging China," *CQ Researcher*, Nov. 11, 2005, pp. 957-980.

15. For background, see Mary H. Cooper, "Exporting Jobs," *CQ Researcher*, Feb. 20, 2004, pp. 149-172.

16. Ji Yongqing, "Dalian Becomes the New Outsourcing Destination," *China Business Feature*, Sept. 17, 2008, www.cbfeature.com/industry_spotlight/news/dalian_becomes_the_new_outsourcing_destination.

17. Thomas L. Friedman, *The World Is Flat: A Brief History of the Twenty-First Century*, updated edition (2006), pp. 24-28, 463-464.

18. Priya Deshingkar and Claudia Natali, "Internal Migration," in *World Migration 2008* (2008), p. 183.

19. Views expressed here are the speaker's own and do not represent those of his employer.

20. Martin Ravallion, Shaohua Chen and Prem Sangraula, "New Evidence on the Urbanization of Global Poverty," World Bank Policy Research Working Paper 4199, April 2007, http://siteresources.worldbank.org/INTWDR2008/Resources/2795087-1191427986785/RavallionMEtAl_UrbanizationOfGlobalPoverty.pdf.

21. For background on the *hukou* system, see Congressional-Executive Commission on China, "China's Household Registration System: Sustained Reform Needed to Protect China's Rural Migrants," Oct. 7, 2005, www.cecc.gov/pages/news/hukou.pdf; and Hayden Windrow and Anik Guha, "The Hukou System, Migrant Workers, and State Power in the People's Republic of China," *Northwestern University Journal of International Human Rights*, spring 2005, pp. 1-18.

22. Wu Zhong, "How the Hukou System Distorts Reality," *Asia Times*, April 11, 2007, www.atimes.com/atimes/China/ID11Ad01.html; Rong Jiaojiao, "Hukou 'An Obstacle to Market Economy,' " *China Daily*, May 21, 2007, www.chinadaily.com.cn/china/2007-05/21/content_876699.htm.

23. "Scientific Outlook on Development," "Full text of Hu Jintao's report at 17th Party Congress," section V.5, Oct. 24, 2007, http://news.xinhuanet.com/english/2007-10/24/content_6938749.htm.

24. Wu Zhong, "Working-Class Heroes Get Their Day," *Asia Times*, Oct. 24, 2007, www.atimes.com/atimes/China_Business/IJ24Cb01.html.

25. "Internal Migration, Poverty and Development in Asia," *Briefing Paper no. 11*, Overseas Development Council, October 2006, p. 3.

26. Clare T. Romanik, "An Urban-Rural Focus on Food Markets in Africa," The Urban Institute, Nov. 15, 2007, p. 30, www.urban.org/publications/411604.html.

27. UN-HABITAT, *op. cit.*, pp. 24-26.

28. "How Shifts to Smaller Family Sizes Contributed to the Asian Miracle," *Population Action International*, July 2006, www.popact.org/Publications/Fact_Sheets/FS4/Asian_Miracle.pdf.

29. Edson C. Tandoc, Jr., "Says UP Economist: Lack of Family Planning Worsens Poverty," *Philippine Daily Inquirer*, Nov. 11, 2008, http://newsinfo.inquirer.net/breakingnews/nation/view/20081111-171604/Lack-of-family-planning-worsens-poverty; Blaine Harden, "Birthrates Help Keep Filipinos in Poverty," *The Washington Post*, April 21, 2008, www.washingtonpost.com/wp-dyn/content/story/2008/04/21/ST2008042100778.html.

30. Kenneth Fletcher, "Colombia Dispatch 11: Former Bogotá Mayor Enrique Peñalosa," Smithsonian.com, Oct. 29, 2008, www.smithsonianmag.com/travel/Colombia-Dispatch-11-Former-Bogota-mayor-Enrique-Penalosa.html.

31. Charles Montgomery, "Bogota's Urban Happiness Movement," *Globe and Mail*, June 25, 2007, www.theglobeandmail.com/servlet/story/RTGAM.20070622.whappyurbanmain0623/BNStory/lifeMain/home.

32. Bus Rapid Transit Planning Guide, 3rd edition, Institute for Transportation & Development Policy, June 2007, p. 1, www.itdp.org/documents/Bus%20Rapid%20Transit%20Guide%20%20complete%20guide.pdf.

33. Project details at www.masdaruae.com/en/home/index.aspx.

34. Awad and Schuler remarks at Greenbuild 2008 conference, Boston, Mass., Nov. 20, 2008.

35. "Green Dreams," Frontline/World, www.pbs.org/frontlineworld/fellows/green_dreams/; Danielle Sacks, "Green Guru Gone Wrong: William McDonough," *Fast Company*, Oct. 13, 2008, www.fastcompany.com/magazine/130/the-mortal-messiah.html; Timothy Lesle, "Cradle and All," *California Magazine*, September/October 2008, www.alumni.berkeley.edu/California/200809/lesle.asp.

36. Shannon May, "Ecological Crisis and Eco-Villages in China," *Counterpunch*, Nov. 21-23, 2008, www.counterpunch.org/may11212008.html.

37. Rujun Shen, "Eco-city seen as Expensive 'Green-Wash,'" *The Standard* (Hong Kong), June 24, 2008, www.thestandard.com.hk/news_detail.asp?we_cat=9&art_id=67641&sid=19488136&con_type=1&d_str=20080624&fc=8; see also Douglas McGray, "Pop-Up Cities: China Builds a Bright Green Metropolis," *Wired*, April 24, 2007, www.wired.com/wired/archive/15.05/feat_popup.html; Malcolm Moore, "China's Pioneering Eco-City of Dongtan Stalls," *Telegraph*, Oct. 19, 2008, www.telegraph.co.uk/news/worldnews/asia/china/3223969/Chinas-pioneering-eco-city-of-Dongtan-stalls. html; "City of Dreams," *Economist*, March 19, 2009, www.economist.com/world/asia/displaystory.cfm?story_id=13330904.

38. Details at www.tianjinecocity.gov.sg/.

39. "LEED Projects and Case Studies Directory," U.S. Green Building Council, www.usgbc.org/LEED/Project/RegisteredProjectList.aspx.

40. Remarks at Greenbuild 2008 conference, Boston, Mass., Nov. 20, 2008.

41. Population Reference Bureau, "Urbanization," www.prb.org; Tertius Chandler, *Four Thousand Years of Urban Growth: An Historical Census* (1987).

42. Lewis Mumford, *The City In History: Its Origins, Its Transformations, and Its Prospects* (1961), pp. 417-418.

43. Frederick Engels, *The Condition of the Working Class in England* (1854), Chapter 7 ("Results"), online at Marx/Engels Internet Archive, www.marxists.org/archive/marx/works/1845/condition-working-class/ch07.htm.

44. Population Reference Bureau, *op. cit.*

45. Robert A. Caro, *The Power Broker: Robert Moses and the Fall of New York* (1975).

46. Jane Jacobs, *The Death and Life of Great American Cities* (1961).

47. Mumford, *op. cit.*, pp. 454-455.

48. Joshua Kurlantzick, "The Big Mango Bounces Back," *World Policy Journal*, spring 2000, www.worldpolicy.org/journal/articles/kurlant.html; UN-HABITAT, *op. cit.*, pp. 74-76.

49. BBC News, "Interactive Map: Urban Growth," http://news.bbc.co.uk/2/shared/spl/hi/world/06/urbanisation/html/urbanisation.stm.

50. Licia Valladares and Magda Prates Coelho, "Urban Research in Latin America: Towards a Research Agenda," MOST Discussion Paper Series No. 4 (undated), www.unesco.org/most/valleng.htm#trends.

51. Jose de Gregorie, "Sustained Growth in Latin America," Economic Policy Papers, Central Bank of Chile, May 2005, www.bcentral.cl/eng/studies/economic-policy-papers/pdf/dpe13eng.pdf.

52. UN-HABITAT, *op cit.*, pp. 70-74.

53. Hernando de Soto, *The Mystery of Capital: Why Capitalism Triumphs in the West and Fails Everywhere Else* (2000), excerpted at http://ild.org.pe/en/mystery/english?page=0%2C0.

54. Deepak Kindo, "Microfinance Services to the Urban Poor," *Microfinance Insights*, March 2007; World Bank, "10 Years of World Bank Support for Microcredit in Bangladesh," Nov. 5, 2007; "Micro Finance Gaining in Popularity," *The Hindu*, Aug. 25, 2008, www.hindu.com/biz/2008/08/25/stories/2008082550121600.htm.

55. Michael Richardson, "Clinton Warns APEC of 'Digital Divide,'" *International Herald Tribune*, Nov. 16, 2000, www.iht.com/articles/2000/11/16/apec.2.t_2.php.

56. Abigail Keene-Babcock, "Study Shows Half the World's Population With Mobile Phones by 2008," Dec. 4, 2007, www.nextbillion.net/news/study-shows-half-the-worlds-population-with-mobile-phones-by-200.

57. "Millennium Development Goals Report 2008," United Nations, p. 48, www.un.org/millenniumgoals/pdf/The%20Millennium%20Development%20Goals%20Report%202008.pdf.

58. Robyn Dixon, "Zimbabwe Slum Dwellers Are Left With Only Dust," *Los Angeles Times*, June 21, 2005, http://articles.latimes.com/2005/jun/21/world/fg-nohomes21.

59. Ewen MacAskill, "UN Report Damns Mugabe Slum Clearance as Catastrophic," *Guardian*, July 23, 2005, www.guardian.co.uk/world/2005/jul/23/zimbabwe.ewenmacaskill.

60. Freedom House, "Freedom in the World 2008: Zimbabwe," www.freedomhouse.org/uploads/press_release/Zimbabwe_FIW_08.pdf.

61. "Crisis Reveals Growing Finance Gaps for Developing Countries," World Bank, March 8, 2009, http://web.worldbank.org/WBSITE/EXTERNAL/NEWS/0,,contentMDK:22093316~menuPK:34463~pagePK:34370~piPK:34424~theSitePK:4607,00.html.

62. "Swimming Against the Tide: How Developing Countries Are Coping with the Global Crisis," World Bank, background paper prepared for the G20 finance Ministers meeting, March 13-14, 2009, http://siteresources.worldbank.org/NEWS/Resources/swimmingagainstthetide-march2009.pdf.

63. "Major Victories for City Priorities in American Recovery and Reinvestment Act," U.S. Conference of Mayors, Feb. 23, 2009, www.usmayors.org/usmayornewspaper/documents/02_23_09/pg1_major_victories.asp.

64. Carter Dougherty, "Chinese Premier Injects Note of Optimism at Davos," *The New York Times*, Jan. 29, 2009, www.nytimes.com/2009/01/29/business/29econ.html?partner=rss.

65. Jamil Anderlini and Geoff Dyer, "Downturn Causes 20m Job Losses in China," *Financial Times*, Feb. 2, 2009, www.ft.com/cms/s/0/19c25aea-f0f5-11dd-8790-0000779fd2ac.html.

66. Joe McDonald, "World Bank Cuts China's 2009 Growth Forecast," The Associated Press, March 18, 2009.

67. "Bill and Melinda Gates Urge Global Leaders to Maintain Foreign Aid," Bill and Melinda Gates Foundation, Jan. 30, 2009, www.gatesfoundation.org/press-releases/Pages/2009-world-economic-forum-090130.aspx.

68. Mark Landler and David E. Sanger, "World Leaders Pledge $1.1 Trillion to Tackle Crisis," *The New York*

Times, April 4, 2009, www.nytimes.com/2009/04/03/world/europe/03summit.html?_r=1&hp.

69. "Is Demolition the Way to Go?" World Bank, www.worldbank.org/urban/upgrading/demolition.html.

70. "What Is Urban Upgrading?" World Bank, www.worldbank.org/urban/upgrading/what.html.

71. For more information, see "Urban Poor Fund," *Shack/Slum Dwellers International*, www.sdinet.co.za/ritual/urban_poor_fund/.

72. Neal R. Peirce, "Gates Millions, Slum-Dwellers: Thanksgiving Miracle?" *Houston Chronicle*, Nov. 22, 2007, www.sdinet.co.za/static/pdf/sdi_gates_iupf_neal_peirce.pdf.

73. "Statement at the African Ministerial Conference on Housing and Urban Development," UN-HABITAT, Abuja, Nigeria, July 28, 2008, www.unhabitat.org/content.asp?cid=5830&catid=14&typeid=8&subMenuId=0.

74. Michael Meyer, *The Last Days of Old Beijing* (2008), pp. 54-55; Richard Spencer, "History is Erased as Beijing Makes Way for Olympics," *Telegraph* (London), June 19, 2006, www.telegraph.co.uk/news/worldnews/asia/china/1521709/History-is-erased-as-Beijing-makes-way-for-Olympics.html; Michael Sheridan, "Old Beijing Falls to Olympics Bulldozer," *Sunday Times* (London), April 29, 2007, www.timesonline.co.uk/tol/news/world/asia/china/article1719945.ece.

75. Meyer, *op. cit.*, pp. 45, 52.

76. Solomon J. Greene, "Staged Cities: Mega-Events, Slum Clearance, and Global Capital," *Yale Human Rights & Development Law Journal*, vol. 6, 2003, http://islandia.law.yale.edu/yhrdlj/PDF/Vol%206/greene.pdf.

77. Slum Rehabilitation Authority, "Dharavi Development Project," www.sra.gov.in/htmlpages/Dharavi.htm; Porus P. Cooper, "In India, Slum May Get Housing," *Philadelphia Inquirer*, Sept. 22, 2008.

78. Mukul Devichand, "Mumbai's Slum Solution?" BBC News, Aug. 14, 2008, http://news.bbc.co.uk/2/hi/south_asia/7558102.stm.

79. Henry Chu, "Dharavi, India's Largest Slum, Eyed By Mumbai Developers," *Los Angeles Times*, Sept. 8, 2008, www.latimes.com/news/nationworld/world/la-fg-dharavi8-2008sep08,0,1830588.story; see also Dominic Whiting, "Dharavi Dwellers Face Ruin in Development Blitz," Reuters, June 6, 2008, http://in.reuters.com/article/topNews/idINIndia-33958520080608; and Mark Tutton, "Real Life 'Slumdog' Slum To Be Demolished," CNN.com, Feb. 23, 2009, www.cnn.com/2009/TRAVEL/02/23/dharavi.mumbai.slums/.

80. Unless otherwise cited, this section is based on "The 2008 Global Cities Index," *Foreign Policy*, November/December 2008, www.foreignpolicy.com/story/cms.php?story_id=4509.

81. T. Ramavarman, "Masdar To Proceed with $15 Billion Investment Plan," *Khaleej Times Online*, March 16, 2009, www.khaleejtimes.com/biz/inside.asp?xfile=/data/business/2009/March/business_March638.xml&section=business&col=; Stefan Nicola, "Green Oasis Rises From Desert Sands," *Washington Times*, Feb. 2, 2009, www.washingtontimes.com/themes/places/abu-dhabi/; Elisabeth Rosenthal, "Gulf Oil States Seeking a Lead in Clean Energy," *The New York Times*, Jan. 13, 2009, www.nytimes.com/2009/01/13/world/middleeast/13greengulf.html.

82. David Teather and Richard Wachman, "The Emirate That Used to Spend It Like Beckham," *The Guardian*, Jan. 31, 2009, www.guardian.co.uk/world/2009/jan/31/dubai-global-recession; Robert F. Worth, "Laid-Off Foreigners Flee as Dubai Spirals Down," *The New York Times*, Feb. 12, 2009, www.nytimes.com/2009/02/12/world/middleeast/12dubai.html; Elizabeth Farrelly, "Dubai's Darkening Sky: The Crane Gods are Still," *Brisbane Times*, Feb. 26, 2009, www.brisbanetimes.com.au/news/opinion/dubais-darkening-sky-the-crane-gods-are-still/2009/02/25/1235237781806.html.

83. Raja Murthy, "Taj Mahal Leads India's Recovery," *Asia Times*, Dec. 3, 2008, www.atimes.com/atimes/South_Asia/JL03Df01.html.

84. Joe Nocera, "Mumbai Finds Its Resiliency," *The New York Times*, Jan. 4, 2009, http://travel.nytimes.com/2009/01/04/travel/04journeys.html.

BIBLIOGRAPHY

Books

Meyer, Michael, *The Last Days of Old Beijing: Life in the Vanishing Backstreets of a City Transformed*, Walker & Co., 2008.
An English teacher and travel writer traces Beijing's history and describes life in one of its oldest neighborhoods as the city prepared to host the 2008 Olympic Games.

Silver, Christopher, *Planning the Megacity: Jakarta in the Twentieth Century*, Routledge, 2007.
An urban scholar describes how Indonesia's largest city grew from a colonial capital of 150,000 in 1900 into a megacity of 12-13 million in 2000, and concludes that overall the process was well-planned.

2007. State of the World: Our Urban Future, Worldwatch Institute, Norton, 2007.
Published by an environmental think tank, a collection of articles on issues such as sanitation, urban farming and strengthening local economies examines how cities can be healthier and greener.

Articles

"The 2008 Global Cities Index," *Foreign Policy*, November/December 2008, www.foreignpolicy.com/story/cms.php?story_id=4509.
Foreign Policy magazine, the Chicago Council on World Affairs and the A.T. Kearney management consulting firm rank the world's most "global" cities in both industrialized and developing countries, based on economic activity, human capital, information exchange, cultural experience and political engagement.

"Mexico City Bikers Preach Pedal Power in Megacity," *The Associated Press*, Dec. 28, 2008.
Bicycle activists are campaigning for respect in a city with more than 6 million cars, taxis and buses.

Albright, Madeleine, and Hernando De Soto, "Out From the Underground," *Time*, July 16, 2007.
A former U.S. Secretary of State and a prominent Peruvian economist contend that giving poor people basic legal rights can help them move from squatter communities and the shadow economy to more secure lives.

Bloom, David E., and Tarun Khanna, "The Urban Revolution," *Finance & Development*, September 2007, pp. 9-14.
Rapid urbanization is inevitable and could be beneficial if leaders plan for it and develop innovative ways to make cities livable.

Chamberlain, Gethin, "The Beating Heart of Mumbai," *The Observer*, Dec. 21, 2008, www.guardian.co.uk/world/2008/dec/21/dharavi-india-slums-slumdog-millionaire-poverty.
Eight boys growing up in Dharavi, Asia's largest slum, talk about life in their neighborhood.

Osnos, Evan, "Letter From China: The Promised Land," *The New Yorker*, Feb. 9, 2009, www.newyorker.com/reporting/2009/02/09/090209fa_fact_osnos.
Traders from at least 19 countries have set up shop in the Chinese coastal city of Guangzhou to make money in the export-import business.

Packer, George, "The Megacity," *The New Yorker*, Nov. 13, 2006, www.newyorker.com/archive/2006/11/13/061113fa_fact_packer.
Lagos, Nigeria, offers a grim picture of urban life.

Schwartz, Michael, "For Russia's Migrants, Economic Despair Douses Flickers of Hope," *The New York Times*, Feb. 9, 2009, www.nytimes.com/2009/02/10/world/europe/10migrants.html?n=Top/Reference/Times%20Topics/People/P/Putin,%20Vladimir%20V.
Russia has an estimated 10 million migrant workers, mainly from former Soviet republics in Central Asia — some living in shanty towns.

Reports and Studies

"Ranking of the World's Cities Most Exposed to Coastal Flooding Today and in the Future," *Organization for Economic Cooperation and Development*, 2007, www.rms.com/Publications/OECD_Cities_Coastal_Flooding.pdf.
As a result of urbanization and global climate change, up to 150 million people in major cities around the world could be threatened by flooding by 2070.

"State of World Population 2007," *U.N. Population Fund*, 2007, www.unfpa.org/upload/lib_pub_file/695_filename_sowp2007_eng.pdf.

A U.N. agency outlines the challenges and opportunities presented by urbanization and calls on policy makers to help cities improve residents' lives.

"State of the World's Cities 2008/2009: Harmonious Cities," UN-HABITAT, 2008.

The biennial report from the U.N. Human Settlements Programme surveys urban growth patterns and social, economic and environmental conditions in cities worldwide.

For More Information

Chicago Council on Global Affairs, 332 South Michigan Ave., Suite 1100, Chicago, IL 60604; (312) 726-3860; www.thechicagocouncil.org. A nonprofit research and public education group; runs the Global Chicago Center, an initiative to strengthen Chicago's international connections, and co-authors the Global Cities Index.

Clean Air Initiative for Asian Cities, CAI-Asia Center, 3510 Robinsons Equitable Tower, ADB Avenue, Ortigas Center, Pasig City, Philippines 1605; (632) 395-2843; www.cleanairnet.org/caiasia. A nonprofit network that promotes and demonstrates innovative ways to improve air quality in Asian cities.

Institute for Liberty and Democracy, Las Begonias 441, Oficina 901, San Isidro, Lima 27, Peru; (51-1) 616-6100; http://ild.org.pe. Think tank headed by economist Hernando de Soto that promotes legal tools to help the world's poor move from the extralegal economy into an inclusive market economy.

Overseas Development Institute, 111 Westminster Bridge Road, London SE1 7JD, United Kingdom; (44) (0)20 7922 0300; www.odi.org.uk. An independent British think tank focusing on international development and humanitarian issues.

Shack/Slum Dwellers International; (+27) 21 689 9408; www.sdinet.co.za. The Web site for the South Africa-based secretariat of an international network of organizations of the urban poor in 23 developing countries.

UN-HABITAT, P.O. Box 30030 GPO, Nairobi, 00100, Kenya; (254-20) 7621234; www.unhabitat.org. The United Nations Human Settlements Programme; works to promote socially and environmentally sustainable cities and towns.

World Bank, 1818 H Street, N.W., Washington, DC 20433, USA; (202) 473-1000; http://web.worldbank.org. Two development institutions with 185 member countries, which provide loans, credits and grants to middle-income developing countries (International Bank for Reconstruction and Development) and the poorest developing countries (International Development Association)

16

Reducing Your Carbon Footprint

Can Individual Actions Reduce Global Warming?

Thomas J. Billitteri

AFP/Getty Images/Jewel Samad

Early-bird shoppers crowd into a Best Buy store in Los Angeles at 5 a.m. on Nov. 28 for post-Thanksgiving bargains. Concern about climate change, coupled with the nation's economic woes, is causing many Americans to ratchet back on their consumption of goods and services. But many environmentalists say government must also do its part to reduce carbon emissions by enacting tough, environmentally friendly policies.

From *CQ Researcher*,
December 5, 2008.

When Karen Larson, a mother of two in Madbury, N.H., took the "New Hampshire Carbon Challenge" she couldn't believe the results.[1]

The statewide effort to help residents reduce their environmental impact includes an online calculator to help consumers measure their "carbon footprint" — the amount of carbon dioxide (CO_2) created by their activities and consumption patterns. Carbon dioxide is the main greenhouse gas (GHG) that scientists believe leads to global warming.[2]

The calculator showed that Larson could save $700 a year and cut her carbon footprint by some 4,400 pounds by taking such actions as replacing her lightbulbs with compact fluorescents, cutting back on showers by a few minutes, putting electronics on power strips that she turned off when not in use, lowering her furnace a few degrees and getting an annual tune-up on her heating system.

"It blew my mind when I finished and it told me how much money per year I could save," Larson says. But the financial savings were only a "side benefit" to a larger objective, she adds. "The main goal was to find out how to be more earth-friendly and not use so many resources."

From voluntary actions like Larson's to emerging federal policies promoted by President-elect Barack Obama, more and more attention is shifting to the impact consumers have on the environment and how individuals can lower the amount of carbon dioxide emitted by the production, transportation, use and disposal of the goods and services they consume.

Studies show the amount of CO_2 emitted by individuals can rival that of industry and commerce. For example:

- American consumers control — directly or indirectly — about two-thirds of the nation's greenhouse-gas emissions, compared to 43 percent for consumers elsewhere.[3] Passenger cars account for 17 percent of U.S. emissions, as do residential buildings and appliances.[4]
- Transportation accounts for a third of carbon output in the United States, 80 percent of it from highway travel.[5]
- The average home creates more pollution than the average car, according to the Environmental Protection Agency (EPA).[6] If every U.S. home replaced a single incandescent lightbulb with a compact fluorescent, it would prevent the equivalent of the GHG emissions from more than 800,000 cars.[7]

"Individual actions can make a significant difference," says Bill Burtis, communications manager for Clean Air-Cool Planet, an environmental group in Portsmouth, N.H. "We're not going to be able to solve the problem without a concerted and unified effort at reaching individuals and changing behavior in residential sectors, whether it's the compact fluorescent lightbulb or changing our use of two-cycle, highly polluting gas-powered lawn equipment."

However, environmental advocates argue that while individual carbon-fighting actions are important, putting too much attention on voluntary actions by individuals can undermine efforts to pass mandatory government policies to control emissions.

"Every time an activist or politician hectors the public to voluntarily reach for a new bulb or spend extra on a Prius, ExxonMobil heaves a big sigh of relief," Mike Tidwell, director of the Chesapeake Climate Action Network, a grassroots group in Takoma Park, Md., wrote last year.[8]

"While . . . we have a moral responsibility to do what we can as individuals, we just don't have enough time to win this battle one household at a time," he wrote. "We must change our laws. I'd rather have 100,000 Americans phoning their U.S. senators twice per week demanding a prompt phase-out of inefficient automobile engines and lightbulbs than 1 million Americans willing to 'eat their vegetables' and voluntarily fill up their driveway and houses with the right stuff."

Nonetheless, experts say concern about climate change, coupled with worries about the faltering economy, are prompting more and more consumers to make environmentally friendly choices. An ABC News/Planet Green/Stanford University poll in July found that 71 percent of respondents said they were trying to reduce their carbon footprint, mainly by driving less, using less electricity and recycling. A fourth of them said saving money was their primary goal, a third said improving the environment was their chief aim and 41 percent said they were motivated by a combination of the two. (*See graph, p. 390.*)[9]

And according to a Harris Poll last spring, more than 60 percent of respondents said they had cut their home energy use to offset their carbon footprint or reduce their emissions, and 43 percent said they'd bought more energy-efficient appliances.[10]

But such findings should be viewed warily, say public opinion researchers. Soaring energy prices, reflected last summer in $4-per-gallon gasoline prices, are likely a stronger motivator for many Americans to cut their carbon appetites than concern about climate change, they say. And only about one in 10 respondents to the Harris Poll said they had ever calculated their personal or household carbon footprint, and more than one in four said they were doing nothing to reduce emissions.

"Experience suggests that we should be somewhat skeptical of claims people make about doing the 'right thing,' " Harris noted in an analysis of its poll results. While the polling company found it "encouraging" that so many Americans feel it is important to reduce their carbon emissions, U.S. energy consumption keeps rising, leading Harris to conclude that "whatever actions people are taking are probably modest ones."[11]

The trend toward larger homes, for instance, spurs greater energy use and higher personal carbon output. The average size of a new single-family home in the United States more than doubled from 1950 to 2005 — from 983 square feet to 2,434 — while the average number of occupants fell 22 percent between 1950 and 2000.[12] Between 1990 and 2006, residential carbon dioxide emissions jumped 26 percent, outpacing the 20 percent growth in population.[13]

For many consumers — even those committed to improving the environment — cutting back on carbon isn't always easy. For one thing, figuring out one's environmental footprint can be a challenge. Should a footprint consist only of carbon-based fossil fuels, such as those that run most cars and power plants, or should it also include non-carbon-based greenhouse gases like nitrous oxide emitted by fertilizers? Should it include methane gas released by landfills and from the digestive tracts of cattle? How should people measure their share of emissions stemming from the manufacture and transportation of consumer goods? And how does one best measure his share of emissions stemming from travel in airplanes or trains?

"Despite its ubiquitous appearance, there seems to be no clear definition of [carbon footprint] and there is still some confusion [over] what it actually means and measures," noted a research firm in Great Britain, where studies on climate change and consumers' environmental choices have been robust for years.[14]

Consumers also must figure out which actions actually help the environment and which only seem to. For example, in a list of 10 "green heresies," *Wired* pointed out that conventional agriculture can be more environmentally friendly than organic farming. "Organic produce *can* be good for the climate, but not if it's grown in energy-dependent hothouses and travels long distances to get to your fridge," the magazine noted.[15]

And of course, the geographic realities of American life — homes and jobs separated by miles — can thwart the good intentions of even the most environmentally conscious consumer.

Big Metro Areas Have Smaller Carbon Footprints

Big-city areas like Los Angeles emit less carbon per capita than smaller areas like Knoxville, Tenn., in large part because urban areas have high-density development patterns or depend on low-polluting mass transit.

Highest and Lowest Carbon Emissions in Metro Areas

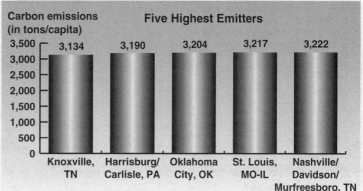

Source: "Shrinking the Carbon Footprint of Metropolitan America," Brookings Institution, May 2008

"When we don't live close to where we work, there's not much choice in the matter," says Duane T. Wegener, a professor of psychological sciences at Purdue University who studies the social aspects of energy policy. The ways most cities are built "tie us into a particular level of energy use, and there's not a lot of control over it."

As consumers and policy makers continue to seek ways to reduce harmful, globe-warming pollutants, here are some of the questions they are asking:

Home Heating and Cooling Emit Most Carbon

Space heating and cooling alone account for more than a third of all the energy used in residential buildings in the United States and more than 450 million metric tons of carbon dioxide.

Carbon Dioxide Emissions From Residential Energy Use
(by percentage and in million metric tons)

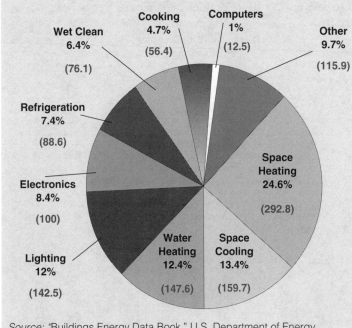

Cooking 4.7% (56.4)

Computers 1% (12.5)

Wet Clean 6.4% (76.1)

Other 9.7% (115.9)

Refrigeration 7.4% (88.6)

Electronics 8.4% (100)

Lighting 12% (142.5)

Water Heating 12.4% (147.6)

Space Cooling 13.4% (159.7)

Space Heating 24.6% (292.8)

Source: "Buildings Energy Data Book," U.S. Department of Energy, November 2008

Are measures of individual carbon emissions valid?

Measuring one's carbon footprint is both art and science.

A growing number of agencies and organizations — from the federal Environmental Protection Agency to the nonprofit Nature Conservancy to the Berkeley Institute of the Environment at the University of California — offer online calculators intended to help people measure their greenhouse-gas emissions.

The calculators gauge the environmental impact of various activities, from heating and lighting a home to flying across the country, and guide consumers on how to reduce their carbon output. But the calculators vary widely in their detail and conclusions.

Air travel, a significant emission source, is a case in point. Calculating a per-passenger carbon footprint is a highly complex task that must take into account such factors as aircraft type, weather conditions, the number of landings and takeoffs along a given route and whether an aircraft flies with a full or partial load.

In June the Montreal-based United Nations International Civil Aviation Organization (ICAO) introduced a carbon calculator that "responds to the wish of many travelers for a reliable and authoritative method to estimate the carbon footprint of a flight," said Robert Kobeh González, president of the ICAO Council.[16]

But the calculator can produce misleading data, according to an official of a company that produces fuel data and supplied the aircraft-performance model used for the ICAO's estimates of airline emissions.

"Producing a single number is crude," Dimitri Simos, a director at Lissys Limited, said. "If you go from Heathrow [airport in London] to Athens, ICAO gives 217 kilograms [478 pounds] of CO_2 [per person]. That hides huge variations. Fly in a full [Boeing] B767 and it's nearer to 160 kg [352 pounds] per person, or in a half-empty [Airbus] A340 it's more like 360 kg [793 pounds]."[17]

While carbon calculators help make people more aware of their individual environmental impact, Daniel Kammen, a professor in the energy and resources group at UC Berkeley, told the *Chicago Tribune* that "the downside is that the methodology is being worked on as we speak."[18]

Still, many see the calculators as useful. Elise L. Amel, director of environmental studies and associate professor of industrial and organizational psychology at the University of St. Thomas in Minneapolis, says that while most carbon calculators are "not very fine-grained," they still serve a valuable purpose.

"Usually the options for somebody's response [to a calculator's questions] don't necessarily perfectly match any one individual situation," she says. Most calculators "use general categories" and give a "rough estimate . . . so it's no wonder they all give you a little something different. But I don't think they should be used to necessarily diagnose each minute activity that you should adjust. They're just to get people aware that, 'Wow, we're using resources at a rate that's really hard to replace.'"

Making people think about their carbon footprint — and doing something about it — is the idea behind the New Hampshire Carbon Challenge and its calculator.[19]

The challenge, which has been online for a year, enables New England residents to measure their carbon footprints, set goals for reducing their emissions and pledge action. Households that take the challenge can be linked to other households through organizations such as churches, schools, civic group and businesses to show the collective action of residents' individual efforts.

So far, about 1,000 households in New Hampshire and Massachusetts have taken the challenge, and thousands more have used the calculator and other Web-based tools developed by the challenge, says Denise Blaha, the organization's codirector. People who have taken the challenge have reduced their home and vehicle energy use by an average of 17 percent, saving roughly $850 annually in fuel and electricity costs, she says. More than 5 million pounds of CO_2 have been pledged for reduction as a result of the challenge, with an energy cost saving of $700,000, she says.

"Households have really been the overlooked sector," Blaha says. "Most of our actions are very, very concrete, real-world and simple to make: changing lightbulbs, putting electronic devices on power strips and turning down your thermostat."

Still, while swapping energy-hogging lightbulbs for miserly compact fluorescents may be simple, some efforts at environmental responsibility can be maddeningly complex.

For example, some consumers try to factor in "food miles" — the distance food travels from farm to table — into their carbon-footprint calculations. Conventional wisdom says food travels an average of 1,500 miles from farm to plate, Jane Black, a *Washington Post* food writer, noted in the online journal *Slate*.[20] But, she wrote, the ubiquitous figure is based on a university study of how far 33 fruits and vegetables grown in the United States traveled to a Chicago produce market. That figure, though it perhaps raises consumers' awareness of the environmental issues involved in food choices, oversimplifies the complex and global nature of the food industry, Black argued.

"If we all think in food miles, the answer is obvious: Buy local. But new studies show that in some cases it can actually be more environmentally responsible to produce food far from home."

Black cited a 2006 report from Lincoln University in New Zealand, which concluded that it was four times more energy efficient for people in London to purchase New Zealand lamb, which feeds on grass, than grain-fed lamb from England.

She also noted that Tesco, the British grocery giant, has begun adding carbon labels to its products. The labels — which reveal how much carbon was emitted by an item's production, transportation and consumption — are now on an initial 20 products, including orange juice and detergent.[21]

But, Black wrote, "Like food miles, these new numbers raise as many questions as they answer."

Should government do more to encourage individuals to reduce their carbon footprints?

While individuals can do much to reduce their carbon consumption and conserve precious resources, environmental advocates say policy makers should provide more incentives to further those efforts.

"Government needs to take a holistic approach to the whole carbon footprint issue," says Eric Carlson, founder and president of Carbonfund.org, a nonprofit organization in Silver Spring, Md., that sells "carbon offsets" — certificates that consumers and businesses can purchase voluntarily to help compensate for their own carbon footprints. The certificates are used to subsidize "green" efforts such as renewable-energy, reforestation and energy-efficiency projects.

Carlson says the federal government has a number of piecemeal consumer programs that provide incentives for greater energy efficiency, such as tax credits for home insulation and the purchase of hybrid cars. But he says

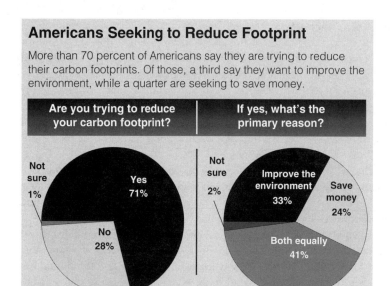

Americans Seeking to Reduce Footprint

More than 70 percent of Americans say they are trying to reduce their carbon footprints. Of those, a third say they want to improve the environment, while a quarter are seeking to save money.

Are you trying to reduce your carbon footprint?

Not sure 1%

Yes 71%

No 28%

If yes, what's the primary reason?

Not sure 2%

Improve the environment 33%

Save money 24%

Both equally 41%

Source: ABC News/Planet Green/Stanford University

Washington should adopt much broader policies aimed at encouraging business and consumers to transition from carbon-emitting energy to green power.

He suggests that the government adopt a "cap-and-trade" climate policy that would set limits on how much carbon dioxide energy producers, utilities or manufacturers could emit. (*See "Current Situation," p. 400.*) The European Union and a handful of other countries have adopted similar carbon-trading systems since 2005.[22] Companies that emit less than allowable amounts could sell their excess "rights to pollute" to others. In some versions of the scheme the rights are auctioned to companies, and the auction revenue is returned to taxpayers or used for renewable-energy development.

Legislation sponsored by Sens. Joseph I. Lieberman, I-Conn., and John W. Warner, R-Va., featured a cap-and-trade system, but the Senate rejected it earlier this year. President-elect Obama favors such a system.

Some have argued that cap-and-trade plans could raise energy costs, squeeze consumers and hurt the economy more than help the environment. Cap-and-trade bills are "nothing short of government re-engineering of the American economy," wrote Ben Lieberman, a policy analyst at the conservative Heritage Foundation.[23]

Yet others say the approach would be good for the environment, partly because it would tend to make goods and services created and transported in carbon-intensive ways more expensive and lower the relative price of items produced in less carbon-intensive ways. Producers and consumers would have incentives to make more environmentally friendly choices.

Including the cost of carbon emissions in the price of goods and services would give consumers an idea of the carbon intensity of the items they buy and motivate them to make greener choices, says Eric Haxthausen, director of U.S. climate change policy for the Nature Conservancy.

He gives the example of buying fruit in a grocery store. "There's a lot of emphasis on buying local," he says. "It's tricky, but the beauty of the cap-and-trade approach is that it's all worked out through the pricing system. You don't have to calculate whether this fruit was shipped on an ocean-going vessel" and therefore had a bigger carbon footprint to reach the grocery shelf than fruit grown closer to home. The price "would distinguish between a high and low" footprint.

Consumers need more information about the environmental consequences of the goods and services they use, according to Rep. Brian Baird, D-Wash., chairman of the House Subcommittee on Research and Science Education. For instance, he told the panel at a hearing last year, households consume more than a third of the energy used in the United States each year, 60 percent of which is used at home. Yet consumers typically lack the information they need to factor energy use into purchases and behaviors, he continued, and both government and industry have "fallen far short in providing the needed information to the public in a way that will result in behavior changes."[24]

At the hearing — on how the social sciences can help solve the nation's energy problems — John A. "Skip" Laitner, senior economist for technology policy at the American Council for an Energy-Efficient Economy, said more funding was needed "to expand our understanding of the social dynamics of energy consumption, energy conservation and energy efficiency." (*See sidebar, p. 396.*) Funding for non-economic social-science research on energy consumption has fallen sharply since the 1980s, he said.[25]

Advocates also say part of the answer to shrinking individual environmental footprints is for government authorities, including those on the state level, to adopt stronger energy-efficiency building standards for such things as thermal insulation and heating and cooling systems. While such codes are "unsexy," Carlson says, they are highly effective in cutting energy consumption and reducing carbon emissions.

Many environmental advocates also call for a wider array of subsidies and tax incentives for energy-efficient appliances and alternative-energy equipment.

California is widely viewed as one the states most determined to promote energy efficiency and clean-energy technology. For example, under a 10-year "Go Solar California" rebate program, the state plans to inject more than $3 billion into financial incentives for Californians to adopt solar energy. The money, which comes from utility bills, is expected to pay for about 3,000 megawatts of new solar energy.[26]

But the program has sparked controversy in some quarters, as the *San Francisco Chronicle* noted last summer. The rebates are aimed at cutting solar prices over time by creating a big and competitive solar industry in the state, the newspaper said, but "solar remains far more expensive than many other forms of power generation."

Moreover, a consumer watchdog group complained that wealthy companies or homeowners have received too many of the rebates, the *Chronicle* reported. Mark Toney, executive director of The Utility Reform Network, said "subsidies should go to the most needy, not the most wealthy."[27]

A spokesman for the California Public Utilities Commission said in mid-November that a program to direct some solar-rebate money to low-income housing was in "the talking stage."

Efforts Focus on Home Energy Use

Nearly two-thirds of Americans are reducing home energy usage, including 43 percent who have bought more energy-efficient appliances, according to a recent survey. Nevertheless, more than a quarter are doing nothing to offset their carbon footprints.

Which of the following have you done in an attempt to offset your carbon footprint or reduce your emissions?

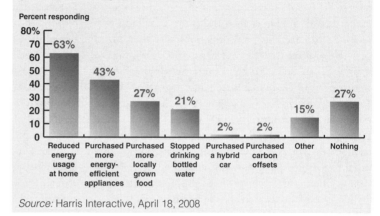

Source: Harris Interactive, April 18, 2008

Can individual action significantly reduce global climate change?

In September the nonprofit Pew Center on Global Climate Change and the Alcoa aluminum company foundation kicked off a partnership aimed at encouraging the company's U.S. employees and their local communities to reduce their energy consumption and shrink their carbon footprints.

Pew and Alcoa initially are rolling out the program at nine of the company's 120 U.S. locations, with plans to expand it. The Make an Impact program, modeled after a similar Alcoa effort in Australia, includes an interactive Web site that offers a carbon calculator and other tools to help employees cut their energy bills and emissions.

Concern about energy efficiency and climate change "is a huge focus inside our business, and this is about translating that into a tool that can be used to increase awareness and actions on climate change" among Alcoa's employees and their communities, says Libby Archell, director of communications for the Alcoa Foundation.

Katie Mandes, vice president of communications at the Pew center, says efforts focused on individuals can make a significant difference in curbing greenhouse-gas emissions. "Are individuals going to be able to solve this problem? No," she says. "But is there a role for individuals? Yes. . . . It's got to be an across-the-board change."

Individuals and households can play a larger role in cutting energy use and emissions than many consumers realize. "U.S. households account for about 38 percent of national carbon emissions through their direct actions, a level of emissions greater than that of any entire country except China and larger than the entire U.S. industrial sector," wrote Gerald T. Gardner, professor emeritus of psychology at the University of Michigan-Dearborn, and Paul C. Stern, director of the Committee on the Human Dimensions of Global Climate Change at the National Research Council.[28] "By changing their selection and use of household and motor-vehicle technologies without waiting for new technologies to appear, making major economic sacrifices or losing a sense of well-being, households can reduce energy consumption by almost 30 percent — about 11 percent of total U.S. consumption."

A study by graduate students at the University of Michigan's School of Natural Resources and Environment a decade ago underscored how cost-effective changes in construction methods and other choices can reduce residential energy consumption.

A conventionally built 2,450-square-foot home in Ann Arbor, Mich., had a global-warming potential equal to 1,013 metric tons of CO_2 over its life span — from construction to use and, ultimately, to demolition — with 92 percent of that occurring during the home's use phase. An energy-efficient home, meanwhile, had a global-warming potential of only 374 metric tons of CO_2 — 67 percent of it occurring in the home's use phase. And although the energy-efficient home cost about 10 percent more than the conventional home, over its lifespan the more efficient home wound up costing about the same as the less efficient one.[29]

And at today's energy prices, the life-cycle costs are substantially lower for the efficient house, according to Gregory Keoleian, codirector of the Center for Sustainable Systems at the University of Michigan.

Refrigeration accounts for 8 percent of average U.S. residential energy consumption.[30] A separate study by the center found that households can save significant costs and energy by replacing refrigerators made before 1995.

Though many consumers still use old energy-hungry appliances, many others own units carrying the government's "Energy Star" label denoting energy efficiency. Last year alone, the joint program between the Environmental Protection Agency and Energy Department helped Americans save enough energy to avoid greenhouse-gas emissions equivalent to the output of 27 million cars while saving $16 billion on utilities.[31]

Even so, the program has drawn some criticism recently. *Consumer Reports* magazine in October questioned the Energy Star label on several refrigerator models and said the program's qualifying standards are lax and its testing procedures dated. Moreover, the article pointed out, the program allows companies to test their own products, with the government relying on manufacturers to evaluate appliances made by competitors and to report suspicious claims about energy use.[32]

Many environmentalists say that while individual carbon-reducing actions, including purchasing energy-efficient appliances, are important in fighting GHG emissions, government action is crucial.

"We need fundamental change and strong policy to drive those changes," says Joe Loper, vice president of policy and research at the Alliance to Save Energy, an advocacy group in Washington. "If I take all those actions and reduce my footprint and no one else does it, it's not going to matter."

Tidwell, of the Chesapeake Climate Action Network, says living a low-carbon lifestyle is good, but "if that becomes the dominant core of our response to climate change, then we fail. We might as well do nothing.

"Instead of wagging our finger at Aunt Betty and telling her she needs to change all her lightbulbs, what we need to do is pass laws in this country so the next time Aunt Betty goes to buy lightbulbs, there are only energy-efficient lightbulbs. You do that with . . . statutory changes that send the right clean-energy market signals to the economy."

In fact, Congress and the Bush administration did take a step in that direction last year with passage of an energy bill that, among other things, requires that traditional incandescent lightbulbs be phased out in favor of more-efficient ones by 2020.[33] But Tidwell asks, "Why wait 12 years?"

He compares the effort to fight GHG emissions to the 1960s-era battle for civil rights, noting that the government didn't rely on voluntary action to curb racial discrimination but instead banned it outright. "When it comes to climate change, we have to start thinking of this as the moral and economic immediate crisis that it is," he says. With that in mind, he adds, "you quickly begin to realize that we have to set strict energy standards that are fair but mandatory."

"We don't have time to green our neighborhoods one house at a time. Nature has chosen its own deadline."

BACKGROUND

Environmental Awakening

Concern about society's impact on the environment has deep historical roots. In ancient Rome, garbage, sewage and the detritus of the metal-working and tanning industries polluted the water and air, turning the fetid atmosphere into what the Romans sometimes called *gravioris caeli* — "heavy heaven."[34]

In the 18th and 19th centuries, economists and political philosophers such as John Stuart Mill worried about the impact of population growth, industrialization and urbanization on food supplies, resources and natural beauty.

In 20th-century America, environmentalism and conservationism grew deep roots, nurtured by the preservation of lands for parks and monuments under Presidents Theodore Roosevelt (1901-1909) and William Howard Taft (1909-1913) and the formation of the Civilian Conservation Corps under President Franklin D. Roosevelt (1933-1945).[35] The Corps put hundreds of thousands of unemployed young adults, mostly men, to work planting trees, developing lakes and ponds, building logging roads, improving campgrounds and performing other tasks.

In the early 1960s, Rachel Carson, an environmental writer and marine biologist with the U.S. Fish and Wildlife Service, became what many regard as the chief catalyst of the modern environmental movement with publication of her 1962 book *Silent Spring*, which described the threats that the pesticide DDT posed to the environment and food chain. In an earlier work, *The Sea Around Us*, published in 1951, she presciently warned of a "startling alteration of climate" and "evidence that the top of the world is growing warmer."[36]

Other environmental milestones in the second half of the 20th century reinforced the idea that choices made not only by government and industry but also by individuals could have a profound influence on climate change, the sustainability of natural resources and, ultimately, the future of the human race. On the first Earth Day, April 22, 1970, millions of Americans participated in rallies and forums that helped thrust environmental concerns onto the political stage.

"It was on that day that Americans made it clear that they understood and were deeply concerned over the deterioration of our environment and the mindless dissipation of our resources," Earth Day's founder, the late Sen. Gaylord Nelson, D-Wis., wrote on the event's 10th anniversary. "That day left a permanent impact on the politics of America."[37]

Throughout the 1970s, a tide of environmental legislation flowed from Washington, including the Clean Air Act of 1970, Endangered Species Act of 1973 and Clean Water Act of 1977. Also, the Environmental Protection Agency was formed in the early 1970s under the Republican administration of President Richard M. Nixon.

But environmentalism was not always as durable as the millions who gathered on Earth Day 1970 had hoped. In the 1980s, for example, the Reagan administration drew heavy criticism for the actions of Interior Secretary James G. Watt and EPA Administrator Anne Gorsuch, both of whom resigned under fire.

Still, the 1980s saw a growing awareness of the consequences of both individual and industrial behavior on the global climate. For example, the Montreal Protocol, an international treaty that the Reagan administration and scores of other governments signed, called for the elimination of ozone-depleting chemicals called chlorofluorocarbons, found in such items as air conditioners and consumer aerosol products.

But getting a handle on carbon emissions has been a much tougher task. Richard A. Benedick, the Reagan administration's chief representative in the Montreal Protocol talks, noted in *The New York Times* in 2005 that tackling carbon dioxide emissions was different than phasing out chlorofluorocarbons, which the newspaper noted were produced by only a small number of companies in a few countries.[38]

"Carbon dioxide is generated by activities as varied as surfing the Web, driving a car, burning wood or flying to Montreal," *Times* environmental reporter Andrew C. Revkin pointed out. "Its production is woven into the fabric of an industrial society, and, for now, economic growth is inconceivable without it."[39]

'Planetary Emergency'

By the late 1990s and early 2000s the scientific community widely agreed that carbon dioxide and other greenhouse gases spelled a global climate catastrophe in the making. But many consumers and policy makers seemed to ignore warnings of catastrophic climate change. Domestic energy use continued to soar at the turn of the 21st century, gas-gulping sport-utility vehicles crowded the highways and conservative pundits belittled the notion that human activity was altering the climate in perilous ways.

In 2001 the Bush administration backed away from the Kyoto Protocol, an international treaty setting targets for industrialized nations to reduce their greenhouse-gas emissions, which the Clinton administration signed but never sent to the Senate for ratification. President George W. Bush complained that the treaty would damage the U.S. economy and that it didn't require developing countries to cut emissions.[40]

Still, other voices have persisted in raising alarms about climate change, none so urgently as former Vice President Al Gore. In a celebrated 2006 documentary film and book — *An Inconvenient Truth* — Gore warned of global peril stemming from rising greenhouse-gas emissions. "Not only does human-caused global warming exist," Gore wrote, "but it is also growing more and more dangerous, and at a pace that has now made it a planetary emergency."[41]

While industry and governments are key targets of Gore's message, individual Americans increasingly are hearing it too — sometimes from unlikely sources, including religious leaders. The Vatican announced this year that "polluting the environment" was among seven new sins requiring repentance.[42] And in March, 44 Southern Baptist leaders, including the current president of the conservative Southern Baptist Convention and two past presidents, signed a declaration supporting stronger action on climate change.[43] The year before, the convention had questioned the notion that humans are mainly responsibile for global warming.

While individual action is widely viewed as important to protecting the environment, many still question how much impact consumers can have on GHG emissions through their daily choices. Research offers competing views.

In a 2007 study provocatively titled "The Carbon Cost of Christmas," European researchers concluded that in the United Kingdom total consumption and spending on food, travel, lighting and gifts over three days of festivities — Christmas Eve, Christmas Day and the traditional U.K. holiday of Boxing Day — could result in as much as 650 kilograms [1,433 pounds] of CO_2 emissions per person — 5.5 percent of Britons' total annual carbon footprint and equivalent to the weight of 1,000 Christmas puddings.[44]

The researchers said consumers could cut their carbon emissions by more than 60 percent by taking such steps as cutting out unwanted gifts, buying "low-carbon" presents such as recycled wine glasses, reusing Christmas cards or using the phone or e-mail to send greetings, composting vegetable peelings when cooking Christmas dinner and reducing holiday lighting.

"In this time of seasonal goodwill, we should all spare a thought for the planet," they wrote.

But another study by researchers at the Massachusetts Institute of Technology pointed to "very significant limits" that voluntary lifestyle choices can have on energy use and carbon emissions. The researchers — MIT students under the direction of Timothy Gutowski, a professor of mechanical engineering — studied 18 different lifestyles ranging from that of a Buddhist monk, a retiree and a 5-year-old to a coma patient, pro golfer and investment banker. They found that even the most constrained lifestyle has an environmental impact far larger than the global average and that none of the lifestyles — including the most modest, that of a homeless person — ever resulted in CO_2 emissions below 8.5 metric tons annually, according to Gutowski.[45]

"The takeaway [from the study] is that we have a very energy-intensive system" that limits how much voluntary actions by individuals can affect climate change, Gutowski says. Still, he says, "I wouldn't want to say people shouldn't take voluntary action. It's a complicated path. Groups may take voluntary action, government agencies notice and then take actions that change the system."

Even so, individuals can sometimes innocently make matters worse by trying to be "green." An example is

C H R O N O L O G Y

1890-1960 *Conservation movement emerges, nurtured by government action, including passage of Air Quality Act in 1967.*

1970s *Under pressure from environmentalists, Congress enacts landmark anti-pollution measures.*

1970 Millions gather for Earth Day on April 22. . . . Congress establishes Environmental Protection Agency and expands Clean Air Act.

1972 Government bans DDT.

1973 Endangered Species Act enacted.

1974 Safe Drinking Water Act enacted.

1978 New York State Department of Health declares public health emergency at Love Canal hazardous waste landfill site.

1979 Accident at Three Mile Island in Pennsylvania, the most serious in the history of U.S. commercial nuclear power plants, leads the Nuclear Regulatory Commission to tighten oversight.

1980s-1990s *Concern about greenhouse gas (GHG) emissions and climate change grows worldwide.*

1983 Interior Secretary James Watt and EPA chief Anne Gorsuch resign amid environmentalists' criticism.

1987 Twenty-four nations initially sign Montreal Protocol, pledging to phase out ozone-depleting chemicals; 169 other countries eventually sign on.

1988 NASA scientist James Hansen warns Congress that global warming is occurring.

1989 Oil tanker *Exxon Valdez* runs aground, contaminating more than 1,000 miles of Alaskan coastline and killing hundreds of thousands of birds and other wildlife in the biggest U.S. oil spill in history.

1992 First international pledge to reduce greenhouse-gas emissions emerges from Earth Summit in Rio de Janeiro, Brazil.

1997 Kyoto Protocol limiting GHG emissions is approved, but wealthy nations are allowed to meet obligations by purchasing "offset" projects in developing countries. United States signs but does not ratify the agreement.

1998 U.S. Green Building Council starts Leadership in Energy and Environmental Design (LEED) program to rate energy-efficient buildings.

2000-Present *Bush administration relaxes strict environmental controls. President-elect Barack Obama pledges to make environmental issues a central focus.*

2001 President George W. Bush reneges on campaign pledge and rejects controls on greenhouse emissions.

2005 Kyoto Protocol takes effect. . . . Members of European Union begin trading carbon credits.

2006 "An Inconvenient Truth," a documentary film featuring former Vice President Al Gore, focuses national attention on global warming.

2007 Federal Trade Commission begins review of environmental marketing guidelines to include carbon offsets, renewable energy certificates and other so-called green products.

2008 British supermarket giant Tesco begins to print "green scores" on some items to show their environmental impact. . . . Senate rejects global warming bill that features a cap-and-trade system on greenhouse gas emissions. . . . Alcoa Aluminum Co. and Pew Center on Global Climate Change form partnership to help Alcoa employees and their communities reduce their carbon footprints. . . . President-elect Barack Obama says his presidency will "mark a new chapter in America's leadership on climate change." He pledges to develop a two-year economic-stimulus plan to save or create 2.5 million jobs, including "green" jobs in alternative energy and environmental technology.

Using Psychology to Influence Consumers' Behavior

Peer pressure proves potent.

Policy makers and environmentalists aren't the only ones trying to figure out how to coax consumers to shrink their carbon appetites. So too are psychologists.

Last summer the American Psychological Association formed a task force to address the role that psychology can play in helping individuals embrace environmentally sustainable practices and cope with the consequences of climate change.

"There hasn't been as much focus on the psychological impacts [of climate change], and we have reason to believe they'll be very serious," says task force member Susan Clayton, a professor of psychology and chair of environmental studies at the College of Wooster in Ohio.

If people have less access to "green, natural, healthy settings," she says, the result could be increased stress and aggression and diminished social interaction. Climate change also could lead to increased competition for dwindling resources such as food and water, sparking social conflict, she says.

When it comes to influencing individual behavior, Clayton says it is "more effective to change the structure of the situation" than trying to change people's minds through preaching.

That might mean providing recycling bins to households, not simply lecturing them on the merits of recycling. Letting consumers know how many of their neighbors are replacing their lightbulbs with compact fluorescents also can be effective, she says. "Social norms matter a lot. . . . Peer pressure never goes away" as an effective catalyst for influencing behavior, she says.

And, Clayton adds, providing the means for feedback on individual behavior — say, putting separate electric meters in apartment houses so tenants can monitor their individual power usage — can give consumers an incentive to conserve.

"Psychologists are increasingly becoming involved in helping alleviate environmental problems," according to Douglas Vakoch, an associate professor in the department of clinical psychology at the California Institute of Integral Studies in San Francisco.

"Most people recognize [that] we face a severe environmental crisis, but it's hard to deal with that head-on because many people feel helpless to do anything about it. . . . Psychologists are very experienced in dealing with denial and in helping to frame messages in ways that people can hear the bad news without being paralyzed by it."

Vakoch says policy experts and government leaders can learn from the psychology field. "The most important lesson . . . is that there is no 'one size fits all' solution to environmental problems," he says. "To create effective public policies, leaders need to recognize that different people are willing to adopt

consumers' efforts to recycle old computers and other electronic items, which contain a toxic brew of chemicals, heavy metals and plastics such as mercury, lead and polyvinyl chloride.

In November the CBS News program "60 Minutes" followed recycled computer parts from Denver to Guiyu, China, which reporter Scott Pelley described as "one of the most toxic places on Earth . . . a town where the blood of the children is laced with lead." In Guiyu, computer parts are stripped, melted down and recycled by impoverished Chinese risking their health and lives for $8 a day.[46]

"This is really the dirty, little secret of the electronic age," said Jim Puckett, founder of the Seattle-based Basel Action Network, a watchdog group named for a treaty intended to keep wealthy nations from exporting toxic waste to poor countries.[47]

One way to keep harmful products from adding toxins to the environment is to produce them in an environmentally conscious way in the first place. Forrester Research Inc. found in a survey of 5,000 U.S. adults that 12 percent were willing to pay more for electronics that consume less energy or are made by an environmentally friendly manufacturer.[48]

In addition, computer manufacturers have been creating products that are more energy efficient than past models, companies are creating software to help older computers use less energy and some are making new models out of recycled materials, *The Wall Street Journal* reported.[49]

Katharine Kaplan, a product manager in the Energy Star program, said computer makers have sought to improve energy efficiency for years and that "the newer focus has been on toxins and recycling."[50]

more environmentally sound behaviors for different reasons. What's compelling for one person will fail for another."

Psychological research has helped shed light on individuals' understanding of environmental issues and their willingness to make changes in their personal consumption habits.

For example, Stanford University psychologist Jon Krosnick found that as people's knowledge about climate change grew, the more concerned about it they became, though political affiliation and trust in science were also important factors.

The link between knowledge about climate change and concern about it "was especially true for respondents who described themselves as Democrats and those who said they trusted scientists," Krosnick told the annual convention of the American Psychological Association in August. "But for Republicans and those who had little trust in scientists, more knowledge did not mean there was more concern."[1]

In another study, Robert B. Cialdini, a professor of psychology and marketing at Arizona State University, found that a small change in message cards asking guests at an upscale Phoenix hotel to reuse their towels had huge potential environmental consequences.

One card exhorted residents to "Help Save the Environment" and was followed by information stressing respect for nature, Cialdini told a House subcommittee last year.[2] Another card stated "Help Save Resources for Future Generations," followed by information stressing the importance of saving energy for the future. A third card asked guests to "Partner With Us to Help Save the Environment," followed by information urging them to cooperate with the hotel in preservation efforts. A fourth card, Cialdini noted, said "Join Your Fellow Citizens in Helping to Save the Environment," followed by information that the majority of hotel guests reuse their towels when asked.

The outcome was striking, he said. Compared with the first three messages, the final one — based on a "social norm," or the perception of what most others were doing — increased the reuse of towels by an average of 34 percent, Cialdini said.

Not everyone in the psychological field agrees that trying to change consumer habits can ever do enough to make a significant difference in nationwide or worldwide carbon emissions or climate change. But many believe the effort is worthwhile.

"There's a huge debate going on among psychologists over whether it's just futile to even bother talking about individual behavior" or whether action by policy makers and corporations is the key to solving the nation's environmental problems, says Elise L. Amel, director of environmental studies and associate professor of industrial and organizational psychology at the University of St. Thomas in Minneapolis.

"I think it's got to be both. This is such a crucial problem coming so quickly that we can't leave any stone unturned."

---

[1] "Climate Change, Global Warming, Among Environmental Concerns Discussed at Psychology Meeting," press release, American Psychological Association, Aug. 15, 2008.

[2] Robert B. Cialdini, testimony to the Subcommittee on Research and Science Education, House Committee on Science and Technology, Sept. 25, 2007.

But corporate efforts to make their products and services green — and to induce consumers to sign on to those efforts — can be challenging. For instance, in Britain, critics say carbon labels on retail goods often confuse consumers or give them information they don't want.[51]

Forum for the Future, a London think tank, expressed concern this year about giving consumers information without proper context. "Only a handful of our focus group participants associated carbon emissions [and climate change] with what they buy in the shops," the group stated in a report. "The majority knew that carbon emissions are linked with cars, airplanes and factories. They made that connection because they can 'see' the emissions, which makes them easy to interpret as being 'bad for the environment.' However, the link between products and climate change was less intuitive to them."[52]

In the United States, most companies are willing to embrace only "incremental change" on carbon labeling, Joel Makower, a sustainability consultant and cofounder and executive editor of Greener World Media Inc., in Oakland, Calif., told *The Christian Science Monitor*.[53]

Wal-Mart, the world's biggest retailer, kicked off a broad sustainability strategy in 2005 aimed at reducing the company's environmental impact. But Matt Kisler, the company's senior vice president of sustainability, told *The Monitor* that he has doubts about current carbon-labeling methodologies and customers' ability to link carbon with consumer goods. "I'm not sure the consumer will ever make a purchase based on the carbon footprint, especially the mass consumer," he told the newspaper.[54]

Green goods can indeed be a tough sell, but some analysts say a more concerted effort by business is

How to Shrink Your Carbon Footprint

Driving habits and home energy use are key factors.

Here are several ways environmentalists, behavioralists and climate scientists say individual consumers can shrink their own carbon footprints:

Alter driving habits. Use of private motor vehicles accounted for more than 38 percent of total U.S. energy use in 2005, according to calculations by Gerald T. Gardner, professor emeritus of psychology at the University of Michigan-Dearborn, and Paul C. Stern, director of the Committee on the Human Dimensions of Global Climate Change at the National Research Council.[1]

Carpooling to work with another person can potentially save as much as 4.2 percent of individual and household energy consumption, they estimated. Other savings can come from avoiding sudden accelerations and stops (up to 3.2 percent); combining errand trips to a half of current mileage (up to 2.7 percent); cutting speeds from 70 to 60 miles per hour (up to 2.4 percent) and getting frequent tune-ups (3.9 percent).

Buying a car that gets an average of 30.7 miles per gallon rather than 20 can save an estimated 13.5 percent of household energy use, the authors estimated.

Reduce home energy consumption. Home space heating accounts for 34 percent of a typical homeowner's utility bill, according to the U.S. Department of Energy. Appliances and lighting account for the same proportion, followed by water heating (13 percent), electricity for air conditioning (11 percent) and refrigeration (8 percent).[2]

Residential buildings not only soak up money for utilities but also emit carbon dioxide. Heating accounts for 25 percent of CO_2 emissions, according to the Energy Department, followed by cooling (13.4 percent), water heating (12.4 percent), lighting (12 percent) and electronics, including color televisions (8.4 percent). Refrigerators and freezers account for more than 7 percent or residential carbon emissions.[3]

Gardner and Stern estimated that replacing 85 percent of all incandescent bulbs with compact fluorescents of equal brightness would reduce total individual and household energy consumption in the United States by 4 percent. Turning down the heat from 72 to 68 degrees during the day and to 65 degrees at night, and turning up the air conditioning from 73 to 78 degrees, would save 3.4 percent of energy consumption, they estimated.[4]

Many other steps can reduce a consumer's energy consumption — and carbon footprint — as well. For example, a home-energy checklist assembled by the American Council for an Energy-Efficient Economy suggests turning down the water-heater temperature to 120 degrees, cleaning or replacing furnace and air-conditioner filters, caulking leaky windows, improving attic and wall insulation and replacing inefficient appliances, among other tips.[5]

Downsize and scale back on consumption. Bigger homes typically use more energy, emit more carbon dioxide and produce more waste (which winds up in methane-emitting landfills) than smaller homes. Bigger cars tend to use more fuel. Bigger consumption patterns, from purchases of furniture, food and electronics to car and airplane travel, add to consumers' emissions.

The median size of a new single-family home rose more than 60 percent between 1970 to 2006, according to data from the National Association of Home Builders, and in 2006 nearly a fourth of new homes contained 3,000 square feet or more. Also that year, more than a fourth of the new homes had three or more bathrooms and about one in five had garage space for three or more cars.[6]

Know that geography can matter. Researchers at the Brookings Institution, a think tank in Washington, found that the carbon footprints of the nation's 100 largest metropolitan areas vary significantly and that development patterns and the availability of rail transit play a key role in the differences.

needed to guide consumers on the merits of environmentally friendly items.

A report by McKinsey & Co. consultants this fall cited a 2007 consumer survey by the trade publication *Chain Store Age*, which found that only 25 percent of respondents reported having bought a green product other than organic foods or energy-efficient lighting. The McKinsey researchers also noted that most green items have small market shares. For instance, green laundry detergent and household cleaners account for less than 2 percent of U.S. sales.[55]

"[R]egions with high density, compact development and rail transit offer a more energy- and carbon-efficient lifestyle than sprawling, auto-centric areas," Brookings said. In addition, it said, while carbon output from urban centers continues to grow, the carbon footprint of a resident of a large metro area is 14 percent smaller than that of the average American and has grown in recent years by only half as much.[7]

Beware of "greenwashing." Along with an avalanche of "green" products and information on how to cut personal carbon emissions has come a steady tide of "greenwashing" — what TerraChoice Environmental Marketing Inc., based in Philadelphia, calls "the act of misleading consumers regarding the environmental practices of a company or the environmental benefits of a product or service."[8]

In a paper titled "The 'Six Sins of Greenwashing,' " the firm said it surveyed six "category-leading big-box stores" and identified 1,018 consumer products making 1,753 environmental claims. Of the total products examined, it said, "all but one made claims that are demonstrably false or that risk misleading intended audiences."[9]

Based on the survey, the firm identified what it said were six patterns of greenwashing:

- **The Sin of the Hidden Trade-Off** — Basing the suggestion that a product is "green" on only one environmental attribute without paying attention to other factors that may be more important, such as impacts on global warming, energy or water use or deforestation. An example is paper marketed as having recycled content without attention to the air, water and global-warming impact of its manufacture.
- **The Sin of No Proof** — Making an environmental claim that can't easily be backed up by supporting information or a reliable third-party certification.
- **The Sin of Vagueness** — Making poorly defined claims or ones that are so broad that consumers are likely to misunderstand the true meaning. An example is claiming a product is "chemical free." "[N]othing is free of chemicals," TerraChoice said. "Water is a chemical. All

plants, animals and humans are made of chemicals as are all of our products."
- **The Sin of Irrelevance** — Making claims that might be true but are unimportant or not helpful. The most common example concerns chlorofluorocarbons, a key factor in depletion of the ozone layer, TerraChoice said. "Since CFCs have been legally banned for almost 30 years, there are no products that are manufactured with it."
- **The Sin of the Lesser of Two Evils** — Claims that, while they may be true within a product category, can distract consumers from the category's broader environmental impact. Organic cigarettes are an example, Terra-Choice said.
- **The Sin of Fibbing** — Making false claims, such as saying a detergent is packaged in "100% recycled paper" but whose container is made of plastic.

---

[1] Gerald T. Gardner and Paul C. Stern, "The Short List: The Most Effective Actions U.S. Households Can Take to Curb Climate Change," *Environment*, September/October, 2008.

[2] "Your Home's Energy Use," U.S. Department of Energy, www1.eere .energy.gov/consumer/tips/home_energy.html.

[3] "2008 Buildings Energy Data Book," U.S. Department of Energy, http://buildingsdatabook.eren.doe.gov/.

[4] Gardner and Stern, *op. cit.*, http://buildingsdatabook.eere.energy.gov/ Table-View.aspx?table=2.4.3.

[5] "Home Energy Checklist for Action," American Council for an Energy-Efficient Economy, www.aceee.org/consumerguide/checklist.htm.

[6] "Selected Characteristics of New Housing," National Association of Home Builders, April 3, 2008, www.nahb.org/page.aspx/category/ sectionID=130.

[7] "Brookings Institution Ranks Nation's 100 Largest Metro Areas for Carbon Footprint," press release, Brookings Institution, May 29, 2008, www.brookings.edu/~/media/Files/rc/papers/2008/05_carbon_foot print_sarzynski/pressrelease.pdf. The report is by Marilyn A. Brown, Frank Southworth and Andrea Sarzynski, "Shrinking the Carbon Footprint of Metropolitan America," Brookings Institution, May 8, 2008, www.brookings.edu/reports/2008/05_carbon_footprint_sarzyn ski. aspx.

[8] "The 'Six Sins of Greenwashing,' " TerraChoice Environmental Marketing Inc., November 2007.

[9] *Ibid.*

"Consumers in the United States and other developed countries have…done little to lighten their carbon footprint," the McKinsey consultants wrote. "Some of this lag between talking and walking could reflect insincerity, laziness or posturing. But much more of it stems from

the failure of business to educate consumers about the benefits of green products and to create and market compelling ones."[56]

Patagonia, the clothing and outdoor-gear retailer, tracks the environmental footprint of more than a dozen of its products and shares both the good and bad on its Web site, though it warns that its environmental examinations are "partial and preliminary."[57]

For example, a down jacket's footprint, from origin of the fiber to the garment's distribution, spans more than 20,000 miles, touching California, Hungary, Japan, China and Nevada. The jacket's manufacture and transportation created nearly seven pounds of CO_2 emissions and enough energy to burn an 18-watt compact fluorescent bulb continuously for 22 days, according to Patagonia.[58]

The jacket uses "high-quality goose down . . . [that] comes from humanely raised geese," and the garment's light shell is made from recycled polyester, Patagonia says. But the zipper is treated with a substance "that contains perfluorooctanoic acid (PFOA), a synthetic chemical that is now persistent in the environment."

Says Patagonia, "We're investigating alternatives to the use of PFOA in water repellents — and looking for ways to recycle down garments."[59]

CURRENT SITUATION

Creating Incentives

Beginning in 2009 buyers of the fuel-sipping Honda Civic hybrid will no longer receive one of the most popular incentives for going green: a tax credit.[60]

Tax incentives on hybrids phase out after an automaker sells 60,000 of them, a benchmark Honda reached in 2007 and that Toyota — maker of the hybrid Prius — hit in 2006. With gas prices having fallen sharply since peaking at $4-plus per gallon last summer, "it's getting a lot more expensive to be an environmentally conscious driver," *The Wall Street Journal* noted in an article about dwindling tax incentives on hybrid cars.[61]

Even so, the tax code remains one of the most powerful tools in the environmental-policy arsenal.

Consumers have long received tax breaks on everything from hybrid cars to energy-efficient appliances and home weatherization. The Emergency Economic Stabilization Act of 2008 — the so-called bailout bill passed this fall to deal with the rapidly deteriorating economy — included, extended or amended several such incentives as well as others aimed at businesses and public utilities.[62] For example, the measure included tax breaks for solar systems to

generate electricity or heat hot water, energy-efficient home improvements and even bicycle commuting.[63]

As legislators and policy analysts contemplate how climate change and energy consumption are likely to unfold in coming decades, they are weighing other ideas to reduce carbon emissions.

"The most efficient approaches to reducing emissions involve giving businesses and individuals an incentive to curb activities that produce CO_2 emissions, rather than adopting a 'command-and-control' approach in which the government would mandate how much individual entities could emit or what technologies they should use," the Congressional Budget Office said in a report on policy options for reducing carbon-dioxide emissions.[64]

Cap-and-trade systems, which can be structured in a variety of ways, provide such incentives. They allow companies that emit CO_2 and other polluting gases to buy (or are allocated) emission credits that allow them to continue emitting a certain amount of the pollutant. Companies emitting less can save their allowances for the future or sell them at a profit to other companies.

So-called "cap-and-dividend" or "cap-and-cash-back" schemes also have been suggested. Entrepreneur and writer Peter Barnes, the author of *Who Owns the Sky?* and *Climate Solutions: A Citizens Guide*, described the idea this way: The caps would be placed "upstream — that is, on the small number of companies that bring carbon into the economy. An upstream cap could be administered without monitoring smokestacks, without a large bureaucracy and without favoring some companies over others. . . . If carbon doesn't come into the economy, it can't go out."

Caps would be auctioned rather than given away free, and the revenue would go to taxpayers to help offset the higher price of fuel and other carbon-intensive products.[65]

"This can be done through yearly tax credits, or better yet through monthly cash dividends wired … to people's bank accounts or debit cards."

Because that income would be taxed, the government would recoup about 25 percent of the money and could use it "as it sees fit," Barnes wrote. "More importantly, ordinary families would get the lion's share of the auction revenue, and get it in a way that rewards conservation. Since everyone would get the same amount back, those

Will President-elect Obama's clean energy plan work?

YES
Bracken Hendricks
Senior Fellow, Center for American Progress

Written for *CQ Researcher*, December 2008

Barack Obama is not yet sworn in as president, and it is far too early to know the details of his policies, let alone their effectiveness. Yet it is clear from the recent election campaign that the energy road map the president-elect and Democrats in Congress have laid out will move the nation forward. After years of inaction and obstruction from the White House, it is time for leadership. We must place clean energy center stage in America's economic renewal.

Clean energy means jobs and hundreds of billions in investment. The country faces a collapsing housing market, record unemployment and a fiscal crisis that hurts communities. New demand for goods and services from an energy transition can stimulate the economy. Retrofitting millions of homes for energy efficiency will put construction workers back on the job. Rebuilding our infrastructure for transit, alternative fuels and a renewable electricity grid will jump-start local economies. And retooling industry to serve the growing market for a new generation of cars and clean technology is our best hope for restoring manufacturing jobs.

A recent study by the Center for American Progress showed that investing $100 billion in smart incentives for energy efficiency and renewable energy would create 2 million jobs. These "green" jobs are in familiar professions in manufacturing and construction, driven by new technology and innovation. Clean and efficient energy investments have more local content and are harder to outsource, and they redirect spending from wasted energy into skilled labor. As a result, they create more jobs at better wages.

Climate solutions also mean fixing broken markets. Inaction in the face of global warming is not costless. Global warming is the biggest market failure the world has ever known, and if left unchecked will cost the economy trillions of dollars in lost productivity. We need smart policies that cap emissions and help businesses respond to the real costs of waste and pollution. Designed properly, smart climate policies can cut energy bills, increase consumer choice and create new markets and desperately needed demand for the ingenuity of American companies and workers.

We cannot drill our way out of our oil dependence, and we cannot deny our way to a stable climate. Barack Obama and congressional leaders instead have offered a vision that invests in innovation, that faces tough challenges squarely and that finds opportunity in crisis. This is real leadership. It is long overdue, and it will put America back to work.

NO
Kenneth P. Green
Resident Scholar, American Enterprise Institute

Written for *CQ Researcher*, December 2008

Barack Obama campaigned on an energy agenda of greenhouse-gas pricing, vehicle-efficiency standards and a fleet of plug-in hybrid cars. His plan is supposed to increase energy independence, lower greenhouse-gas emissions and create 5 million "green" energy jobs. Will it work? It's doubtful.

Obama's proposed cap-and-trade program, which would reduce greenhouse-gas emissions 80 percent by 2050, is 10 percent more stringent than the variation (S 2191) proposed by Sens. John Warner and Joseph Lieberman that was killed in 2007. The Congressional Budget Office estimated their proposal would cost $1.2 trillion from 2009-2018. The Environmental Protection Agency projected it would raise gas prices by $0.53 per gallon and hike electricity prices 44 percent in 2030. Economists at CRA International estimated the Warner-Lieberman proposal would result in a loss of 4 million jobs by 2015, growing to 7 million by 2050. In the face of a global financial crisis and a long, deep recession, passage of such a plan is both unlikely and undesirable.

The Obama plan also calls for reducing oil imports by tightening vehicle fuel-economy standards and subsidizing a fleet of 1 million plug-in hybrid vehicles. But there's a problem here: U.S. automakers are teetering on the brink of bankruptcy, Americans are strapped for cash, and plug-in hybrids are considerably more expensive than currently available cars. The National Renewable Energy Laboratory estimates the additional costs of plug-in hybrid vehicles that slightly outperform conventional hybrids at between $3,000 and $7,000. For the really fuel efficient ones, the laboratory estimates a premium of $12,000-$18,000.

Hybrids also cost more to insure. Are Americans going to shell out that kind of cash in a recession? I don't think so. Government fleets might buy some, but even they are strapped for cash and will have to cut costs elsewhere to afford plug-in hybrids.

As for creating green jobs, "job creation" is simply a myth. Governments do not create private-sector jobs, or wealth. They can only curtail jobs in one way (through taxation or regulation) and generate other jobs with subsidies and incentives. But since they impose costs in "managing" such programs — and because the market has already rejected the goods that the government is pushing (or there would be no need for intervention) — there are invariably fewer jobs and less wealth creation at the end of the day.

Markets create jobs, as markets create wealth: All the government can do is move it about to suit its priorities.

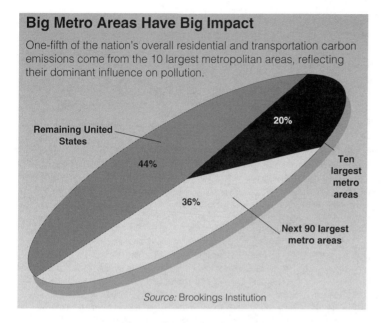

Big Metro Areas Have Big Impact

One-fifth of the nation's overall residential and transportation carbon emissions come from the 10 largest metropolitan areas, reflecting their dominant influence on pollution.

Remaining United States

44%

20%

36%

Ten largest metro areas

Next 90 largest metro areas

Source: Brookings Institution

who use the most carbon would lose, and those who use the least would gain — their dividends would exceed what they pay in higher prices."

The approach would benefit low-income families because they use less energy than wealthier ones and would pay little if any tax on their dividends, Barnes argued.

Barnes wrote that while a carbon cap would raise fuel prices "for years to come," the cap-and-dividend approach would protect families' finances "by permanently linking dividends to carbon prices. As carbon prices rise, so — automatically — do dividends. If voters scream about rising fuel prices . . . politicians can truthfully say, 'How you fare is up to you. If you guzzle, you lose; if you conserve, you gain.'"

The Chesapeake Climate Action Network's Tidwell favors the cap-and-dividend idea and says that while it would lead to higher prices for fossil-based fuels, it would create an incentive for industry and consumers to conserve and switch to alternative forms of energy. The cap-and-dividend approach, he argues, would "lead to more energy-efficient cars, the discontinued use of energy-inefficient lightbulbs, and it will overnight [provide an incentive for] energy audits for homes, weatherization" and other carbon-reducing actions.

"Let's let the invisible hand of [economist] Adam Smith take over, but first we make carbon fuels more expensive

as they come out of the ground and give the money back to all Americans in a way that protects the poor, so it's fair and effective and market-driven."

Roger W. Stephenson, executive vice president for programs at Clean Air-Cool Planet, says his group also favors an approach that generates dividends, but rather than sending a check to individuals, he says it would be better to recycle the revenue through the tax system for such purposes as reducing payroll taxes, corporate tax rates and other business and consumer taxes and modifying the earned-income tax credit for low-income working people. Some money could also be used for social services, such as giving a "carbon boost to the food-stamp program" to help compensate poor people who are outside the tax system but nonetheless affected by higher energy costs.

"We have to be sure that non-taxpayers are taken into account — the bottom quintile of the population — people who don't pay taxes but who can be affected disproportionately" by a carbon-capping system, he says.

While the cost of the emission allowances would raise consumers' energy prices, Stephenson says, the government could limit the auction price of the credits to avoid extremes in consumers' utility bills.

Stephenson says in addition to a carbon-cap regime, new technology and greater energy efficiency — especially in transportation — also are needed to solve the climate-change problem. "Transport is 30 percent or more of emissions," he noted. "We cannot and are not going to stop driving. Therefore technology and efficiency solutions are essential."

Obama's Plan

President-elect Obama's energy plan calls for a wide range of efforts to reduce the carbon footprints of both industry and consumers, including an "economy-wide cap-and-trade program" aimed at cutting greenhouse-gas emissions by 80 percent by 2050. Under the plan, all pollution credits would be auctioned and the proceeds used for investments in "clean" energy, habitat protection and "rebates and other transition relief for families."[66]

The Obama plan also calls for other energy- and carbon-cutting efforts, including weatherizing at least a million low-income homes annually for the next decade, cutting electricity demand 15 percent from projected levels by 2020, putting a million plug-in hybrid cars on the road by 2015 and requiring a fourth of the nation's electricity to be generated by renewable sources by 2025.[67]

Obama even wants the federal government to practice what he is preaching. He has said he would transform the White House fleet to plug-in vehicles within a year of his inauguration "as security permits" and ensure that half of the new vehicles the government purchases are plug-ins and battery-powered cars by 2012, according to *The New York Times*.[68]

In mid-November Jason Grumet, the Obama campaign's chief energy and environment adviser, told a conference on carbon trading that Obama "will move quickly on climate change" but offered no specifics.[69]

Obama said during his campaign that spending $150 billion over the next decade to increase energy efficiency would help to create 5 million "green-collar" jobs in such industries as insulation installation, wind-turbine production and construction of energy-efficient buildings.[70] But some policy experts have questioned the job-creation number and underlying assumptions behind it, and the numbers have even been debated by the president-elect's own advisers.[71] "U.S. automakers are teetering on the brink of bankruptcy, Americans are strapped for cash and plug-in hybrids are considerably more expensive than currently available cars," according to Kenneth P. Green, resident scholar at the American Enterprise Institute. "As for creating green jobs," he continues, " 'job creation' is simply a myth." (*See "At Issue," page 401.*)

While many details of Obama's energy plan remain to be fleshed out, he is wasting no time in making climate change and cutting CO_2 emissions a priority. In November, in a four-minute video message to the Governors' Global Climate Summit, he said his presidency would "mark a new chapter in America's leadership on climate change."[72] Separately, he also pledged to develop a two-year economic-stimulus plan to save or create 2.5 million jobs, including "green" jobs in alternative energy.

"I promise you this: When I am president, any governor who's willing to promote clean energy will have a partner in the White House," Obama told the climate summit.

"Any company that's willing to invest in clean energy will have an ally in Washington. And any nation that's willing to join the cause of combating climate change will have an ally in the United States of America."[73]

Fred Krupp, president of the Environmental Defense Fund, said Obama was "clearly rejecting the timid, business-as-usual approach" to climate and energy issues.[74]

Eileen Claussen, president of the Pew Center on Global Climate Change, said in a statement that the president-elect's remarks were "exactly the kind of leadership the country and the world have been waiting for . . . . We urge the bipartisan leadership in Congress to work closely with the new president to quickly enact an economy-wide cap-and-trade system."

As policy makers in Washington turn their attention toward shaping a new energy strategy, concerns about a global climate catastrophe stemming from carbon-dioxide and other GHG emissions have never been stronger. Greenhouse-gas emissions from developed nations rose 2.3 percent between 2000 and 2006, according to the United Nations Framework Convention on Climate Change.[75] And individuals, as well as industry, are the source of those emissions.

The U.N. also reported in November that a poisonous, brownish haze of carbon dust from cars, coal-fired power plants, wood-burning stoves and other sources was blocking out sunlight in Asia and other continents, altering weather patterns and making people sick.[76] The thick, atmospheric blanket sprawls from the Arabian Peninsula to the Yellow Sea, goes past North and South Korea and Japan in the spring, and sometimes drifts to California.[77]

"We used to think of this brown cloud as a regional problem, but now we realize its impact is much greater," said Professor Veerabhadran Ramanathan, the leader of the U.N. scientific panel. "When we see the smog one day and not the next, it just means it's blown somewhere else."[78]

Achim Steiner, executive director of the U.N. Environment Program in Beijing, said, "The imperative to act has never been clearer."[79]

OUTLOOK

Economic Imperatives

Several developments could lead to significant changes — both positive and negative — in how consumers use energy and leave their carbon imprints on the environment.

One is the financial crisis. In the short term, many consumers are feeling pressured financially to lower their thermostats, forgo vacations and take other steps to conserve money — all of which help reduce carbon emissions.

In its annual Thanksgiving survey, AAA said the number of Americans planning to travel by car, plane, bus or train was down 1.4 percent from last year's record and marked the first decline since 2002. Air travel was expected to fall by 7.2 percent, the group said.[80]

The nation's economic woes also are causing consumers — even the wealthiest ones — to ratchet back on consumption of goods and services. "People are saying, 'We are going to save money, and we are going to save the environment,'" said Wendy Liebmann, chief executive of WSL Strategic Retail, a consulting firm in New York.[81]

But the economic slowdown could also drive up consumers' carbon emissions. With state and federal budgets under severe pressure, less money may be available for everything from tax credits for energy-saving home retrofitting to subsidies for mass-transit services.

In November, financial pressure stemming from the credit crisis propelled officials from 11 major transit agencies — including those in New York, Boston, Washington, Los Angeles and Chicago — to seek help from Congress to avoid cuts in bus and subway service for millions of riders.[82]

In New York, the Metropolitan Transportation Authority called for raising fare and toll revenues by 23 percent in 2009 because of a huge budget gap, and it also drafted proposals for service reductions.[83]

Such moves would likely put more cars on the road and lead to greater emissions.

In the longer term, the incoming Obama administration could make dramatic changes to U.S. energy policy that would affect not only industry but consumers as well.

"Few challenges facing America, and the world, are more urgent than combating climate change," Obama told the governors' climate summit.[84]

Still unknown is whether Congress will pass a climate bill and how such legislation might be structured to provide incentives for consumers to lower their carbon emissions. "I think we have to wait and see" what Congress and the Obama administration do on energy policy, says Haxthausen of the Nature Conservancy.

"A reasonable conjecture is that [the administration] would want to get out in front of this issue, whether with legislation or a set of principles. In both houses of Congress, members are ready to work on this issue. They'll probably be looking to the administration for signals."

NOTES

1. See New Hampshire Carbon Challenge, http://carbonchallenge.sr.unh.edu/.

2. For background, see Marcia Clemmitt, "Climate Change," *CQ Researcher*, Jan. 27, 2006, pp. 73-96; and Colin Woodard, "Curbing Climate Change," *CQ Global Researcher*, February 2007, pp. 27-50.

3. Jeffrey Ball, "A Big Sum of Small Differences," *The Wall Street Journal*, Oct. 2, 2008.

4. *Ibid.*

5. Marilyn A. Brown, Frank Southworth and Andrea Sarzynski, "Shrinking the Carbon Footprint of Metropolitan America," Brookings Institution, May 2008, p. 8, www.brookings.edu/reports/2008/05_carbon_footprint_sarzynski.aspx.

6. "What's your EnviroQ? Answer Page," Environmental Protection Agency, www.epa.gov/epahome/enviroq/index.htm.

7. "Compact Fluorescent Light Bulbs," www.energystar.gov/index.cfm?c=cfls.pr_cfls.

8. Mike Tidwell, "Consider Using the N-Word Less," *Grist*, Sept. 4, 2007, http://grist.org/feature/2007/09/04/change_redux/index.html.

9. Gary Langer, "Fuel Costs Boost Conservation Efforts; 7 in 10 Reducing 'Carbon Foot-print,'" ABC News, Aug. 9, 2008, http://abcnews.go.com/PollingUnit/story?id=5525064&page=1.

10. Harris Interactive, "For Earth Day: Two-thirds of Americans Believe Humans are Contributing to Increased Temperatures," The Harris Poll #44, April 18, 2008, www.harrisinteractive.com/harris_poll/index.asp?PID=898.

11. *Ibid.*

12. Center for Sustainable Systems, University of Michigan, "Residential Buildings," http://css.snre.umich.edu/css_doc/CSS01-08.pdf.

13. *Ibid.*

14. Thomas Wiedmann and Jan Minx, "A Definition of 'Carbon Footprint,' " *ISA* [UK] *Research & Consulting*, June 2007, www.isa-research.co.uk/docs/ISA-UK_Report_07-01_carbon_footprint.pdf.

15. Joanna Pearlstein, "Surprise! Conventional Agriculture Can Be Easier on the Planet," in "Inconvenient Truths: Get Ready to Rethink What It Means to Be Green," *Wired*, May 19, 2008, www.wired.com/science/planetearth/magazine/16-06/ff_heresies_intro.

16. "Universal, Neutral and Transparent Method for Estimating the Carbon Footprint of a Flight," press release, International Civil Aviation Organization, June 18, 2008, www.icao.int/icao/en/nr/2008/PIO200803_e.pdf.

17. Gerard Wynn, "Critics [say] air travel carbon offsetting too crude," Reuters, Aug. 21, 2008, www.reuters.com/article/environmentNews/idUSLK20281120080821.

18. Tim DeChant, "Calculating footprint often uses fuzzy math; Results vary, but give idea of environmental impact," *Chicago Tribune*, Aug. 10, 2008, p. 1.

19. The New England Carbon Estimator is at http://carbonchallenge.sr.unh.edu/calculator.jsp.

20. Jane Black, "What's in a Number?" *Slate*, Sept. 17, 2008, www.slate.com/id/2200202/.

21. See Eric Marx, "Are you ready to go on a carbon diet?" *The Christian Science Monitor*, Nov. 10, 2008, www.csmonitor.com/2008/1110/p13s01-wmgn.html.

22. For background, see Jennifer Weeks, "Carbon Trading," *CQ Global Researcher*, November 2008, pp. 295-320.

23. Ben Lieberman, "Beware of Cap and Trade Climate Bills," *Web Memo No. 1723*, Heritage Foundation, Dec. 6, 2007, www.heritage.org/Research/Economy/wm1723.cfm.

24. Opening Statement of Chairman Brian Baird before the Subcommittee on Research and Science Education, House Committee on Science and Technology, Sept. 25, 2007, http://science.house.gov/publications/OpeningStatement.aspx?OSID=1293.

25. Testimony of John A. "Skip" Laitner before the Subcommittee on Research and Science Education, House Committee on Science and Technology, Sept. 25, 2007, p. 11, http://science.house.gov/publications/Testimony.aspx?TID=7922.

26. David R. Baker, "State rebates lead more people to go solar," *San Francisco Chronicle*, July 15, 2008, www.sfgate.com/cgi-bin/article.cgi?f=/c/a/2008/07/14/BUNL11OVEF.DTL.

27. Quoted in *ibid.*

28. Gerald T. Gardner and Paul C. Stern, "The Short List: The Most Effective Actions U.S. Households Can Take to Curb Climate Change," *Environment*, September/October 2008, www.environmentmagazine.org/Archives/Back%20Issues/September-October%202008/gardner-stern-full.html.

29. Steven Blanchard and Peter Reppe, "Life Cycle Analysis of a Residential Home in Michigan," University of Michigan Center for Sustainable Systems, September 1998, www.umich.edu/~nppcpub/research/lcahome/homelca.PDF. The project was submitted in partial fulfillment of requirements for the degree of master of science in natural resources.

30. "Residential Buildings," Center for Sustainable Systems, University of Michigan, http://css.snre.umich.edu/css_doc/CSS01-08.pdf.

31. "About Energy Star," www.energystar.gov/index.cfm?c=about.ab_index.

32. "Energy Star has lost some luster," *Consumer Reports*, October 2008, p. 24.

33. Paul Davidson, "A new twist for light bulbs that conserve energy," *USA Today*, April 22, 2008, p. 10B, www.usatoday.com/money/industries/energy/environment/2008-04-21-light-bulbs_N.htm.

34. Environmental timeline, Radford University, www.runet.edu/~wkovarik/envhist/about.html.

35. Jennifer Weeks, "Buying Green," *CQ Researcher*, Feb. 29, 2008, pp. 193-216.

36. Rachel Carson, *The Sea Around Us*, Illustrated Commemorative Edition, (Oxford University Press, 2003), pp. 223, 225.

37. Gaylord Nelson, "Earth Day '70: What It Meant," *EPA Journal*, April 1980, www.epa.gov/history/topics/earthday/02.htm.

38. Andrew C. Revkin, "On Climate Change, a Change of Thinking," *The New York Times*, Dec. 4, 2005.

39. *Ibid.*

40. "Q&A: The Kyoto Protocol," BBC News, Feb. 16, 2005, http://news.bbc.co.uk/2/hi/science/nature/4269921.stm.

41. Al Gore, *An Inconvenient Truth* (2006), p. 8.

42. Daniel Stone, "The Green Pope," *Newsweek*, April 17, 2008, www.newsweek.com/id/132523.

43. Neela Banerjee, "Southern Baptists Back a Shift on Climate Change," *The New York Times*, March 10, 2008, www.nytimes.com/2008/03/10/us/10baptist.html?scp=1&sq=southern%20baptists%20back%20a%20shift&st=cse.

44. Gary Haq, Anne Owen, Elena Dawkins and John Barrett, "The Carbon Cost of Christmas," Stockholm Environment Institute, 2007, www.climatetalk.org.uk/downloads/CarbonCostofChristmas2007.pdf.

45. Timothy Gutowski, *et al.*, "Environmental Life Style Analysis," IEEE International Symposium on Electronics and the Environment, May 19-20, 2008. All other authors were graduate or undergraduate students at the Massachusetts Institute of Technology during the 2007 spring term.

46. "Following The Trail Of Toxic E-Waste," "60 Minutes," CBS News, Nov. 9, 2008, www.cbsnews.com/stories/2008/11/06/60minutes/main4579229.shtml.

47. *Ibid.*

48. Joseph De Avila, "PC Movement: How Green Is Your Computer?" *The Wall Street Journal*, Sept. 4, 2008, http://online.wsj.com/article/SB122048465164497063.html.

49. *Ibid.*

50. *Ibid.*

51. Marx, *op. cit.*

52. *Ibid.*

53. *Ibid.*

54. *Ibid.*

55. Sheila M. J. Bonini and Jeremy M. Oppenheim, "Helping 'green' products grow," *The McKinsey Quarterly*, October 2008.

56. *Ibid.*

57. "Environmentalism: Leading the Examined Life," *Patagonia*, www.patagonia.com/web/us/contribution/patagonia.go?assetid=23429&ln=267. See also www.patagonia.com/web/us/footprint/index.jsp.

58. *Ibid.*

59. www.patagonia.com/web/us/footprint/index.jsp.

60. Mike Spector, "The Incentives to Buy Hybrids Are Dwindling," *The Wall Street Journal*, Nov. 6, 2008, http://online.wsj.com/article/SB122593537581103821.html.

61. *Ibid.*

62. U.S. Department of Energy, "Consumer Energy Tax Incentives," www.energy.gov/taxbreaks.htm. See also, "P.L. 110-343/The Emergency Economic Stabilization Act of 2008: Energy Tax Incentives," www.energy.gov/media/HR_1424.pdf.

63. Ashlea Ebeling, "The Green Tax Gusher," *Forbes*, Nov. 24, 2008, p. 150.

64. "Policy Options for Reducing CO2 Emissions," Congressional Budget Office, February 2008, p. vii, www.cbo.gov/ftpdocs/89xx/doc8934/02-12-Carbon.pdf.

65. Peter Barnes, "Obama's 'number 1 priority,' Reuters, http://blogs.reuters.com/great-debate/2008/11/11/obamas-number-1-priority/.

66. "New Energy for America," http://my.barackobama.com/page/content/newenergy_more#emissions.

67. *Ibid.*

68. Jim Motavalli, "The Candidates' Clean Car Plans," *The New York Times*, Oct. 23, 2008, http://wheels.blogs.nytimes.com/2008/10/23/the-candidates-clean-car-plans/.

69. Deborah Zabarenko, "Obama will act quickly on climate change: adviser," Reuters, Nov. 12, 2008, www.reuters.com/article/environmentNews/idUSTRE4AB84K20081112?feedType=RSS&feedName=environmentNews.

70. Jeffrey Ball, "Does Green Energy Add 5 Million Jobs? Pitch Is Potent; Numbers Are Squishy," *The Wall Street Journal*, Nov. 7, 2008, p. A13.

71. *Ibid.*

72. The Associated Press, "Obama Promises Leadership on Climate Change," *The New York Times*, Nov. 18, 2008, www.nytimes.com/aponline/washington/AP-Obama-Climate-Change.html?sq=climate%20summit&st=nyt&scp=2&pagewanted=print.

73. Quoted in *ibid.*

74. *Ibid.*

75. Richard Black, "Obama vows climate 'engagement,'" BBC News, Nov. 18, 2008, http://news.bbc.co.uk/2/hi/science/nature/7736321.stm.

76. Andrew Jacob, "Report Sees New Pollution Threat," *The New York Times*, Nov. 14, 2008, www.nytimes.com/2008/11/14/world/14cloud.html?hp.

77. *Ibid.*

78. *Ibid.*

79. *Ibid.*

80. Oren Dorell and Alan Levin, "Economy sets travel back a bit for holiday," *USA Today*, Nov. 18, 2008, www.usatoday.com/travel/news/2008-11-18-aaa-holiday-travel-forecast_N.htm.

81. Jennifer Saranow, "Luxury Consumers Scrimp for Sake of Planet, and Because It's Cheaper," *The Wall Street Journal*, Nov. 4, 2008, http://online.wsj.com/article/SB122575617614495083.html.

82. Lena H. Sun, "U.S. Transit Agencies Ask Congress for Help in Averting Service Cuts," *The Washington Post*, Nov. 19, 2008, p. 2D, www.washingtonpost.com/wp-dyn/content/article/2008/11/18/AR2008111803174_pf.html.

83. William Neuman, "M.T.A. Said to Plan 23% Increase in Fare and Toll Revenue," *The New York Times*, Nov. 19, 2008, www.nytimes.com/2008/11/19/nyregion/19transit.html.

84. Juliet Eilperin, "Obama Addresses Climate Summit," *The Washington Post*, www.washingtonpost.com/wp-dyn/content/article/2008/11/18/AR2008111803286.html.

BIBLIOGRAPHY

Books

Barnes, Peter, *Climate Solutions: A Citizens Guide*, Chelsea Green Publishing, 2008.
An entrepreneur and writer blames global warming on market failure and misplaced government priorities.

Brower, Michael, and Warren Leon, *The Consumer's Guide to Effective Environmental Choices*, Three Rivers Press, 1999.
Though nearly a decade old, this book by veteran environmental experts helps consumers determine what impact their decisions will have on the environment, backed by research from the Union of Concerned Scientists.

Gore, Al, *An Inconvenient Truth*, Rodale, 2006.
The former vice president and Nobel Peace Prize winner argues that exploding population growth and a technology revolution have transformed the relationship between humans and the Earth.

Articles

Ball, Jeffrey, "Six Products, Six Carbon Footprints," *The Wall Street Journal*, Oct. 6, 2008, http://online.wsj.com/article/SB122304950601802565.html.
Companies calculate the carbon footprints of their products in different ways, making it hard for consumers to compare goods.

Bonini, Sheila M., and Jeremy M. Oppenheim, "Helping 'green' products grow," *The McKinsey Quarterly*, October 2008.
Two consultants contend that the failure of business to educate consumers about green products has helped discourage them from doing more to reduce their carbon footprints.

Ebeling, Ashlea, "The Green Tax Gusher," *Forbes*, Nov. 24, 2008, www.forbes.com/forbes/2008/1124/150.html.
As part of this fall's $700 billion bailout bill, Congress enacted a new round of tax breaks that benefit consumers who embrace energy-saving home improvements and alternative energy.

El Nasser, Haya, " 'Green' efforts embrace poor," *USA Today*, Nov. 23, 2008, www.usatoday.com/news/nation/2008-11-23-green-poor_N.htm.
Cities and community groups are trying to help low-income households reduce their energy consumption.

Gardner, Gerald T., and Paul C. Stern, "The Short List: The Most Effective Actions U.S. Households Can Take to Curb Climate Change," *Environment*, www.environmentmagazine.org/Archives/Back%20Issues/September-October%202008/gardner-stern-full.html.
A professor emeritus of psychology and the director of the National Research Council's Committee on the Human Dimensions of Global Climate Change argue that households often lack accurate and accessible information on how to reduce carbon emissions and mitigate climate change.

Knight, Matthew, "Carbon dioxide levels already a danger," *CNN*, www.cnn.com/2008/TECH/science/11/21/climate.danger.zone/index.html.
International scientists — led by James Hansen, director of NASA's Goddard Institute for Space Studies — conclude

that carbon dioxide concentrations in Earth's atmosphere are in the danger zone, threatening food shortages, more intense storms and other calamities.

Specter, Michael, "Big Foot," *The New Yorker,* **Feb. 25, 2008.**
An excessive carbon footprint is the modern equivalent of a scarlet letter, but calculating one's environmental impact of modern life "can be dazzlingly complex," the writer says.

Wald, Matthew L., "For Carbon Emissions, a Goal of Less Than Zero," *The New York Times,* **March 26, 2008, www.nytimes.com/2008/03/26/business/businessspecial2/ 26negative.html?scp=14&sq=carbon&st=cse.**
Researchers around the world are searching for so-called carbon-negative technologies that remove carbon dioxide from the atmosphere.

Reports and Studies

"Policy Options for Reducing CO$_2$ Emissions," Congressional Budget Office, February 2008, www
.cbo.gov/ftpdocs/89xx/doc8934/02-12-Carbon.pdf.
The congressional agency analyzes incentive-based options for reducing greenhouse-gas emissions, especially carbon dioxide.

Brown, Marilyn A., Frank Southworth and Andrea Sarzynski, "Shrinking the Carbon Footprint of Metropolitan America," Brookings Institution, May 2008, www.brookings.edu/reports/2008/~/media/ Files/rc/reports/2008/05_carbon_footprint_sarzynski/ carbonfootprint_report.pdf.
The researchers quantify transportation and residential carbon emissions for the 100-largest U.S. metropolitan areas.

Haq, Gary, Anne Owen, Elena Dawkins and John Barrett, "The Carbon Cost of Christmas," Stockholm Environment Institute, 2007, www.climatetalk.org .uk/downloads/CarbonCostofChristmas2007.pdf.
Three days of Christmas festivities could result in 650 kilograms (1,433 pounds) of carbon dioxide emissions per person, according to the United Kingdom-based researchers.

For More Information

Alliance to Save Energy, 1850 M St., N.W., Suite 600, Washington, DC 20036; (202) 857-0666; www.ase.org. Group that promotes energy efficiency worldwide.

American Council for an Energy-Efficient Economy, 529 14th St., N.W., Suite 600, Washington, DC 20045-1000; (202) 507-4000; www.aceee.org. Fosters energy efficiency to promote economic prosperity and environmental protection.

Carbonfund.org, 1320 Fenwick Lane, Suite 206, Silver Spring, MD 20910; (240) 247-0630; www.carbonfund.org. Provides certified carbon offsets to help individuals, businesses and organizations reduce their carbon footprints.

Chesapeake Climate Action Network, P.O. Box 11138, Takoma Park, MD 20912; (240) 396-1981; www .chesapeakeclimate.org. Grassroots group that fights global warming in Maryland, Virginia and Washington, D.C.

Clean Air-Cool Planet, 100 Market St., Suite 204, Portsmouth, NH 03801; (603) 422-6464; www .cleanair-coolplanet.org. Nonprofit organization that

partners with businesses, colleges and communities in the Northeast to reduce carbon emissions.

Energy Star, US EPA, Energy Star Hotline (6202J), 1200 Pennsylvania Ave., N.W., Washington, DC 20460; (888) 782-7937; www.energystar.gov. Environmental Protection Agency and Department of Energy program promoting energy-efficient products.

Nature Conservancy, 4245 North Fairfax Dr., Suite 100, Arlington, VA 22203-1606; (703) 841-5300; www .nature.org. Conservation organization that works worldwide to protect ecologically sensitive land and water.

New Hampshire Carbon Challenge, 8 College Road, CSRC, Morse Hall, Durham, NH 03824; (603) 862-3128; http://carbonchallenge.sr.unh.edu. Works to help households and communities reduce their energy consumption.

Pew Center on Global Climate Change, 2101 Wilson Blvd., Suite 550, Arlington, VA 22201; (703) 516-4146; www.pewclimate.org. Seeks new approaches to dealing with climate change.

Socially Responsible Investing

17

Can Investors Do Well by Doing Good?

Thomas J. Billitteri

Solar panels generate electricity from the roof the state capitol in Salem, Ore. Many socially responsible investors seek out firms that address environmental concerns, such as climate change and energy conservation. But critics say social investments tend not to perform as well as traditional ones.

From *CQ Researcher*,
August 29, 2008.

When Ann B. Alexander and her husband sold their natural-foods store in Durham, N.C., they pocketed a tidy sum. But they didn't want their profits to simply sit in a typical investment fund, even if they did grow in value. They also wanted their money to do good.

So the Alexanders chose a financial adviser who specializes in "socially responsible investing," an increasingly popular approach that combines investors' financial goals with a desire to improve society and hold corporations accountable.

"To have these corporations pay attention and be more responsible, it's good for all of us — not just the people who have money," Mrs. Alexander says.

She's not alone in that view. Socially responsible investing — sometimes called "sustainable investing," "ethical investing" and simply SRI — involves screening investments according to social or environmental, as well as financial, criteria, plus other strategies.

From a small niche of the financial world during the protest era of the 1960s and '70s, SRI has evolved into a complex and controversial trillion-dollar global industry spanning mutual funds, pension plans and big institutional and private holdings.

Many investors are attracted to social investing for moral or ethical reasons stemming from such concerns as climate change, workers' rights, workplace diversity, skyrocketing CEO pay and corporate political influence. Others may choose it out of a practical belief that companies that treat their employees well and protect the environment will be more profitable, more open about their operating methods and less prone to lawsuits and regulatory sanctions.

Social Investments Did Well Over Long Term

Over its 18-year life span, the Domini 400 Social Index (DS400) — composed of companies regarded as socially responsible — has outperformed the Standard and Poor's 500 (S&P 500). At one-, three-, five- and 10-year intervals, however, the DS400 lagged behind the S&P.*

DS400 Performance Statistics

	Total returns as of July 31, 2008							
	July 08	Last Qtr	Year to Date	One Year	Three Year	Five Year	10 Year	Since 5/1/90
KLD's DS400 Index	0.24%	-3.84	-12.67	-11.49	1.32	5.49	2.46	10.40
S&P 500	-0.84%	-2.73	-12.65	-11.09	2.85	7.03	2.91	9.91

* The S&P 500 is a broad stock market index (indicator) representing 500 publicly traded companies.

Source: KLD Research and Analytics

But the SRI movement also has vocal critics. They argue that stock-screening methods used by SRI mutual funds are ill-defined and highly subjective, that social investments tend not to perform as well as traditional ones and that SRI techniques have no place in the public pension world, where money managers have a fiduciary duty to maximize investment returns. (*See "At Issue," p. 425.*)

Despite the controversy, the movement is growing.

According to the Social Investment Forum, the industry's main trade group, assets under management using at least one of three core SRI strategies — investment screening, shareholder advocacy and "community investing" in areas underserved by traditional financial institutions — totaled $2.71 trillion last year, more than four times the amount in 1995 and about one of every nine dollars under professional management in the United States.[1]

As it has grown, social investing has evolved from its roots of simply avoiding tobacco, alcohol and other so-called sin stocks. It is placing more and more weight on companies that pay close attention to environmental, social and corporate-governance issues — "ESG" factors in investing parlance.

That emphasis is spreading to the broader financial world. For example, Goldman Sachs Group developed a 179-page report last year that recommended 44 companies based on a formulation that included ESG performance.[2] Most SRI assets — an estimated $1.9 trillion, according to the forum — are in accounts professionally managed for big institutional and high-net-worth clients.[3] Institutions embracing the responsible-investing approach include religious groups, private foundations, hospitals and labor unions and some of the nation's biggest public pension funds.

"In the old days, most people associated with SRI were just taking tobacco, alcohol and gambling stocks out of portfolios," says Paul Hilton, director of advanced equities research at the Calvert Group, a major SRI mutual fund company. "More recently, there's an understanding that what we're trying to do is use environmental, social and governance factors as another way to identify risk and better-quality management. That's getting through to institutional investors in a way that it hasn't before."

Even so, some critics argue that social investments have tended to produce mediocre returns compared to the broader stock market, causing investors to give up potential profits. The Domini 400 Social Index, a main SRI benchmark, has lagged the broader Standard & Poor's 500 index at the last one-, three-, five-, and 10-year intervals, although it is ahead of the S&P 500 over its entire 18-year life. (*See chart above.*)[4]

The use of socially responsible investing approaches in public pensions has been especially controversial. In California, which adopted a so-called double-bottom-line approach eight years ago, the $239 billion California Public Employees' Retirement System — the nation's largest pension fund — missed $400 million in gains by forgoing investments in China and other countries, according to *Business Week*. And a ban on tobacco stocks cost the $172 billion California State Teachers' Retirement System $1 billion, though returns for both pension funds still outpaced the S&P 500 over the past five years, the magazine said.[5]

California's pension system reportedly is considering reversing its tobacco policy, which critics of socially responsible investing say could undermine support for the movement.[6] (*See "Current Situation," p. 424.*)

And while the SRI mutual fund industry's stock-picking methods have become more refined over the years, they have occasionally led to stumbles. This summer Pax World Management Corp., a venerable SRI firm founded in 1971, settled charges with the Securities and Exchange Commission that it violated its own guidelines against buying shares in such businesses as alcohol, gambling, tobacco and defense.[7]

The company agreed to a $500,000 fine. The actions took place from 2001 through 2005 in two Pax funds, though not in its World Balanced Fund, which held over 90 percent of Pax World assets at the time, the company said.[8]

"We regret and take full responsibility for what occurred," said Joseph F. Keefe, who became CEO in 2005 after the SEC began its investigation.

Despite such missteps, supporters say responsible investing is fast moving into the mainstream. While investors may be attracted to it for different reasons, "the remarkable thing is the coming together of these investors with literally, now, trillions of dollars to press companies, or encourage companies, to act in a responsible way," says Timothy Smith, senior vice president of the environment, social and governance group at Walden Asset Management, an SRI firm in Boston.

Concerns over global climate change and natural-resource sustainability have provided perhaps the biggest impetus for the recent growth. So-called green funds new to the market include the Global Alternative Energy fund, started by the Calvert Group in May 2007, the Winslow Green Solutions Fund, started last November, and the five-month-old Global Green Fund, part of Pax World Funds.

Climate concerns are heavily reflected not only in mutual funds but also in shareholder activism. In this year's proxy season, 54 global-warming resolutions were filed with U.S. companies by public pension funds, labor and religious groups, and other institutional investors, according to the *Environmental Leader*, an online trade publication. Many of the investors belong to the Investor Network on Climate Risk, an alliance of 60 institutional investors with combined assets of more than $5 trillion.[9]

Also driving the responsible-investing movement are concerns about corporate behavior. Scandals at Enron and Tyco International earlier this decade, outrage over child exploitation in overseas sweatshops making goods for the West, and corporate links to human-rights hotspots like Sudan have led some investors to look carefully at which companies are in their stock portfolios.

Social investing is getting a boost, too, from the ongoing intergenerational transfer of wealth that is putting unprecedented sums of inherited money into the hands of baby boomers — a group weaned on '60s-era activism.[10]

"More and more and more people are asking themselves the question: What do I want to profit from?" says Cliff Feigenbaum, managing editor of the *Green Money Journal*, a Santa Fe, N.M., publication that promotes social and environmental values in investing.

As socially responsible investing attracts more followers and gains more scrutiny, here are some key questions that proponents and skeptics are asking:

Do socially screened investment funds perform as well as regular ones?

Meir Statman, a professor of finance at Santa Clara University in California, has done several studies examining the performance of socially responsible mutual funds and ones that don't apply social screens.

"The performance is about the same," he says. "Not better or worse."

Scores of other studies have reached similar conclusions. Yet they have not convinced everyone.

"The use of social criteria may be fine for the affluent who gamble their own money as a feel-good, vanity investment — but for those who can't afford to take a chance, SRI is a bad bet," argues Jon Entine, an adjunct fellow at the conservative American Enterprise Institute and a longtime critic of socially responsible investing.[11]

But Mercer, the international consulting firm, along with the United Nations Environment Program Finance Initiative, said last year after reviewing 20 academic studies on the subject, "Investors incorporating environmental, social and corporate-governance factors within their investments don't have to give up returns." Mercer formed a "Responsible Investment" business unit in 2004.[12]

Likewise, researchers at Lehigh University in Pennsylvania studied SRI-screened portfolios and concluded that "there is no cost to being good."[13]

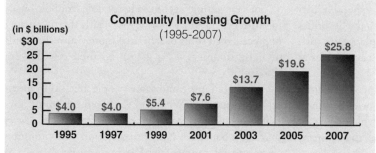

Community Investing Continues to Expand

The amount of money that social investors directed to communities that are underserved by traditional financial services grew sixfold since 1995, and 32 percent from 2005 to 2007.

Community Investing Growth
(1995-2007)

(in $ billions)

1995	$4.0
1997	$4.0
1999	$5.4
2001	$7.6
2003	$13.7
2005	$19.6
2007	$25.8

Source: "2007 Report on Socially Responsible Investing Trends in the United States," Social Investment Forum Foundation, 2008

And Alex Edmans, a finance professor at the University of Pennsylvania's Wharton School, found that stocks of companies on *Fortune* magazine's "Best Companies to Work For" list robustly outperformed the overall market.[14]

Yet, other evidence has fed doubts about the correlation between socially responsible investing and stock performance. Robert F. Stambaugh, a finance professor at Wharton, and two colleagues found that when returns are adjusted for risk, index-style investors don't give up much when they use socially responsible funds, but investors who select actively managed funds can lose more than 3.5 percentage points of return a year.[15]

"Sometimes being socially responsible is costly, and sometimes it isn't," Stambaugh said. "It depends on what kind of investor you are."[16]

And research published in the *Harvard Business Review* this year found only a minor correlation between corporate social responsibility and good financial results.[17]

Some of the most spirited debate over social investing has involved SRI barometers like the benchmark Domini 400 Social Index. The index limits or avoids stocks in some industries that have performed well in recent years, such as gambling and military weapons, and is significantly weighted in industries like banking and information technology that hit rough patches.

Peter D. Kinder, president of KLD Research & Analytics, a Boston SRI research firm that owns the Domini 400 Social Index, says a key reason the index has lagged the S&P 500 is that it generally excludes stocks in companies that extract natural resources, such as oil-drilling and mining. Those sectors have soared recently.

"When industries we typically either exclude or underweight are doing very well, we will perform less well," Kinder says. On the other hand, "we are going to perform better relative to the broad market indexes when these companies underperform."

Amy Domini, who as a KLD partner helped develop the index before starting her own company, Domini Social Investments, says social investing has been hurt by market shifts stemming from the post-Sept. 11 "war economy."

"People look for hard assets — gold, copper, oil — and those are not very robust for social investment funds to find opportunities in," she says.

Domini's flagship Social Equity Fund was down nearly 12 percent for the year ending July 31 and, partly because of the bursting of the tech bubble earlier this decade, has lagged the S&P 500 over time.

Amy Domini has felt the pressure that so-so stock performance can bring to bear. Since the Social Equity Fund's inception in 1991 it had tracked the Domini 400 index, but two years ago she changed that strategy. Now the fund is actively managed through a computer-driven stock-picking strategy.

"Emotionally, this is difficult," Domini said in 2006, after making the hard decision to change the fund's management approach. "But I can't ask my shareholders to be patient forever."[18]

Entine argues that the lag in benchmarks like the Domini 400 Social Index points to a larger flaw in the social-investment approach. To avoid investing in defense, energy and tobacco — areas viewed negatively in the SRI world but that have soared in recent years — social funds made unwise bets on banking and technology, only to be slammed by speculative bubbles, he charges.

"They overweighted based on social prejudices, not on metrics," he says. And "when you overweight, essentially you're gambling with someone's money."

Squabbling over SRIs' performance compared with conventional investments isn't likely to be settled soon. "This debate is going to continue," insists Walden Asset Management's Smith, a 40-year veteran of SRI. "There's no definitive knockout blow on either side."

Are screening methods used by mutual funds sound?

Walden Asset Management bills itself as "a leader in socially responsive investing," but that doesn't mean it shuns one of the world's most controversial industries. For years Walden has owned shares in British Petroleum (BP).

Smith calls the company a "work in progress" on health and safety issues and expresses discouragement over a deadly BP refinery explosion in Texas in 2005 and an oil spill in Alaska in 2006. Still, he says, the company continues to be a leader among energy producers on the climate-change issue.

"In some industries, companies stand out, and therefore there are some positive reasons to approach them," he says.

Walden's approach is known as "best in class," one of the strategies that SRI funds use to choose which stocks fit their missions.

The Domini Social Equity Fund holds stock in McDonald's — a company that, over the years, has been accused of everything from destroying rainforests for food production to polluting the environment with packaging waste. But Amy Domini says McDonald's has answered the critics, setting standards for beef production, addressing the packaging issue and participating in a coalition to improve international labor standards.

"Am I going to throw out the company that has been highly responsive to consumers and third parties time after time?" she asks.

Screening industries and companies has always been tricky for social investors, leaving the field open to charges that it compromises too much, lacks uniform screening standards and uses methods that are confusing or even misleading to people who want to invest with a conscience.

Cold cash totaling $12 million awaits the winner of the World Series of Poker main event at Harrah's in Las Vegas. Investment funds that screen for gambling stocks had more than $41 billion in assets under management in 2007.

The research can be "shoddy," Entine charges, adding, "It depends on how well companies manage their reputations, more or less."

Social-investment advocates strongly disagree, however. "We are more sophisticated than we've ever been, and we'll be more sophisticated tomorrow," says the *Green Money Journal's* Feigenbaum. "We're screening as thoroughly as we possibly can."

When it comes to SRI mutual funds, investment companies use different approaches to decide which industries and stocks to avoid, embrace or include with reservations.

In *Socially Responsible Investing: Making a Difference and Making Money*, a 2001 overview of the field, Domini noted that "[r]esponsible investors struggle with how best to deal with . . . troubling industries. Many have come down on the side of avoiding problematic industries completely. Others decide to underweight some industries while avoiding others."[19]

Social funds may also use their shares to exert pressure on companies through shareholder resolutions or direct talks. "There are companies that we invest in that on balance have a positive record but still have some problems, and then we use our shareholder voice to engage those companies and urge them to change," Smith says.

In addition, SRI portfolios may shift to keep up with changing times or circumstances.

Last November *The Wall Street Journal* reported that Pax World discontinued its policy of rejecting alcohol- and gambling-related investments after the policy led it three years ago to divest from Starbucks Corp. — widely regarded as socially progressive — for licensing the Starbucks name to a coffee liqueur. "Now, while they still decline to invest in weapons and tobacco makers, portfolio managers . . . weigh potential investments based on a mix of financial and ESG metrics, including corporate governance, community relations, product integrity, human rights and climate change," the newspaper said.[20]

The Calvert Group, based in Bethesda, Md., traditionally has avoided the nuclear-power industry. But Hilton, the advanced equity research director, says the company's two newest funds — Global Alternative Energy and International Opportunities — are willing to invest in certain companies that have existing nuclear plants but are doing excellent work in wind or solar power. "That's not so much of an endorsement of nuclear as the fact that we want to invest in companies pushing ahead in alternative energy," he says.

SRI proponents regard such attempts at nimbleness as signs that the SRI field is maturing. But the field's methods have also invited sharp criticism, some of it from people dedicated to the cause of sustainability and social responsibility.

In 2004, environmentalist and entrepreneur Paul Hawken — cofounder of the Smith & Hawken garden-supply business and executive director of the Natural Capital Institute, a research group in California — wrote a stinging critique of the SRI mutual-fund industry, saying he wanted to help it better respond to investors seeking "to invest with a conscience and a purpose."[21]

Hawken charged, in part, that SRI mutual funds lack common standards, definitions and codes of practice; that taken as a whole, the investment portfolio of the combined SRI fund industry "is virtually no different" than that of conventional mutual funds and that the field is marred by a "lack of transparency and accountability in screening and portfolio selection." He also said the language used to describe social-investment mutual funds is "vague and indiscriminate" and that "fund names and literature can be deceptive."

This year he was quoted in a British newspaper as saying that "the situation has not got better. [Financial] performance has become the primary driver. They [ethical funds] are doing everything they can to be acceptable to the broadest possible clientele, and with that has come dilution of meaning and standards."[22]

Julie N. W. Goodridge, president of Northstar Asset Management, a Boston social-investment firm, wants the SRI industry to adopt stronger disclosure policies to give investors a better understanding of what funds are investing in, and why.

"The consumer doesn't necessarily know that different funds do different things," she says. "I worry that consumers . . . find that Coca-Cola is sitting right smack in the middle of their fund, and they would never purchase the product, let alone invest in it."

Many social investors have avoided Coke for reasons as varied as its use of precious water resources in developing countries and its sponsorship of this year's Olympics in China, which has been accused of human-rights abuses.

Joe Nocera, a business columnist for *The New York Times*, last year criticized KLD Research & Analytics, upon whose analysis many SRI funds rely for guidance. Nocera said KLD had only two dozen researchers monitoring 3,000 companies and that they relied on media reports, blogs, interactions with activist groups and conversations with companies themselves. "That hardly seems like enough to make a decision on whether a company is good or bad," Nocera wrote.[23]

But Kinder, KLD's president, says his company "takes full advantage of Internet tools" to research companies and that "companies don't fundamentally change that often.

"We've been looking at these companies for 20 years. I feel as confident as any researcher can be in our research."

Can socially responsible investors influence corporate behavior?

Last year, activists led by Christian Brothers Investment Services — an investment advisory firm for Catholic dioceses, hospitals and other organizations — put forth a shareholder resolution seeking an independent examination of the environmental and social impact of Newmont Mining Co.'s operations around the globe.

But the Denver-based gold-mining company didn't fight the resolution — in fact, it endorsed it before the shareholder vote was taken.[24]

Social-investment proponents point to that endorsement as a sign of the SRI movement's growing influence, although the resolution's ultimate effect on Newmont's operations remains to be seen.

"It's hard to say at this point what the results will be," says Julie Tanner, assistant director of socially responsible investing at Christian Brothers and a member of a panel reviewing the issue. "We look forward to the board's recommendations," due before Newmont's 2009 annual meeting, she says.

As the SRI movement matures, it has shifted more and more of its focus toward engaging companies directly on a variety of issues, ranging from CEO pay to human rights in overseas factories.

"Responsible investors are no longer simply avoiding stocks in companies they have problems with," says Smith of Walden Asset Management. "They are, with trillions of dollars behind them, engaging companies, writing letters, talking with them, voting their proxies, filing shareholder resolutions and voting and debating at stockholder meetings. These are not investors that are seeking an illusory, pure corner of the marketplace, they are investors who are engaged in the [real] marketplace."

Shareholder resolutions are the most visible manifestation of that engagement. From 2005 through the first half of 2007, a total of 1,065 resolutions were filed on social, environmental and governance issues. Support for resolutions filed in the first half of 2007 stood at 15.4 percent of votes cast, compared with 9.8 percent in 2005 and 13.3 percent in 2006.[25]

Even so, shareholder activism can be a tough road. Companies often resist resolutions, especially those that are costly or disruptive. Even resolutions that pass may not lead to real change.

"A company can do something because of a shareholder resolution," says Tanner, "but the real question is,

Socially Responsible Investments Near $3 Trillion

The amount of money in socially responsible investments (SRI) in the United States has increased more than fourfold since 1995, to $2.7 trillion. The number of funds offering SRI has also increased from 55 in 1995 to 260 in 2007 (not shown).

Socially Responsible Investing in the United States, 1995-2007
(in $ billions)

Year	Social screening	Shareholder advocacy	Community investing	Total*
1995	$162	$473	$4	$639
1997	$529	$736	$4	$1,185
1999	$1,497	$922	$5	$2,159
2001	$2,010	$897	$8	$2,323
2003	$2,143	$448	$14	$2,164
2005	$1,685	$703	$20	$2,408
2007	$2,098	$723	$26	$2,711

* The sum of the three columns for each year is less than the total shown because overlapping assets involved in screening and advocacy are subtracted to avoid double counting.

Source: "2007 Report on Socially Responsible Investing Trends in the United States," Social Investment Forum Foundation, 2008

how robust and substantive is the company's implementation going to be?"

SRI advocates often prefer to try engaging in direct talks with management on issues of concern before going the resolution route. "The preferable outcome is negotiated settlement," says KLD's Kinder. "It doesn't start with a proxy resolution. A proxy resolution in the context of socially responsible investing typically represents failed negotiations."

In the Newmont Mining case, Tanner says, "We were in dialogue with the company, we had been speaking to them for a while; it wasn't necessarily that we were having a failed negotiation [but] I don't think things had gotten to where the company was going to take action to the extent we wanted them to."

Jeanne M. Logsdon, a professor of business ethics, and Harry J. Van Buren III, an assistant professor of business and society, both at the University of New Mexico, study shareholder activism. In one study they classified more than 1,700 shareholder resolutions filed over seven years with companies listed on the New York Stock Exchange by members of the Interfaith Center on Corporate Responsibility (ICCR), an association of 275 faith-based institutional investors. More than 40 percent of the resolutions — 707 in all — related to "justice" issues, with employment and economic-development concerns the most prevalent.[26]

Logsdon says that while few such resolutions are approved by the majority of a corporation's stockholders, they can be effective in raising the public profile of an issue.

"You don't do it because you're going to win, you do it to bring attention to an issue — to take a stand and to signal to corporate leaders that the companies should be dealing with the issue."

Van Buren, who engages in shareholder activism on behalf of the Episcopal Church, says the church doesn't do much stock screening but has been an active filer of shareholder resolutions.

For years, he says, the church has been raising red flags about predatory and sub-prime lending, issues now at the center of Wall Street upheaval. "If you look at shareholder resolutions that are being filed today, in many cases they represent leading-edge issues that corporations or other stakeholders pick up in years to come," Van Buren says.

In a study published this summer, Logsdon and Van Buren looked at how social activists influence companies through direct dialogues with management. They studied nearly 1,200 new and ongoing dialogues initiated by ICCR members from 1999 through 2005 with such firms as Citigroup (20 dialogues), Wal-Mart Stores (19), Coca-Cola (15) and Bank of America and Target (14 each).[27]

Van Buren says no researchers have systematically tracked whether shareholder activists are getting concrete results from such talks. But one "imperfect" measure of such progress, he says, is the number of shareholder resolutions that are withdrawn. Such withdrawals may occur if a company makes enough of an effort to change to satisfy the concerns of activists.

"The real work of shareholder activism gets done in ongoing corporate dialogue," Van Buren says. "You file a resolution, you may withdraw the resolution, but with a condition of withdrawal the company will agree to do something."

According to the Social Investment Forum, of the 1,065 resolutions filed from 2005 to mid-2007, 319 were withdrawn, and 561 were voted on.[28]

BACKGROUND

Early Activism

Socially responsible investing boasts a venerable history. In an 18th-century sermon titled "The Use of Money," Methodism's founder, Anglican minister John Wesley, warned against engaging in liquor production, industries that pollute, such as tanning, and practices such as bribery.[29] "We ought not to gain money at the expense of life," he declared.[30]

Quakers and Methodists in early America refused to invest in ways that might help the slave trade.[31] And during the 19th century, many religious investors refused to put money into such activities as liquor and tobacco production, gambling and weapons manufacturing.

Mutual funds that practice social screening emerged in the 20th century. The Pioneer Fund began screening out "sin" stocks in 1950 to serve Christian investors.[32] Then, in 1971, two United Methodist Church activists in peace, housing and employment issues started the Pax World Fund, which the company calls the first broadly diversified, publicly available mutual fund to use social as well as financial criteria in making investment decisions.[33]

During the late 1960s and early '70s intense social activism by advocacy and religious groups began placing more and more emphasis on corporate responsibility.

The Interfaith Center on Corporate Responsibility (ICCR) was founded in 1971 partly as a result of opposition to the Vietnam War. Progressive clergy questioned whether churches were profiting from the war, such as by holding the stocks of arms manufacturers. The questioning spread further, to such issues as production of nuclear weapons, sale of military goods abroad, creation of space weapons and apartheid in South Africa.[34]

"One of the things our founding members looked at was, if we're going to be endowed institutional investors, do we choose to own companies that make money on some of the precise social ills we're trying to counteract?" says Laura Berry, executive director of the ICCR.

South Africa's system of racial separation was becoming a highly visible cause among many socially conscious investors during this period. In 1971 the Episcopal Church filed the first church-sponsored shareholder resolution, which called on General Motors (GM) to close down operations in South Africa because of the government's apartheid policy.

That same year, Leon Sullivan, an African-American Baptist minister, was elected a GM director. He used his influence on the GM board to campaign against apartheid and later in the decade developed a code of conduct for companies doing business in South Africa, known as the Sullivan Principles.

By the 1980s, growing anti-apartheid fervor, including pressure from Sullivan, helped lead a number of churches, pension plans and other institutional investors to either divest themselves of stock in U.S. corporations operating in South Africa or use their shares to urge companies to withdraw from the country. By the end of the 1980s scores of companies were leaving South Africa.

"The anti-apartheid campaigns of the 1980s provided a galvanizing moment in the history of SRI," according to the Social Investment Forum.[35]

Apartheid was not the only concern that led social investors to examine their stock portfolios. Environmental catastrophes such as a poison-gas leak at a Union Carbide factory in Bhopal, India, that killed thousands in 1984, a nuclear disaster in Chernobyl, in what is now Ukraine, in 1986 and the *Exxon Valdez* oil spill in Alaska in 1989 "served as flashpoints for investor concerns over pollution and corporate responsibility," the forum noted.[36]

By the 1990s and early 2000s, social investing had percolated into the mainstream investment world, leading to the creation of an entire infrastructure of investing vehicles.

Tobacco and Alcohol Are Top Social Screens

Among funds that screen for social and environmental concerns, tobacco is the most prevalent criterion incorporated into portfolio management. It was employed by 166 of the 260 funds identified, with more than $174 billion in total net assets under management, or more than 86 percent of the total assets of socially screened funds. Alcohol is the second most predominant social screen, with more than $158 billion managed by 125 funds.

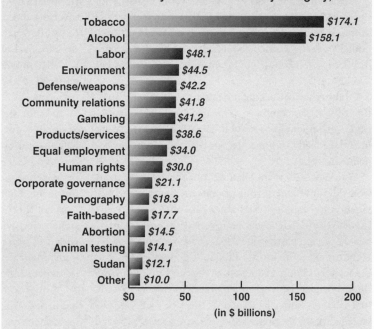

Total Net Assets of Socially Screened Funds by Category, 2007

Source: "2007 Report on Socially Responsible Investing Trends in the United States," Social Investment Forum Foundation, 2008

SRI indexes — for example, the Domini 400 Social Index, Dow Jones Sustainability Indexes (started in 1999) and London-based FTSE4Good indexes (2001) — have helped track the financial performance of companies and provided yardsticks for corporations to measure their social progress against their peers.

And as social investing grew, it also evolved and diversified. "We started looking for positive things to invest in," says Goodridge of Northstar Asset Management, "and we then moved into shareholder activism in a big way, and then into community investing."

Key Rulings

Several key government rulings have added to the movement's momentum.

In 2003, the Securities and Exchange Commission (SEC) required all mutual funds to disclose their proxy votes on corporate issues — the kind of disclosure that had long been sought by social investors.[37] In 1999 Domini Social Investments says it became the first mutual fund manager to publicly disclose its proxy votes.

Critics of the SEC decision said it would be costly for shareholders and wouldn't help them much, but SRI proponents hailed the new regulation. "What [mutual funds] are really worried about is the end of all their conflicts of interest," said Tim Grant, then-president of Pax World Funds.[38]

Another key government decision involved the question of whether company-sponsored "defined contribution" retirement plans, such as 401(k) plans, could include socially responsible funds among their investment choices.

A decade ago, many plan administrators shied away from SRI investments out of concern that they would violate the Employee Retirement Income Security Act (ERISA) rule that assets be invested for the exclusive benefit of plan participants.[39]

However, William M. Tartikoff, general counsel of the Calvert Group, requested an advisory opinion from the Labor Department, which said in a May 1998 letter that ERISA's fiduciary standards don't rule out socially screened funds as long as "the investment was expected to provide an investment return commensurate to investments having similar risks."[40]

"The letter gave us great optimism," Tartikoff said on the 10th anniversary of the opinion. "It gave us comfort and validated what we were doing and what we were saying was correct and that, in fact, there was no inherent reason for Calvert's socially screened funds not to be chosen on a 401(k) platform and as a plan option."[41]

SRI has made inroads into the retirement-plan field since then. A survey last year by Mercer found that 19 percent of defined-contribution retirement-plan sponsors had an SRI option, and an additional 41 percent expected to be doing so within three years.[42]

As the responsible-investing movement moves more and more into the mainstream, investors and money managers are increasingly using environmental, social and corporate governance factors — or ESG principles — to guide their investment decisions, a trend driven in no small way by concern over climate change.

Traditional investing firms have picked up on the trend. For instance, State Street Global Advisors has been using some ESG-applied research over the past couple of years, according to *The Wall Street Journal.* "We are seeing these factors start mainstreaming," said William Page, chairman of State Street's ESG Team. He said the team uses ESG research for accounts of some wealthy investors and private institutional investors.[43]

While ESG principles are consistent with the traditional moral or ethical impulses that have long driven the SRI movement, many institutional investors are embracing them because they help shape financial performance.

In 2006, the United Nations issued a set of voluntary "Principles for Responsible Investment" that aimed to help institutional investors and others integrate environmental, social and governance factors into their investment decisions and ownership practices with the goal of improving long-term returns.

As of this past May, the principles had garnered 362 signatories, with European investors leading the way with 148 signatories representing $9.7 trillion in assets under management. North America counted 70 signatories with $2.3 trillion.[44]

"The great majority of new signatories continue to be mainstream pension funds, insurance companies and investment managers, with a minority coming from the dedicated socially responsible investment sector," according to the latest progress report on the effort.[45]

The U.N. principles are not the only effort to frame investment decisions around corporate responsibility. For example, the Carbon Disclosure Project, a British nonprofit group, serves institutional investors with some $57 trillion in assets under management globally. It analyzes how major companies around the world are responding to climate change and examines the commercial risks and opportunities it presents to corporations. This year it sought emissions data from 3,000 companies.

Investment-consulting firms also are adopting ESG principles in their work. For example, Mercer's "responsible investment" unit employs 16 people around the world

C H R O N O L O G Y

1700s-1950 *Roots of socially responsible investment movement are developed.*

1700s Methodism founder John Wesley warns against making money "at the expense of life."

1800s Religious investors shun liquor and tobacco production, gambling and weapons-making.

1950 Pioneer Fund becomes first mutual fund to screen for "sin" stocks.

1970s *Opposition to South African apartheid and Vietnam War helps establish social-investing practices.*

1971 Interfaith Center on Corporate Responsibility is founded. . . . Pax World launches nation's first publicly available socially responsible investment fund. . . . Episcopal Church files first church-sponsored shareholder resolution.

1977 African-American pastor Leon H. Sullivan develops code of conduct on human rights and equal opportunity for companies operating in South Africa, eventually helping to dismantle apartheid; in the late 1990s Sullivan announces successor set of principles for industries worldwide.

1980s-1990s *Anti-apartheid campaigns help spur companies to leave South Africa, and socially responsible investing moves increasingly into the mainstream investment world.*

1982 Calvert Social Investment Fund becomes the first U.S. mutual fund to prohibit investments in companies doing business in South Africa.

1985 Social Investment Forum launched.

1989 Ceres, a national network of investors and others addressing environmental sustainability, is formed.

1990 Domini 400 Social Index, first index to measure performance of broad group of socially responsible stocks in the U.S., is created.

1998 U.S. Department of Labor tells Calvert Group federal law doesn't preclude use of "socially responsible" funds in retirement plans.

1999 Domini Social Investments becomes first mutual fund manager to publish its proxy-voting record. . . . Dow Jones Sustainability Indexes are launched.

2000-Present *Public pensions impose tobacco and other screens; climate change rises as social-investment priority.*

2000 Nation's largest pension fund, the California Public Employees' Retirement System (CalPERS), announces it will sell its tobacco holdings.

2001 London-based FTSE4Good indexes begin tracking social investing.

2003 Securities and Exchange Commission requires mutual funds to disclose proxy votes.

2004 Environmentalist and entrepreneur Paul Hawken issues report critical of the socially responsible investment mutual fund industry, arguing it "has no standards, no definitions and no regulations other than financial regulations." . . . Mercer, a worldwide consulting firm, starts a "responsible investment" unit.

2006 United Nations launches "Principles of Responsible Investment" to help large investors integrate environmental, social and governance factors into their investment decisions. . . . Sudan Divestment Task Force established to persuade states, universities and other groups to restrict Sudan-linked investments.

2007 Mercer survey finds that 19 percent of defined-contribution retirement plans include a social-investment option, and 41 percent plan to add one over the next three years. . . . *Los Angeles Times* publishes stories critical of investment policy of Bill & Melinda Gates Foundation.

2008 California public pensions reportedly re-examining policy on tobacco investments. . . . New Hampshire orders its public pensions to divest Sudan-related investments, sparking resistance from pension officials. . . . Social Investment Forum identifies $2.7 trillion in assets under management using socially responsible investing strategies.

Big Financial Institutions Are Going 'Green'

Investors say profits and principles go together.

Climate change may be bad for planet Earth, but it's good for the bottom line, Deutsche Bank decided last fall, launching its new "climate change investment initiative."

The big German-based institution was one of the latest mainstream financial institutions and socially responsible investment companies to focus on issues like global warming, renewable fuels and environmental sustainability.

"Companies and investors are quickly realizing that climate change is not merely a social, political or moral issue — it is an economic and business issue as well," declared Kevin Parker, Deutsche Bank's global head of asset management.[1]

Environmental funds have been attracting stronger investor interest lately than some other social-investing vehicles. Inflows of money into so-called green funds totaled $766 million for the year ending May 31, compared with net outflows among religion-based funds of $37 million for that 12-month period, according to Morningstar, which tracks both categories under the heading of socially responsible investing.[2]

"It's not just tree huggers" who are concerned about global warming, Holly Isdale, managing director and head of wealth advisory at Lehman Brothers, the big New York investment bank, told *The Wall Street Journal* last year. "There's money to be made, and people want to know how to make it."[3]

New green funds have sprouted at longtime social-investment companies such as Pax World, whose five-month-old Global Green Fund invests in companies focused on "mitigating the environmental impact of commerce," and Calvert, which launched the Calvert Global Alternative Energy Fund last year and plans to launch a Global Water Fund later this year.[4]

And new funds are popping up at traditional financial firms in Europe, the United States and elsewhere.

"In a little more than two years," Parker of Deutsche Bank wrote in March, "we estimate retail investors all over the world have pumped around $66 [billion] into more than 200 newly launched mutual funds and exchange-traded funds investing in companies that help to mitigate or adapt to climate change."[5]

Last November Deutsche Bank, through its U.S. retail unit, launched the DWS Climate Change Fund, which focuses on alternative energy, energy-efficient products and companies that deal with damage to the environment.

One overarching factor is behind the growth in socially responsible investing, says Peter Kinder, president of KLD Research & Analytics, a social-investing research firm: "There's always more than one answer [but] the short answer is, global warming is driving this. There is in my mind no question that is true."

who rate institutional money managers according to how they use environmental, social and governance factors in their investment decisions. That information helps guide Mercer clients, such as pension-plan administrators and private foundations, in selecting investment managers.

"It's a big growth area for us," says Craig Metrick, U.S. head of responsible investments for Mercer Investment Consulting. "Increasing numbers of institutional investors are coming to look at social responsibility as way to manage risk and look for opportunities in their portfolios long term."

Sharp Critiques

As responsible investing has become a more prominent part of the investing scene, it also has invited sharp

critiques — particularly where mutual funds and stock screening are concerned.

In his 2004 report, for example, environmentalist Hawken took issue with screening methods among SRI funds, saying they "allow practically any publicly held corporation to be considered as an SRI portfolio company."[46] The single most important criterion for a company, he argued, is whether "its services and products [are] helpful to the world we inhabit. . . . What does it matter if one fast-food company is singled out as 'best in its class. . . ?" he posited. ". . . [I]f you are going the wrong way, it doesn't matter how you get there."[47]

In his influential book *The Market for Virtue: The Potential and Limits of Corporate Social Responsibility,*

KLD's Global Climate 100 Index, which the company bills as the first global index focused on climate change, marked its third anniversary this summer. It holds companies engaged in renewable energy, clean technology and "future fuels."[6]

In an analysis of social-screening trends among all investment funds, the Social Investment Forum found that environmental concerns were more frequently incorporated into fund management than either labor-friendly criteria or alcohol restrictions, though the 146 funds that incorporate environmental criteria managed fewer assets overall than either labor-friendly or alcohol-restricted funds.[7]

As environmental issues heat up, social investors are joining with others in public-policy advocacy efforts.

This summer, for example, an investor group that includes officials from a number of social-investment companies, pension funds, foundations and other organizations, called on the U.S. Senate to extend tax credits for renewable energy and energy-efficiency projects that are set to expire at the end of the year.[8] In May investors urged senators to enact legislation dealing with global warming.[9]

The efforts were organized by Ceres, a 19-year-old Boston-based network of investors, environmental groups and other public-interest groups, and its Investor Network on Climate Risk, made up of more than 70 institutional investors with roughly $6 trillion in assets.

Ceres also started the Global Reporting Initiative, based in Amsterdam, which provides a framework for companies and other organizations to report their environmental, social and governance performance. The framework is used by more than 1,200 companies worldwide.[10]

"While U.S. policy makers are running in place on climate change," Ceres President Mindy S. Lubber wrote in a blog this month, "global investors are moving quickly to make money from its far-reaching risks and opportunities."[11]

---

[1] Deutsche Asset Management, "Investing In Climate Change," October 2007, p. 1.

[2] Elizabeth O'Brien, "Green Investing — The Gold Rush," *Financial Planning*, Aug. 1, 2008.

[3] Jilian Mincer, "Why 'Green' Investing Has Gained Focus," *The Wall Street Journal*, June 21, 2007.

[4] O'Brien, *op. cit.*

[5] Kevin Parker, "Investment is key in climate change battle," *Financial Times*, March 24, 2008.

[6] "KLD Global Climate 100 Index Marks Three Year Anniversary," www.kld.com/newsletter/archive/press/pdf/1216311527_GC100%20 3rd%20Anniv%20release_Final%20(2).pdf.

[7] Social Investment Forum, "2007 Report on Socially Responsible Investing Trends in the United States," p. 11.

[8] Ceres, "Investors with $1.5 Trillion in Assets Call on Congress to Extend Renewable Energy and Energy Efficiency Tax Credits," press release, July 29, 2008, www.ceres.org/NETCOMMUNITY/Page.aspx?pid=923&srcid=705.

[9] Ceres, "Investors Managing $2.3 Trillion Call on Congress to Tackle Global Climate Change," press release, www.incr.com/NET COMMUNITY/Page.aspx?pid=900.

[10] Ceres, www.ceres.org/NETCOMMUNITY/Page.aspx?pid=415& srcid=552.

[11] Mindy S. Lubber, "Climate Change: Investors' Next Global Mega-Trend? Harvard Business Publishing, Aug. 1, 2008, http://blogs.harvard business.org/leadinggreen/2008/08/climate-change-investors-next.html.

David Vogel — a professor of both business ethics and political science at the University of California, Berkeley — questioned some of the underlying tenets of the SRI movement.

Vogel noted, for example, "that the social-investment community was no more able than any other investors to identify the failures of corporate governance that created such massive shareholder losses at the beginning of the twenty-first century." He pointed out that Enron and WorldCom, among other troubled companies, were widely held by SRI funds.[48]

"Implicit in the very existence of SRI is the claim that it is possible to identify which firms are more or less responsible," Vogel also wrote. "Not only is this claim questionable, but the selection criteria employed by SRI fund managers and researchers can be criticized on several grounds."[49]

In *The Atlantic* magazine last fall, former Wall Street stock analyst Henry Blodget concluded, in the words of the article's subtitle, that "socially responsible investing is neither as profitable nor as responsible as advertised."[50]

Blodget himself is no stranger to Wall Street controversy, as he acknowledges in the article. In 2003 regulators banned him from the securities industry and fined him $4 million for putting out misleading stock research in violation of federal laws.[51] He has since become a prominent financial writer.

In *The Atlantic* article Blodget argued that "the central dilemma for most socially responsible investors" is that "virtue can cost you."

Should Foundations Screen Their Investments?

Some experts question the benefit to society.

With some $39 billion in assets, the Bill & Melinda Gates Foundation is the nation's largest private foundation — and a revered leader in global philanthropy. But early last year, the *Los Angeles Times* questioned the huge philanthropy's investment practices.

The newspaper concluded that the foundation "reaps vast financial gains every year from investments that contravene its good works" and had holdings in companies that had "failed tests of social responsibility because of environmental lapses, employment discrimination, disregard for worker rights, or unethical practices."[1]

In one example, it noted that the foundation had spent $218 million on polio and measles immunization and research worldwide, including in the Niger Delta, but that it had invested $423 million in companies responsible for most of the oil-plant flares that spew pollution in the delta.[2]

The articles touched off a spirited debate in philanthropy circles about whether foundations, which together control some $670 billion in assets, should do more to align their investments with their charitable missions.[3]

"Creating an immutable firewall between investments and grants is nonsensical, a strategy worthy of ostriches, not leaders," Allison H. Fine, a senior fellow at Demos: A Network for Ideas and Action, said at a panel discussion last year prompted by the articles on the Gates Foundation.[4]

But some philanthropy experts question whether efforts by foundations to screen their investment holdings can do very much to help society.

"There is definitely a virtue to aligning the investments of the foundation with its mission," says Mark Kramer, founder and managing director of FSG Social Impact Advisors, a nonprofit consulting organization in Boston. But, he adds, while shareholder advocacy "can absolutely have a demonstrable impact" on companies, "it's not clear that screening public equities has any demonstrable social impact."

"It would be very hard to have enough volume [of shares] to truly affect the [stock] price of a large publicly traded company, and even if you did, in a sense what you might be doing is simply giving a better investment to someone who's less ethically inclined because they can pick up a valuable stock at a lower price.

"So there is something around the advocacy piece — the pressure that . . . foundations can put on companies — but whether they are not buying tobacco stocks, or are buying stocks in companies they believe are more sustainable, it's a good thing to do in terms of aligning your values, but it's not clear that it's a high impact strategy."

Gates keeps its grant making and investment management separate. It says Bill and Melinda Gates guide the foundation's endowment managers in voting proxies in a manner "consistent with the principles of good governance and good management" and that "they have defined areas in which the endowment will not invest, such as companies whose profit model is centrally tied to corporate activity that they find egregious." The foundation singles out tobacco stocks.[5]

On the issue of investments in Sudan, the foundation says the Gateses have directed the investment team "to be consistent with the approach taken by the endowment managers for Harvard, Yale and Stanford universities." The Gates endowment "no longer has any holdings in the companies identified by these institutions in their investment policy statements on Sudan."

If Gates does not apply broader screens to its investments, it is hardly alone among foundations. A *Chronicle of Philanthropy* survey of the 50 wealthiest private foundations in 2006 found that 34 applied no screens or declined to comment. At least 13 foundations said they screened for tobacco, and several of those screened other investments, such as those in alcohol, firearms or gambling.[6]

The newspaper also found that most foundations don't do much to influence shareholder votes. The survey found that 30 of the 50 big foundations allow their money managers to make all decisions about proxy voting.[7]

Even so, some foundations are strengthening efforts to bring their philanthropic aims into closer alignment with their endowments.

A study last year of 92 foundations by FSG Social Impact Advisors found that the number of philanthropies involved in mission-related investing doubled in the previous decade. Making loans to charities, a process known as program-related investments, was the most common approach.[8]

The F. B. Heron Foundation, Annie E. Casey Foundation and Meyer Memorial Trust have challenged other grant makers to dedicate at least 2 percent of assets to mission-related investments.[9]

Also, the W.K. Kellogg Foundation last year earmarked $100 million for social and mission-driven investments in Africa and the United States. "Few foundations have fully realized the potential of what is commonly referred to as 'double-bottom-line investing,' " Sterling Speirn, the foundation's CEO, said. "We want to maximize our social return on the investment front."[10]

The John D. and Catherine T. MacArthur Foundation formalized a policy for voting shareholder proxies "to reduce or eliminate a substantial social injury caused by a company's actions," according to the *Los Angeles Times*.[11]

The William and Flora Hewlett Foundation screens tobacco stocks from its portfolio but says it generally is not attracted to investment screening. Instead, it is attracted to proxy voting, which "appears to be having an increasing influence on management decisions" and "is unlikely to degrade investment returns."

"We believe that we can be most effective in voting proxies that implicate climate change or forestry practices," the foundation says, adding that it "may selectively choose to exercise proxies when doing so is seen to have a particularly beneficial impact."[12]

Philanthropy experts say that actions like proxy voting remain the exception rather than the rule among philanthropies, though.

"Proxy voting is a basic first step in aligning investments and mission," according to As You Sow, a nonprofit shareholder-advocacy organization in San Francisco engaged in social and environmental issues. Yet, it adds, "when it comes to using the proxy process, most foundations still passively follow management recommendations even when they are not aligned with the foundations' own mission and values."[13]

Bill and Melinda Gates, with investor Warren Buffett, don't invest in companies whose profit is tied to "egregious" corporate activity, such as tobacco production.

Still, Larry Fahn, the group's executive director, says that "slowly but surely," more and more foundations are trying to align their investments with their missions by voting their proxies, engaging companies in dialogue on social and environmental issues and publishing information showing which proxy issues are important to them and how they are voting.

He says that while a foundation may dedicate its annual payout, typically 5 percent of assets, to a mission-related cause, what is needed is a "95 percent solution" in which the rest of the assets are used "proactively" through shareholder engagement.

[1] Charles Piller, Edmund Sanders and Robyn Dixon, "Dark cloud over good works of Gates Foundation," *Los Angeles Times*, Jan. 7, 2007.

[2] *Ibid.*

[3] Figure based on Foundation Center data.

[4] Ian Wilhelm, "Philanthropy Experts Debate Merits of Socially Responsible Investments," *The Chronicle of Philanthropy*, Feb. 22, 2007. The panel was organized by the Hudson Institute's Bradley Center for Philanthropy and Civic Renewal.

[5] Bill and Melinda Gates Foundation, "Our Investment Philosophy," www.gatesfoundation.org/AboutUs/OurWork/Financials/RelatedInfo/OurInvestmentPhilosophy.htm.

[6] "Stock-Investment Policies at the 50 Wealthiest Private Foundations," *The Chronicle of Philanthropy*, May 4, 2006.

[7] Harvy Lipman, "Meshing Proxy With Mission," *The Chronicle of Philanthropy*, May 4, 2006.

[8] Ian Wilhelm, "Foundations Seek to Tie Investments to Their Charitable Missions," *The Chronicle of Philanthropy*, April 19, 2007. The study is Sarah Cooch and Mark Kramer, "Compounding Impact: Mission Investing by US Foundations," Social Impact Advisors, March 2007, www.fsg-impact.org/images/upload/Compounding%20Impact(5).pdf.

[9] Social Investment Forum, "2007 Report on Socially Responsible Investing Trends in the United States," 2008, p. 22.

[10] "W.K. Kellogg Foundation Launches Mission-Driven Investing," press release, Oct. 23, 2007, www.wkkf.org.

[11] Charles Piller, "Foundations align investments with their charitable goals," *Los Angeles Times*, Dec. 29, 2007.

[12] "The William and Flora Hewlett Foundation Social Investment Policy," www.hewlett.org.

[13] Michael Passoff, "Proxy Season Preview 2008," As You Sow, www.asyousow.org.

Getty Images/William Thomas Cain

Noted shareholder activist Evelyn Davis addresses Ford directors at the company's annual meeting in Wilmington, Del., on May 8, 2008. Agenda items included discussion of a proposed report on global warming sought by the Free Enterprise Action Fund, which owns Ford shares. Many socially responsible investors seek to influence workplace and corporate governance issues.

He points out that an investor who put $1,000 in the S&P 500 in 1957 would have ended up with $124,000 in 2003, but about 5 percent less than that amount if tobacco company Philip Morris (now Altria) — the index's single best-performing stock for 46 years through 2003 — had been excluded. If the investor had put the entire $1,000 into Philip Morris alone, the ending figure would have been $4.6 million.

Blodget's critique touched on a litany of other points as well, but ended on a conciliatory note.

"[S]ocially responsible investing certainly deserves to go mainstream," he wrote. "Capital allocation decisions *can* help shape behavior. Even with different investors emphasizing different priorities, there is usually some common ground. And we need to stop insisting that SRI should be both socially *and* financially superior to traditional alternatives. It is unlikely to be both. . . .

"A lifetime of investing in SRI funds might cost you a lot more than organic milk and hybrid cars," he concluded.

"But as SRI investors become both cannier and more numerous, the sacrifice involved need not amount to the 5 percent you might have lost by boycotting Philip Morris. Perhaps, even if SRI returns are no higher than can be achieved through traditional investing — or even a bit less — the practice can be its own reward."

CURRENT SITUATION

Taking Stock

Generalizing about the performance of the SRI field in this year's volatile stock market is difficult, partly because of the field's broad array of holdings and investment strategies.

"Socially responsible funds are such a diverse group that it's hard to make a judgment on how well they do," David Kathman, an analyst at Morningstar who follows SRI mutual funds, noted earlier this year. "Comparisons are not all that helpful."[52]

The Domini 400 Social Index provides at least a clue, though. Through the first seven months of 2008 the 400 corporations charted by the Domini index performed about the same as the S&P 500 — roughly a negative 12.7 percent for the period. The Domini index did better than the S&P 500 in July, rising 0.24 percent compared with a 0.8 percent decline in the broader index.

Business Week noted this spring that some socially responsible funds were seeing inflows of money as investors shifted out of riskier investments and looked for ways to make money in energy.[53]

SRI managers didn't think the market downturn had shaken investor confidence in socially responsible investing, the magazine said. Geeta Aiyer, a portfolio manager at Boston Common Asset Management, said a bigger test of investors' willingness to stick with SRI has occurred during the past five years, as energy-extraction and defense stocks — shunned by social investors — have been among the strongest categories.

"For years we've been addressing the opportunity costs — what you give up by being a social investor," Aiyer said. "Now we see there's opportunity from being a social investor."[54]

Still, this year's collapse of financial stocks has left the social-investing field open to sharp criticism from those who argue that it has put too much faith in banking stocks, just as it invested heavily in technology shares early this decade.

Almost every company "that went south in the '01 collapse [in communication technology], and almost every major financial company that has imploded during the current bubble was ranked high by social investors," says Entine at the American Enterprise Institute (AEI).

AT ISSUE

Should public pensions engage in socially responsible investing?

YES

Timothy Smith
Senior Vice President,
Walden Asset Management

Written for *CQ Researcher*, August 2008

Public pension funds have a fiduciary duty to protect the financial interests of their beneficiaries and are obliged to keep their eye on that target. However, a responsible fiduciary is also obliged to consider a range of non-financial factors that can and do affect shareholder value.

Increasingly, it is understood that environmental, social and governance (ESG) issues have a bottom-line impact. Thus, it would be a limited and ideologically rigid perspective to flatly state these ESG issues have no place in the investment process because they are "social." In fact, pension funds around the globe increasingly have integrated ESG as a necessary ingredient, not as an afterthought.

Investment responsibility, as practiced by pension funds with literally trillions of dollars under management, includes a wide variety of strategies. Critics often define ESG or Responsible Investing as "all about screening" — thus setting up a convenient but historically inaccurate premise.

In fact, active, responsible investors are involved in:

- Voting their proxies thoughtfully and conscientiously;
- Discussing issues with companies seeking forward-looking policies and practices;
- Filing shareholder resolutions urging companies to improve environmental, social or governance practices; and
- Integrating ESG into their investment process.

Globally, investors managing more than $14 trillion in assets have endorsed the U.N.'s "Principles for Responsible Investing," which hold that part of an investor's fiduciary duty is making ESG a part of their investment process. Several issues are before such pension funds, including:

- Governance Reforms — Pension funds from unions to states and cities promote governance reforms as a means of building accountability to investors and improving value for investors.
- Climate Change — Future global devastation from climate change is well documented. As a result, investors globally with more than $50 trillion in assets under management have supported the Carbon Disclosure Project, which urges companies to reveal and reduce their carbon emissions. Hundreds of companies have stepped up as climate leaders — not for narrow, "green" reasons but because it makes long-term business sense.

Should responsible pension funds engage in ESG investing? If hundreds of companies agree that it is in their business interests to be good corporate citizens, forward-looking pension funds are simply ensuring this is reflected in their investment philosophy.

NO

Alicia H. Munnell
Director, Center for Retirement
Research, Boston College

Written for *CQ Researcher*, August 2008

Social investing by public pension plans is generally ineffective and potentially dangerous. Federal law prohibits private plans from introducing social considerations into their investment decisions. And the case for public plans is even stronger.

Advocates suggest that screening out, say, tobacco stocks will hurt tobacco companies, thereby reducing cigarette supplies. But the evidence suggests the opposite. With efficient capital markets, the price of the stock is equal to the present discounted value of future cash flows. Boycotting tobacco stocks may result in a temporary fall in the stock price, but as long as some buyers remain they can swoop in, purchase the stock and make money. So ridding portfolios of "sin" has no impact on the targeted firm.

The same is true for stock prices of "good" companies. But how does one identify desirable firms? Corporations that pass ideological litmus tests today with respect to good governance, protecting the environment or having diverse workforces may turn into tomorrow's disasters. Enron had independent directors; Arthur Andersen prided itself on quality. And Odwalla was dedicated to producing healthy juices, but flaws in its systems led to injuries and a death.

Private pension plans hold virtually no investments in companies that have been screened for either "good" or "bad" characteristics. From the beginning, the Department of Labor has stringently enforced the duties of loyalty and prudence set by ERISA (Employee Retirement Income Security Act). In 1994, the Labor Department reminded fiduciaries that they are prohibited from "subordinating the interests of participants and beneficiaries . . . to unrelated objectives." And the department has just reiterated this warning, so social investing remains a public pension fund phenomenon.

But public pension funds are particularly ill-suited to social investing. First, in many instances, the environment surrounding public pension fund investing is politically charged, and encouraging fund trustees to take "their eyes off the prize" of the maximum return for any given level of risk is asking for trouble. Public pension funds have made serious mistakes in the past when they have introduced social concerns into their investment decisions. Second, the decision-makers and the stakeholders are not the same people. The decision-makers are either the fund boards or the state legislatures. The stakeholders are tomorrow's beneficiaries and/or taxpayers. If social investing produces losses either through higher administrative costs or lower returns, tomorrow's taxpayers will have to ante up, or future retirees will receive lower benefits. Public plans should just say "No" to socially responsible investing.

Walden Asset Management, which calls itself "a leader in socially responsive investing," owns shares of British Petroleum despite its mixed environmental record. Walden says BP has been a leader among energy producers on the climate-change issue. Above, BP's refinery in Grangemouth, Scotland.

Yet some SRI managers did see trouble coming. *Business Week* noted that Patrick McVeigh — president of Boston-based Reynders, McVeigh Capital Management, which engages in social investing — sent a letter to *Business Ethics* magazine back in 2004 after it named government-supported mortgage-funding provider Fannie Mae the most socially responsible company that year. The letter attacked Fannie Mae for having become a big hedge fund by taking on piles of debt as it grew into new markets, *Business Week* said. Fannie's stock has plummeted this year amid huge credit losses.

"*Business Ethics* said Fannie was in the business of building dreams," McVeigh said. "We said they're in the business of building nightmares by putting people in homes they can't afford."[55]

As the SRI movement becomes larger and more prominent, it can expect to become a bigger target for critics, including those who do not think it belongs in the pension world.

AEI's Entine is among the most vocal. "In some states and municipalities, including California, New York and New York City, elected and appointed politicians responsible for overseeing public retirement funds are embracing highly controversial social and environmental criteria to decide on which companies to invest in or publicly lobby against," he wrote in a 2005 book examining public pensions and socially responsible investing.[56]

He went on to say that "social investors and advocacy groups have allied themselves with union leaders and sympathetic politicians, introducing ideology into the management of public pension funds with a stated goal of more directly influencing corporate and public policy."

The California Public Employees' Retirement System (CalPERS) and State Teachers' Retirement System (CalSTRS) have been under scrutiny for a "double bottom line" approach to investing that was introduced in 2000 by former state Treasurer Philip Angelides. His idea, according to *Business Week*, was to take pension fund money out of two asset classes that were performing poorly back then — tobacco stocks and emerging markets — and reinvest in businesses and real estate in low-income California neighborhoods. The aim was to produce both a social return and a healthy financial return.[57]

But a recent CalSTRS report revealed that the $172-billion fund would have had $1 billion more had it not shunned tobacco stocks, *Business Week* said. What's more, real estate investing by the $239 billion CalPERS has been "particularly painful," it said, adding: "Among other bad deals, it faces a loss of nearly $1 billion on one land investment alone.

"The performance of the double bottom line plan illustrates the potential drawbacks of socially responsible investing," *Business Week* wrote. "While it's fine for individual investors to vote their conscience by putting money into the growing number of socially responsible mutual funds, they should know that it could lead to weaker investment performance. . . . Like it or not, people do gamble, smoke and buy expensive nuclear-powered war machines."

Pension Problems

How long the California public pension system plans to stick with its investing approach is in question, however. The system reportedly is considering reversing its anti-tobacco policy. The State Teachers' Retirement System could vote on such a move as early as this fall, and the California Public Employees' Retirement System is also monitoring the issue, according to a report in August in *The New York Sun*.[58]

Entine at AEI says a retreat by the California funds would be a "watershed" event and a sign that "from a fiduciary standpoint [SRI] doesn't hold water."

In New Hampshire the legislature passed a law requiring its two public pension systems to sell its investments supporting Sudan, which has been accused of genocide, but pension officials said the law may be unconstitutional, The Associated Press (AP) reported in August.[59] Pension officials were studying whether to challenge the law after Democratic Gov. John Lynch refused a request to veto the bill.

In a letter obtained by the AP, a pension system lawyer told Lynch the divestiture provision violated the state constitution "because it would require the board to make investment decisions for [a] purpose other than providing benefits to members and beneficiaries and divest assets in order to further the foreign policy objectives of the New Hampshire legislature."

Public-pension squabbles aren't alone in stirring up controversy on the social-responsibility front. Shareholder activism by religious investors and union pension funds also has been controversial.

Last year the Securities and Exchange Commission asked for public comment on proposals to limit the right of shareholders to file proxy resolutions and participate in nominating corporate board members. The curbs were supported by business groups and opposed, in part, by thousands of investors through a Web site formed by the Social Investment Forum and Interfaith Center on Corporate Responsibility (ICCR).[60] The SEC dropped the proxy-resolution idea, but the board-nomination proposal remains on hold, according to Smith at Walden Asset Management.

Meanwhile, the SEC, in a shift of policy, told corporations this year they have to let shareholders vote on a proposal for universal health insurance coverage — a topic of great concern to many social investors.[61] The proposal, backed by such groups as the ICCR, urges companies to adopt "principles for comprehensive health-care reform" such as those reported by the Institute of Medicine, part of the National Academy of Sciences.

The SEC said it was appropriate for shareholders to tell companies what they think by voting on "significant social policy issues" that go beyond day-to-day business matters.[62]

> The integration of environmental, social and governance standards into investment analysis is already widespread in Europe, and many SRI advocates believe the approach will continue to gain favor in the United States among professional money managers.

Religious groups and labor unions have submitted the same basic health-care proposal to several dozen corporations, according to *The New York Times*. "We are working for a national policy that provides universal access to health care, and we do hold more than 30,000 shares of General Electric stock," Barbara Kraemer, a Roman Catholic nun who is national president of the School Sisters of St. Francis, told the newspaper. "As we pursued the proposal with G.E., the company requested a dialogue in lieu of the shareholder resolution, so we withdrew it. The dialogue was productive, resulting in G.E.'s public endorsement of the Institute of Medicine principles."[63]

A shareholder campaign by the AFL-CIO that includes the health-care proposal has met stiff resistance from the U.S. Chamber of Commerce. Last year the chamber asked the Labor Department to weigh in on whether a shareholder campaign by the labor group on health care and other issues was in line with ERISA, the federal retirement-security law. The opinion, issued in December, said pension trustees risk running afoul of their fiduciary duty when they "attempt to further legislative, regulatory or public policy issues through the proxy process when there is no clear economic benefit to the plan."[64]

The letter "sends a clear message that union pension trustees need to put workers' retirement security first, instead of any political agenda," Chamber President Thomas J. Donohue said. "Union pension savings belong to beneficiaries and retirees and must not be tapped to advance goals other than the economic enhancement of those funds."[65]

But Daniel Pedrotty, director of the AFL-CIO's Office of Investment, argues that the advisory opinion broke no new ground and does not preclude investor groups proposing shareholder resolutions that affect their members and beneficiaries.

That, he says, was borne out by the SEC's recent change in policy allowing shareholders to vote on the health-care principles.

"The No. 1 competitive concern for companies is the cost of health care," he says. "We felt like it was a legitimate issue."

OUTLOOK

Continued Growth?

As global warming, health-care access, employee rights in overseas factories and other issues continue to emerge on the public-policy scene, the appeal of socially responsible investing seems likely to grow.

But the sector also faces challenges in convincing skeptics that its screening methods are sound and that financial returns on social investments are at least competitive — if not superior — to those found in conventional ones.

The integration of environmental, social and governance standards into investment analysis is already widespread in Europe, and many SRI advocates believe the approach will continue to gain favor in the United States among professional money managers.

"To the degree that we can demonstrate that this integration approach can improve performance, that's where you're going to attract more investors" to SRI, says Hilton of the Calvert Group. "In some ways, if we're successful, it will almost put us out of a job, meaning if we're really good at integrating ESG performance and we have better performance, then at some point everybody's going to start doing that."

SRI advocates point to the fact that progress reports on environmental sustainability have become standard among many of the nation's biggest corporations as climate and energy issues have gained prominence. While some of those companies may engage in "green-washing" — substituting public relations spin for substance in promoting their environmental efforts — the reports are nonetheless significant, SRI advocates say.

"Yes, [the corporate sustainability reports] might not be as honest as they could be," the *Green Money Journal's* Feigenbaum says. But "they're putting it in print, they're putting themselves out there in a way they have never done before. There is an unstoppable movement [among] every board in this country" to address the sustainability issue.

Besides environmental issues, a range of other concerns is likely to animate the SRI movement in coming years, either through investing practices or shareholder activism. Those issues range from excessive CEO pay, which many SRI proponents see as a social as well as governance concern, to divestment in stocks of companies profiting from commerce in Sudan.

As advocates of socially responsible investing look back on the movement's early roots, they express the hope that it continues to move into the mainstream.

"This industry is a lot like the natural-foods industry," says Alexander, the former grocery store owner in Durham, N.C. "When we started, people thought we were just a bunch of wacko hippies. Twenty or 25 years later, at the convenience store they're selling soy milk. I'm hoping socially responsible investing is like that."

NOTES

1. Social Investment Forum, "2007 Report on Socially Responsible Investing Trends in the United States," 2008, p. iv.

2. Carolyn Cui, "For Money Managers, A Smarter Approach to Social Responsibility," *The Wall Street Journal*, Nov. 5, 2007.

3. Social Investment Forum, *op. cit.*, p. v.

4. KLD Research & Analytics Inc., www.kld.com/ indexes/ds400index/performance.html.

5. Christopher Palmeri, "CalPERS: The Price of Good Intentions," *Business Week*, Aug. 11, 2008, p. 54.

6. Julie Satow, "Big Funds Eye Reinvesting in Tobacco Firms," *The New York Sun*, Aug. 14, 2008.

7. John Hechinger, "Pax Funds Strayed From Its Mission," *The Wall Street Journal*, July 31, 2008, p. 1C.

8. Pax World, "A Letter From the President and CEO of Pax World Funds," http://paxworld.com/ homepage/2008/07/30/a-letter-from -the-president-and-ceo-of-pax-world-funds.

9. "Investors File 54 Global Warming-Related Shareholder Resolutions," *Environmental Leader*, March 9, 2008. Investors withdrew 14 of the 54 resolutions after the companies agreed to disclose potential impacts from emerging climate regulations and strategies for curbing greenhouse-gas emissions, *Environmental Leader* said.

10. For background, see Alan Greenblatt, "Aging Baby Boomers," *CQ Researcher*, Oct. 19, 2007, pp. 865-888.

11. Jon Entine, "Delusional Goodwill," in "The Debate Room: SRI: Invest With Your Heart and Soul," *Business Week*, July 2007, www.businessweek.com/debateroom/archives/2007/07/invest_with_you.html.

12. "Responsible investment doesn't hurt returns: UNEP & Mercer report reveals," Oct. 24, 2007, www.mercer.com. University of California, Berkeley, "New Study on Employee Satisfaction and Long-Run Stock Performance Wins 2007 Moskowitz Prize for SRI Research," Nov. 5, 2007.

13. Ann-Marie Anderson and David H. Myers, "The cost of being good," *Review of Business*, fall 2007.

14. Alex Edmans, "Does the Stock Market Fully Value Intangibles? Employee Satisfaction and Equity Prices," 2008, http://papers.ssrn.com/sol3/papers.cfm?abstract_id=985735.

15. Knowledge@Wharton, "Risks and Costs of Socially Responsible Investing," Aug. 13, 2003, http://knowledge.wharton.upenn.edu/article.cfm?articleid=831. The study is Christopher Geczy, Robert F. Stambaugh and David Levin, "Investing in Socially Responsible Mutual Funds," *Social Science Research Network*, July 22, 2003, revised, Feb. 15, 2006, http://papers.ssrn.com/sol3/papers.cfm?abstract_id=416380.

16. Knowledge@Wharton, *ibid*.

17. Joshua D. Margolis and Hillary Anger Elfenbein, "Do Well by Doing Good? Don't Count on It," *Harvard Business Review*, January 2008, pp. 19-20.

18. Lauren Young, "A Social Fund's Strategic Shift," *Business Week*, May 26, 2006.

19. Amy Domini, *Socially Responsible Investing: Making a Difference and Making Money* (2001), p. 57.

20. Cui, *op. cit.*

21. Paul Hawken and the Natural Capital Institute, "Socially Responsible Investing: How the SRI industry has failed to respond to people who want to invest with conscience and what can be done to change it," 2004.

22. Proinsias O'Mahony, "Ethical living: Profits and principles," *The Guardian*, Feb. 21, 2008, p. 18.

23. Joe Nocera, "Well-Meaning but Misguided Stock Screens," *The New York Times*, April 7, 2007.

24. Ben Arnoldy, "US mining company agrees to 'green' review," *The Christian Science Monitor*, April 26, 2007.

25. Social Investment Forum, *op. cit.*, p. 27. Data source is RiskMetrics Group.

26. Jeanne M. Logsdon and Harry J. Van Buren III, "Justice and Large Corporations," *Business and Society*, online publication, May 2, 2008.

27. Logsdon and Van Buren, "Beyond the Proxy Vote: Dialogues Between Shareholder Activists and Corporations," *Journal of Business Ethics*, online publication, June 18, 2008.

28. Social Investment Forum, *op. cit.* Data source is RiskMetrics Group.

29. Domini, *op. cit.*, p. 29.

30. *Ibid.*, p. 28.

31. Social Investment Forum, *op. cit.*, p. 4.

32. *Ibid.*

33. Pax World Funds, www.paxworld.com/about/pax-history.

34. "About ICCR: FAQ," Interfaith Center on Corporate Responsibility, www.iccr.org/about/faq.php.

35. Social Investment Forum, *op. cit.*

36. *Ibid.*

37. Jonathan Glater, "S.E.C. Adopts New Rules For Lawyers and Funds," *The New York Times*, Jan. 24, 2003.

38. *Ibid.*

39. Jennifer Byrd, "Socially conscious investing blossoms with DOL's blessing," *Investment News*, June 23, 2008.

40. *Ibid.*

41. "Calvert Celebrates 10-Year Anniversary of DOL letter," Calvert Group, www.calvertgroup.com.

42. "U.S. Defined Contribution & Socially Responsible Investing Survey," Mercer, June 5, 2007, www.mercer.com.

43. Cui, *op. cit.*

44. United Nations, "Principles for Responsible Investment Report on Progress 2008," www.unpri.org/report08, p. 4.

45. *Ibid.*

46. Hawken, *op. cit.*, p. 18.

47. *Ibid.*, p. 27.

48. David Vogel, *The Market for Virtue* (2005), p. 38.

49. *Ibid.*, p. 39.

50. Henry Blodget, "The Conscientious Investor," *The Atlantic*, October 2007.

51. Clark Hoyt, "Taint by Association," *The New York Times*, Nov. 11, 2007

52. Nancy Stancill, "Investing with Conscience," *Charlotte Observer*, May 11, 2008.

53. David Bogoslaw, "Socially Responsible Funds Hang Tough," *Business Week*, May 14, 2008.

54. *Ibid.*

55. *Ibid.*

56. Jon Entine, ed., *Pension Fund Politics* (2005), pp. 1-2.

57. Christopher Palmeri, "The Golden State's not-so-golden goose," *Business Week*, July 16, 2008.

58. Satow, *op. cit.*

59. Norma Love, "NH funds mull Sudan divestment," The Associated Press, Aug. 11, 2008, www.boston.com.

60. www.SaveShareholderRights.org.

61. Robert Pear, "S.E.C. Backs Health Care Balloting," *The New York Times*, May 27, 2008.

62. *Ibid.*

63. *Ibid.*

64. Letter from Robert J. Doyle, director of regulations and interpretations, U.S. Department of Labor, to Thomas J. Donohue, president and CEO, U.S. Chamber of Commerce, Dec. 21, 2007, and Pear, *ibid.*

65. U.S. Chamber of Commerce, "Chamber Applauds DOL Union Proxy Activity Decision," Jan. 3, 2008.

BIBLIOGRAPHY

Books

Domini, Amy, *Socially Responsible Investing: Making a Difference and Making Money*, Dearborn Trade, 2001.
The founder of Domini Social Investments writes, "by integrating deeply held personal or ethical concerns into the investment decision-making process, investors can bring about a world that values and supports human dignity and environmental sustainability."

Entine, Jon, ed., *Pension Fund Politics: The Dangers of Socially Responsible Investing*, AEI Press, 2005.
Five critics, including Alicia H. Munnell, director of the Center for Retirement Research at Boston College, analyze the history, strategy and risks of retirement-fund involvement in socially responsible investing.

Vogel, David, *The Market for Virtue: The Potential and Limits of Corporate Social Responsibility*, Brookings Institution Press, 2005.
A professor of business ethics and political science at the University of California, Berkeley explores the corporate-responsibility movement, including the claims and performance record surrounding socially responsible investing.

Articles

Blodget, Henry, "The Conscientious Investor," *The Atlantic*, October 2007.
A former securities analyst argues that socially responsible investing is not as profitable as many advocates contend.

Chatterji, Aaron, and Siona Listokin, "Corporate Social Irresponsibility," *Democracy*, winter 2007, www.democracyjournal.org/article.php?ID=6497.
Only governments and multilateral agreements can solve social injustices, and progressives should "end their fixation with corporate social responsibility."

Feigenbaum, Cliff, "Essays on the Future: Looking Ahead," *GreenMoneyJournal.com*, www.greenmoney-journal.com/index.mpl?newsletterid=41.
A publication focusing on environmental sustainability presents essays from key figures in the SRI movement.

Nocera, Joe, "Well-Meaning but Misguided Stock Screens," *The New York Times*, April 7, 2007, http://select.nytimes.com/2007/04/07/business/07nocera.html.
A business journalist analyzes socially responsible investing, including the work of KLD Research & Analytics, a Boston company that constructs widely used SRI

indexes, and says social investing "oversimplifies the world."

Palmeri, Christopher, "CalPERS: The Price of Good Intentions," *Business Week*, **Aug. 11, 2008.**
The SRI strategy of the California Public Employees' Retirement System and the California State Teachers' Retirement System has hurt fund returns, even though the funds have outpaced the S&P 500 over the past five years.

Piller, Charles, Edmund Sanders and Robyn Dixon, "Dark cloud over good works of Gates Foundation," *Los Angeles Times*, **Jan. 7, 2007, www.latimes.com/ news/nationworld/nation/la-na-gatesx07jan07,0, 6827615.story.**
Part of a series on the Bill & Melinda Gates Foundation, the article says the nation's biggest private foundation invested "in many companies that have failed tests of social responsibility because of environmental lapses, employment discrimination, disregard for worker rights or unethical practices."

Satow, Julie, "Big Funds Eye Reinvesting in Tobacco Firms," *The New York Sun*, **Aug. 14, 2008, www .nysun.com/business/big-funds-eye-reinvesting-in-tobacco-firms/83867/.**
A newspaper says California's public system is considering reversing its policy of shunning tobacco stocks, in what could be a "potentially devastating blow" to the socially responsible investing movement.

Reports and Studies

"The language of responsible investment," *Mercer LLC*, **Jan. 3, 2008, www.mercer.com/ridictionary.**
An international business-consulting firm offers a guide to key organizations in the SRI field and defines such insider terms as "eco-efficiency" ("the ratio between goods produced or services rendered and the resources consumed or waste produced").

"2007 Report on Socially Responsible Investing Trends in the United States," *Social Investment Forum*, **2008, www.socialinvest.org/pdf/SRI_Trends_ ExecSummary_2007.pdf.**
Socially responsible investing is growing at a faster rate than the broader universe of investment assets under professional management, says this biennial report from the trade group of the social-investing field.

Hawken, Paul, "Socially Responsible Investing: How the SRI industry has failed to respond to people who want to invest with conscience and what can be done to change it," *Natural Capital Institute*, **2004, www .responsibleinvesting.org/database/dokuman/SRI% 20Report%2010-04_word.pdf.**
An ecologist and entrepreneur offers a stinging critique of the socially responsible mutual fund industry, arguing among other things that it lacks transparency and that the language surrounding socially responsible investing is vague.

For More Information

American Enterprise Institute, 1150 17th St., N.W., Washington, DC 20036; (202) 862-5800; www.aei.org. A private, nonpartisan, not-for-profit institution dedicated to research and education on issues of government, politics, economics, and social welfare.

As You Sow, 311 California St., Suite 510, San Francisco, CA 94104; (415) 391-3212; www.asyousow.org. Shareholder advocacy group that seeks to bring about change within public companies.

Calvert Group, 4550 Montgomery Ave., Suite 1000N, Bethesda, MD 20814; (800) 368-2748. Mutual fund company engaged in socially responsible investing.

Ceres, 99 Chauncy St., 6th Floor, Boston, MA 02111; (617) 247-0700; www.ceres.org. Coalition of investors, environmental organizations and other public-interest groups working with companies and investors to address "sustainability" issues such as climate change.

Domini Social Investments, P.O. Box 9785, Providence, RI 02940; (800) 762-6814; www.domini.com. Mutual fund company engaged in socially responsible investing.

Interfaith Center on Corporate Responsibility, Room 1842, 475 Riverside Dr., New York, NY 10115; (212) 870-2295; www.iccr.org. Association of 275 faith-based institutional investors.

KLD Research & Analytics, 250 Summer St., 4th Floor, Boston, MA 02210; (617) 426-5270; www.kld.com. Independent investment research firm focusing on social investing.

Natural Capital Institute, 3 Gate Five Road, Suite A, Sausalito, CA 94965; (415) 331-6241; www.naturalcapital .org. Research group on social responsibility, corporate accountability and environmental issues.

Pax World Funds, 30 Penhallow St., Suite 400, Portsmouth, NH 03801; (800) 767-1729; www.paxworld.com. Mutual fund company engaged in socially responsible investing.

Social Investment Forum, 1612 K St., N.W., Suite 650, Washington, DC 20006; (202) 872-5361; www.social invest.org. Trade group for socially responsible investment industry.

Walden Asset Management, One Beacon St., Boston, MA 02108; (617) 726-7155; www.waldenassetmgmt.com. A division of Boston Trust & Investment Management Co. that specializes in socially responsive investments.

Supporting researchers for more than 40 years

Research methods have always been at the core of SAGE's publishing program. Founder Sara Miller McCune published SAGE's first methods book, *Public Policy Evaluation*, in 1970. Soon after, she launched the *Quantitative Applications in the Social Sciences* series—affectionately known as the "little green books."

Always at the forefront of developing and supporting new approaches in methods, SAGE published early groundbreaking texts and journals in the fields of qualitative methods and evaluation.

Today, more than 40 years and two million little green books later, SAGE continues to push the boundaries with a growing list of more than 1,200 research methods books, journals, and reference works across the social, behavioral, and health sciences. Its imprints—Pine Forge Press, home of innovative textbooks in sociology, and Corwin, publisher of PreK–12 resources for teachers and administrators—broaden SAGE's range of offerings in methods. SAGE further extended its impact in 2008 when it acquired CQ Press and its best-selling and highly respected political science research methods list.

From qualitative, quantitative, and mixed methods to evaluation, SAGE is the essential resource for academics and practitioners looking for the latest methods by leading scholars.

For more information, visit **www.sagepub.com**.